Data
Presentation

Interpretation

CLINICAL BIOCHEMISTRY

PRINCIPLES · METHODS
APPLICATIONS
2

Series Editors
H. Ch. Curtius
M. Roth

Walter de Gruyter
Berlin · New York
1989

Data Presentation Interpretation

Editors
H. Keller
Ch. Trendelenburg

W DE G
Walter de Gruyter
Berlin · New York
1989

Editors

Professor
Dr. Herbert Keller
Institut für Klinische Chemie und
Hämatologie des Kantons St. Gallen
Frohbergstrasse 3
CH-9000 St. Gallen
Switzerland

Professor
Dr. med. Christian Trendelenburg
Chefarzt des Instituts für Laboratoriumsmedizin
Städtisches Krankenhaus Frankfurt/M.-Höchst
Gotenstraße 6–8
D-6230 Frankfurt a. M. 80
Fed. Republic of Germany

Series Editors

Dr. Hans-Christoph Curtius
Head, Division of Clinical Chemsitry
Department of Pediatrics
University of Zurich
Steinwiesstrasse 75
CH-8032 Zurich
Switzerland

Dr. Marc Roth
Director, Central Laboratory of
Clinical Chemistry
Cantonal Hospital
University of Geneva
CH-1211 Geneva 4
Switzerland

Publishers

Walter de Gruyter & Co.
Genthiner Straße 13
D-1000 Berlin 30
Fed. Republic of Germany
Telephone (030) 26005-0
Telefax (030) 26005-251 · Telex 184027

Walter de Gruyter, Inc.
200 Saw Mill River Road
Hawthorne, N.Y. 10532
USA
Telephone (914) 747-0110
Telefax (914) 747-1326 · Telex 646677

Library of Congress Cataloging-in-Publication Data

Data presentation/interpretation.
(Clinical biochemistry ; 2)
Includes index.
1. Diagnosis, Laboratory – Statistical methods. 2. Diagnosis, Laboratory – Graphic methods. 3. Medical protocols. 4. Expert systems (Computer science)
I. Keller, H. (Herbert), 1925–. II. Trendelenburg, Ch. (Christian), 1945–. III. Series: Clinical biochemistry (Berlin, Germany) ; 2.
RB38.3.D38 1989 616.07′5 89-11959
ISBN 0-89925-550-7 (U.S.)

Deutsche Bibliothek Cataloging-in-Publication Data

Data presentation, interpretation / ed. H. Keller ; Ch. Trendelenburg. – Berlin ; New York : de Gruyter, 1989
(Clinical biochemistry ; 2)
ISBN 3-11-010735-X (Berlin)
ISBN 0-89925-550-7 (New York)
NE: Keller, Herbert [Hrsg.]; GT

Typesetting: Arthur Collignon GmbH, Berlin. – Printing: Gerike GmbH, Berlin. – Binding: Dieter Mikolai, Berlin. – Cover design: Hansbernd Lindemann, Berlin.

Foreword

The medical laboratory is undergoing a fundamental change. Whereas the main emphasis used to be in the areas of analytical techniques, instrumentation and error analysis, the modern laboratory is increasingly faced with problems of patho-biochemistry and pathophysiology, and especially with the questions of diagnostic decision-making techniques. The number of parameters which a modern diagnostic laboratory makes available to the practicing physician is so large that the practice of reporting numerical data, supplemented only by indicators of normal ranges, is no longer adequate. Quite often, physicians who receive laboratory reports do not understand their full implications, because their knowledge of all the wide spectrum of modern laboratory diagnostic techniques is not adequate.

In this situation, new forms of presentation and interpretation of laboratory data must be found. In addition to numerical data, graphic presentations, which have the advantage of more flexibility, are becoming more common. A well designed graph can convey the essential information at a glance, and permits reduction of data without loss of information. Special methods of representation can be used to emphasize essential facts, temporal courses and multiple parameters.

Self-explanatory graphic presentations, which may be available on a color-graphics computer monitor, are still in the experimental stage. For the time being, clinical chemical laboratories will continue to report their findings in numerical form, supplemented by reference intervals from appropriate comparison groups.

Interpretations of laboratory findings are always important when the diagnosis is difficult, with respect to clinical decisions; they are also important when the consequences are grave for the patient. On the other hand, only those laboratory findings should be interpreted which are significantly related to the problem at hand. In other words, the laboratory should not distract the physician with non-essential secondary findings. Only those findings which are important in the context of the clinical problem should be interpreted, not those which might draw attention away from the primary problem.

In order to answer a wide variety of such questions competently, the editors have sought out experts in a number of areas to contribute to this book. They were aware that discussion of the themes "data presentation and interpretation" by so many authors could not be condensed into a concise work. The intention was, instead, to present as many facets of the picture as possible, so that the competent reader could be as well informed as possible about the complexities of the subject.

July 1989

H. Keller
Ch. Trendelenburg

Contributors

Ian O. Angell, Dr.
Professor of Information Systems
The London School of Economics
and Political Science
University of London
Hougthon Street
GB-London WC2A 2AE
Great Britain

Claus Badke
Medizinisches Fachrechenzentrum
Labordatenverarbeitung
Medizinische Hochschule Hannover
Konstanty-Gutschow-Str. 8
D-3000 Hannover 61
West-Germany

Susanne Bothung
Medizinisches Fachrechenzentrum
Labordatenverarbeitung
Medizinische Hochschule Hannover
Konstanty-Gutschow-Str. 8
D-3000 Hannover 61
West-Germany

Johannes Büttner, Dr. med., Dr. rer. nat.
Professor für Klinische Chemie
Institut für Klinische Chemie I
Medizinische Hochschule Hannover
Konstanty-Gutschow-Str. 8
D-3000 Hannover 61
West-Germany

Paul D. Clayton, Ph. D.
Professor of Information Science
in Medicine
Director of the Center
for Medical Informatics
Columbia University College
of Physicians and Surgeons
162 Fort Washington Ave AP1310
New York, N.Y. 10032
U.S.A

Donald P. Connelly, M. D., Ph. D.
Associate Professor
Department of Laboratory Medicine
and Pathology
Box 511 UMHC
University of Minnesota
Minneapolis, Minnesota 55455
U.S.A.

M. Eggstein, Dr. med.
Professor für Innere Medizin
Eberhard-Karls-Universität
Medizinische Klinik
Abt. Innere Medizin IV
Otfried-Müller-Straße 10
D-7400 Tübingen 1
West-Germany

R. Scott Evans, Ph. D.
Research Assistant Professor
Division of Infectous Diseases
Department of Medicine
University of Utah
LDS Hospital
325 8th Ave
Salt Lake City, Utah 84143
U.S.A.

Reed M. Gardner, Ph. D.
Professor of Medical Informatics
University of Utah
LDS Hospital
325 8th Ave
Salt Lake City, Utah 84143
U.S.A.

Willi Gerhardt, M. D., Ph. D.
Department of Clinical Chemistry
Lasarettet
S-251 87 Helsingborg
Sweden

Erwin Hansert, Dr.
Max-Planck-Institut für Psychiatrie
Biostatistik
Kraepelinstraße 2–10
D-8000 München 40
West-Germany

Peter J. Haug, M. D.
Assistant Professor of
Medical Informatics
University of Utah
LDS Hospital
325 8th Ave
Salt Lake City, Utah 84143
U. S. A.

B. Keller, Dipl.-Math. ETHZ
State Institute of Clinical Chemistry
and Haematology
Frohbergstraße 3
CH-9000 St. Gallen
Switzerland

Herbert Keller, M. D., Ph. D.
Professor of Clinical Chemistry
State Institute of Clinical Chemistry
and Haematology
Frohbergstraße 3
CH-9000 St. Gallen
Switzerland

R. Maulbetsch, Dr. rer. nat.
Eberhard-Karls-Universität
Medizinische Klinik
Abt. Innere Medizin IV
Otfried-Müller-Straße 10
D-7400 Tübingen 1
West-Germany

Michael McNeely, B. Sc., M. D.,
FACB, FRCPC
President and Chief Executive Officer
Island Medical Laboratories
4489 Viewmont Avenue
Victoria B. C. V8Z 5K8
Canada

Dorothea Nagel, Dr. rer. nat.,
Dipl.-Math.
Institut für Klinische Chemie
Klinikum Großhadern
der Universität München
Marchioninistraße 15
D-8000 München 70
West-Germany

Michael Oellerich, Dr. med.
Professor für Klinische Chemie
Institut für Klinische Chemie I
Medizinische Hochschule Hannover
Konstanty-Gutschow-Str. 8
D-3000 Hannover 61
West-Germany

P. E. Politser, M. D., Ph. D.
Associate Professor of
Health Decision Sciences
Depts. of Health Policy and
Management and Biostatistics
School of Public Health
Harvard University
677 Huntington Avenue
Boston, Massachusetts 02115
U. S. A.

Albert J. Porth, Dr. rer. nat.,
Dipl.-Math.
Professor für Medizinische Informatik
Medizinisches Fachrechenzentrum
Labordatenverarbeitung
Medizinische Hochschule Hannover
Konstanty-Gutschow-Str. 8
D-3000 Hannover 61
West-Germany

T. Allan Pryor, Ph. D.
Professor of Medical Informatics
University of Utah
LDS Hospital
325 8th Ave
Salt Lake City, Utah 84143
U. S. A.

W. Rehpenning, Dr. rer. nat.
Dozent für Mathematik in der Medizin
Institut für Mathematik und
Datenverarbeitung in der Medizin
Universitäts-Krankenhaus Eppendorf
Martinistraße 52
D-2000 Hamburg 20
West-Germany

W. Renn, Dr. rer. nat.
Eberhard-Karls-Universität
Medizinische Klinik
Abt. Innere Medizin IV
Otfried-Müller-Straße 10
D-7400 Tübingen 1
West-Germany

Linda J. Reynolds, BSc., Msc.
Tutor
Department of Graphic Design
Royal College of Art
Kensington Gore
GB-London SW7 2EU
Great Britain

Harald Sator, Dr. biol. hum., Dipl.-Inf.
Institut für Klinische Chemie
Klinikum Großhadern
der Universität München
Marchioninistraße 15
D-8000 München 70
West-Germany

Berthold Schneider, Dr. rer. nat.
Professor für Biometrie
Institut für Biometrie
Medizinische Hochschule Hannover
Konstanty-Gutschow-Str. 8
D-3000 Hannover 61
West-Germany

Dankwart Stamm, Dr. Dr.
Professor für Klinische Chemie
Max-Planck-Institut für Psychiatrie
Institut für Klinische Chemie
Kraepelinstraße 2–10
D-8000 München 40
West-Germany

Christian Trendelenburg, Dr. med.
Professor für Klinische Chemie
Institut für Laboratoriumsmedizin
Städtisches Krankenhaus
Frankfurt a. M.-Höchst
Gotenstraße 6–8
D-6230 Frankfurt a. M. 80
West-Germany

Wolfgang Vogt, Dr. med.
Professor für Klinische Chemie
Vorstand des Instituts für Klinische
Chemie und Laboratoriumsmedizin
Deutsches Herzzentrum München
des Freistaates Bayern
Lothstraße 11
D-8000 München 2
West-Germany

Homer R. Warner, M. D., Ph. D.
Professor and Chairman
Department of Medical Informatics
University of Utah
University Medical Center
Salt Lake City, Utah 84112
U. S. A.

Per Winkel, overlaege, Doc. Med. Sci.
Department of Clinical Chemistry
University Hospital of Copenhagen
9 Blegdamsvej
DK-2100 Copenhagen
Denmark

Michael Worzyk, Dipl.-Math.
Medizinisches Fachrechenzentrum
Labordatenverarbeitung
Medizinische Hochschule Hannover
Konstanty-Gutschow-Str. 8
D-3000 Hannover 61
West-Germany

Abbreviations and Symbols

Abbreviations

A	alcoholism, individual with alcoholism
ABS	non-alcoholic patients
AI	artificial intelligence
ALK	patients with alcoholism
ALP	*see* AP
AMI	acute myocardial infarction
AP	alkaline phosphatase, serum (EC 3.1.3.1)
Apo	apolipoprotein
ASAT	aspartate aminotransferase
AST	*see* GOT
BASIC	beginners
BFCI	Bayes formula based on conditional independence
BG	blood glucose level
BGT	target blood glucose level
Bil	bilirubin
bit	binary digit
BUN	Blood urea nitrogen
CAD	computer aided design
CAM	computer assisted manufacture
CEA	carcinoembryonic antigen
CFD	class fraction of diseased (individuals)
CFR	class fraction of reference (individuals)
CGA	colour graphics adaptor
CH	cholesterol
CK	creatine kinase (activity)
CK-MB	creatine kinase isoenzyme MB
CPD	class prevalence of diseased (individuals)
CPR	class prevalence of reference (individuals)
CV	coefficient of variation
D	discriminant region
D^+, D^-	region containing the "positive", or "negative" test results
DA	discriminant analysis
DAP	name of a program for digital and analogous data presentation
DIS	diseased (persons)
DM	diabetes mellitus
EDP	electronic data processing
EGA	enhanced graphics adaptor
EKG	electrocardiogram
GES	healthy blood donors
GGT	γ-glutamyl-transferase (activity), γ-glutamyl-transpeptidase
GH	growth hormone
G.K.S.	graphical kernel system

GLDH	glutamate dehydrogenase, serum (EC 1.4.1.3)
GOT	glutamate-oxaloacetate transaminase (aspartate aminotransferase)
GPT	glutamate-pyruvate transaminase (alanine aminotransferase, serum)
GT	γ-glutamyl transferase
HB	haemoglobin
HbA1	glycolized haemoglobin
HCaCL	Cleveland patients suffering from hypercalcinosis
H.C.I.	human computer interaction
HDL	high density lipoprotein
HPT	prim. hyperparathyreoidism
IVGTT	intravenous glucose tolerance test
LDF	linear discriminant function
LDH	lactate dehydrogenase
LDH1	lactate dehydrogenase isoenzyme 1 (H4)
LDL	low density lipoprotein
LIS	laboratory information system
LR	likelihood ratio
MCHC	mean corpuscular haemoglobin concentration
MCV	(erythrocyte) mean cellular volume, mean corpuscular volume
MIST	maximal insulin stimulating test
NA	non-alcoholism, individual not with alcoholism
NEG	negative (result)
NONLIN	name of a program for nonlinear regression
OGTT	oral glucose tolerance test
PAC	prostatic adenocarcinoma
PALI	programmed accelerated laboratory investigation
PAP	prostatic acid phosphatase
PAT	patient group
PL	phospholipids
POS	positive (result)
PV	predictive value
PVP	predictive value of the positive test result
PVpos	predictive value of a positive result
PVneg	predictive value of a negative result
REF	reference (group or persons)
RIA	radioimmunoassay
ROC	receiver operating characteristic
ROM	read only memory
S-Ca	serum calcium
SCE	Scandinavian Committee on Enzymes
S-K	serum potassium
SENS	sensitivity
S.I.G.	special interest group
SPEZ	specivity
$T_{(k)}$	simple discriminant region of the kth order

T3	triiodothyronine
T4	thyroxine
TBG	thyroxine binding globulin
TG	triglycerides
TRH	thyrotropic releasing hormone
TSH	thyroid stimulating hormone
VAR	variance-covariance operator
VDT	video display terminal
VGA	video graphics adaptor

Symbols

A	information loss, information decrease
α	sensitivity
β	1-specifity
C	test capacity, channel capacity
D	input probability, events
E	expectation operator
H	entropy
I	transinformation, information gain
k	order of a simple discriminant region
m	number of variables or dimensions
p, P	probability (function)
Φ	normal distribution function
r	Pearson correlation coefficient
s_R	Rand's measure
s_{RC}	Rand's measure corrected for chance
T	output probability, test result
U	output probability, test result (in multiple tests)
w	probability density functions
x, X	given or measured value of a variable
$\setminus$	residue sets (set theory)
$\cup$	union (set theory)
$\cap$	intersection (set theory)
$\supset$	contains (set theory)
$\subset$	is contained in (set theory)

Contents

1 Introduction

Data Presentation and Interpretation: Problems, Tools and Goals (H. Keller) 3

2 Data Presentation

Chapter 2.1 How to Make Laboratory Information More Informative: Psychological and Statistical Considerations (P. E. Politser) 11

1 Introduction 11
2 Background 11
3 Basic Concepts 12
3.1 Filtering 12
3.2 Simplification 14
3.3 Coding 14
3.4 Grouping 14
3.5 Recognition 18
3.6 Segregation 20
4 Discussion 25
Appendix 26
Acknowledgements 30
References 30

Chapter 2.2 Result Reports from Large Centralized Laboratories (A. J. Porth, C. Badke, S. Bothung and M. Worzyk) 33

1 Introduction and Basic Aspects 33
2 Internal Laboratory Reports 37
2.1 Working place lists, result entry lists and evaluation lists 37
2.2 Criticism for plausibility checks 37
2.2.1 Extreme value check 37
2.2.2 Double result list 40
2.2.3 Extreme previous value deviations 40
2.2.4 Criticism of striking trends 40
2.2.5 Striking result constellations 41
2.3 Result processing 43
2.3.1 Calculation printout 43
2.3.2 Remainder list of the specimen data 45
2.3.3 Remainder list of requests 45
3 Findings Reports for the Attending Physician 45
3.1 General remarks on contents and form 45
3.2 Cumulative findings 49

3.3 Preliminary findings (collective findings) 52
3.4 Job-oriented individual findings 54
3.5 Special findings 55
3.5.1 Microbiological findings 55
3.5.2 Medical report 56
4 Requirements for Efficient Findings Reports 58
4.1 System design 59
4.2 Data base design 60
4.3 Time behaviour 60
References 61

Chapter 2.3 Graphical Data Display and Medical Decision Making (D. P. Connelly) 63
1 Introduction 63
1.1 Communicating at the semantic level 63
1.2 Aiding the decision maker 65
1.3 Graphical presentation as a decision aid 65
2 Graphical Displays in the Laboratory 66
2.1 Referential displays 66
2.2 Monitoring displays 67
2.3 Diagnostic displays 69
2.4 Laboratory operations and management displays 71
3 The Effectiveness of Graphical Displays in Aiding Decision Making 71
3.1 Graphical aids for decision makers: Studies of effectiveness 72
3.2 Human perception and graphical display 72
3.3 The effectiveness of graphical displays in laboratory reporting 74
4 Summary 76
Acknowledgement 77
References 77

Chapter 2.4 Digital and Analogous Presentation of Laboratory Results in Dependence of Time (W. Renn, R. Maulbetsch and M. Eggstein) 81
1 Introduction 81
2 Linear Interpolation 83
3 Model Formulation 92
4 Model Identification 101
4.1 Theoretical identifiability 102
4.2 Parameter estimation 103
4.3 Practical identifiability 105
4.3.1 Goodness of fit 106
4.3.2 Examination of residuals 106
4.3.3 Accuracy of parameter estimates 107
5 Nonlinear Interpolation 109
6 Digital and Analogous Presentation 112

7 Conclusion ... 118
References ... 119

Chapter 2.5 Computer Graphics for Data Presentation (I. O. Angell) ... 121
References ... 131

Chapter 2.6 Use of Colour in Tables and Graphic Displays (L. Reynolds) ... 133
1 Introduction ... 133
2 Visual Coding ... 134
2.1 The concept of visual coding ... 134
2.2 Visual coding in computer-generated information ... 135
3 General Principles in the Use of Colour ... 136
3.1 Colour and legibility ... 136
3.2 Colour as a means of coding information structure ... 137
3.2.1 Colour as a grouping cue ... 137
3.2.2 Colour for emphasis ... 137
3.2.3 Colour coding ... 139
3.2.4 Flashing ... 139
4 Presentation of Tabular Information ... 140
4.1 The structure of tables ... 140
4.2 Layout in tables ... 140
4.3 Colour in tables ... 143
5 Graphic Displays ... 146
5.1 A comparison with tabular formats ... 146
5.2 The impact of computer graphics ... 147
5.3 Charts versus graphs ... 147
5.4 Pie charts ... 147
5.5 Bar charts ... 148
5.5.1 Structure ... 148
5.5.2 Detail ... 148
5.5.3 Use of colour ... 150
5.6 Graphs ... 151
5.6.1 Structure and detail ... 151
5.6.2 Use of colour ... 153
6 Conclusions ... 155
References ... 157

3 Interpretation of Laboratory Results

Chapter 3.1 The Bayes Approach: Systematic Graphic Evaluation of Diagnostic Tests (W. Gerhardt) ... 161
1 Terminology ... 161
2 Function of a Diagnostic Test ... 162
2.1 Study groups ... 162

2.2 Anterior prevalence of disease 162
2.3 2 × 2 tables 163
2.4 Discriminations 164
3 Systematic Graphic Evaluation of Diagnostic Test: S-Myoglobin at Suspected AMI 166
3.1 Samples drawn 8–12 hours after symptoms 166
3.2 Samples drawn 4–8 hours after symptoms 171
4 Anterior Prevalence, CPD, and Predictive Values 173
5 Selection of Sample Groups for Clinical Studies 177
6 Comment Corriger la Fortune 178
6.1 How to improve test results 179
6.2 How to impair test results 179
7 Proper Reporting of Test Evaluations 181
7.1 2 × 2 tables for presentation and reconstruction of test data 182
8 Diagnostic Strategies 183
8.1 Two-class tests in sequence 184
8.2 The clinician as the first test in a diagnostic strategy 186
8.3 Two-class tests in parallel 188
8.4 Tests for screening 190
Appendix 191
Bayes theorem 191
Likelihood ratios, class prevalence of disease (CPD), predictive values, posterior odds of disease, and risk quotients (RQ) 192
Receiver operating characteristic (ROC) 195
Efficiency 197
Comparison of likelihood ratio, post. odds DIS, ROCs, and efficiency with double histograms with plots of test quantities 197
Double histograms with plots of test quantities 198
Synonym dictionary of some current concepts in test evaluation 200
Acknowledgements 201
References 202

Chapter 3.2 Deterministic Methods for Computerized Interpretation of Laboratory Tests (M. D. D. McNeely) 205
1 Introduction 205
2 Hardware and Software Considerations 206
2.1 Microcomputer vs. minicomputer 206
2.2 Language 206
2.3 Commercial programming aids 207
2.4 Commercial deterministic interpretive reporting systems 207
2.5 Essential features 207
2.6 Composing the comment 207
2.7 Information sources 208

3 "Canned" Comments ... 208
3.1 Definition ... 208
3.2 Word processor applications ... 208
3.3 Menu driven and canned comments ... 209
4 Logic Techniques ... 211
4.1 Branching logic ... 211
4.2 Octal coding ... 213
4.3 Binary branching tree ... 213
4.4 Matrix logic ... 213
4.5 Other approaches ... 216
5 Conclusion ... 216
References ... 216

Chapter 3.3 Information Theoretical Model of a Clinical Chemical Test (J. Büttner) ... 219
1 Introduction ... 219
2 Information Theoretical Model of a Test ... 219
3 Quantification of Uncertainty ... 220
4 Sophistication of the Information Theoretical Model of a Clinical Chemical Test ... 221
5 Calculation of the Information Theoretical Quantities ... 224
6 Extension of the Model to Tests with Quantitative Results ... 227
7 Treatment of Test Combinations (Multivariate Model) ... 228
8 Relationship between the Information Theoretical and the Probability Theoretical Model of a Test ... 233
9 Practical Applications ... 236
9.1 Quality criterion of a test ... 236
9.2 Optimizing the decision limit of a test ... 236
9.3 Adaptation of a test to a given input probability ... 238
9.4 Optimization of decision limits in combined tests ... 239
9.5 Selection of a decision scheme in the combination of several tests 240
9.6 Calculation of the "optimum sets" in multiple tests ... 241
9.7 Cost-benefit analysis of a test ... 241
9.8 Performance of a clinical chemical test in the framework of the entire diagnostic process ... 241
10 Prospectus ... 243
Appendix: Calculation of the test capacity (channel capacity) ... 244
Acknowledgements ... 245
References ... 245

Chapter 3.4 The Multivariate Approach (P. Winkel) ... 247
1 Introduction ... 247
2 One Observation per Patient ... 247
2.1 Predefined clinical classes presumed ... 247

2.1.1 Probabilistic diagnosis ... 247
2.1.1.1 Bayes theorem for multiple variates ... 247
2.1.1.2 Choice of probabilistic mode ... 250
2.1.1.3 Application of Bayes theorem ... 255
2.1.1.4 Assessing the performance of probabilistic diagnosis 255
2.1.2 Studying the relationships between variates and between variates and clinical classes ... 256
2.1.3 The effect of random analytical variation ... 259
2.2 Constructing new clinical classes ... 260
2.2.1 Numerical taxonomy ... 260
2.2.1.1 The techniques of numerical taxonomy ... 260
2.2.1.2 Unresolved problems in numerical taxonomy ... 263
2.2.1.3 Considerations when using numerical taxonomy ... 265
2.2.2 Multivariate prognostic stratifications ... 267
3 Multiple Observations per Patient (Time Series Analysis) ... 268
References ... 272

Chapter 3.5 Cluster Analysis as a Help in Interpretation of Complex Data (W. Vogt, D. Nagel and H. Sator) ... 277
1 Introduction: On the Theory of Diagnosis and Diagnosing ... 277
1.1 Taxonomy in medicine ... 277
1.2 Diagnosis as basis of action ... 278
1.3 Medical action and diagnostic procedure ... 278
1.4 How clinical chemistry can help ... 279
2 Data Reduction as an Aid in Interpretation ... 279
2.1 Commonly used methods ... 279
2.2 Pattern cognition by cluster analysis ... 280
3 A Cluster Based Model as a Help in Clinical Judgement ... 281
3.1 Collection of data ... 281
3.2 Selection and number of the clinical chemical parameters ... 282
3.3 Transformation of data ... 283
3.3.1 Linear transformation ... 283
3.3.2 Nonlinear transformation ... 283
3.4 Choice of classification method ... 286
3.5 Classification: Description of reality or product of change? ... 287
3.5.1 Internal stability of the structure found ... 288
3.5.2 Measuring the external stability ... 292
3.5.3 Reclassification: Reliability despite of analytical errors? ... 294
3.5.4 Choosing the number of clusters ... 297
3.6 Representation of the groups ... 299
3.7 Clinical significance ... 301
3.7.1 Comparison of the results of classification with the laboratory diagnosis ... 302
3.7.2 Comparison of the results of classification with the final diagnosis ... 303
3.7.3 Comparison of the laboratory diagnosis with the final diagnosis ... 304

4 A Diagnostic Strategy 304
4.1 Outline of the problem 305
4.2 A simple example 306
4.3 Problems of practical application 310
4.4 Reduction of clinical chemical parameters needed 311
4.4.1 Stepwise diagnostic strategy 311
4.4.2 Diagnostic strategy using combinations of parameters 315
4.4.3 Considering analytical errors in the strategy 315
5 Experience with the Model in Clinical Practice 316
5.1 Design of the study 316
5.2 Subjective estimation of the usefulness of the knowledge of cluster membership 318
5.3 Objective evaluation of the usefulness of the knowledge of cluster membership 322
6 Outlook 322
6.1 Dependence of the model on patient sample used 322
6.2 Dependence of the model on analytical methods 323
6.3 Aspects for the future 323
Appendix: Comparison of Classifications 323
References 326

Chapter 3.6 Quasi Distribution-Free Assessment of Multivariate Data of Clinical Chemical Tests with the Aid of the X-transformation (W. Rehpenning) 329
1 Introduction 329
2 The X-Transformation of van der Waerden 331
2.1 The multivariate application of the X-transformation 335
2.2 The effect of the X-transformation 337
2.3 Structural investigations with the aid of the X-transformation 340
2.4 The role of the X-transformation in multivariate data assessment for diagnostic purposes 347
2.5 The correction for age 349
2.6 The empirical distribution of the test quantities 352
2.7 Deviation patterns and group-conformity behaviour 357
3 Conclusions 361
References 365

Chapter 3.7 Data Driven Interpretation of Laboratory Results in the Context of a Medical Decision Support System (P. D. Clayton, R. S. Evans, T. A. Pryor, R. M. Gardner, P. J. Haug and H. R. Warner) 367
1 Introduction 367
1.1 Philosophical issues in automated decision assistance 367
2 Functional Implementation of the HELP System 370
2.1 Interface to the clinical laboratory system 375
2.2 Experience with the decision-making aspects of the system 376

3 Summary 378
References 379

Chapter 3.8 Expert Systems in Clinical Chemistry (Ch. Trendelenburg) 381
1 Introduction 381
2 Structure and Mode of Operation of Expert Systems 382
2.1 Characteristic properties of expert systems 382
2.1.1 Separation of knowledge bases and inference mechanism 384
2.1.2 Capability of symbolic inference 384
2.1.3 Capability of explaining the inferences 384
2.2 Prolog (programming in logic) as efficient implementation language 385
2.3 A brief review of medical expert systems 386
3 Requirements Made of a Knowledge-Based System for Supporting Laboratory Medicine Diagnostics 387
3.1 Implementation of an inference engine suitable for medical question complexes 388
3.2 Combination of numerical and non-numerical information 388
3.3 Weighting 388
3.4 Screen representation and user interface 389
3.5 Properties of the inference engine 390
3.6 Storing and reconsultation 391
3.7 Imaging of the time concept 391
3.8 Fact input 392
3.9 Display of conclusions and correction 392
3.10 Explanation facility 392
3.11 Connection to the outer world 393
3.12 Range facility 393
3.13 Indication output 393
3.14 Structure of the knowledge base, rule types 393
3.15 Language concept, rule maintenance 395
3.16 Extension of the language scope of PRO.M.D. 396
4 Applications 397
5 Further Expert Systems for Clinical Chemistry Question Complexes 398
6 Limits and Possibilities of Knowledge Representation by Expert Systems in the Field of Laboratory Medicine 399
6.1 Limits 399
6.2 Possibilities 399
6.3 Objectives 400
References 400

Chapter 3.9 Single-Point Method and Bayesian Approach for Individualizing Theophylline Dosage (M. Oellerich and B. Schneider) 403
1 Introduction 403
2 Single-Point Method 404
2.1 Pharmacokinetic basis 404

2.2 Application for dosage prediction 406
2.2.1 Sampling time and population constant $\bar{\Omega}$ 406
2.2.2 Computer simulations 408
2.2.3 Nomogram for dosage prediction 409
3 Bayesian Method 411
3.1 Pharmacokinetic parameter estimation 411
3.2 Drug dosing program for theophylline 414
4 Validation of Methods 417
4.1 Single-point methods 417
4.2 Bayesian method 420
References 422

Chapter 3.10 Discriminant Analysis: A Special Procedure for the Detection of Alcoholism (E. Hansert and D. Stamm) 425
1 Introduction 425
2 The Problem of Detecting and Excluding Alcohol Abuse on the Basis of Clinical Chemical Parameters 425
3 Description of the Alcoholism Study 426
4 Problems in Applying Discriminant Analysis to Clinical Chemical Measurements 427
5 Decision Limits and Simple Discriminant Regions 431
6 The Computer Program for Finding Simple Discriminant Regions 434
7 "Nesting" of Discriminant Regions 437
8 Discussion of the Procedure 441
9 Discussion of the Results 442
References 444

Chapter 3.11 A New Distribution-Free, Multivariate Discriminating Method First experiences gained in the discrimination of patientgroups suffering from coronary heart disease, alcoholism, and primary hyperparathyreoidism (B. Keller and H. Keller) 447
1 Outline of the Method 447
2 Efficiency Tests 451
2.1 Textbook model: Discriminant analysis for the assessment of survival in cases of icterus hemolyticus neonatorum (2 parameters available) 451
2.2 Bivariate clinical study: Discriminant analysis for the assessment of the coronary status (8 parameters available) 453
2.3 Trivariate clinical study: Discriminant analysis for the detection of alcoholism (24 parameters available) 458
2.4 Tetravariate clinical study: Discriminant analysis for the identification of primary hyperparathyreoidism (8 parameters available) 462
Appendix 468
References 468

Subject Index 471

1 Introduction

Data Presentation and Interpretation: Problems, Tools and Goals

H. Keller

The book in hand is concerned with a topic, which is one of the most difficult in laboratory medicine. Many different forms of data presentation are described in the literature and articles concerning interpretative laboratory reports are numerous. This variety shows that we have still not achieved a consensus about how laboratory data should be presented and interpreted. Therefore even the 17 contributions of this volume give no cardinal principles for optimal data presentation and interpretation. Nevertheless they indicate concrete possibilities to reach the intended goal, to give the physician in his patients treatment an effective help for his decision making.

The volume consists of two parts: First the *Presentation* of data, introduced by a paper by P. E. Politser, which exemplifies impressively how optimal presentation could be. The actual presentation used in clinical laboratories today strays far from it.

Modern data presentation as it is used in a big central laboratory is described by A. J. Porth et al. in a detailed description; D. P. Conelly and W. Renn, R. Maulbetsch and M. Eggstein deal with the special aspect of graphical displays and suggest many original solutions.

The concluding papers by J. O. Angell and L. Reynolds inform about hardware tools and software tools used for graphical displays.

The second part of the volume is concerned with *Interpretation* of laboratory data, i. e. its logical basis. The first three articles deal with the most important basic approaches: the conditional probabilities corresponding to the Bayes Theorem (Willie Gerhardt), the information-theoretical treatment (J. Büttner) and the deterministic methods (M. D. D. McNeely).

The next three contributions are especially concerned with the evaluation of laboratory data: an introductory survey by P. Winkel is followed by detailed observations of clustering (W. Vogt et al.) and X-transformation (W. Rehpenning), and the third contribution is about expert systems, explained by P. D. Clayton et al. and Ch. Trendelenburg.

The last section describes the following special applications: drug monitoring (M. Oellerich and B. Schneider), detection of alcoholism (E. Hansert and D. Stamm) and detection of coronary heart disease, alcoholism and primary hyperparathyroidism (B. and H. Keller). The methods applied are not restricted to this special problem, but can be transferred to answer other questions: drug monitoring for example to estimate endocrinological function tests, or the screening of alcoholism for other addictions.

The contribution of Keller and Keller also includes as an example the separation of alcoholics from non-alcoholics. The interested reader can note, that the data used for this example is the one provided by Hansert and Stamm. Thus advantages and disadvantages of different discriminating analyses can be compared.

Data presentation and interpretation are closely tied: if e. g. the concentration specification of a serum component is supplemented with the age- and sex-specific reference interval, a simple form of help for interpretation is given. Symbols (*, ↑, ↓, different printing inks etc.) to enhance single data, i. e. to focus the attention of the reader on it, are also of interpretative value.

In a clinical laboratory one has to distinguish between intra-laboratory and external presentation and interpretation. *Intra-laboratory presentation* concerns the listing of received samples, lists for each place of work, so-called lists of remainders, extreme value lists, deltacheck lists etc. Other important examples for intra-laboratory presentation are graphical displays of permanent quality controls and result presentation, before the results are transmitted to check them or after they are transmitted to cope with further inquiries from physicians, e. g. over the telephone.

Intra-laboratory interpretation concerns first of all the evaluation of the validity of tests or test combinations. Therefore a large laboratory for example has to evaluate a newly described test before introducing it, to assure that to its ordering physicians an effective help for their decision making is given. Only carefully calculated cost/benefit assessments based on the data of the ordering clinics, its special wards and scientific interests, allow rational decisions.

External presentation means display of laboratory results for the ordering physician. It differs depending on the situation: the result display for an intensive-care unit, which has to receive the results of certain parameters as quickly as possible, is necessarily different from the result presentation for an endocrinological function test, performed for an ambulant patient. The same is true for the *external interpretation*, which should provide the physician with the requested help for decision making, but avoid superfluous advice: the medical management of a hepatological expert division should not receive an interpretative text about the meaning of the result of transaminases; whereas the practising physician is grateful to receive competent dosage advice along with the monitoring of antiepileptic drugs.

All contributions of this book take it for granted explicitly or implicitly, that presentation and interpretation of laboratory data are performed by electronic data processing. Or in other words, for all authors of this volume A. E. Rappoport's question "... Should laboratories have their own computers? ..." is answered positively.

At the moment computer technology is undergoing drastic changes. Computer systems ordered today are already old-fashioned at the time of their installation and operation. The situation is made more complicated, because the manufacturers of measuring instruments and autoanalyzers use more and more "intelligent interfaces" for their configurations. The intersecting areas are considered easily adjustable for on-line data transmission. The hardware intersecting area may be standardized (e. g. V 24), but the form of data records differs greatly depending on product and manufacturer (coding of method number, measuring value, dimension, additional informations etc.). As a result, major difficulties very often arise when a connection to the computer system is attempted. If several "multichannel analyzers" transmit large amounts of information simultaneously to the central processing unit, the system may break down. Therefore several laboratories use self-developed "intelligent interfaces" with an integrated storage for measured values of appropriate capacity. Corresponding to its disposal the computer allows the data to be transmitted and avoids narrow passages during processing. Additional protection by an external data buffer against power failure and a breakdown of the central processing unit prevents the loss of data. To prevent a breakdown of the calculating system and the possibility of loosing data partially or totally, duplex-EDP systems are used, an effective, but expensive solution. – Commercial computer networks are applied in an increasing number but whether they will be of use in clinical laboratories can not be answered yet. Until now no logical PC solutions have been established in bigger laboratories.

Not only is the hardware technology subject to constant changes – one only needs to think of the permanent "new generations" of PC's – but also for software, especially for the operating system and program languages, a consolidation is not in sight. No one can predict nowadays, with which operating systems and program languages laboratory EDP systems will work in the year 1990 or even 2000. This prognostic insecurity is reflected in the contributions to this volume: no author gives concrete recommendations concerning hardware or software or even recommendable periphery systems. From the economical point of view it might be considered a shortcoming, but it includes the advantage that data presentation and interpretation will keep being a challenge to the creativity of clinical laboratories for a long time. All of the following articles prove, how original individual self-developments can be compared to ready-made solutions, which are being offered commercially today. Obviously ready-made systems can only provide a "basic supply" of presentation and interpretation. In fact they just serve as electronical card-index boxes and list generators, most often completed with bookkeeping programs. These are the reasons why a description of an average commercial system was dispensed with.

Difficulties similar to forecasting the future development of hardware-specific and computer-specific software arise for the logical procedures of interpretation of laboratory results for the next decade. Which one will dominate? Will the expert systems push aside the deterministic methods? How far will information-theoretical considerations spread? It is not possible to make a statement at the moment. Even less so, because the manifold logical procedures do not exclude, but often complement each other.

And there is one other aspect to consider before reading this volume: the interpretative presentation is the last stage of a working process – today generally referred to as documentation – which contains five stages: listing, formatting, storing, refinding and presentation of data.

To list data the clinical laboratory has once more to distinguish between internal and external data listing. External data concern the order for examination in a broad sense, basically the identification of the ordering physician, patient, patient sample and order for examination. At the best, physical records can be obtained through an on-line connection to a hospital information system, which allows access to all identification data. The most complicated solution is the manual input of all necessary data over the keyboard of a terminal. This process – still performed in many laboratories – can be simplified by corresponding masks. Modern interim solutions are barcode-ciphered data and information in machine-readable non-coded writing, e. g. OCR-writing, which can be adapted by special reading units. Mark-sense-cards or mark-sense-sheets are widespread today.

With internal data listing the situation is almost the same: analyzers with corresponding big data output are preferably connected on-line to EDP systems. The problematic nature of on-line connections – with or without additional interface – were already mentioned. Working places with small data production work often off-line with keyboard terminals. In any case a basic difficulty must be overcome: the data of the results have to be connected securely to the identification of the patient material. Today an "indirect identification of samples" is still in use (the position of a sample holder or processing vessel in a rotation plate, resp. chain, magazine etc. gives the identification), but has been replaced by the "direct identification" with barcode-marked processing vessels. For a secure and efficient operation a matching laboratory organisation is required. Besides these possibilities single laboratories use mark-sense-cards and/or mark-sense-sheets for their internal transmission. In summary, commercial hardware is available for data listing which can cover the basic needs of a laboratory.

The technical aspect of data listing is crucial for security: data listing via keyboard and terminal provides the user with the highest flexibility, but includes no protection against wrong input. But also the on-line data transfer can be subject to malfunctioning. All wrong, mutilated or missing information leads to consequences for the output.

Listing of data in a laboratory requires a definition of which data must be put in. Basically there are three data blocks: the individual data of the patients, the data of the ordering physician and the specific laboratory data. There are no rules, not even recommendations, how detailed these blocks must or should be. The individual data of the hospital-patients are generally registered by the administration. This leads to a missing of important information for the laboratory, e. g. body-height, weight, body-surface, affiliation to a group with special disease disposition etc. If these data are required for the interpretation, the laboratory has to find a way to get them. Frequently the ordering physician is not correctly identified. Instead of a personal identification, a division of the hospital is named and instead of an aimed, well-founded question, e. g. "suspicion for prostata carcinoma," an overall request like "enzyme block" is made.

Many unanswered questions concern the laboratory-specific data. Is the time between bloodletting and processing in the laboratory to be noted? Does it have to be stated, with which reagent and/or from whom, with which analyzer and when a certain analysis was performed? Should it be documented, whether it is venous or arterial blood and which additives were used? Is the specific alteration of the material, e. g. lipemic, hemolytic, jaundiced, stated in every case? Facts, which are not noted, cannot be processed and therefore possible consequences for analytical results cannot be considered.

The storage of data for the basic documentation takes place in a predefined format, which must consider the special needs of a clinical laboratory. An unsuitable format might lead to a long response time of the system and greatly restrict the efficiency and/or the user friendliness.

For the physical storage of data, a multitude of media are at disposal: band cassettes, big band stations, diskettes of different sizes, mass-storage-units and many others. The development of electronical storage media is moving at a great pace, and it cannot be estimated which solutions will succeed in the future. One thing is for sure, every electronical storage medium contains the danger of an irreversible loss of data. To minimise this danger, two physically independent data storages are frequently used. It is possible to feed them at the same time (parallel) or in corresponding, short time intervals, i. e. to transfer the data from one current storage media to the other for long time storage.

The organisation of data in a computer decides the capacity and flexibility of the system software. Originally laboratory systems were equipped with so-called request files, worksheet files etc. The structure of data in these computer files, analogous to the processing in the laboratory, offered a comfortable way to get to the data for every specific function. But this structure restricted the use of the computer system to the possibilities prepared by the software engineer. If the system were applied for something different, which the software engineer did not provide for, the structure of the fixed data files set up insurmountable limits, i. e. the access to the data could only be guaranteed over the specific program. Any other access demanded every time the installation of a new program, for which eventually even the data structure had to be altered. A considerable increase of flexibility was reached with the introduction of database systems. Here we can distinguish between non-relational and relational databases. The non-relational databases request fixed datapaths, and the relational databases allow dynamic datapaths. For unfixed data relationships, a non-relational database is useful as it makes a short access time possible. The advantages of the slower relational base are that variable conditions can be freely formulated, e. g. "bigger than," "smaller than," "different from," "similar to" etc. Considering the up-to-date techniques, it seems possible that combinations of relational and non-relational databases will be developed, which will specially suit the medical laboratories.

To find single data again among millions of others requires the possession of suitable hardware tools and software tools for listing and storage. But if because of unsuitable tools the access times for the data base just last a few minutes, rational use of the EDP system is made impossible. A modern database system should be planned so that it is not only useful for the presentation of daily reports and cumulative reports, but that also scientific questions can be dealt with, e. g. comparing therapy effects, seasonal deviations, changes in the behaviour of the ordering physicians and many others.

Three hardware alternatives for the presentation of data are possible: screens, printers and plotters. Competent authors recommend colour screens, because the information on the screen is representable in a special easily comprehensible way (see Reynolds and Angell). Unchanged will be the use of printers in the future, but many different technologies will be used: matrix printers, inkjet printers, thermo printers and increasingly often laser printers. All printing systems have their advantages and disadvantages, concerning speed and quality, but most of all the costs. Screens as well as printers are used in the laboratory and many modern laboratories also provide terminals and/or printers for their ordering physicians, to quicken the transfer of data. General recommendations are impossible to give for external or intra-laboratory use. A thorough analysis of demand should be performed for every single issuing place, before it is decided, whether and in which form a data output is required.

The use of plotters, which normally allow the production of colour graphics, seems to be generally restricted to intra-laboratory use.

Everything said in the following chapters is based upon an optimal configuration of listing-, ciphering-, storage- and refinding-algorithms for the specific necessities. Everything not listed, wrongly ciphered, lost in the storage or mutilated, and for which no suitable orders exist, cannot be presented or interpreted.

2 Data Presentation

Chapter 2.1
How to Make Laboratory Information More Informative: Psychological and Statistical Considerations

P. E. Politser

1 Introduction

In recent years, laboratory data have become more numerous and more difficult to evaluate. Automatic analyzers and computer storage capabilities have multiplied the potential volume of test information, and new diagnostic technologies have increased its complexity. Simultaneously, the growing proportion of elderly patients with multiple organ system failure has made data interpretation more difficult. Thus, it is no surprise to find many published studies (reviewed in (1)) documenting failures to recognize important findings and diagnostic possibilities.

Some of these errors may arise from deficiencies in laboratory reports, which often seem to complicate rather than simplify information. Irrelevant data in them can distract us from significant results, as a magician's gesture can divert us from a trick. Both exploit the limitations of human attention.

Fortunately, these and other deceptions may be preventable if we design laboratory reports with human psychological limitations in mind. This paper illustrates how this may be done. It provides a framework, based on psychological research, for improving standard laboratory reports and computer data displays.

2 Background

Creating better laboratory reporting methods may help compensate for human limitations and support decision making in ways that other decision support methods cannot. Computer-assisted diagnostic methods all have serious deficiencies. Although Bayesian methods can be used to automatically interpret complex laboratory data (i. e. compute disease likelihoods), they typically do not assist in the early stages of diagnosis, in discriminating multiple co-existing disorders, or in monitoring changes in disease. Such applications are possible in theory, but in practice the possible disease spectra often are too numerous and the data insufficient. Artificial intelligence programs, derived from expert judgment rather than a database, may enlarge the range of applications. However, they usually do not assist perceptual recognition of patterns to monitor disease progression or treatment effects, and they also require simple patient categorizations, as do algorithms and most other decision support methods.

Graphic display techniques are free of these deficiencies. They can assist perception of abnormalities or relationships very early in the diagnostic process, help monitor subtle changes, summarize or reduce data, and even suggest the presence of coexisting diseases. They also leave the usual judgment tasks to the physician. As a result, they are often more clinically acceptable than other decision support methods.

Many different display techniques have been proposed (reviewed in (2)), including computer-generated time plots, multiple graphs related to a common system or organ, summary statistics, and displays only of abnormal results. These methods, however, can hinder as well as help. They sometimes overwhelm the clinician with too many results or omit important findings in an effort to reduce information overload. Some designers of computer display systems have begun to acknowledge the need to consider "human factors" (2, 3). Others (4) have developed general guidelines for the choosing graphic techniques in the presentation of research data. However, no one has yet adequately demonstrated the applicability of these new techniques to the improvement of laboratory data presentations. Also, the need for many additional techniques, motivated by other psychological literature, has been ignored. Most seriously, no adequate framework has existed specifically to guide selection of methods for display of laboratory data. This paper makes a contribution toward filling these gaps.

3 Basic Concepts

In the following sections, I present guidelines to show how to better tailor routine laboratory and computer-generated reports to the requirements of human perception. By pre-editing laboratory reports – as humans do when they read the reports – less human effort in extracting information from them is needed. These guidelines illustrate the use of this principle to assist six natural stages of perception: filtering, simplification, coding, grouping, recognition, and segregation (5, 6). Later, I discuss further guidelines to help decide which editing techniques to use when.

3.1 Filtering

Humans initially edit information by filtering. For example, physicians screen data in a written laboratory report to locate key findings. With experience, they may become reasonably proficient at this; however, superfluous data may still be highly distracting, and filtering them may require cognitive effort that would be better spent elsewhere. For example, the report in Figure 1a includes unnecessary numbers

a.

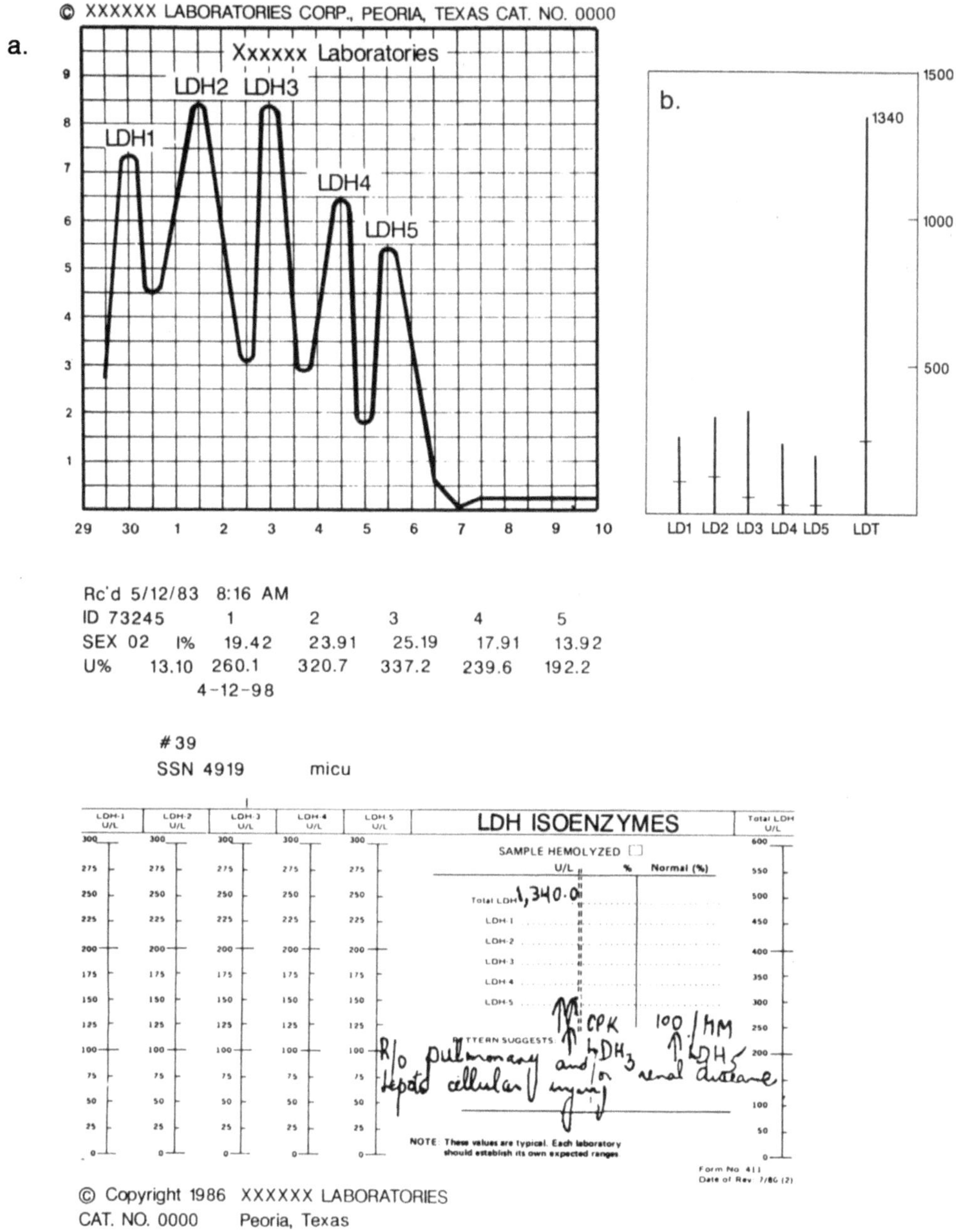

Fig. 1. In report *a*, a typical LDH isoenzyme report, I% refers to isoenzyme (1 – 5) proportions, U% refers to isoenzyme (1 – 5) values. These terms are not explained in the graph and we have discovered (informally) that many clinicians do not understand them. In report *b*, a simplified bar chart represents the essential information in *a*. The graph in *b* shows the LDH isoenzyme (1 – 5) and total (T) with a slash for the upper range of normal. (Reprinted by permission from Clinical Chemistry, 1986, Volume 32, pages 1510 – 1517.)

and grids and repetitious reports of normal ranges. It repeats the copyright number and the name and location of the company producing the sheet several times. Regrettably, similar distractions typically occur on many slips throughout patients' charts. (On many slips, in fact, the most prominently displayed piece of information is the name of the chairman of the department!)

These unnecessary distractions can be easily eliminated, and guidelines for doing so have already been published (7). Figure 1b illustrates the results of applying such guidelines. The isoenzyme proportions and total activity of lactate dehydrogenase (LDH) are given as bar graphs with a slash for the upper range of normal. This display omits unnecessary markings, repetitions, and legends, yet captures all the important information in the previous figure. This "pre-filtering" reduces the need for human perceptual editing.

3.2 Simplification

Humans also edit data by mentally simplifying them – for example, by rounding numbers. Unfortunately, this may require considerable effort, e. g., when laboratory data are reported to four digits as in Figure 1a. We do not need and cannot remember all of these numbers. Thus, we may want to simplify such data, perhaps by rounding them to two significant digits when additional ones are unlikely to improve judgment.

Nonnumerical, graphic reports of data also can be simplified. Cleveland and McGill (8) have suggested classifying graphical displays hierarchically, from the simpler (e. g., point graphs, bar charts) to the more complex (e. g., pie charts, Chernoff faces). As they demonstrated, the more complex the display, the less accurate people are in discriminating differences in values. For example, whereas some investigators have suggested that we report multiple test results by representing each as a different feature in a drawing of a human face (a Chernoff face) and varying their sizes or shapes according to the magnitudes of the result, discriminating the differences in the sizes and shapes of facial features is perceptually complex. The user must remember the meaning of each feature, and irrelevant variations in one feature can bias judgments about another (9). Thus, despite claims that such displays are more memorable, empirical evidence suggests we generally should avoid them and choose simpler methods (8).

3.3 Coding

To further reduce quantitative data, people often remember them only qualitatively. They mentally "code" data as changes from a reference point, e. g., as above or below normal, more or less than expected, etc. (5). Unfortunately, they may choose reference points improperly, by comparing laboratory findings with the normal range for an inappropriately general population. For example, failure to consider sex in using the hematocrit to regulate blood transfusion and similar omissions are discussed in ref. 1. To correct such errors, we commonly normalize reference values with respect to age, sex, lean body mass, etc. Still other forms of coding may involve comparing different test results and noting whether or not inequalities occur – e.g., is LDH 1 > LDH 2? Many common clinical heuristics take this form. If enough prior data points are available, we may even consider the patient's own past average of results. Because patients often have different homeostatic set points regulating their own normal values, recomputing the normal range based on previous results, when feasible, may avoid erroneous judgments about the significance of change (1).

Another common problem in coding results as normal or abnormal arises from the problem of multiple testing. Obviously the likelihood of an abnormal test result for a normal individual is not merely 0.05 (the frequency for an individual test) but depends on the number of tests done. If we perform 20 independent tests, the likelihood of a false-positive rises to 0.64. Thus the question arises: should we alter the normal range on the basis of the number of tests involved to ensure that the likelihood of an abnormal result from a normal patient still remains below some threshold (e.g., 0.05)? (This could be done automatically, with computer lab data displays, and perhaps with the Bonferroni inequality or, ideally, methods considering inter-test correlations (10).) Such a correction might have disadvantages, when physicians are accustomed to working with standard normal ranges for a common battery of tests; however, an adjustment in the normal range might be advisable in less common cases, especially when a large number of tests are done. If used to supplement rather than replace traditional reporting methods, this could help the physician "see" the data in another perspective. It could also diminish the frequency of unnecessary repeat tests to rule out false positives.

3.4 Grouping

Humans also perceptually edit data by crudely separating or grouping them (11). However, this can prove difficult if data are improperly arranged. To illustrate, try to identify the letter F, first in Figure 2a and then 2b. People are much slower and less accurate in Figure 2a because the letter F is not separated from the other group of symbols as it is in Figure 2b (11); in the first panel they must examine details within the group to see the letter.

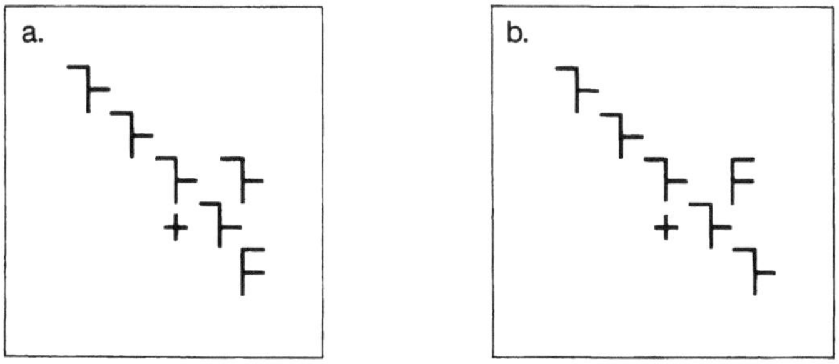

Fig. 2. The perceptual effects of grouping abstract symbols (adapted from ref. *12*): *a.* distinctive element (F) displayed together with others; *b.* distinctive element displayed separately. (Reprinted by permission from Clinical Chemistry, 1986, Volume 32, pages 1510–1517.)

a

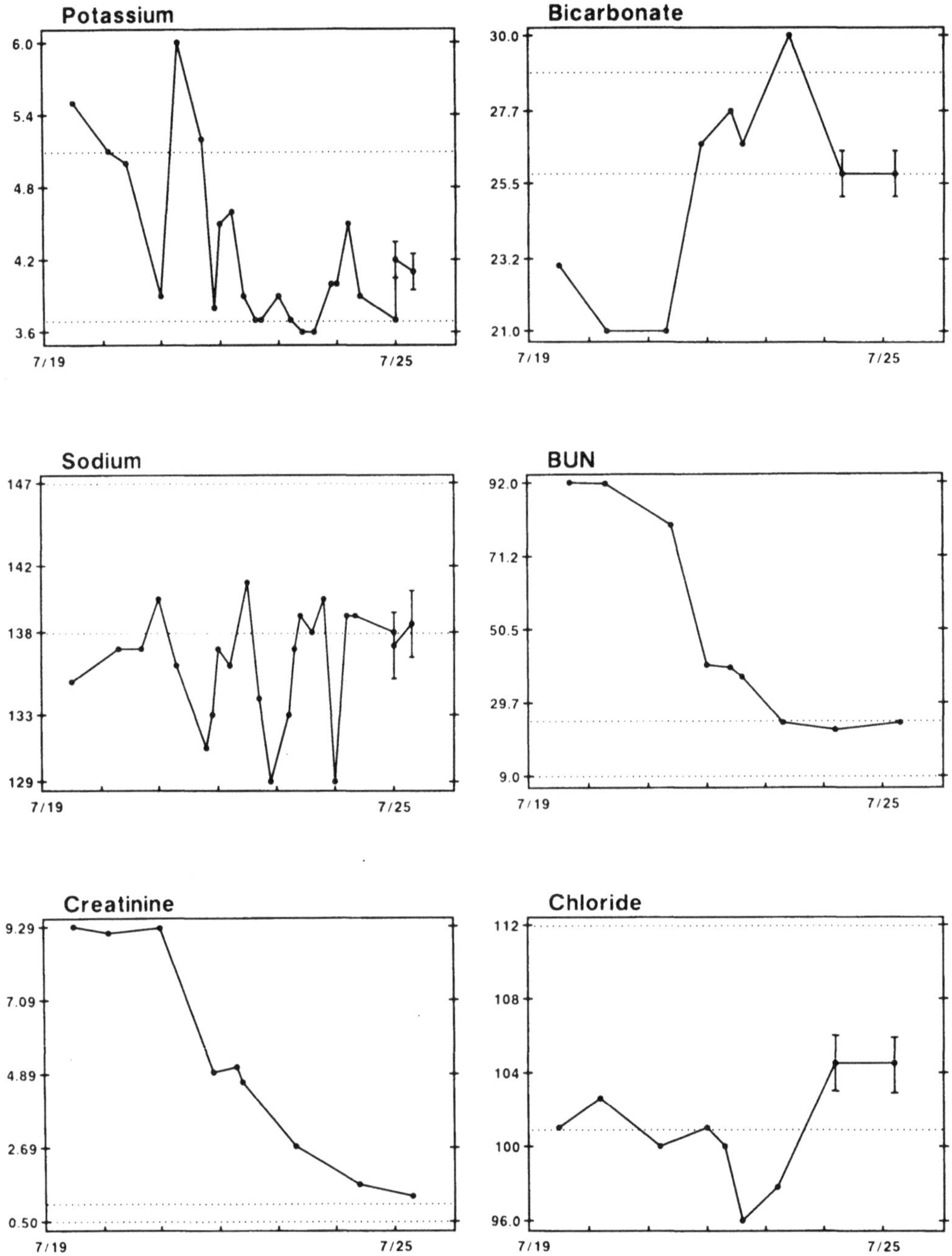

Fig. 3. Effect of grouping methods on information conveyed (adapted from ref. *2*): *a*. a disorganized panel of kidney function test profiles; *b*. the same panel, with related profiles linearly organized (BUN = blood urea nitrogen). (Reprinted by permission from Clinical Chemistry, 1986, Volume 32, pages 1510–1517.)

b

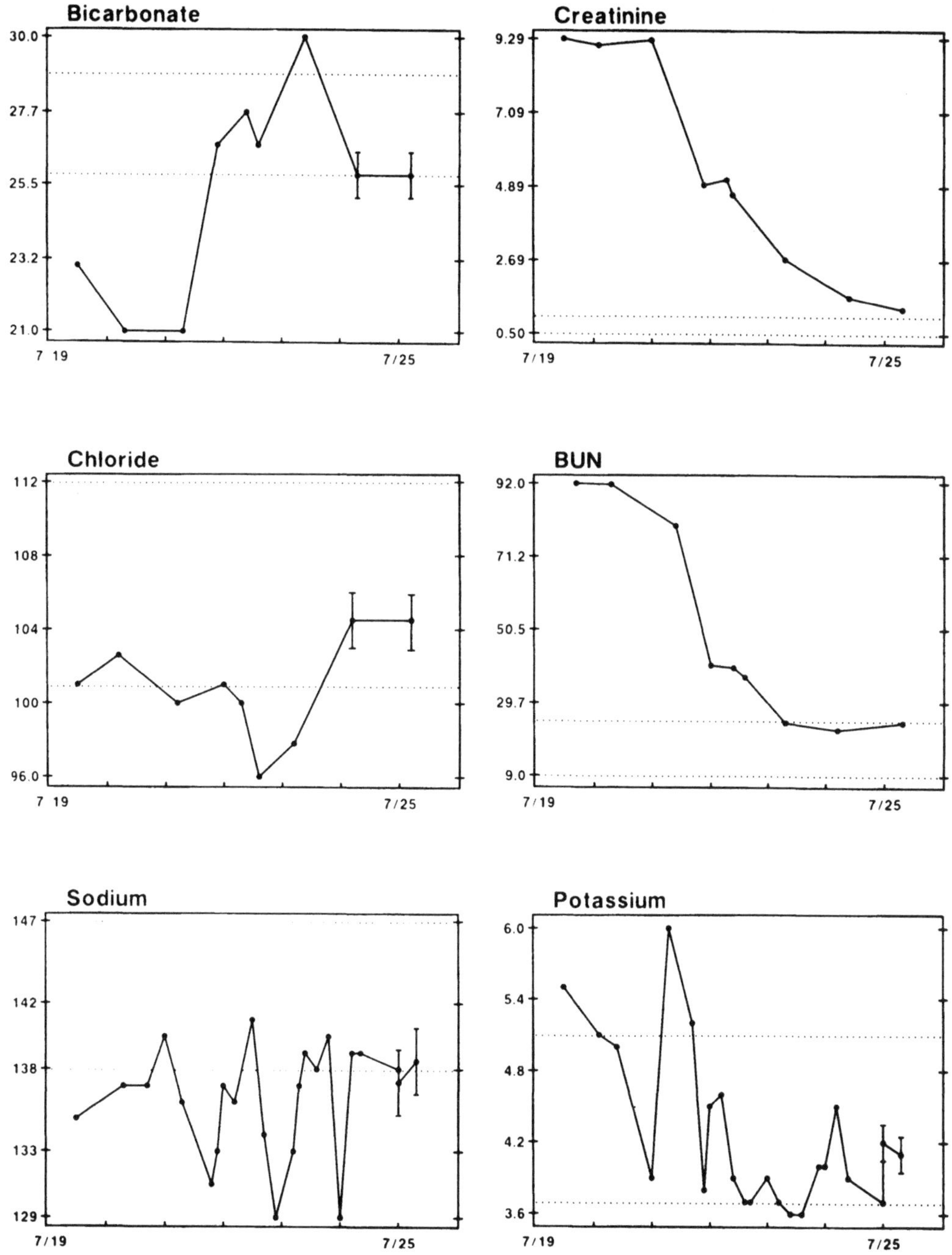
Bicarbonate
30.0
27.7
25.5
23.2
21.0
7 19
7/25
Creatinine
9.29
7.09
4.89
2.69
0.50
7/19
7/25
Chloride
112
108
104
100
96.0
7 19
7/25
BUN
92.0
71.2
50.5
29.7
9.0
7/19
7/25
Sodium
147
142
138
133
129
7/19
7/25
Potassium
6.0
5.4
4.8
4.2
3.6
7/19
7/25

Similarly, when lab reports are not properly grouped in a patient's chart, we may miss related or unusual findings. However, we can mechanically rearrange the data to make perception easier. For example, Connelly et al. developed a computer system to automatically group related findings such as results of renal or liver-function tests. We can also regroup related results at an even more detailed level. To illustrate, suppose we intentionally disorganize a panel of renal function tests in the system Connelly developed (Fig. 3a). Here, the synchronous patterns for blood urea nitrogen (BUN) and serum creatinine are almost indiscernible. However, rearranging the related elements in a single row or column (Fig. 3b) helps us to much more readily see the pattern correlations – the improvements in renal function (urea and creatinine) synchronously with the declining potassium.[1] Elsewhere, Connelly et al. (3) provided similar examples showing synchronous increases in urea nitrogen and creatinine, suggesting a renal rather than pre-renal cause. We notice related trends much more easily with such arrangements because perception naturally flows across a single row or column, or in some other continuous linear direction, as previously illustrated in Figure 2 (12).

3.5 Recognition

At a later stage of perception, we often must recognize less obvious findings, such as changes in a patient's condition. This, however, may require the ability to ignore irrelevant similarities and focus attention on differences in successive test results. Because this is a difficult task for unaided perception, especially when irrelevancies are prominent (13), assistance may be needed.

For example, Figures 4 (top and middle) shows that isoenzyme patterns with similar shapes on successive days may be scarcely distinguishable. However, by subtracting the raw values on day 1 from those on day 2 and thus removing the similar pattern features, an important change is clearly apparent: a statistically significant increase in LDH 1, often evidence of recurrent acute myocardial infarction. Thus, even when a characteristic sign of myocardial infarction (LDH 1 > LDH 2) escapes notice in the daily patterns, it may become obvious when the change between them is emphasized (ΔLDH 1 > ΔLDH 2 in Fig. 4 bottom). Such subtraction could, of course, be misleading when analytical variability is high, but when it is justified (e. g., when the differences are statistically significant), it could eliminate irrelevant similarities that distract us from important differences in successive results. Thus, it relieves the human mind of a difficult editing task.

Transformations other than subtraction might also help eliminate irrelevancies and clarify subtle differences. Displaying ratios, logarithms, reciprocals, or other transformations can, in theory, remove irrelevant curvilinear features of patterns in time plots of serial results (similar to those shown in Fig. 3). Reciprocal transformations have been advocated to linearize patterns for serial determinations of plasma creatinine, so as to distinguish changes due to renal transplant rejection

---

[1] This example is intended only to illustrate the perceptual effects of grouping, not to advocate that we always group analyses measuring the same function. In some cases the clinician may wish to group tests from different organs to examine their associations.

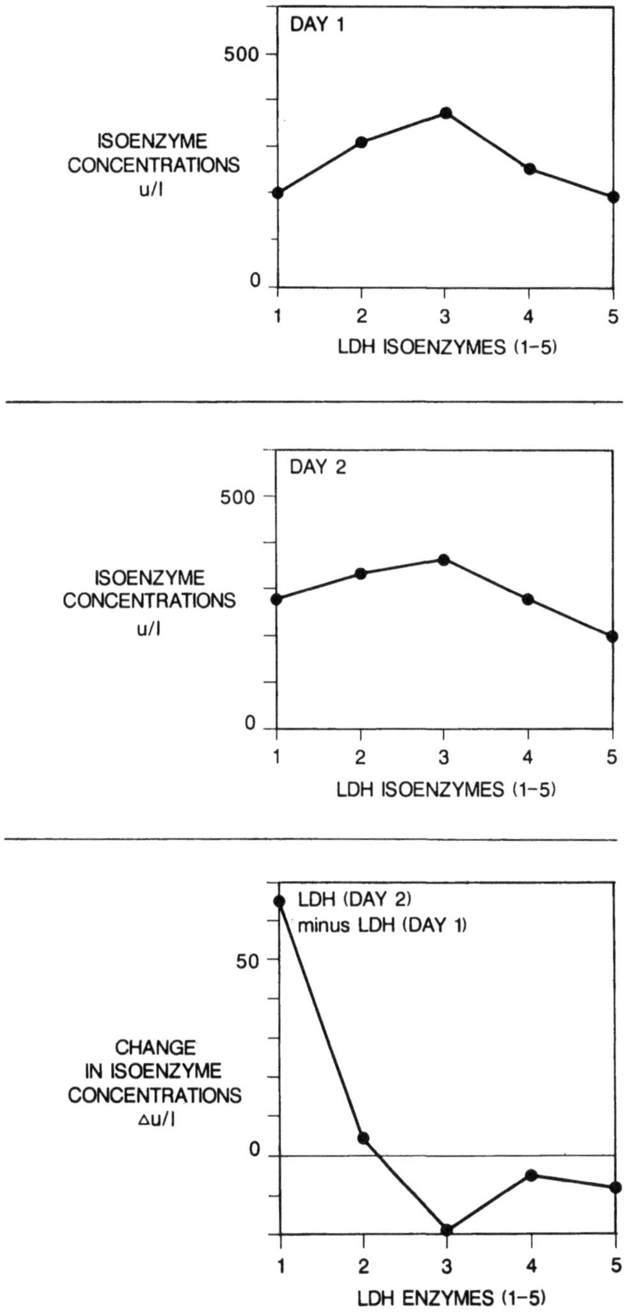

Fig. 4. Organization of information to facilitate recognition of changes: LDH isoenzyme on day 1 *(top)* and day 2 *(middle)* and the change in isoenzyme concentrations from day 1 to day 2 *(bottom)*. (Reprinted by permission from Clinical Chemistry, 1986, Volume 32, pages 1510–1517.)

(14). Unfortunately, patients are often too heterogeneous for any single transformation method to linearize patterns for all of them; choosing a transformation (e.g., a logarithm rather than a reciprocal) after the data have been obtained would be unreasonable and could lead to spurious results (missing a rejection when it occurs or seeing one when it does not). Thus these methods have often had limited utility in practice.

Other transformations, however, that assume less about the homogeneity of patients can minimize irrelevant changes or highlight important ones. For example, weighted averages of serial tests sometimes can remove short-term variability and help detect long-term trends (15). Conversely, a summation of successive results may quickly detect important short-term changes (16). Even altering the measurement scale to "fill" the graph, as shown in Figure 3 and as is done automatically in computer graphing systems, can magnify trends in serial tests. Unfortunately, this attempt to highlight important changes can also accentuate irrelevant ones. For example, the variations in serum sodium (Fig. 3) appear large only because the scale is considerably expanded. Thus, the choice of scale and transformation method requires a careful balancing of the risks of false positives and false negatives.

The choice of appropriate scales or transformations may also be influenced by the need to detect relationships between graphs (Fig. 3) as well as changes within a particular graph. Because different tests naturally have different means and variabilities, some changes may appear more salient than others, even when this conclusion is not warranted. Equalizing the error ranges in Figure 3 is one way to overcome this problem. Standardizing different tests (subtracting their means and dividing by their standard errors) also can aid comparisons and sometimes improve interpretation (17).

Overall, the issue of which scale to choose is a complex one. We should try to strike a balance between different communication goals, depending on the clinical problems likely to occur. Any single simplistic principle (e.g., "always change the scale to fill up the graph") is likely to fail. Also, when a single graph (as in Fig. 3) cannot satisfy all the requirements simultaneously, separate ones (individual and joint plots) may be preferred.

3.6 Segregation

Beyond recognizing changes due to a single disease or event (e.g., transplant rejection), sometimes we must detect co-existing disorders. Often we intuitively attempt to segregate their effects, trying to decide which individual diseases have caused the observed test abnormalities and to what extent. Unfortunately, in complex cases with failures of multiple organ systems, the effects of one disorder may conceal those of another (18). Thus, when possible, methods that mechanically segregate the effects of separate diseases may be useful.

To illustrate how co-existing diseases may be overlooked, suppose we construct four different LDH isoenzyme profiles: one by graphing their normal mean activity concentrations, another by adding a contribution from the heart, another by including one from the liver, and a final one by adding a contribution from both organs, according to the reported proportions of isoenzymes in these tissues (19).

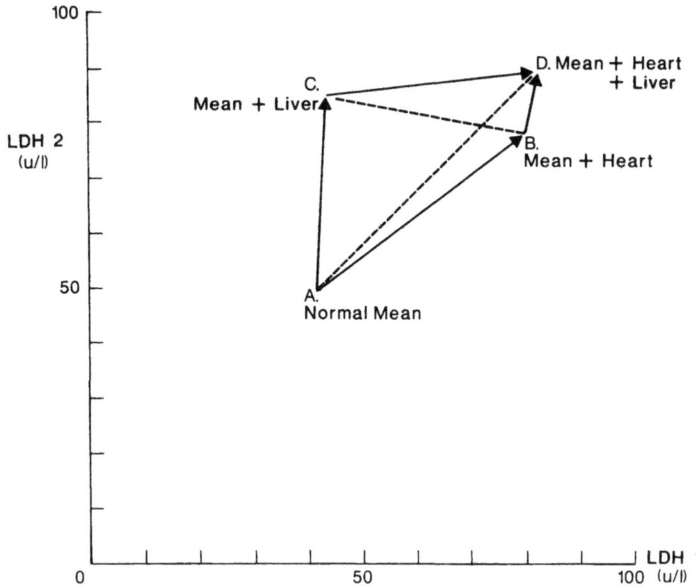

Fig. 5. Construction of the four LDH isoenzyme profiles. The arrows note that, to the normal mean values (point A) we add a contribution from the heart, which has high LDH 1 and low LDH 2 (point B). To this we also add a contribution from liver, which has low LD 1 and high LD 2 (point D). The arrows from A to C and C to D represent a similar sequence of additions, but in reverse order (first a contribution from the liver and then one from the heart). (Reprinted by permission from Clinical Chemistry, 1986, Volume 32, pages 1510–1517.)

(The graph in Fig. 5 illustrates this construction method for two isoenzymes, but the conclusions that follow apply to all five.) Such profiles actually do occur in patients with acute myocardial infarction, congestive heart failure, or coexisting diseases (myocardial infarction complicated by congestive heart failure). Suppose next we ask: in theory, how easy to distinguish should these profiles be – for example, how easy should it be to detect co-existing disorders (to distinguish their profile from the normal one)? The answer is that, statistically, profiles for co-existing diseases and for normal values are further apart (Fig. 5) and so they clearly should differ more than the other pair of profiles (heart alone and liver alone). In fact, this is true for any proportions of contributions representing different severities of congestive heart failure and acute myocardial infarction, and virtually any common measure of similarity.[2] So, when we display the entire set of LDH profiles

[2] This holds true in five dimensions (although only two are pictured in Fig. 5) for any proportions of contributions from the liver or heart, representing different severities of congestive heart failure and acute myocardial infarction, and according to an infinite variety of statistical distance measures – Euclidean, city block, or any Miskowski metric (20) with a parameter between zero and infinity. It also is true for the Mahalanobis distance, computed from the data in ref. 21.

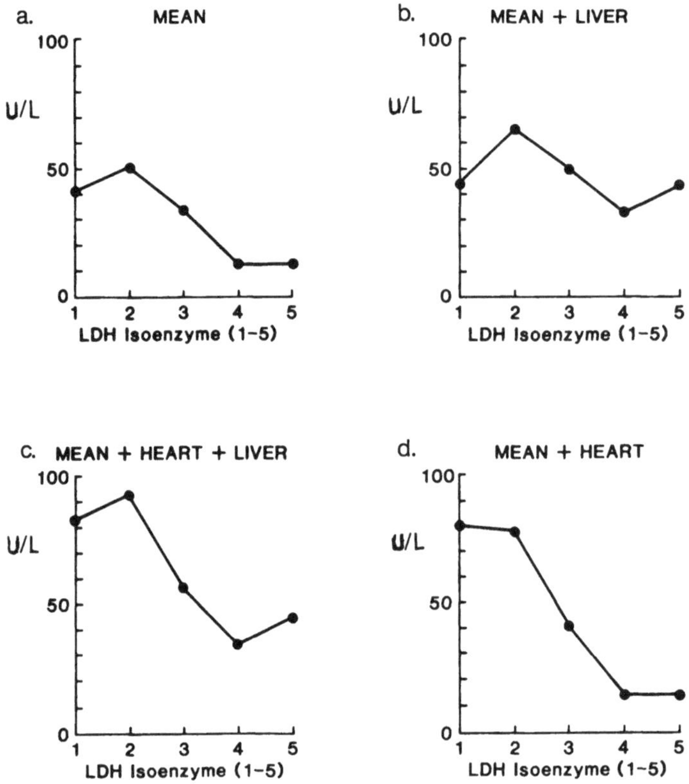

Fig. 6. Actual data from the four LDH isoenzyme profiles whose construction was illustrated in Figure 5. (Reprinted by permission from Clinical Chemistry, 1986, Volume 32, pages 1510–1517.)

(as in Fig. 6), it should be easier to discriminate coexisting disorders (Fig. 6c) from normal results (Fig. 6a) than to distinguish individual abnormalities from the heart (Fig. 6d) and from the liver (Fig. 6b).

Curiously, however, our perceptions of the actual profiles so constructed seem to violate the predictions of statistical theory. The pair we expected to be *less* similar (Figs. 6a and 6c) appear *more* similar, and the pair we expected to be more similar (Figs. 6b and 6d) look quite different. In the latter pair, our attention naturally focuses on their obvious dissimilarities in shape (due to the LDH "flip," LDH 1 > LDH 2, characteristic of acute myocardial infarction, in the heart profile, and the increased proportion of LDH 5, characteristic of liver abnormalities, in the other). However, in the combined profile (Fig. 6c), the liver contributions mask those of the heart; by increasing LDH 1 relative to LDH 2, they prevent the LDH flip. Thus the combined profile (Fig. 6c) resembles the normal profile (Fig. 6a).

This similarity, however, is an illusion. The similar shapes of the profiles first catch our eyes and distract us from their statistically important differences (the graph in Fig. 6c is elevated compared to that of Fig. 6a). Indeed, with traditionally reported proportions or electrophoretic profiles (as in Fig. 1a), even this difference would vanish.[3]

To remove this illusion, my colleagues and I have developed a method to estimate the separate contribution from each organ (21). We first defined five iso-enzymatically similar types of organs through a cluster analysis. We then estimated (through solving a system of linear equations or by regression analysis) the unknown amounts that each "type" of organ contributed to the total LDH. A detailed description of this method can be found in the appendix and in a separate paper (21).

Illustrative example

Figure 7 demonstrates the application of this technique to the data of a patient admitted with clinical and laboratory evidence of myocardial infarction. Two days after admission, this patient developed recurrent chest pain and clinical signs of mitral valve prolapse, including pulmonary edema. Subsequently, he underwent surgery for coronary artery bypass with mitral valve replacement, had cardiac respiratory failure, and died. The right side of the graph showing the organ contributions clearly indicates that most of the LDH came from the heart, liver, and lungs (or from iso-enzymatically similar organs).

Interestingly, the increases in lung and liver LDH clearly mirror the development of congestive heart failure after mitral prolapse. From the raw data, the pathologist noted possible liver abnormalities, because of the increase in LDH 5, but failed to consider lung congestion. More importantly, clear evidence of recurrent infarction appears in the transformed data from the last three days (Fig. 7b). The clinical staff considered re-infarction unlikely because the electrocardiographic and clinical findings (chest pain relieved by nitroglycerin) were ambiguous, there was no report of creatine kinase MB[4] isoenzyme, and there was no new LDH 1:2 flip (LDH 1 > LDH 2). Undoubtedly, the lung and liver contributions increased LDH 2 relative to LDH 1 and prevented the flip. The estimates of separate organ contributions, however, clearly reveal the previously hidden heart abnormalities.[5]

The validity of this transformation was confirmed by autopsy, which revealed evidence of a new infarction in the anterior papillary muscle, undoubtedly the cause of the mitral prolapse. An experimental test of this new method with 73 patients

---

[3] Although bar graphs as in Figure 1b might clarify differences between Figures 6a and c, they have their own drawbacks. Cleveland and McGill (8) have pointed out, the longer the bars, the more difficult it is to distinguish differences within a profile (e. g., between the isoenzymes 1 – 5 *within* Fig. 6a or 6c). So bar graphs might help correct one problem, but would create another. They also would not elucidate the sources of the abnormalities in Figure 6c (heart and liver). Moreover, even if point graphs (e. g., Figs. 6a and c) are suboptimal, they are often used in automated systems for graphing laboratory data (e. g., Figs. 3a and b); thus our example serves to illustrate serious problems with these commoner forms of display. Failure to use the same scales for the axes or merely reporting isoenzyme percentages, as laboratories sometimes do, further compounds these problems.

[4] Actually the CK MB was never determined because the total CK activity was not high enough to fractionate the isoenzymes. A closer inspection of the data revealed, in fact, the presence of a recurrent peak in the total CK, but this was overlooked by the clinical staff.

[5] Because erythrocyte isoenzyme proportions are similar to those from the heart, we might have also considered hemolysis (e. g., hemolytic anemia) as a potential source of LDH abnormalities. However, no evidence supported this (the patient was not anemic).

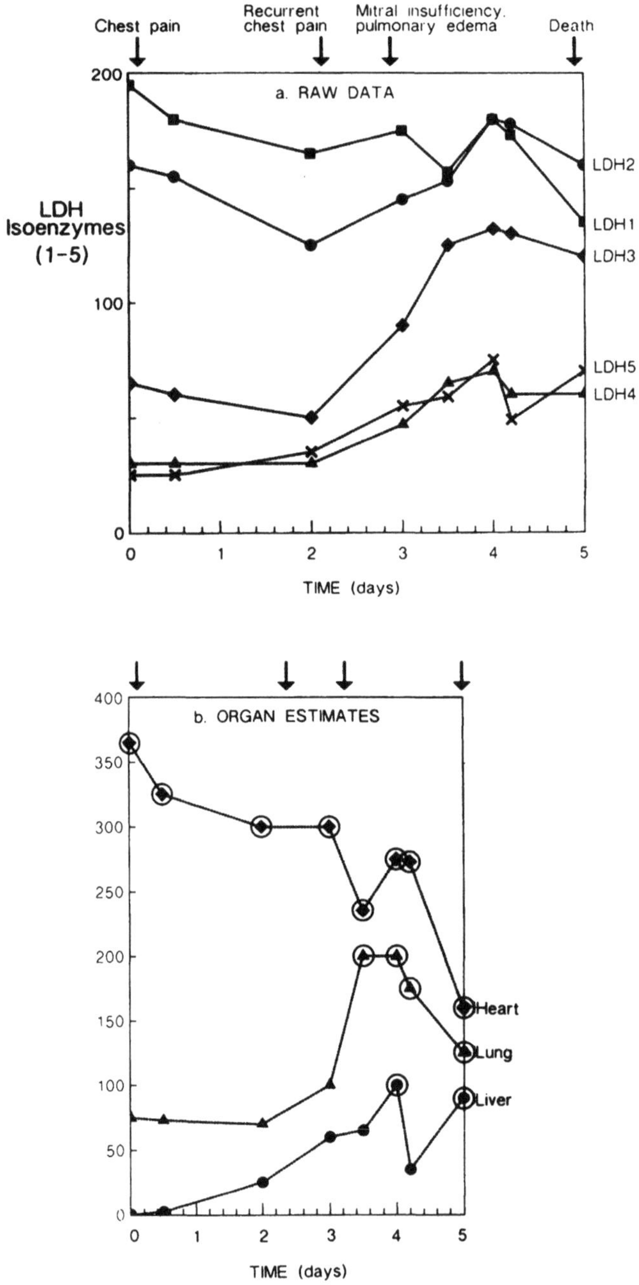

Fig. 7. Interpretive display of changes in serial LDH isoenzymes in a patient with mitral prolapse: *a*. Raw data (uncorrected actual activities of LDH isoenzymes in serum). *b*. Display of estimates of the amounts of total LDH (from the data in *a*) attributable to separate clusters of organs (the arrows in *b* (unlabelled) refer to the same clinical events described at the corresponding times in *a*). (Reprinted by permission from Clinical Chemistry, 1986, Volume 32, pages 1510–1517.)

in the intensive care unit also revealed gains in the detection of acute myocardial infarction and other disorders (such as pulmonary embolism). We performed a split-half cross-validation, taking half of the cases, determining the optimal threshold based on discriminant analysis, and applying it to the other half of the cases to determine sensitivity and specificity. The order of analysis was then reversed and the results from both analyses were summed to calculate overall sensitivity and specificity. The result was that for the detection of acute MI, the test had 96% sensitivity and 100% specificity. It significantly outperformed unaided pathologists' judgments and accepted indices (LDH 1:2, LDH 1, total LDH, LDH 1:total LDH) for interpretation of LDH isoenzymes (21). Moreover, the cases in which this approach did better were almost always ones in which other diseases or complications concealed the effects of acute myocardial infarction. Thus, the uncovering of hidden disorders by estimating separate organ contributions was not isolated to the case in Figure 7 but appeared useful for other patients as well. A more complete discussion of the empirical validation of this methodology is reported elsewhere (21).

4 Discussion

I have discussed human psychological limitations at different stages of perception, and have suggested possible methods for displaying laboratory data to attenuate these vulnerabilities. Natural psychological editing skills serve as a model for many of the methods.

We can speculate about how and when each may be useful. Filtering and simplification techniques seem the most likely to be applied first, because they are the first steps naturally required to perceive important results and their use does not omit important information; they could be used routinely in laboratory reports or computer displays. Depending on the context, humans may naturally switch between other editing mechanisms, such as coding and grouping (22), which might best be used interactively through computer displays (2, 3). With these, any data temporarily suppressed or condensed could easily be retrieved. Recognition methods, such as rescaling or subtracting serial results, probably should apply when more subtle distinctions are critical or are most likely to be obscured. For example, we may need to rescale results in routine reports to communicate the impression of change more forcefully when serum constituents are tightly regulated. An interactive computer system might even be programmed to display changes automatically when they are most likely to be obscured – e.g., when two successive multivariate profiles (as in Figs. 4a and b) are highly correlated or perceptually similar by other standards. Similar principles could guide use of segregated displays like the estimates of organ-specific LDH, but only under special circumstances, when appropriate methodologies exist. Inevitably, in the presence of noise there is a tradeoff concerning what differences (or segregated components) should be reported. Statistical tests of their significance, perhaps weighted by their clinical importance, should help guide their selection.

Most importantly, the physicians themselves (more likely to be the end users of the data rather than pathologists) must decide what displays are needed for patient management and must have some flexibility in choosing them. Display needs may vary in different settings. In intensive care, with serious time constraints, for example, physicians may need simpler, more familiar forms of test display (perhaps even numeric rather than graphic). Also, a physician actively searching for expected information may require different forms of editing than one who is passively receiving unexpected findings. The clinician's needs may also differ depending on whether the data are for monitoring patients or for diagnosis. In the latter case, clinical priorities for the display of information should center on the clinician's diagnostic hypotheses. Such methods require extensive knowledge of disease-finding relationships, as embodied in artificial intelligence systems. Elsewhere I will discuss a system that provides further editing mechanisms based on hypotheses, whereby the clinician selects, at a terminal, different segments of a data set on the basis of various hypotheses (23).

Clearly, further research is needed to answer many other questions and to refine the guidelines just presented. Studies must determine when the benefits of adding edited displays to the raw data outweigh the increased information burdens they create. We must also ask whether new displays produce patterns with unclear relevance or eliminate information needlessly; in some cases, redundant information may actually improve human judgment (1, 8). Editing may also be less useful for physicians with more expertise or for those more familiar with a case (e.g., consultants vs house staff) (12). Given that natural perceptual editing techniques appear to change with increased expertise or familiarity (24, 25), different physicians may require different degrees of prior simplification.

Once we better understand the skills of specialists in editing complex patterns of laboratory data, we may even discover new, more effective strategies. By modeling and applying these refined expert editing skills to data, we may see new information that is informative even to the experts themselves.

Appendix

The following section discusses three methods that can be used to derive organ-specific estimates of LDH isoenzyme abnormalities. Each method was based on a cluster analysis that define isoenzymatically similar and distinct "types" of organs. Correlations between the published proportions of isoenzymes in each pair of organs (19) provided a measure of similarity between them. Five enzymatically similar organ types were defined:

Cluster 1: Pancreas, lymph, adrenal, thyroid
Cluster 2: Liver muscle, tongue mucus (liver/muscle-related type)
Cluster 3: Heart, kidney, erythrocyte (heart-related type)
Cluster 4: Leukocyte, platelet
Cluster 5: Lung, colon, gallbladder, esophagus, duodenum, brain, thymus, prostate, spleen (lung-related type)

The first method to estimate the contribution of each organ type to the total LDH (which we call "organ-specific estimates") simply solved a set of linear equations for the unknown contributions.

To illustrate this method heuristically, suppose for example that we measured only two isoenzymes and that only two organ types (cluster 2 and 3) were potential contributors (this assumption is made only to simplify the explanation and is not necessary, as we shall see later).

Then to estimate the contribution of each organ cluster to the total LDH, we could write the following equations:[6]

$$\begin{aligned}\text{Total L1} &= (\%\ \text{L1 in cluster 2}) \cdot (\text{Total LDH due to cluster 2}) \\ &\quad + (\%\ \text{L1 in cluster 3}) \cdot (\text{Total LDH due to cluster 3}) \qquad (1)\end{aligned}$$

$$\begin{aligned}\text{Total L2} &= (\%\ \text{L2 in cluster 2}) \cdot (\text{Total LDH due to cluster 2}) \\ &\quad + (\%\ \text{L2 in cluster 3}) \cdot (\text{Total LDH due to cluster 3}) \qquad (2)\end{aligned}$$

Note then that if we divide all terms in the first equation by the percentage of L1 in cluster 3, divide all terms in the second equation by the percentage of L2 in cluster 3, and then subtract the second equation from the first, we obtain:

$$\frac{\text{Total L1}}{\%\ \text{L1 in cluster 3}} - \frac{\text{Total L2}}{\%\ \text{L2 in cluster 3}} = \frac{\%\ \text{L1 in cluster 2}}{\%\ \text{L1 in cluster 3}} - \frac{\%\ \text{L2 in cluster 2}}{\%\ \text{L2 in cluster 3}} \cdot (\text{Total LDH due to cluster 2})$$

Then if we divide both sides by the term in brackets, we obtain:

$$\frac{(\text{Total L1}/\%\ \text{L1 in cluster 3}) - (\text{Total L2}/\%\ \text{L2 in cluster 3})}{(\%\ \text{L1 in cluster 2}/\%\ \text{L1 in cluster 3}) - (\%\ \text{L2 in cluster 2}/\%\ \text{L2 in cluster 3})} = \text{Total LDH due to cluster 3}$$

All of the quantities on the left side of this equation are measured or known from published data; thus we can estimate from these the total *LDH ascribable due to cluster 2*. Likewise, by subtracting (percentage of L1 in cluster 2) · (Total LDH due to cluster 2) from both sides of equation (1) and then dividing by the (percentage of L1 in cluster 3), we obtain:

$$\text{Total L1} - (\%\ \text{L1 in cluster 2}) \cdot (\text{Total LDH due to cluster 2}) = \text{Total LDH due to cluster 3}$$

Since we have estimated the total LDH due to cluster 2 and have measures or estimates of the other quantities on the left side of the equation, we now can use this to compute the other previously unknown quantity, *the total LDH due to cluster 3*. We have solved two equations for two unknown quantities.

[6] In the method we have developed, we actually use corrected values for total L1 and L2 – we first subtract the mean L1 values from total L1 and the mean L2 from total L2. The percent of L1 and L2 in each cluster are averages of the published isoenzyme proportions across each organ in the cluster.

We can analogously do this by writing five equations (for the total values of each isoenzyme 1 – 5) to solve for five unknowns (the total LDH due each isoenzymatically similar cluster of organs); i.e., we assume:

$$L = X\beta + \mu$$

where

$L = \{L_i\}$ indicates the observed value of each isoenzyme, $i = 1, \ldots, 5$

$X = \{X_{ij}\}$ represents the proportion of isoenzyme i in organ cluster j (as defined in the Methods section), $i = 1, \ldots, 5$ and $j = 1, \ldots, 5$.

and $\mu = \{\mu_i\}$ is the population mean value for each isoenzyme. Then, given the observed values of L and published or derived values of X and μ,[7] we solved for $\beta = (\beta_1, \beta_2, \beta_3, \beta_4, \beta_5)$ which is the amount of total LDH contributed by organ clusters 1 to 5. This method has the clinical advantage of allowing crude estimation of all five organ clusters but does not provide confidence intervals for them.

A second method to derive organ-specific estimates is through regression analysis. Since we have only five measurements (isoenzymes) at each time, we decided to try to estimate and provide confidence intervals for three clusters, at most. In our cases, the heart, liver, and lung-related types accounted for nearly all of the variance and thus seemed sufficient; however, the investigator may use any three of the five organ clusters.

To estimate the organ sources we then assume:

$$L = \mu + X\beta + \varepsilon$$
$$X = (X_1, X_2, X_3)$$
$$O = (O_1, O_2, O_3)$$
$$E(\varepsilon) = O \quad V(\varepsilon) = \sigma^2 I \text{ and } \varepsilon \sim N(O, \sigma^2 I)$$

Although intuitively a weighted regression analysis would seem potentially preferable to this method, to more adequately account for interdependencies between the residuals, we have found empirically that assuming independence provides much more adequate predictions in terms of the total variance accounted for by the regression. This may be due to differences in the correlations between errors for different types of cases (myocardial infarction, congestive heart failure, etc.). Better estimates of these could eventually provide further improvements, but such improvements are likely to be minor in view of the current performance levels.

A more important reason for caution in the interpretation of the results of the regression is that the confidence intervals are probably somewhat narrower than the true intervals, which would really include the possibility of errors in the published parameter estimates between and within individuals. This is not a serious flaw because the potential error from false positives is generally less than that from false negatives and because clinical judgment ultimately must be used in interpreting the

---

[7] Based on our cluster analysis and published data (19) we used the values;

$$X = \begin{bmatrix} 13.75 & 12.00 & 44.00 & 12.00 & 7.83 \\ 21.25 & 12.75 & 43.33 & 50.50 & 19.17 \\ 60.00 & 16.25 & 11.00 & 33.00 & 34.25 \\ 0.00 & 15.00 & 0.67 & 4.00 & 25.92 \\ 5.00 & 44.75 & 0.33 & 0.00 & 12.00 \end{bmatrix} \quad \mu = \begin{bmatrix} 41.3 \\ 50.3 \\ 33.8 \\ 12.8 \\ 12.8 \end{bmatrix}$$

clinical significance of the results anyway. In the future, however, as better estimates of the errors in our parameter estimates become available, we may be able to obtain more accurate confidence intervals through errors-in-variables methods (26).

A final reason for caution arises from the possibility of negative results for the organ-specific estimates. We recognized that these might have been due to systematic parameter error – the use of average values for the proportions of each isoenzyme in each organ cluster rather than proportions for an individual organ (e.g. the exact published proportions of each isoenzyme in the heart when solely the heart is expected to contribute). However, perhaps because of other, nonsystematic individual variations in parameters, and because multiple organs in a given cluster may simultaneously contribute, we found empirically that models with parameters based on individual organs performed more poorly. Thus, we decided to recommend use of the present model, with two important safeguards. A "danger signal" from the computer program can be emitted (a) if negative results occur beyond a (conservatively) significant range of analytic variability or (b) if the correlation between actual and predicted total LDH falls below a predefined criterion. The clinician can then choose to ignore the results.

Fortunately, our model seldom performed poorly. To test its adequacy, we computed the correlation between the "predicted" and actual total LDH as the multiple correlation between the mean isoenzyme proportions in each organ cluster and the observed total LDH, in excess of the normal mean (with the regression intercept term set equal to zero). By this criterion, a poor fit of the original regression model ($r < 0.8$) occurred in less than 4% of all the (nearly 400) measurements we analyzed.

One might criticize our use of correlations as an index of the adequacy of our model because they are not very sensitive to deviations in the "beta weights" (here the organ-specific estimates). Also, the way we have chosen to compute them (by setting the intercept equal to zero) does not enable us to interpret them as indicating the ability of the model to make predictions, beyond those of the mean total LDH values in our population (which would be possible if an intercept were used). While we did correct the total LDH to be predicted for the *normal* mean levels before the analysis, we intentionally decided not to include an intercept (and thus correct also for the *overall* means). Our reason for not doing this was that the overall means would be influenced by different disease types and extents in different cohorts of patients and thus would be of little use in making predictions in future cases. Thus, computing a correlation (with an intercept) that could be interpreted as the model's ability to improve upon predictions from the overall means would have had little significance. Hence, we believe that our method of computing correlations was the fairest indicator of model performance. Also, we provided even more compelling evidence of the adequacy of our model. We showed that it outperformed any existing standard diagnostic indices in the actual diagnosis of acute myocardial infarction.

Recognizing that the mere possibility of negative values in organ-specific estimates, despite their lack of effect on model performance, still could disturb some clinicians, we developed yet a third method for computing them. The derivation of this method had a less clear interpretation but it almost assured elimination of potentially negative values for the organ-specific estimates.

If we define

$$P_i = \frac{L_i - \mu_i}{L_T - \mu_T} \text{ where } L_T = \sum_{i=1}^{5} L_i$$

and then suppose

$$log \frac{P_i}{1 - P_i} = X_i\beta' + \varepsilon_i \qquad i = 1, ..., 5$$

and if we use regression to estimate

$$\hat{\beta}' = \{\hat{\beta}'_j\} \qquad j = 1, ..., 5$$

then

$$\hat{\beta}_j = \left| \frac{e^{\hat{\beta}j}}{1 + e^{\hat{\beta}j}} \right| (L_T - \mu_T)$$

is an estimate of the contribution of organ cluster j lying in the interval $[O, L_T - \mu_T]$.

This method produces results indistinguishable from the previous two in terms of overall accuracy. In the future, we will need to see whether it will produce better or worse discrimination than the alternative methods on a larger sample of cases.

Acknowledgements

I thank David Chou, Hang-Yat Tam, G. S. Kumar, Timothy Clark, Stephen Powell, and Robert Galen for their comments and assistance. This research was supported in part by grants LM04132, LM03306 and LM04086 from the National Library of Medicine. Dr. Politser is recipient of NIH Research Career Development Award LM0080 from the National Library of Medicine. This paper was adapted from two manuscripts published in Clinical Chemistry (32: 1510-17 and 1517-24, 1986).

References

1. Politser, P. E. (1982) Decision analysis and clinical judgement: a re-evaluation. Med. Decision Making *1*, 368–89.
2. Connelly, D. P., Lasky, L. C., Keller, R. & Moore, A. A. (1980) Graphical representation of clinical laboratory data. In: Computer applications in medical care (O'Neill, J. T., ed.) pp. 841–848, IEEE Computer Society Press, Washington.
3. Connelly, D. P., Lasky, L. C., Keller, R. M. & Morrison, D. S. (1981) A System for graphical display of clinical laboratory data. Am. J. Clin. Path. *78*, 729–37.
4. Tufte, E. A. (1983) The visual display of quantitative information. Graphics Press, Cheshire, CT.
5. Kahneman, D. (1973) Attention and effort, Prentice-Hall, Englewood Cliffs, N.J.
6. Kahneman, D. & Tversky, A. (1979) Prospect theory: an analysis of decision under risk. Econometrica *47*, 263–91.
7. Tukey, J. W. (1977) Exploratory data analysis. Addison-Wesley Publishing Co., Boston.
8. Cleveland, W. S. & McGill, R. (1983) Graphical perception: theory, experimentation, and application to the development of graphical methods. J. Am. Stat. Assoc. *79*, 531–54.
9. Pachella, R. G., Somers, P. & Hardzinski, M. (1981) A psychophysical approach to dimensional integrality. In: Auditory and

visual pattern recognition (Getty, D. J. and Howard, J. H. Jr., eds.) Lawrence Erlbaum Associates, Hillsdale, N.J.

10. Ingelfinger, J. A., Mosteller, F., Thibodeau, L. A. & Ware, L. H. (1983) Biostatistics in clinical medicine. Macmillan Publishing Company, New York.
11. Cooper, L. A. (1980) Recent themes in visual information processing: a selected overview. In: Attention and performance (Nickerson, R. S., ed.) Lawrence Erlbaum Associates, Hillsdale, N.J.
12. Prinzmetal, W., Banks, W. P. (1977) Good continuation affects visual detection. Perception Psychophysics *11*, 542–50.
13. Tversky, A. & Gati, I. (1982) Similarity, separability, and the triangle inequality. Psychol. Rev. *89*, 123–54.
14. Gore, S. M. (1981) Assessing methods – transforming the data. Brit. Med. J. *283*, 348–500.
15. Jacquez, J. A. (1972) Compartmental analysis in biology and medicine. Elsevier, New York.
16. Peterson, P. H., Groth, T. & Hjelm, M. (1979) Characterization of increased synthesis of acute phase proteins form plasma concentration measurements: correction of sampling errors and of disturbing variations of plasma volume, exchange fluxes, and catabolic rates in the clinical situation. J. Clin. Comp. *8*, 180–201.
17. Parkerson, G. R. Jr. (1978) Labstand: a computerized system for reporting clinical laboratory data in standard units. J. Family Practice *6*, 611–20.
18. The principles and practice of medicine, 20th ed. (1980) (Harvey, A. M., Johns, R. J., McKusick, V. J., Owens, A. H. & Ross, R. S., eds.) Appleton-Century Crofts, New York.
19. Roberts, R. (1981) Diagnostic assessment of myocardial infarction based on lactate dehydrogenase and creatine kinase isoenzymes. Heart Lung *10*, 486–506.
20. Coombs, C. H., Dawes, R. M. & Tversky, A. (1970) Mathematical psychology: an elementary introduction. Prentice-Hall, Englewood Cliffs, N.J.
21. Politser, P. E., Powell, S. & Fink, J. (1986) A new method for reporting the sources of abnormal activities of lactate dehydrogenase in serum. Clin. Chem. *32*, 1517–24.
22. Payne, J. W. (1982) Contingent decision behavior. Psychol. Bull. *92*, 382–402.
23. Politser, P. E., Gastfriend, D., Bakin, D. & Nguyen, L. (1987) An intelligent display system to aid psychiatric diagnosis in primary care. Medical Care (in press).
24. Johnson, M. D. & Russo, J. E. (1984) Product familiarity and learning new information. J. Consumer Res. *11*, 542–50.
25. Larkin, J., McDermott, J., Simon, D. P. & Simon, H. A. (1980) Expert and novice performance in solving physics problems. Science *208*, 1335–42.
26. Fuller, W. (1980) Properties of some estimators for the errors in variables model. Ann. Statistics *8*, 407–22.

Chapter 2.2
Result Reports from Large Centralized Laboratories

A. J. Porth, C. Badke, S. Bothung and M. Worzyk

1 Introduction and Basic Aspects

Even before introduction of laboratory computers in the sixties the presentation of measurement results and findings data from a medical laboratory played a very significant part. Although from the start particular significance was given to the communication interface between the laboratory and the attending physician, the limits of what could be expected from the technical personnel were soon reached because the mechanization and automation of laboratory test stations led to amounts of data which could no longer be transferred rapidly enough without loss of quality by writing them down manually. Consequently, apart from supporting the internal processes in the laboratory, an essential objective of the introduction of computers into laboratory work was to facilitate the report presentation in appropriate manner.

Any good result and findings report, whether directed to the technical laboratory personnel or the attending physician, should have the following attributes: informative, clear, complete, easy to handle, clearly structured, as free from errors as possible and, last but not least, rapid availability.

If the historical development of computer-aided laboratory findings is studied, as set forth in the fundamental publications of Bock and Eggstein (1), Delbrück (2), Büttner (3), Knedel (4) and Porth (5), it is immediately noticed that most of the basic components of reporting were already present from the start in the various laboratory computer systems. The present state of the art will be illustrated below with the aid of examples from the laboratory information system QUADROLAB ("quality-controlling universally applicable data maintenance and real-time oriented laboratory information system") which was developed in the seventies on the basis of many years system experience at the Hanover Medical University and has been constantly updated on the basis of new knowledge. For this purpose the practical experience of 5 different central laboratories will be evaluated and incorporated (2 × regional hospitals, 1 × municipal hospital, 1 × large joint medical practice and 1 × medium joint medical practice).

The essential trends in recent years can be outlined as follows: Whereas the first reports usually consisted only of numerical results possibly amplified by an asterisc in the case of pathological values, today the results may consist selectively of numbers with amplifying text or of text on its own, these texts or auxiliary texts (comments) being established with the aid of complex logic structures or being explicitly added by the laboratory physician. These interpretative text amplifications are an important aid for the attending physician; in particular, they support the introduction of new analysis techniques and the consistency of interpretation when the medical personnel changes. In Section 2 various types of internal laboratory reports will be explained and discussed and in Section 3 reports which leave the laboratory; firstly, however, below a few basic aspects will be explained.

As regards the configuration of the reports, very different types of representation are encountered from system to system and unfortunately do not always follow the aforementioned criteria but are very frequently influenced by individual concepts of the system analysists and programmers. This can reach such an extent that the laboratory is "sold" as essential obviously nonsensical methods of presentation, as will become clear from the following example:

Of course, any numerical statement simultaneously implies a rough indication of its accuracy; weight figures in milligrams, grams, kilograms or tonnes imply in each case different information on the range within which the measurement made can vary. It follows from this that when giving analysis results the number of decimal points used should correspond to the accuracy of the analysis measurements. This basic accuracy must be individually definable for each parameter. However, in computer printouts it is repeatedly found that for example the transaminases GOT (AST) and GPT (ALT) are presented with 3 digits after the decimal point where an integer presentation would be logical and appropriate.

The representation accuracy of numerical results is not only significant as regards the interpretation of their accuracy in the findings presentation but also as regards proper storage space administration. In practice this means in computer-aided laboratory operation that for each biological parameter for which numerical results are to be administered and stored a result unit is to be defined as base. The results are then stored as multiples of this result unit and in findings reports printed out with the decimal point number thus defined.

This will be illustrated with the aid of a few biological parameters:

1. Sodium in serum with result unit 1 mmol/l, i. e. the results are output as integers;
2. potassium in serum with result unit 0.1 mmol/l, i. e. the results are output with one decimal digit;
3. cholinesterase determination (activated with acetylthiocholine) in the serum with result unit 10 U/l, i. e. the results are output as multiples of 10, that is always with a zero as the last digit;
4. cholinesterase determination (activated with butyrylthiocholine) in the serum with result unit 0.01 kU/l, i. e. the results are output with two digits after the decimal point.

As important as such definitions are for unitary representation of results and as an aid in evaluation, exceptions to this rule must be allowed. In example 2 for potassium the result unit is specified as 0.1 mmol/l. This is no doubt expedient for the assessment by the attending physician but for internal laboratory quality control it is better to work with two digits after the decimal point because the measuring methods have the corresponding accuracy. It may also happen that an analysis apparatus fundamentally supplies the data with more decimal places than warranted by the measurement accuracy. In such a case as well the laboratory director must be able to stipulate that in an individual case or generally exceptions to the basic rule are allowed, both in laboratory internal reports and in findings reports which leave the laboratory. Furthermore, clinical-chemical parameters exist (e. g. carbamazepine) which in the lower measurement range can be processed with a higher accuracy (one decimal place more) than is possible with extremely high values; this is important when the decision range requires greater accuracy and the latter is also justified by the method precision.

Attention must be paid not only to a systematic representation of the results but also the uniform use throughout of the other features and representation forms; this makes the data contents themselves more informative and the column heading is needed less as aid for clear interpretation.

The following rules always applied are taken from the QUADROLAB system.

1. *Date:* Each data specification has the form dd.mm.yy; dd stands for day, mm for month and yy for year and they are always given with 2 digits (i. e. where necessary with a leading zero). Other forms such as 3.9.86 or 86-9-3 or 86/3/9 or 9/3/86 for 03. 09. 86 do not appear in the output. For the sake of unity each date should always be entered in this form but sometimes it may be expedient to allow shortened entries as long as clarity is not impaired.
2. *Time-of-day:* The time-of-day is put in the form hh:mm:ss; hh stands for hour, mm for minute and ss for second; here as well the representation is always in 2 digits; specification of the seconds may sometimes be omitted.
3. *Result field:* The size of a result field should be made such that firstly all the result order of magnitude occurring and brief texts have enough room and secondly as many results as possible fit into one line and thus onto one findings sheet. A good well proven compromise are 8 characters for the result (in digit representation or as short text) and in addition 2 characters for supplemented qualification symbols (e. g. + high, – low, + + extremely high, – – extremely low) or additional result-linked text abbreviations which can then be explained in detail in a footnote.
4. *Designation of the biological parameters:* For internal laboratory use it suffices to work with 6-position mnemotechnical abbreviations (Table 1). Digits as code numbers involve the risk of confusion and should be avoided. In findings reports intended for the attending physician it must be possible to cmploy long texts with the possibility of selecting their length in the computer. For display reasons at the Hanover Medical University a length of 18 characters has been adopted; this field contains a one to two-place investigation material code (S = serum, U = urine, P = plasma, ...) and the parameter designation follows, separated by a hyphen; the measured value unit then appears right-justified (Fig. 8).
   It is obvious that in the case of all these particulars the complete ASCII character font (in particular uppercase and lowercase) is available and must be applicable.

Table 1. The most frequently used abbreviations of biological parameters in this chapter

CGLU	Glucose in capillary blood
Dxxxxx	Secretion of Uxxxxx pro die (= per 24 h)
KONKRE	Concrement analysis
SAP	Alkaline phosphatase in serum
SA1G	a_1-globuline of SEEL
SA2G	a_2-globuline of SEEL
SBG	Beta-globuline of SEEL
SBIL	Bilirubin, total in serum
SCA	Calcium in serum
SCHE	Cholinesterase in serum (substrate: butyrylthiocholine)
SCHO	Cholesterol in serum
SCO2	CO_2 total in serum
SCK	Creatine kinase in serum
SCKMB	Creatine kinase MB isoenzyme in serum
SCL	Chloride in serum
SEAL	Albumin of SEEL
SEEL	Serumelectrophoresis
SGG	γ-globuline of SEEL
SGGT	γ-glutamyltransferase in serum
SGLD	Glutamatdehydrogenase in serum
SGLU	Glucose in serum
SGOT	AST in serum
SGPT	ALT in serum
SHST	Urea in serum
SIP	Phosphate in serum
SK	Potassium in serum
SKRE	Creatinine in serum (by Jaffe)
SKRW	Creatinine in serum (after absorption by Fuller's earth, used for calculating XKCL)
SLDH	Lactate dehydrogenase in serum
SNA	Sodium in serum
SPAL	Prealbumin of SEEL
SPRO	Protein in serum
UDS	Drug-Screening
UGLU	Glucose in urine
UHST	Urea in urine
UK	Potassium in urine
UKRE	Creatinine in urine
UNA	Sodium in urine
UP	Phosphate in urine
UPRO	Protein in urine
USDH	Duration of urine collection in hours
UVMS	Vanillylmandelic acid in urine
UVOL	Volume of urine collection in ml
UVOLST	Volume of urine per hour
XKCL	Clearance rate for creatinine
YKGR	body size (in m)
YKGW	body weight (in kg)
YK	Potassium in special material
YNA	Sodium in special material
YP	Phosphate in special material

2 Internal Laboratory Reports

2.1 Working place lists, result entry lists and evaluation lists

A good laboratory computer system is not only distinguished by offering as many system components as possible to the user but by leaving him free to decide whether or not he wants a specific component implemented in its application. An example of this is the working list component. The laboratory organization at Hanover Oststadt Hospital is set up so that working lists are superfluous and consequently have been eliminated in all software and executive phases in the computer. If however the laboratory director has decided to prepare working lists it must first be established whether in the working list previous values relating to the patient from earlier examinations are to be printed out. Making available previous values can lead to delays in printing out the working lists because firstly a patient-oriented linking between the new job and the result data base must be established and secondly the previous jobs must be searched for corresponding results. Here as well, it is important whether or not the component "search for previous values" is to be used, in accordance with the laboratory's wishes.

To avoid continuously confronting the technical personnel on change of working stations within the laboratory with new list types a form should be offered which is applicable substantially uniformly to all laboratory units. In QUADROLAB it has proved possible in a single list type to implement all working lists, result entry lists (for devices not connected on-line) and evaluation lists of the most varied on-line analysis measuring stations (Fig. 1). If quality control results appear in a list they are represented at the end of a list on a separate sheet in clear check-card form (Fig. 1). A criticism list for plausibility checks can also be appended here to discover implausibilities as early as possible and check them.

In particular in laboratories with a high data throughput, by good data presentation the computer must contribute to drawing the attention of the technical personnel very rapidly and reliably to dubious and possibly erroneous results so as to ensure by rapidly adopted control actions good data quality and the transmission of findings without delay (6).

2.2 Criticisms for plausibility checks

2.2.1 **Extreme value check**

For each biological parameter a result interval can be defined outside which extremely low or high values are located which are marked by two symbols (e.g. −− and ++) affixed to the result. These limits can define as required the biologically possible range or a statistical range from the frequency distribution of the biological parameter (e.g. the 0.1% and 99.9% quantile).

On each data transfer to the computer (off-line via display terminal or on-line from analyzers) and on computer-internal storage transfer from the measuring-station-oriented data file to the patient-oriented data base, extreme results are checked so that, possibly with the aid of a criticism list, the extreme values can be sent to an additional control.

```
LISTE: 194/228 am 30.10.86  14:25 P=P23   B=***   A=30.10.86  11:24:04
G=23  U R I N - S M A          AE=        F=      E=30.10.86  11:24:06
      MI-SATZ#  8022 ... 8730

1                 UNA         UK          UCL          UGLU         UP           UKRE         UGX
  287/30.10.86           14         53           41          0.4         20.5          1.9         0.07
2                 UNA         UK          UCL          UGLU         UP/          UKRE         UGX
  297/30.10.86            1         85           48          0.7         46.2          6.6         0.13
3                 UNA         UK          UCL          UGLU         UP/          UKRE         UGX
  298/30.10.86           16        145           74          1.1         46.2          9.2         0.19
4                 UNA         UK          UCL          UGLU         UP           UKRE         UGX
  310/30.10.86           61         16           63         <0.1          8.3          2.3        <0.01
5                 UNA         UK          UCL          UGLU         UP           UKRE         UGX
  311/30.10.86           62         14           64          0.0          7.0          2.3         0.00
6                 UNA         UK          UCL          UGLU         UP           UKRE         UGX
  317/30.10.86           72         20           78          0.3          5.2          3.2         0.05
7                 UNA         UK          UCL          UGLU         UP           UKRE         UGX
  323/30.10.86           56         33           45          0.1         13.0          6.6         0.03
8                 UNA         UK          UCL          UGLU         UP           UKRE         UGX
  324/30.10.86           76         21           78          0.2         11.5          2.3         0.03
9                 UNA         UK          UCL          UGLU         UP           UKRE         UGX
 9001/30.10.86           <1          0            1          0.0          0.1          0.0         0.00
10                UNA         UK          UCL          UGLU         UP           UKRE         UGX
 9001/30.10.86            0          0            0          0.0          0.0          0.0         0.00
11                UNA         UK          UCL          UGLU         UP           UKRE         UGX
 8562/30.10.86          136         74          147         13.4         24.8         12.2         2.42
12                UNA         UK          UCL          UGLU         UP           UKRE         UGX
 9999/30.10.86            2         <1           <1          0.3          0.4         <0.1         0.00

LISTE: 194/228 am 30.10.86  14:25 P=P23    Qualitaetskontrolle
                                                            WERT    MW/SW      ABWCH.      ..........2....3....>4

P  13 8562/30.10.86  Nxxxxxx URINKONTROLLE 299    UNA    (   136 -   132 ) /  2.64 = 1.5   ++++++++           3 %
P  13 8562/30.10.86  Nxxxxxx URINKONTROLLE 299    UK     (    74 -    72 ) /  1.89 = 1.1   ++++++             3 %
P  13 8562/30.10.86  Nxxxxxx URINKONTROLLE 299    UCL    (   147 -   148 ) /  2.52 =-0.4   --                -1 %
P  13 8562/30.10.86  Nxxxxxx URINKONTROLLE 299    UGLU   (  13.4 -  13.0 ) /  0.50 = 0.8   ++++               3 %
P  13 8562/30.10.86  Nxxxxxx URINKONTROLLE 299    UP     (  24.8 -  24.7 ) /  0.50 = 0.2   +                 <1 %
P  13 8562/30.10.86  Nxxxxxx URINKONTROLLE 299    UKRE   (  12.2 -  12.1 ) /  0.21 = 0.5   +++                1 %
P  13 8562/30.10.86  Nxxxxxx URINKONTROLLE 299    UGX    (  2.42 -  2.34 ) /  0.09 = 0.9   +++++              3 %
```

Fig. 1. Result list with appended quality control printout
(WERT = value, MW/SW = mean value/target value, ABWCH. = deviation).

If a calculated parameter is being formed from several biological parameters implausible findings constellations may be determined via their extreme value position. Examples of this are

- de-Ritis quotient XDRQ = SGOT/SGPT
- Schmidt quotient XSCQ = (SGOT + SGPT) – creatinine deviation SKRABW = (SKRE – SKRW) · 100/SKRW
- urine volume per hour UVOLST = UVOL/USDH
- ionogram value, i. e. the difference between cations and anions in the serum SIONO = SNA + SK + 2 · SCA – SCL – SCO2 – 1.8 · SIP – 0.241 · SPRO

Figure 2 shows a histogram of 20 000 ionogram values and with the aid of the 0.1% and 99.9% quantile values to be derived therefrom the warning limits were fixed at –11 and 27. If these limits are violated more than one to two times a day this may be an indication that the results of one of the underlying parameters include a systematic error.

```
 Q01= -11        Q1= -6         Q50= 3         Q99= 14        Q999= 27

Frequency distribution       07.03.86 .... 10.06.86      20000 Values

 SIONO       NUMBER   %     ----------!---------!---------!---------!
                            !         10        20        30        40  %
 .. -11 :        22   0.11 !
  .. -8 :        42   0.21 !
  .. -5 :       332   1.66 !**
  .. -2 :      1775   8.87 !*********
   .. 1 :      5141  25.71 !**************************
   .. 4 :      6291  31.45 !*******************************
   .. 7 :      3963  19.82 !********************
  .. 10 :      1615   8.07 !********
  .. 13 :       539   2.70 !***
  .. 16 :       170   0.85 !*
  .. 19 :        49   0.24 !
  .. 22 :        18   0.09 !
  .. 25 :        18   0.09 !
  .. 28 :        15   0.08 !
  .. 31 :         2   0.01 !
  .. 34 :         2   0.01 !
  .. 37 :         5   0.02 !
  .. 40 :         0   0.00 !
  .. 43 :         1   0.01 !
  .. 46 :         0   0.00 !
   > 46 :         0   0.00 !
```

Fig. 2. Histogram with quantiles for the ionogram value
SIONO = SNA + SK + 2.0 · SCA – SCO2 – SCL – 1.8 · SIP – 0.241 · SPRO
(Q01 = 0.1% – quantile, Q1 = 1% – quantile, Q50 = mediane, Q99 = 99% – quantile, Q999 = 99.9% – quantile).

2.2.2 Double result list

If on entry of the results into the data base the program detects that results were already present then the old and new result in each case is submitted in a list to the responsible laboratory assistant for a decision.

2.2.3 Extreme previous value deviations

The program for determining extreme previous value deviations lists from selected analyses methods from the current daily production of the laboratory the 5 most extreme previous value deviations. This can be very helpful in obtaining for newly introduced methods warning information even before the data base contains enough data for statistical evaluations (7).

2.2.4 Criticism of striking trends

The program carried out (possibly several times) daily for criticizing striking trends checks all results newly stored in the data base on this day against existing previous values and prints out those for which the result deviations lie outside predetermined intervals.

These intervals are defined by the laboratory per biological parameter on the basis of statistical investigations and explorative data analyses. To take appropriate account of the time difference between the previous value and current value criticism limit values are formed for the following 4 time difference groups:

1. 0, 1 or 2 days,
2. 3 or 4 days,
3. 5, 6 or 7 days and
4. more than 7 days.

To obtain in routine operation criticisms of the best possible informative value the particulars listed in Table 2 are required. The deviation measure employed in this method "beta check" (an improvement over the known "delta check") and the scientific background are described in (7).

Table 2. Components of a trend criticism with the aid of an example

MEDP 43	Code of the sender of the previous specimen
SGOT	Abbreviation of the biological parameter
185+	Previous value (+ means "high")
15	Current value
−680	Beta check value according to Porth (7)
— — — tt	Warning field according to Porth (7)
653/30.06.86	Job number of the current specimen
MEDP 16	Code of the sender of the current specimen
Nxxxx, Stefan	Name of the patient
31.10.57	Date of birth of the patient

2.2.5 Striking result constellations

If it is desired to track down implausible constellations with results on various biological parameters from the same specimen, then fundamentally the procedure described in Section 2.2.1 using additionally calculated parameters can be adopted. Another way, which is simpler in practical implementation and very efficient, is based on bivariate frequency distributions as described in (8). One check rule proceeds initially for the two biological parameters from 16 result intervals in each case which are taken equidistant in the case of symmetrical and progressively

SKRE \ ---SHST	0.6	1.4	2.9	5.0	7.5	10.3	13.5	16.9	20.7	24.7	29.0	33.5	38.3	43.2	48.4	>48.4
>1410	.	.	.	.	.	.	.	.	.	.	.	.	.	.	.	.
1410	.	.	.	.	.	.	.	.	.	P	P	P	.	.	.	.
1240	.	.	.	.	.	.	.	.	P	P	P	P	P	.	.	.
1070	.	.	.	.	.	.	.	.	P	P	P	P	P	P	.	.
920	.	.	.	.	.	.	.	P	P	P	P	P	P	P	.	.
780	.	.	.	.	.	.	P	P	P	P	P	P	P	P	.	.
650	.	.	.	.	.	.	P	P	P	P	P	P	P	P	.	.
530	.	.	.	.	.	P	P	P	P	P	P	P	P	P	.	.
416	.	.	.	.	P	P	P	P	P	P	P	P	P	P	P	.
321	.	.	.	P	P	P	P	P	P	P	P	P	P	P	.	.
238	.	.	P	P	P	P	P	P	P	P	P	P	P	.	.	.
168	.	P	P	P	P	P	P	P	P	P	P	.	.	.	.	.
111	.	P	P	P	P	P	P	P	P	.	.	.	.	.	.	.
68	P	P	P	P	P	P	P	.	.	.	.	.	.	.	.	.
40	.	P	P	P	P	.	.	.	.	.	.	.	.	.	.	.
29	.	.	.	P	.	.	.	.	.	.	.	.	.	.	.	.

Fig. 3. Definition of bivariate plausibility ranges with the aid of result intervals using as example the biological parameters urea (SHST) and creatinine (SKRE) in serum; result combinations which do not lie in the fields marked by P are considered implausible with an error probability of 1%.

Rxxxxxx, Hxxxxx IDNR=35693 *04.12.27 M MEDP42

PRNR= 26/24.10.86 PRDA= 23.10.86 PRUZ= 15:00 EINS= MEDP42

	143	=SNA	5	=SBIL		=SEEL	11	=SGOT
	4.4	=SK	7.1+	=SCHO	62	=SEAL	13	=SGPT
	105	=SCL	2.42!T	STG	3	=SA1G	187	=SAP
	2.45	=SCA	5.9+	SGLU	8	=SA2G	34+	=SGGT
+	0.97	=SIP	497+	=skre	11	=SBG	6.33	=SCHE
+	73	=SPRO	4.7	=shst	16	=SGG	218	=SLDH

!T : The reference range for triglycerides applies only for 12-hour alimentary abstinence

Fig. 4. Screen display of results for an examination job with an implausible findings constellation of creatinine (SKRE) and urea (SHST) in serum; the criticized parameters are output in lowercase letters and at the start of the line there is a plus sign for emphasis; a missing equals sign in front of the analysis abbreviation denotes an unrequested result which was obtained within the scope of a profile.

B E R E C H N U N G S P R O T O K O L L   27.10.86 14:48:00

	alter Wert	neuer Wert	Kenngr.	Ausgangswerte
397/23.10.86		[NERB]	=UVOLST	UVOL= FEHLT USDH= FEHLT
		[NERB]	=XKCL	UKRE= 42.0 SKRW= 59 UVOL= FEHLT USDH= FEHLT
449/23.10.86		19	=SKRABW	SKRE= 124+ SKRW= 104+
		126	=XKCL	UKRE= 8.6 SKRW= 104+ UVOL= 2200 USDH= 24
281/24.10.86		123	=XKCL	UKRE= 8.9 SKRW= 64 UVOL= 1400 USDH= 24 VKGW= 82.00 VKGR= 1.58
286/24.10.86		23-	=XKCL	UKRE= 3.1 SKRW= 294+ UVOL= 3100 USDH= 24
287/24.10.86		51-	=XKCL	UKRE= 4.0 SKRW= 81+ UVOL= 1500 USDH= 24
421/24.10.86		57-	=XKCL	UKRE= 4.9 SKRW= 143+ UVOL= 2400 USDH= 24
		11.8	=DKRE	UKRE= 4.9 UVOL= 2400 USDH= 24
470/24.10.86		18-	=XKCL	UKRE= 5.1 SKRW= 161+ UVOL= 780 USDH= 24 VKGW= 56.60 VKGR= 1.65
471/24.10.86	174+	116	=XKCL	UKRE= 5.0 SKRW= 45- UVOL= 1500 USDH= 24
	77++	18	=SKRABW	SKRE= 53 SKRW= 45-
472/24.10.86		31-	=XKCL	UKRE= 4.2 SKRW= 249+ UVOL= 2600 USDH= 24 VKGW= 60.50 VKGR= 1.65
481/24.10.86		56-	=XKCL	UKRE= 6.5 SKRW= 75 UVOL= 680 USDH= 24 VKGW= 40.00 VKGR= 1.45
484/24.10.86		51-	=XKCL	UKRE= 2.0 SKRW= 78 UVOL= 900 USDH= 14 VKGW= 24.40 VKGR= 1.28
487/24.10.86	204+	209+	=XKCL	UKRE= 14.6 SKRW= 68 UVOL= 1450 USDH= 24 VKGW= 64.30 VKGR= 1.78
		18	=SKRABW	SKRE= 80 SKRW= 68
1483/24.10.86		26.5+	=DKRE	UKRE= 9.8 UVOL= 2700 USDH= 24
		70	=DK	UK= 26 UVOL= 2700 USDH= 24
		294+	=DNA	UNA= 109 UVOL= 2700 USDH= 24
		3.5+	=DGLU	UGLU= 1.3 UVOL= 2700 USDH= 24
		0.65+	=DGX	UGX= 0.24 UVOL= 2700 USDH= 24
		84.0+	=DP	UP= 31.1 UVOL= 2700 USDH= 24
		192	=DCL	UCL= 71 UVOL= 2700 USDH= 24
		135	=XKCL	UKRE= 9.8 SKRW= 136+ UVOL= 2700 USDH= 24
1490/24.10.86		8.2-	=DKRE	UKRE= 2.1 UVOL= 3900 USDH= 24
		62	=DK	UK= 16 UVOL= 3900 USDH= 24
		254	=DNA	UNA= 65 UVOL= 3900 USDH= 24
		[NERB]	=DGLU	UGLU= <0.1 UVOL= 3900 USDH= 24
		[NERB]	=DGX	UGX= <0.02 UVOL= 3900 USDH= 24
		56-	=XKCL	UKRE= 2.1 SKRW= 102+ UVOL= 3900 USDH= 24
1534/24.10.86		3.4-	=DKRE	UKRE= 1.7 UVOL= 2000 USDH= 24
		26	=DK	UK= 13 UVOL= 2000 USDH= 24
		46--	=DNA	UNA= 23 UVOL= 2000 USDH= 24
		606.6++	=DGLU	UGLU= 303.3++ UVOL= 2000 USDH= 24
		109.32++	=DGX	UGX= 54.66++ UVOL= 2000 USDH= 24
		6-	=XKCL	UKRE= 1.7 SKRW= 373+ UVOL= 2000 USDH= 24

Fig. 5. Calculation Log
(alter Wert = old value, neuer Wert = new value, Kenngr. = parameter, Ausgangswerte = underlying values, FEHLT = missing. NERB = A result cannot be calculated because the values of the necessary parameters do not exist or do not permit a calculation).

distant in the case of asymmetrical (e. g. log normal) frequency distribution of the biological parameter. For each of the 16 × 16 = 256 "result squares" it is defined whether the result combinations lying in the square are plausible or not. These determinations can fundamentally be made arbitrarily but as in other methods explorative data analyses are very helpful as a basis. Figure 3 contains the data of the parameter combination (SKRE, SHST) valid for the Hanover Medical University. The criticisms shown in Figure 4 are an example of a display representation in which the essential information is particularly emphasized.

2.3 Result processing

2.3.1 Calculation printout

With the storing of requests parameters whose results are to be calculated may also be initiated. In the course of the investigations the starting parameters come from the various laboratories to the data base and with the aid of stored formulae the corresponding parameter values are calculated. Together with the starting parameters and any previously stored value the result appears in the calculation printout (Fig. 5). If in the starting parameters not only purely numerical results

```
R E S T E L I S T E  -  PROBENDATEN                          23.10.86   09:39:48
LNR        1/01.10.86  -  4000/23.10.86

   239/01.10.86  FALSCHER ABRECHNUNGSSCHLUESSEL XX          (WRONG INVOICING KEY XX   )
   572/05.10.86  VERSICHERER BKKU  UNBEKANNT                (INSURER UNKNOWN          )
    93/20.10.86  KEIN RESULTAT                              (NO RESULT                )
   587/21.10.86  KEIN RESULTAT
   656/21.10.86  KEIN RESULTAT
     1/22.10.86  KEIN RESULTAT
    29/22.10.86  KEIN RESULTAT
   177/22.10.86  FALSCHE VERSICHERTENGRUPPE  FU             (WRONG INSURANCE GROUP FU)
                 KEIN GEBURTSDATUM VORHANDEN                (NO DATE OF BIRTH ON FILE)
   996/22.10.86  NICHT ZUGEORDNET                           (NOT ASSIGNED TO PATIENT )
                 KEIN ABSENDER                              (NO SENDER                )
                 KEINE ANFORDERUNG                          (NO REQUESTS ARE STORED   )
   997/22.10.86  NICHT ZUGEORDNET
                 KEIN ABSENDER
                 KEINE ANFORDERUNG
   998/22.10.86  NICHT ZUGEORDNET
                 KEIN ABSENDER
                 KEINE ANFORDERUNG
   999/22.10.86  NICHT ZUGEORDNET
                 KEIN ABSENDER
                 KEINE ANFORDERUNG
   554/23.10.86  NICHT ZUGEORDNET
                 KEIN ABSENDER
                 KEINE ANFORDERUNG
   658/23.10.86  KEINE ANFORDERUNG
   659/23.10.86  KEINE ANFORDERUNG
   660/23.10.86  KEINE ANFORDERUNG
   661/23.10.86  KEINE ANFORDERUNG
```

Fig. 6. Remainder list – specimen data.

occur but text results (such as "LACKING"), qualitative (such as "NEGATIVE") or range results (such as "< 0.01"), the test result "NERB" is allocated to the parameter to be calculated and initiates in the findings printout the following footnote: "A result cannot be calculated because the values of the necessary parameters do not exist or do not permit a calculation".

```
R E S T E L I S T E  -  ANFORDERUNGEN                         23.10.86   09:39:48
LNR      1/01.10.86  -  4000/23.10.86

  396/09.10.86 *UMETHQ -UHY
  449/09.10.86  SSPEZ
  325/13.10.86  DVMS    UVMS
  326/13.10.86  DVMS    UVMS
  415/13.10.86  DVMS    UVMS
  782/13.10.86  SHBSAL
  152/14.10.86  DVMS    UVMS
  314/14.10.86  DVMS    UVMS
 1360/14.10.86  KONKRE
 1377/14.10.86  KONKRE
   48/15.10.86  SHBSAL
  162/15.10.86  SICT    XIHWZ   XICLR   XIVTV
  170/16.10.86  DVMS    UVMS
  268/16.10.86  SCOFT   XCHWZ   XCCLR   XCVTV
  290/16.10.86  SPAP    SEKONZ
  369/16.10.86  SHBSAL
   39/17.10.86  UDS
  281/17.10.86  SALLOB  SAMOB   SAPROB  SBARB   SBRAL   SBUTAL  SCROT   SCYCB
                SPROP   SSECB   SSECO   SVINY
  379/17.10.86  DVMS    UVMS
  487/17.10.86  DVMS    UVMS
 1331/17.10.86  KONKRE
   25/20.10.86  DVMS    UVMS
   93/20.10.86  UAMPD   UBARD   UBENZE  UMETD   UMORD   UCOD    UDS
  325/20.10.86  SCOFT   XCHWZ   XCCLR   XCVTV
  356/20.10.86  DVMS
 1324/20.10.86  KONKRE
   55/21.10.86  SCOFT   XCHWZ   XCCLR   XCVTV
  577/21.10.86 *YP
  587/21.10.86  UAMPD   UBARD   UBENZE  UMETD   UMORD   UCOD    UDS
  656/21.10.86  VCIC    VIMM
    1/22.10.86  UAMPD   UBARD   UBENZE  UMETD   UMORD   UCOD    UDS
   29/22.10.86  UAMPD   UBARD   UBENZE  UMETD   UMORD   UCOD    UDS
  258/22.10.86  DVMS    UVMS
  435/22.10.86 *SEAL   *SA1G   *SA2G   *SBG    *SGG     SEEL
  453/22.10.86  SEEL    SEAL    SA1G    SA2G    SBG     SGG
  460/22.10.86 *SZN
  504/22.10.86  SCOFT
  505/22.10.86  SCOFT
  569/22.10.86  SXY1    SXY2    UXY5    XYT
    7/23.10.86  SVP
   56/23.10.86  SICT    SCOFT
```

Fig. 7. Remainder list – requests.

2.3.2 Remainder list of the specimen data

In the "remainder list specimen data" attention is drawn to erroneous or non-existent specimen administrative data (Fig. 6):

– wrong or non-existent billing key, insurance code or insurance group code,
– missing patient data for a specimen.

In addition warnings are printed when for a sample no result whatever has been stored for existing requests or no request exists for existing results.

2.3.3 Remainder lists of requests

In the "requests remainder list" in an overall list for the laboratory the following reminders appear by ascending specimen number (Fig. 7):

– requests not (yet) dealt with,
– biological parameters marked by "*" have results for which no request is present,
– biological parameters marked with "–" have been incorrectly entered and have not (yet) been deleted.

3 Findings Reports for the Attending Physician

3.1 General remarks on contents and form

The presentation of the results of an investigation job in clear, orderly and understandable manner is one of the most important tasks of a laboratory and laboratory data processing. Those responsible must meet the demands of the state of the art, not only as regards contents but also as regards form, and thereby ensure the use of a flexible and easily adaptable system.

The findings components explained below are related through their numbering to the explanation numbers of Figures 8 to 17:

(1) Preparation of the cumulative findings report by the Laboratory Data Processing Specialist Computer Centre for the Institute of Clinical Chemistry of the Hanover Medical University.
(2) Printout date and time-of-day to designate the up-to-date printout.
(3) Designation of the submitter (ward, out-patients, policlinic, external submitter).
(4) Summary of the contents of points (1) and (2) (only on the auxiliary sheet).
(5) Pneumatic post number of the sender (for sending findings printouts).
(6) Sheet numbering unique for the patient (and not only for a treatment case and certainly not only for a printout time).
This numbering is made within the findings sheet groups (cf. point (7)) in two parts:

The number in front of the stroke – ascending from left to right – the time sequence of specimen taking (cf. point (12)), the number after the stroke refers – ascending from the top to the bottom – to the sequence of the biological parameters (cf. point (14)).

Sketch showing the arrangement of various numbered findings sheets:

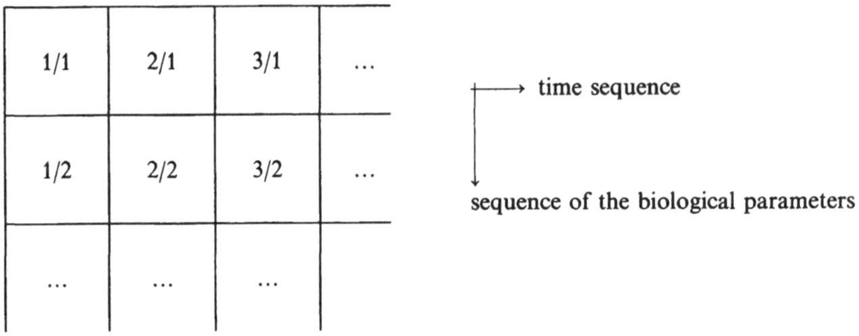

(7) Designation of the findings sheet groups.
The findings may be presented in one or up to 16 different finding sheet groups; at the Central Clinic of Hanover Medical University these are the following:

a) blood,
b) urine and other materials,
c) drugs I,
d) drugs II,
e) calibration dialyzates
f) HBs antigen

Within a findings sheet group the biological parameters are arranged by diagnostic-therapeutical categories.

(8) Surname, christian name, of the patient.

(9) Date of birth of the patient.

(10) Sex of the patient

F = female
M = male
U = unknown

With patients whose date of birth is 1899 or earlier the abbreviation for the sex is followed by the code digit 8.

(11) Identification number of the patient (QUADROLAB can handle up to 30 different types simultaneously).

(12) All results for an examination job are printed out in a column containing as heading the following particulars:

a) Date of taking (from the request form; if not given in the request form the date of receipt is assumed),
b) time-of-day of taking (from the request form; if not given in the request form the time of receipt in the laboratory, rounded off to the nearest hour, is assumed),

c) specimen number (is allocated internally in the laboratory on receipt of the specimen); it consists of a 4-digit consecutive number and the date of receipt.

(13) The last findings column is followed by a column with the reference ranges which are specified corresponding to the sex (and possibly age) of the patient.

(14) Designation of the biological parameters.
The material code is prefixed to the designation of a biological parameter and the measured value unit is suffixed to the parameter.

Material codes are:

A	= arterial blood	P	= plasma
B	= blood	PW	= plasma water
C	= capillary blood	S	= serum
ED	= calibration dialyzate	SP	= saliva
F	= faeces	SW	= serum water
K	= concrement	U	= urine
L	= liquor	V	= venous blood
M	= gastric juice	Y	= special specimen material

Within a findings sheet group the biological parameters are combined into groups by clinical aspects and separated from each other by an empty row.

(15) Each result consists of up to eight characters (numbers, number ranges or abbreviated texts) and up to two suffix qualification symbols; the most common of these are:

−	beneath the reference range (according to point (13))
+	above the reference range (according to point (13))
−−	beneath extreme value range
++	above the extreme value range
K	result has been separately checked; its value was concerned on repetition of the analysis
!U	following a numerical result initiates the following footnote: "Too little material for examination; new submission recommended!"

(16) Instead of a result an 8 character short text entered by the laboratory can appear.
Examples:

ARTER.	= material is arterial blood
ERRON.TK	= during or after the analysis it is found that the sender did not obtain the material properly
ALKAL.	= material is alkaline (e. g. gastric juice)
BLOODY	= material is bloody
MISSING	= sender has not given the necessary particulars (e. g. urine volume, urine collection duration)
COAGUL.	= coagulated capillary blood
HEMOLYT.	= serum is hemolytic
ICTER.	= serum exhibits icterus
NO MAT.	= no material submitted with request
LIPEMIC	= serum is lipemic

M.WRONG	=	material is rejected immediately on receipt of specimen because a wrong monovette was used or incorrect specimen storage (e.g. capillary glucose, urine sediment) does not permit a reliable analysis
NUMEROUS	=	qualitative result (e.g. urine constituents)
NEGATIVE	=	qualitative result
N.DETERM	=	not determinable, analysis cannot be carried out properly because for example the nature of the specimen matrix interferes
N.NECESS	=	determination not necessary (e.g. bilirubin dir. a. indir.)
NORMAL	=	qualitative result
N.DETECT	=	a parameter is not detectable
POSITIVE	=	qualitative result
T.FAULT	=	technical fault, the material having been lost or used up in the repetition without being able to obtain a reliable result
INVALID	=	an already passed on result is subsequently declared invalid
VENOUS	=	material is venous blood
XANTHOC.	=	liquor is xanthochromic
T.L.PROT	=	too little protein for carrying out the urine or liquor electrophoresis
CEN.VEN.	=	material is central venous blood
T.L.MAT.	=	too little material for an analysis; this may also mean that the result of the first measurement had to be discarded and a repetition was not possible because not enough material was available

(17) A requested result which has not however been obtained is given 2 asteriscs "**" in its field.

(18) All values newly stored in the printout are marked by an appended stroke, i.e. values not provided with a stroke were already contained in an earlier findings printout.

(19) A single dot in a result column is only to make the results easy to read and therefore itself has no meaning.

(20) Indication of the health insurance.

(21) For microbiological findings: enumeration of the germs.

(22) For microbiological findings the result marking can be chosen freely; in this case 0 = resistant, + = moderately sensitive, ++ = sensitive.

(23) Free text on the findings.

(24) Summary of the findings (to be separated and stuck in the patient's file at the sender's).

(25) Measured value unit (unless indicated in the long name of the biological parameter (cf. point (14)).

(26) Medical report, made up of text segments with segment insertions and free text.

(27) Specimen number under which the medical report text is to be stored in the data base.

(28) Abbreviation under which the medical report text is to be stored in the data base.
(29) Name of the text segment from the text preservation file.
(30) Insertion for the previously listed text segment.
(31) Separators of the individual insertions.
(32) End symbol of the text segment insertions.
(33) Free text.
(34) Start symbol of the text preservation file.
(35) Definition of the delimiter symbols for segment names, control characters and insertions.
(36) Definition of the end symbol for a text segment.
(37) Name of the text segment.
(38) Symbol for the following continuation line of the text segment.
(39) Insertion of a data element from the data base (here: PRDA = sample taking date).
(40) Symbol for feed control and character repetition; meanings: / = start new line, //nn = insert nn blank lines, /nnz = the character z is to be inserted nn times.
(41) Other text segments can also be called up in a given text segment; a text may even consist solely of a succession of other text segments.
(42) Definition of the points for insertions (see point (30)).

3.2 Cumulative findings

In the cumulative findings report the attending physician is given not only the results with comments, that is the findings, for a current job but these findings are reported in relationship with findings data previously taken from the same patient so that the variations of the different biological parameters are apparent (Figs. 8 and 9). Depending on the requirements (e. g. for certain sender groups) for QUADROLAB several report forms can be freely defined and used. The form used in Oststadt Hospital and in the Central Clinic of Hanover Medical University have proved best.

Each finding sheet consists of a main section having the dimensions of a standard letter (in Germany DIN A4) with filing holes on the left margin and an auxiliary section which for insertion into the patient's file can be folded backwards. Both sections contain all the particulars necessary for clear assignment in the heading (in case the auxiliary section is unintentionally separated).

On the main section of the findings sheet there is room for 5 findings columns and on the auxiliary section 3 findings columns. For up to 3 examination times the reference range (cf. point (13)) appears on the main section and when this limit is exceeded the reference range is printed out on the auxiliary section. Consecutive columns may on the one hand contain results for the same specimen taking date and on the other hand results for specimen taking dates which are far apart, the column always containing the results of one specimen.

```
MEDIZINISCHE  HOCHSCHULE  HANNOVER             S t a t i o n :  (3)  12B
Institut  fuer  Klinische  Chemie 1            Rohrpost : 1666
Direktor: Prof. Dr.Dr.med. Buettner         (5) Blatt:              1 / 1
                                                                    (6)
                                                (7)
                                               ***********************
K L I N I S C H - C H E M I S C H E   B E F U N D E   *  BLUT / URIN  *
(8)                                            ***********************
                          (11)           (1)                       (2)
Nxxxxxxxx, Axxxx
*14.08.72  M  I-Zahl: 14.08.72-5910    Labor-EDV-Ausdruck: 22.10.86, 15:30
(9)       (10)
 Entnahmedatum: 02.10.86    04.10.86    22.10.86  (12)   Referenz-
 Entnahmeuhrz.:    08:00       08:00       08:00         bereich    (13)
 Probennummer :   222/02      110/04      100/22         (Kinder)
(14)              --------    --------    --------       --------------

S-Kalium     mmol/l    4.2           . (19)   3.9  /      3.6 - 5.4
S-Natrium    mmol/l    135           .        138  /      132 - 155
S-Chlorid    mmol/l     99           .         99  / (18) 97 - 108
S-Calcium    mmol/l   2.23           .       2.00- /      2.15 - 2.60
S-Phosphat   mmol/l   1.34           .       1.19  /      0.83 - 1.67
S-Kreatinin  umol/l     80        737+         55  /      bis 115
S-Harnstoff  mmol/l    4.6        24.0+       5.0  /      3.3 - 6.7
S-Harnsre.   umol/l    190    Z.W.MAT. (16)   205  /      149 - 208
                  -----------------------------------
S-Alk. P'ase    U/l   246K (15)      .        203  /      130 - 385
                  -----------------------------------
P-Haemoglobin mg/l      64           .   (17)  **  /      bis 100
=========================================================================
U-Status                             .          .
U-pH-Wert                5           .          .         5 - 7
U-Protein              +++           .          .
U-Glucose          NEGATIV           .          .
U-Ketonkoerper     NEGATIV           .          .
U-Haemoglobin          +++           .          .
                   ---------------------------------

U-Sediment                           .          .
U-Leukozyten          5-10           .          .
U-Erythrozyten         >20           .          .
U-Plattenepithel.      1-4           .          .
U-Nierenepithelien     1-4           .          .
U-Granul. Zylinder    5-10           .          .

!! Ein hierzu frueher ausgedrucktes     !! This makes a corresponding
   Befundblatt wird hiermit                previously printed findings
   ungueltig !!                            sheet invalid !!
=========================================================================

INDIKATION:                              INDICATION:
Seit einigen Tagen Behandlung mit        Treatment for some days with
Isocillin-Saft wegen eitriger            isocillin syrup because of
Tonsillitis, Bauchschmerzen,             purulent tonsilitis, stomach
Schmerzen im Nierenlager,                pains, pains in the kidney area,
Blutdruck: 125/70 mm Hg                  blood pressure: 125/70 mm Hg
```

Fig. 8. Cumulative findings printout (with 3 specimens).

MEDIZINISCHE HOCHSCHULE HANNOVER
Institut fuer Klinische Chemie 1
Direktor: Prof. Dr.Dr.med. Buettner

Poliklinik : (3) NEUPO2
Blatt: 1 / 1 (7) (6)

*** M H H ***
8 1 1 0 / 8 1 3 0
(22.10.86 09.16) (4)

Poliklinik : NEUPO2
Blatt: 1 / 1

K L I N I S C H - C H E M I S C H E B E F U N D E (1) (11) (8)

KLINISCH-CHEMISCHE BEFUNDE

* PHARMAKA * (2)

Nxxxxxxxxx, Sxxxxxx
*27.09.42 F (10) I-Zahl: 27.09.42-5924 Labor-EDV-Ausdruck: 22.10.86, 09:16

Nxxxxxxxxx, Sxxxxxx
*27.09.42 F (12) I-Zahl: 27.09.42-5924

(9)									
Entnahmedatum:	21.03.86	10.04.86	11.04.86	05.06.86	28.06.86	08.08.86	12.09.86	21.10.86	Referenz-
Entnahmeuhrz.:	08.00	09.00	08.00	10.00	10.00	09.30	10.00	10.30	bereich (13)
Probennummer :	3366/21	3224/10	3328/11	3232/05	3200/28	3264/08	3068/12	3202/21	(Frauen)
(14) S-Antiepileptika							(15)		
S-Carbamazep. mg/l	.	.	.	.	.	5.3	5.2	4.9	4.0 - 10.0
S-Phenobarb. mg/l	17.2	14.2-	24.0	23.3	20.1	23.3	18.3	22.7	15.0 - 25.0
S-Primidon mg/l	11.2	11.8	16.7!D	14.1!D	10.3	10.3	6.5	9.8	5.0 - 12.0

!D : Die obere Grenze des therapeutischen Bereichs von Primidon kann in bestimmten Faellen erst bei 15.0 mg/l liegen.

The upper limit of the therapeutical range of primidone may in specific cases be 15.0 mg/l.

!! Ein hierzu fruener ausgedrucktes Befundblatt wird hiermit ungueltig !!

!! This makes a corresponding previously printed findings sheet invalid !!

==

Es ist zu beachten, dass die angegebenen Referenzbereiche nur als grobe Rahmenempfehlung zu verstehen sind. Die Interpretation der Serumspiegel dieser Pharmaka kann sinnvoll nur unter Beruecksichtigung des klinischen Befundes erfolgen. Die aufgefuehrten Empfehlungen entbinden daher den behandelnden Arzt nicht von der Verpflichtung der eigenen Ueberpruefung im Einzelfall.

It should be noted that the reference ranges given are only a rough general guide. The interpretation of the serum level of these drugs can take place properly only when account is taken of the clinical findings. The recommendations listed thus do not free attending physician from the obligation of conducting his own check in each specific case.

Fig. 9. Cumulative findings printout (with 8 specimens).

3.3 Preliminary findings (collective findings)

For an acute diagnosis or therapy it may be important to the attending physician to have results sooner than is possible for organizational or lack of time for the cumulative findings report. In particular in the case of results with emergency indication from the standby laboratory a rapid presentation of the data is more important than the context with the complete findings. Figure 10 shows an example of collective findings with the necessary patient data for unique assignment. This findings list is suitable for various paper formats for printout in the wards: single-column with a format for 18 characters/line, four-column for 72 or 80 characters/line and seven-column for 126 or 132 characters/line.

```
MEDIZINISCHE  HOCHSCHULE  HANNOVER
Krankenhaus Oststadt - Zentrallabor
Direktor:  Prof. Dr. med. Delbrueck



V o r a b b e f u n d e  des Bereitschaftslabors vom 04.09.86 um 17:45
-----------------------------------------------------------------------

     24A          '      ACHP11      '       1F         '       2412       '
Mxxxxxxxx,Pxxx    'Mxxxxxx,Fxxxxxxxx'Sxxxxxxxxx,Exxx   'Cxxxxxxx,Lxxxxxxx'
  18.09.47-1234   '  56342           '  5439876         '   2345/03.09.86  '
*18.09.47     M   ' *25.08.75     F  ' *17.10.97    M8  ' *29.02.48     M  '
                  '                  '                  '                  '
PrNr 2109/04.09.' PrNr 2457/04.09.' PrNr 2468/04.09.' PrNr 2345/03.09.'
Entn-Dat: 04.09.' Entn-Dat: 04.09.' Entn-Dat: 03.09.' Entn-Dat: 02.09.'
Entn-Zt.: 11:45 ' Entn-Zt.: 14:35 ' Entn-Zt.: 23:15 ' Entn-Zt.: 22:40 '
                  '                  '                  '                  '
    157+  =PNA    '     65   =PPROT  '   2.34-  =PK     '     432+  =PGPT  '
    106   =PCL    '     7.6+ =PHST   '    148   =PNA    '     145+  =PCK   '
    106   =PCL    '     345+ =PKREA  '    108   =PCL    '      99+  =PCKMB '
                  '     5.6  =PCHOL  '                  '                  '
                  '                  '                  '                  '
-----------------------------------------------------------------------

                  '      24A         '      24A         ' (3)              '
            (8)   'Exxxxxxx,Pxxxx    'Rxxxxxxx,Rxxxxxx  '                  '
                  '  17.04.25-4321   '  43276           ' (11)             '
            (9)   ' *17.04.25     M  ' *23.08.17     F  '                  '
                  '                  '                  ' (10)             '
                  ' PrNr 2458/04.09.' PrNr 2469/04.09.'                    '
            (12)  ' Entn-Dat: 04.09.' Entn-Dat: 03.09.'                    '
                  ' Entn-Zt.: 17:35 ' Entn-Zt.: 21:15 '                    '
                  '                  '                  '                  '
            (15)  '     148  =PNA    '    433+ =PGPT    ' (14)             '
                  '     115+ =PCL    '                  '                  '
                  '          =PKREA  '                  '                  '
                  '          =PHST   '                  '                  '
                  '                  '                  '                  '
                  '                  '                  '                  '
=======================================================================
```

Fig. 10. Preliminary findings of the standby laboratory
(PrNr = Specimen number, Entn-Dat = Date taken, Entn-Zt = Time taken).

Praxisgemeinschaft
Dr. med. E. H[redacted]
Arzt für Laboratoriumsmedizin,
Mikrobiologie und Infektionsepidemiologie
Priv.-Doz. Dr. med. P. M[redacted]
Arzt für Nuklearmedizin

Postfach 36[redacted] K[redacted] 17
4[redacted] B[redacted] 1, den 07.05.85
Telefon [redacted] 6[redacted] / 6[redacted]

Dr. med. E. H[redacted], Priv.-Doz. Dr. med. P. M[redacted] · Postfach 36[redacted] · 4[redacted] B[redacted] 1

HOERH

(3)

Herrn Dr. med.
H[redacted]-R[redacted] H[redacted]
Arzt f. Allgemeinmedizin
D[redacted]str. 10

4[redacted] B[redacted] 1

(2)

Verbindlichen Dank für Ihre Überweisung. Wir erhoben folgenden Befund.

Tages Nr./Eingangsdatum	Patient			Befunddatum	Dr. E. H[redacted] · PD Dr. P. M[redacted]
424/02.05.85 (12)	L[redacted], C[redacted] (8)			07.05.85	Postfach 36[redacted] · 4[redacted] B[redacted] 1 Tel. [redacted] 6[redacted] / 6[redacted]
Entnahmedatum/-uhrz.	**Geb.-Datum**	**Geschlecht**	**Versicherer**	**Bemerkungen:**	
02.05.85 00:00	25.08.33 (9)	M (10)	AOKBI (20)		

MATERIAL: Rachenabstrich (Throat smear)

Folgende(r) Keim(e) wurde(n) nachgewiesen : (Following germ(s) was(were) detected:)

(21) STRGG STREPTOCOCCUS, GRUPPE G
CAALB CANDIDA ALBICANS

	STRGG	CAALB
AMPICILLIN/AMOXYCILLIN	++	
AZLOCILLIN	++	
CEPHALOTIN	++	
CEFOXITIN	++	
CHLORAMPHENICOL	++	
ERYTHROMYCIN	++	
GENTAMYCIN	0	
CLINDAMYCIN	++	
NALIDIXIN-/PIPEMIDSAEURE	0	
NITROFURANE	++	
PENICILLIN - G	++	
PENICILLINASEFESTE PENIC	++	
NORFLOXACIN	++	
SULFONAMIDE	0	
TETRACYCLINE	0	
TRIMETHOPRIM/SULFAMETH.	++	
NYSTATIN		++
FLUCYTOSIN		++ (22)
AMPHOTERICIN B		+
CLOTRIMAZOL		++
ECONAZOL		++
NATAMYCIN		+

==

Mit freundlichen Gruessen !

Fig. 11. Microbiological findings
(from the joint practice Dr. Haubold/PD Dr. Mariß, Bielefeld).

3.4 Job-oriented individual findings

The job-oriented individual findings are used whenever for organizational reasons (e. g. outside senders) a systematic and clear patient identification cannot be thoroughly carried out because for example the necessary expenditure would be out of all proportion to the benefit of a cumulation.

Clear and informative individual findings can sometimes also be a good solution as shown by Figures 11 to 14.

Praxisgemeinschaft
Dr. med. E. H[redacted]
Arzt für Laboratoriumsmedizin,
Mikrobiologie und Infektionsepidemiologie
Priv.-Doz. Dr. med. P. M[redacted]
Arzt für Nuklearmedizin

Postfach 36[redacted], K[redacted] 17
4[redacted] B[redacted] 1, den 07.05.85
Telefon ([redacted]) 6[redacted] / 6[redacted]

Dr. med. E. H[redacted] Priv.-Doz. Dr. med. P. M[redacted] · Postfach 36[redacted] · 4[redacted] B[redacted] 1

BEOB (3)

Dres. med.
G[redacted] und D[redacted] E[redacted]
Aerzte f. Gynaekologie
D[redacted]str. 24

4[redacted] B[redacted] 1

Verbindlichen Dank für Ihre Überweisung. Wir erhoben folgenden Befund. (2)

Tages Nr./Eingangsdatum	Patient			Befunddatum	Dr. E. H[redacted] PD Dr. P. M[redacted]
(12) 433/02.05.85	K[redacted], E[redacted] (8)			07.05.85	Postfach 36[redacted] · 4[redacted] Tel. ([redacted]) 6[redacted] / 6[redacted]
Entnahmedatum/-uhrz.	**Geb.-Datum**	**Geschlecht**	**Versicherer**	**Bemerkungen:**	
02.05.85 00:00	22.07.53 (9)	W (10)	AOKBI (20)		

(23) MATERIAL: Vaginalabstrich
Neisseria gonorrhoeae nicht nachgewiesen.
Pilze (Candida-species) nicht nachgewiesen - Bebruetungsdauer: 2Tage
Mykoplasmen nachgewiesen.
Therapeutisch sind folgende Antibiotika zu empfehlen : Tetracyclin

==

Mit freundlichen Gruessen !

(MATERIAL: Vaginal smear
Neisseria gonorrhoeae not detected,
mycosis (candida species) not detected - incubation duration: 2 days
mycoplasma detected,
therapeutically the following antibiotics are recommended: tetracyclin)

Fig. 12. Laboratory findings
(from the joint practice Dr. Haubold/PD Dr. Mariß, Bielefeld).

```
Dres. med. K[redacted]                          2[redacted] G[redacted], den 09.06.86
Aerzte fuer                                     Postfach 1[redacted]
Laboratoriumsmedizin                            Tel. 0[redacted]/3[redacted]
                                                                        (2)

                                 R3HA        (8)
   Herrn Dr. med.
   G[redacted] R[redacted]          (3)    Pat. : [redacted]  (9)
   Prakt. Arzt                             ------
   C[redacted] B[redacted]damm 6           Geb.Datum   :10.04.66  F (10)
                                           Entn.Datum  :05.06.86
   2[redacted] H[redacted] 80                              (12)

A N A L Y S E N B E F U N D E               fuer Labor-Nr.:  141/05.06.86
(14)
                           Resultat   (25)        Referenzbereich
                             (15)                                (13)
```

	Resultat		Referenzbereich
THYRECIDEA			
T3-BINDUNGSKAPAZIT.	28.0	%	22.1 - 33.7
THYROXIN T4	5.5	UG/DL	5.0 - 13.0
THYROXIN-JOD	3.6	UG/DL	3.2 - 8.5
PROTEINGEB.JOD	3.2	UG/DL	2.9 - 8.0
T7-WERT	1.54	UG/DL	1.10 - 4.35
TSH VOR TRH-INJEKT.	0.9	MU/L	0.3 - 4.0
TSH NACH TRH-STIM.	10.6	MU/L	2.0 - 25.0

```
(24)                                 (8)
 141/05.05.86          [redacted] [redacted]
THYR=   T3B= 28.0    TY4= 5.5    TY4J= 3.6    PBJ= 3.2    TY7= 1.54
TSH0= 0.9    TSH3= 10.6
==========================================================================
```

Fig. 13. Analysis findings
(from the laboratory of the physicians Dres. med. S. and D. Kramer, Geesthacht)
(Geb. Datum = Date of birth, Entn. Datum = Date taken, Resultat = Result, Referenzbereich = Reference Range).

3.5 Special findings

3.5.1 Microbiological findings

Microbiological antibiograms may also be stored in the same data base and with the same structures as the clinical-chemical (quantitative or qualitative) results.

A special feature here is the data type "relation result": An antibiotic A is linked to a germ K by the relation x (e.g. $x = 0$ for "resistant", $x = +$ for "moderately sensitive", $x = ++$ for "sensitive"). The findings printout is as in Figure 11 and the results for up to 6 germs can be printed out on one sheet.

Praxisgemeinschaft
Dr. med. E. H[redacted] · Dr. med. R. M[redacted]
Ärzte für Laboratoriumsmedizin,
Mikrobiologie und Infektionsepidemiologie
Priv.-Doz. Dr. med. P. M[redacted]
Arzt für Nuklearmedizin

Postfach 3[redacted], W[redacted] 20
4[redacted] B[redacted] 1, den 11.06.86
Telefon (0[redacted]) 6[redacted]/17[redacted]

Dr E H[redacted] · Dr. R. M[redacted] · PD Dr. P. M[redacted] · Postfach 3[redacted] · 4[redacted] B[redacted] 1

ELCE

(3)

Frau u. Herrn Drs. med.
D[redacted] u. F[redacted] E[redacted]
Aerzte f. innere Medizin
D[redacted]str. 417-419

4[redacted] B[redacted] 1

Verbindlichen Dank für Ihre Überweisung. Wir erhoben folgenden Befund.

(2)

Tages Nr./Eingangsdatum	Patient			Befunddatum	Dr. E. H[redacted] · Dr. R. M[redacted] PD Dr. P. M[redacted] · 4[redacted] B[redacted] 1 Tel. (0[redacted]) 6[redacted]/ 17[redacted]
(12) 617/05.06.86	H[redacted], W[redacted] (8)			11.06.86	
Entnahmedatum/-uhrz.	Geb.-Datum	Geschlecht	Versicherer	Bemerkungen:	
05.06.86 00:00	13.05.28	W	BKKUC		

(14) (9) (10) RESULTAT (20) (25) REFERENZBEREICH

THYROXIN, GES., T4	(S)	14.4+	UG/DL	4.5 - 12.5
TRH-TEST				
TSH, BASAL	(S)	0.38	UU/ML	0.1 - 4.0
TSH, STIMULIERT	(S)	2.9	UU/ML	2.5 - 25.0

(15) (13)

Mit freundlichen Gruessen !

Fig. 14. Findings
(from the joint practice Dr. Haubold/Dr. Malottke/PD Dr. Mariß, Bielefeld)
(RESULTAT = RESULT, REFERENZBEREICH = REFERENCE RANGE).

3.5.2 Medical report

To enable the findings component "medical report" to be used firstly a text preservation file must be created which contains frequently recurring texts which differ from each other only by insertions related to the individual cases (Fig. 17). To generate the medical report in current daily operation the crude text with the current insertions and free texts of the findings then has to be inserted (Fig. 16) and is then stored in the result data base. The findings print module then prints out the medical report (Fig. 15).

Praxisgemeinschaft
Dr. med. E. H[redacted] · Dr. med. R. M[redacted]
Ärzte für Laboratoriumsmedizin,
Mikrobiologie und Infektionsepidemiologie
Priv.-Doz. Dr. med. P. M[redacted]
Arzt für Nuklearmedizin

Postfach 36[redacted] W[redacted]20
4[redacted] B[redacted] 1, den 11.06.86
Telefon (0[redacted]) [redacted]/6[redacted]

(2)

ELGE

Frau u. Herrn Drs. med. (3)
D[redacted] u. F[redacted] E[redacted]
Aerzte f. innere Medizin
D[redacted]str. 417-419

4[redacted] B[redacted] 1

TagesNr./Eingangsdatum:
617/05.06.86

(8) (9) (12)

Betr.: H[redacted], W[redacted]

geb. 13.05.28
M[redacted] STR. 39
4[redacted] B[redacted] 1

(26)

Dear Mr. Exxxxx,

Many thanks for kindly referring the above-mentioned patient to me who came here on June 5, 1986 for a thyroid gland examination.

ANAMNESTIC PARTICULARS:
The patient has lost about 8 kg weight recently and also lost some hair; moderate vitality, heat intolerance.

CLINICAL FINDINGS:
Neck circumference 35/36 cm, palpatory 1.5 times enlarged diffused soft thyroid gland, no specific eye symptoms, pulse 80/min, regular, no tremor, warm-dry skin.

INTRAVENOUS TRH TEST:
TSH basal concentration: 0.38 uU/ml (normal up to 4 uU/ml), 25 min, after i.v. injection of 200 ug TRH 2.9 uU/ml (normal up to 25 uU/ml).

SCINTIGRAM after i.v. injection of 1.0 mCi Tc-99m-pertechnetate:
The scintigraphic image shows a 1.5 times enlarged thyroid gland of regular form and position. The contours are smoothly bordered. The storage pattern is without pathological findings; in particular, there are no indications of hot or cold nodes.

CONCLUSION:
Struma diffusa I with limit-value euthyroid-hyperthyroid metabolic condition, the normal triiodothyronine level and the positive TRH test are at present contraindicative to a decompensated hyperthyroid or a decompensated autonomic adenoma. Consequently, I do not consider initiation of a thyrostatic treatment absolutely essential at present but the in-vitro parameters should be checked in about 4-6 weeks or of course earlier on any signs of hyperthyroid aggravation.

Yours truly,

Fig. 15. Medical report (from the joint practice Dr. Haubold/Dr. Malottke/PD Dr. Mariß, Bielefeld) generated with the aid of the text preservation file (cf. Fig. 17) from the raw text (cf. Fig. 16).

```
   (27)
//PR 617/05.06.86
//KB LETTER (28) (29)
[t01]Exxxxx\\[t10]The patient has lost about 8 kg weight
recently and also lost some hair; moderate vitality, (30) (31)
heat intolerance.[t06]35\36\1.5 times\80\\[t32]0.38\200\2.9\\ (32)
[t48]1.0\\[t51]1.5\\[ul4]Struma diffusa I with limit-value
euthyroid-hyperthyroid metabolic condition, the normal
triiodothyronine level and the positive TRH test are at
present contraindicative to a decompensated hyperthyroid (33)
or a decompensated autonomic adenoma. Consequently, I do
not consider initiation of a thyrostatic treatment
absolutely essential at present but the in-vitro parameters
should be checked in about 4-6 weeks or of course earlier
on any signs of hyperthyroid aggravation.[z01]
```

Fig. 16. Entered raw medical report text stored under the patient
(from the laboratory system of the joint practice Dr. Haubold/PD Dr. Marіß, Bielefeld).

4 Requirements for Efficient Findings Reports

An essential factor in a laboratory information system (see also ref. (9)) is the module for flexible presentation of laboratory findings. The examples illustrated and explained in the previous sections show the great variety of possibilities provided by the present state of the art in computer-aided laboratory medicine. It should be emphasized here that in the QUADROLAB system only a single findings module exists with which all the findings reports which leave the laboratory are generated. The examples in Section 3 show the high flexibility of the program product which is not only definable and executable completely form-controlled and parameter-controlled but also in spite of the multiplicity of the functions provided good maintenance and adaptability to modified support conditions.

Below the essential aspects and system components representing the important boundary conditions for the findings module will be discussed.

```
(34)          (35)                    (36)
TEXTSEGMENTE  [  ]   BEGRENZUNGSZEICHEN  \      *** SEG.HM ***

                              (40)
[a01 ][/]Dear Mr. [],[//02]\

[a10 ][//01]ANAMNESTIC PARTICULARS :[/]\

[a11 ][//01]CLINICAL FINDINGS :[/]\                                 (38)

[a15 ]Many thanks for kindly referring the above-mentioned patient \-
(37)   to me who came here on [PRDA] for a thyroid gland examination. \

[t01 ][a01][a15]\ (41)          (39)                      (42)

[t06 ][a11]Neck circumference [] / [] cm, palpatory [] \-
       enlarged diffused soft thyroid gland, no specific \-
       eye symptoms, pulse []/min, regular, no tremor, \-
       warm-dry skin. \

[t10 ][a10]\

[t32 ][//01]INTRAVENOUS TRH TEST:[/]TSH basal concentration: [] \-
       uU/ml (normal up to 4 uU/ml), 25 min. after i.v. injection \-
       of [] ug TRH [] uU/ml (normal up to 25 uU/ml).

[t48 ][//01]SCINTIGRAM after i.v. injection of [] mCi \-
       Tc-99m-pertechnetate:[/]\

[t51 ][/]The scintigraphic image shows a [] times enlarged \-
       thyroid gland of regular form and position. The contours \-
       are smoothly bordered. The storage pattern is without \-
       pathological findings; in particular, there are no \-
       indications of hot or cold nodes. \

[u14 ][//01]CONCLUSION:[/]\

[z01 ][//02]Yours truly[/] \
```

Fig. 17. Definitions in the text preservation file
(from the laboratory system of the joint practice Dr. Haubold/PD Dr. Mariß, Bielefeld).

4.1 System design

Each data element occuring in the system has an extensive and unique description which is stored in a central data dictionary and which all the programs access. With the aid of central project variables both time sequences as well as data flows and specific program functions are controlled. Uniform data input routines, output routines and structures as well as an extensive subroutine pool ensure that both the programmer and the user employ standardized formats and representation forms. An efficient multiprogramming operating system ensures that all the necessary activities can take place independently of each other, in so far as expedient, parallel to each other without impairing time-critical execution sequences. The operating system also supports different communication networks for data exchange with other computer systems, the necessary data safeguarding and saving precautions ensuring that no unauthorized access to or even destruction of the highly sensitive medical data stocks is possible. By thorough separation between programs, structural data and user data applicability to other laboratories has been achieved and already put into practice several times in the last 9 years.

4.2 Data base design

For storing, maintenance and information recovery of patient and result data a hierarchically structured data base system exists on 3 levels which with the aid of key files ensures rapid data access even with large amounts of data (100 000 – 200 000 patients with 500 000 – 800 000 jobs and more than 5 000 000 results). An important feature is that there is direct access at any time to all the data for the past 1 to 1 1/2 years and these data are therefore available without restriction for all functions of a laboratory information system (in particular for cumulative findings in the case of outpatients or explorative data analyses on unselected data material). Not only the contents but also the structural definitions of the data base can easily be varied because the necessary data elements are locally fixed in each component of the data base (such as patient master data, job master data and result data) and can be updated, amplified or deleted at any time.

4.3 Time behaviour

The time behaviour of a laboratory information system is of major significance to its users. EDP systems are still being offered on the market and bought for laboratories which do not even fulfil the minimum requirements as described by Trendelenburg et al. (10).

The most important points are the following:

1. Laboratory and EDP processes must be able to take place parallel to a great extent.
2. The EDP system must hold in readiness fundamentally all functions for routine operation.
3. The findings printout must be able to take place both continuously and batchwise (sorted by criteria to be defined such as senders, job numbers or patient names). At any time (certainly at least once at the end of a day) it must be possible to print out all findings which are (still) incomplete.
4. Independently of the regular findings printout mode the system must be able to prepare preliminary findings for selected wards with definable selection criteria:
   - at specific times of day and
   - for individual biological parameters.

5. The patient-oriented linking of all job and result data must ensure that over the entire access period of the data base at any time previous values are available within a few seconds (or even fractions of a second).
6. The response time behaviour must as a rule be such that the system reaction is 95% immediate (in less than one second) for all dialogs necessary for routine execution and for all requests from the data base for laboratory-relevant criteria such as patient name, identification number and job number.
7. Even with a very large data base and an extensive test system of plausibility checking it must be ensured that the findings module (even in multi-programming mode) operate so fast that a printer with a throughput capacity of 600 lines/minute can work at its maximum speed.

Long years of practical experience have made the authors aware that many aspects of the presentation and demands made in this publication can only be implemented with high expenditure on conception, programming and current system maintenance; however, the expenditure must not be judged by economic criteria alone because the benefit to the patient and his attending physician often cannot be quantified and must be considered of very high importance when weighing up the advantages and disadvantages.

References

1. Diagnostik-Informationssystem (1970) (Bock, H. E. and Eggstein, M., eds.) Springer, Berlin, Heidelberg, New York.
2. Delbrück, A. (1971) On-line-Datenerfassung und -verarbeitung im klinisch-chemischen Zentrallabor. IBM-Form K12-1012-0, 49–59.
3. Büttner, J. (1972) Klinisch-chemische Analytik und elektronische Datenverarbeitung. Z. klin. Chem. klin. Biochem. *10*, 474–478.
4. Knedel, M. (1971) Datenverarbeitung im klinisch-chemischen Institut am Städtischen Krankenhaus Harlaching. Siemens AG, Erlangen.
5. Porth, A. J. (1975) Laboratoriumsdaten, in: Handbuch der medizinischen Dokumentation und Datenverarbeitung (S. Koller and G. Wagner, eds.) pp. 663–701, Schattauer, Stuttgart, New York.
6. Porth, A. J. (1976) Untersuchungen und Verfahren zur Plausibilitätskontrolle im computerunterstützten klinisch-chemischen Laboratorium, Habilitationsschrift Medizinische Hochschule Hannover.
7. Porth, A. J. (1986) EVA und ATAM – zwei einfache Verfahren zur computerunterstützten Plausibilitätskontrolle im medizinischen Laboratorium, in: Perspektiven der Informationsverarbeitung in der Medizin – Kritische Synopse der Nutzung der Informatik in der Medizin, Göttingen September 1986 (C. T. Ehlers and H. Beland, eds.) pp. 365–369, Springer-Verlag, Berlin, Heidelberg, New York, Tokyo.
8. Porth, A. J. (1988) A method for checking striking result constellations based on bivariate frequency distributions (in preparation).
9. Porth, A. J. & Mieth, I. (1978) Data structures and data base problems in Clinical Chemistry information systems, in: Medical Informatics Europe 78, Proc. Cambridge, England 1978 (J. Anderson, ed.) pp. 489–496, Springer-Verlag, Berlin, Heidelberg, New York.
10. Trendelenburg, Ch., Heidrich, R., Hesseling, B., Luthe, H., Porth, A. J. & Weiss, U. (1985) Forderungen an ein Labor-EDV-System hinsichtlich Zeitverhalten, Meßwerterfassung, Behandlung von Notfalleinsendungen und Systempflege, Dt. Ges. f. Klin. Chemie e. V., Mitteilungen *4/85*, 173–180.

Chapter 2.3
Graphical Data Display and Medical Decision Making

D. P. Connelly

1 Introduction

At times the modern clinical laboratory appears to generate more data than can be used effectively in medical problem solving. Important laboratory results may be overlooked as they lie hidden in a mass of data competing for the physician's attention. Results may be misinterpreted because important associations cannot be conveyed in a conventional laboratory report. Computer systems in the clinical laboratory have been primarily directed at the first, or technical, level of communication which deals with the accurate transfer of data (1). Transcription errors have been reduced and reports made more timely, legible and complete. However, laboratory computers have contributed little to the second, or semantic, level of communication which emphasizes the communication of the significance of a message. Computer-based graphical display techniques for data aggregation and presentation have been advanced as a means of more effectively conveying the significance of laboratory results so as to promote better medical decisions (2, 3). Because trends in computer technology have made the use of graphical displays a feasible supplement to conventional reporting techniques, there is growing interest in graphical reporting of laboratory data. This chapter presents the rationale for the use of graphics, describes the functional uses of graphical displays for laboratory reporting, and reviews what evidence there is regarding the effectiveness of graphical display techniques for decision making.

1.1 Communication at the semantic level

In general, communication of the meaning of a message calls for a complex interplay between the received symbols, the receiver's knowledge regarding the context of the symbols and an inference process performed by the receiving individual. In a medical setting *evidence* from a particular patient (the set of symbols transmitted) is related by *reasoning* (i. e., inference) to a background of relevant medical knowledge (4). For instance, while the symbols communicated to a physician on the telephone may be "Mrs. Jones' glucose is 20 mmol/l", the physician must bring his own domain and patient-specific knowledge to bear through a reasoning process if the meaning and significance of the message is to be conveyed. Thus the physician brings from memory that Mrs. Jones, a newly hospitalized, insulin-dependent diabetic has had a fasting blood glucose ordered. The value being reported is the result of that order and is higher than desirable. Based on the physician's knowledge about diabetes, the laboratory test, and this patient, it is quickly and almost effortlessly decided that an adjustment in therapy is required. The receiver's knowledge underlying this decision is a product of years of training and unique personal experience. The inference process underlying the decision can only be vaguely described.

Clearly communication at the semantic level depends on the message receiver as well as the sender. Figure 1 presents an abstraction of the cognitive process in which semantic transfer of patient data (i. e., evidence) culminates in a clinical decision. Data may be presented to the physician in myriad forms often arriving simultaneously. The information's presence must first be perceived allowing the decision maker to then selectively direct attention from other competing data sources. The intensity of medical practice and a relative data deluge confronting the physician may affect semantic transfer as part of messages are overlooked, misread or too superficially interpreted. Important laboratory results are often overlooked because they are hidden in a mass of data competing for the physician's attention (5, 6). Transference of message meaning can be adversely affected if the receiver's knowledge stored in long-term memory is incomplete, poorly recalled or incorrect. Since the decision maker uses knowledge from past experiences to decide what data sources are important in particular situations, faulty knowledge can also result in misdirected attention. Semantic transfer may be impaired if external knowledge forms (e. g. books, medical records, colleagues) are inaccessible, incorrect or misunderstood. Faulty inference on the part of the decision maker as data and knowledge are processed can similarly disrupt transfer of the significance of a message. There is growing evidence that our inference ability is prone to significant

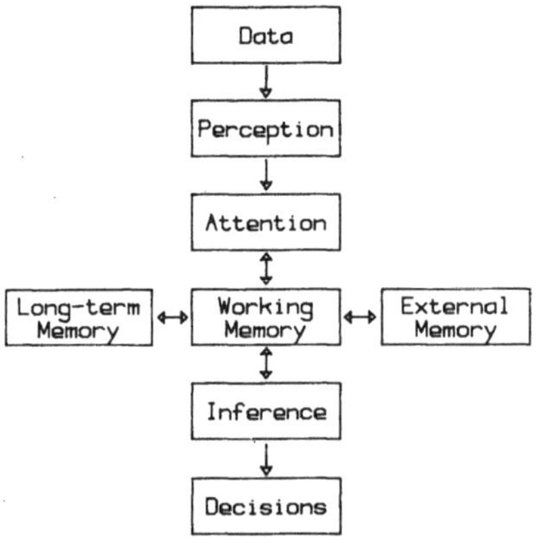

Fig. 1. The presence of patient-specific data in the form of observations, measurements, responses, reports, etc. are perceived and attention actively directed towards the information of interest. The patient-specific data and knowledge of the medical domain stored in the decision maker's memory or available in printed or other external media are brought together and through an inference process, conclusions are drawn and decisions formulated.

and consistent types of errors that can be expected to adversely affect communication at the semantic level (7). The complex and uncertain nature of most medical problems coupled with an imperative to act frequently demands intuitive decision-making as compared to a purely analytic and rational approach. How such intuitive approaches affect semantic transfer is difficult to predict. In short, limitations of the receiver's memory, inference, and attention can adversely affect communication of the significance of laboratory results.

1.2 Aiding the decision maker

While clinicians readily accept and use devices that support their sensory limitations (e.g. stethoscope, EKG machines), they have and use few resources other than external memory aids to support effective semantic transfer of patient data. How might the laboratory computer be employed to improve communication at the semantic level assuming the technical level has been adequately managed? If it can act to supplement the limitations of the message recipient's attention, memory, and inference skills, the laboratory computer can transcend its traditional role as a data transferring device to that of a decision support system. Feinstein, a noted researcher in clinical decision making, has suggested, "... one of the best uses for computers in medical decision making is to display the data in imaginative new ways that can help the physician making the decisions" (8). Graphical display techniques for presenting clinical laboratory data can be directed at each of the areas of limitations and thereby can perhaps provide a useful mechanism of decision support.

1.3 Graphical presentation as a decision aid

Patient management decisions often hinge on a time series of laboratory results. Graphical presentation of such data might be a more effective means of communication than is the conventional tabular form of reporting laboratory results. Graphic displays can aid in information transfer. The brain can perceive a picture at a glance with little concentration effort through a process called preattentive perception (9). A Cartesian graph can be perceived as a simple picture if the axes and scales conform to expectations and need not be read. Simple time-trend graphs can highlight outlying data, trends and data interrelationships in an efficient way not possible with tabular presentation methods.

The key advantages of graphics over more conventional tabular data presentation methods as summarized by Schmid in his *Handbook of Graphic Presentation* (10) and reviewed by Fienberg (11) are: 1. In comparison with other types of presentation, well-designed charts are more effective in creating interest and in appealing to the attention of the reader. 2. Visual relationships, as portrayed by charts and graphs, are more clearly grasped and more easily remembered. 3. The use of charts and graphs saves time, since the essential meaning of large masses of statistical data can be visualized at a glance. 4. Charts and graphs can provide a comprehensive picture of a problem that makes possible a more complete and better-balanced understanding than could be derived from tabular or textual forms of presentation. 5. Charts and graphs can bring out hidden facts and relationships and can stimulate as well as aid, analytical thinking and investigation. In short, graphical displays should not only aid the decision maker's attention, memory and inference skills, but also speed communication and even promote discovery of unforeseen relationships.

2 Graphical Displays in the Laboratory

Despite the potential of computer-based graphical display techniques to convey the meaning of data more effectively, application in the clinical laboratory has been limited. In this section some of the uses of graphical displays for laboratory communication are reviewed. This review concentrates on the clinical use of graphical displays, and only briefly touches on their use in clinical laboratory education and research.

The clinical uses may be functionally classified into 1. referential data presentation, in which the emphasis is on conveying results relative to reference values, 2. patient management or monitoring displays, in which the emphasis is on the ongoing course of a patient with known disease and therapy, 3. diagnostic presentations, in which an effect on diagnostic inference is intended, and 4. laboratory operations or management displays that aim to support high quality patient care services.

The first three categories are ranked, approximately, by the degree of inference promoted by the graphical display. In the referential display, a sense of abnormality and its degree can be conveyed. In monitoring, the display is oriented so that the time course of the abnormality can be readily perceived. A display suggesting diagnostic categories promotes the highest degree of inference. Many of the applications reviewed involve more than one of these functional classes.

2.1 Referential displays

Proposed methods for referential graphical display usually have emphasized the highlighting, at a glance, of abnormal results and degree of abnormality, taking care that important information not be overlooked. A multichannel analysis related to the SMA-12 (Technicon Instruments Corp., Tarrytown, New York) series of analyzers resulted in a display of this type and is well known to laboratory workers and clinicians. Because of its utility and familiarity, Altshuler used this display technique in the Programmed Accelerated Laboratory Investigation (PALI) system, including variables not measured on the SMA-12 (12). The PALI report also included computer-generated graphical plots of densitometric recordings.

In another form of referential display, the scaling factor and origin of all variables of concern are adjusted so that the reference range widths appear of equal size and are spatially contiguous on the graph. This allows values outside the reference

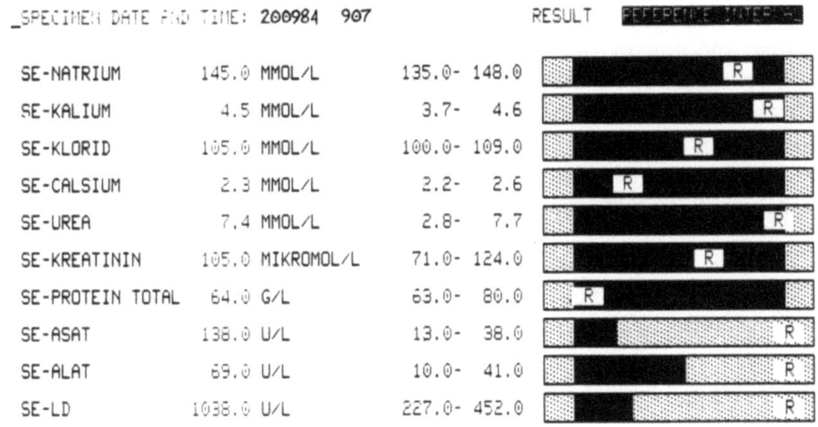

Fig. 2. A form of reference display in which reference ranges are aligned vertically. The value for each analyte is indicated by an R. Results outside the reference range cause the reference interval to be compressed, drawing attention to the out-of-bounds result. (Courtesy of H. E. Solberg, from the Clinical Laboratory Display System – CLDS, see ref. 14).

range to be seen readily. Rushmer proposed normalizing variables by dividing by the standard deviation of the reference population's distribution on each variable and expressing values in standard deviation units (13). The reference ranges of displayed variables were positioned to be spatially contiguous in a vertical column. With Rushmer's approach, 1. widely deviant values would become immediately obvious, 2. attention could be focused immediately on the abnormal finding which would no longer be obscured in a mass of normal results, and 3. the relative significance (i.e., degree of abnormality) of aberrant values would be obvious immediately, without reliance on the clinician's memory of reference ranges. Solberg adapted such a format (Fig. 2) and further enhanced abnormal values through the use of color highlights (14). Van Egmond and Wieme (15) used referential display techniques to display on a single page the results of 46 analytes and their change over time.

Rather than arrange the contiguous reference ranges of multiple variables in a column or row, Williams et al. (16) arranged them in a circular fashion so that the contiguous reference ranges formed an annulus with the variables plotted on scalar "rays" emanating from the center of the ring. Zimmer (17) and later Love (18) subsequently adapted and expanded on the work of Williams et al. to microcomputer implementations. Besides the referential function of this technique, the shape of the polygon formed by connecting the measured values of the variables may provide recognizable patterns useful for diagnosis (16).

2.2 Monitoring displays

Graphical displays to support patient management decisions focus on time trends of laboratory results as the patient responds or reacts to known therapy and disease.

Simple time-trend graphs can highlight outlying data, trends, and data interrelationships, and convey the effect of analytic variability in a way not possible with tabular presentation methods. The utility of this form of display could be extremely important because many patient management decisions hinge on time-related laboratory results.

Although a number of time-trend display systems have been described for looking retrospectively at large data banks for research purposes (19, 20), attempts at

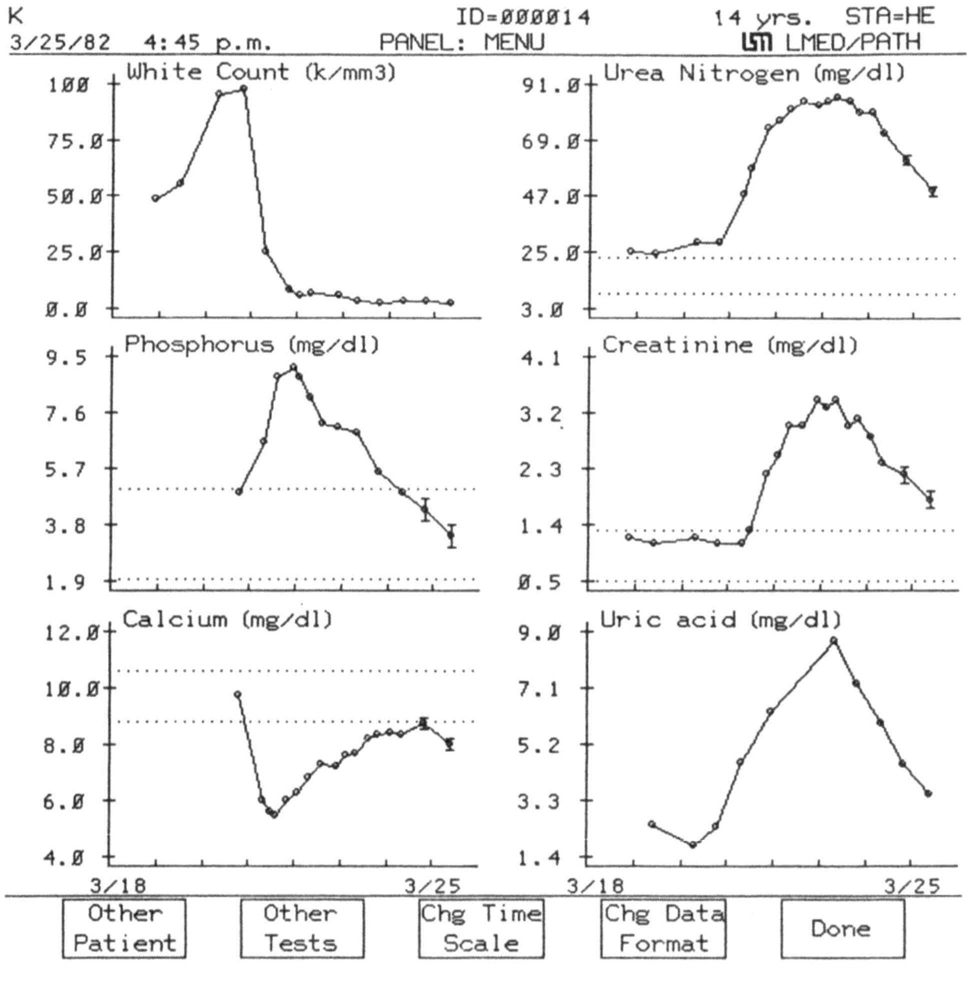

Fig. 3. A monitoring display showing the time course of six analytes in a 14 year old male with leukemia treated with chemotherapy. The dotted lines represent the age-sex-specific reference range. The last two values of four of the analytes have associated error bars that indicate the analytic variability of the procedure. In this case the patient's massively elevated white count drops abruptly on the third hospital day as a result of therapy. The cytolosis is reflected in an elevation in uric acid and phosphorus with the phosphorous change tracked by an inverse change in calcium. Even though well hydrated throughout therapy, acute and resolving renal impairment is indicated by urea nitrogen and creatinine.

integrating such displays into the clinical care process are of relatively recent origin (21). Williams et al. (16) discribed a method for plotting time-trends using the high resolution graphics and "user friendliness" of the PLATO system (Control Data Corporation, Minneapolis, Minnesota). The author and colleagues adapted William's approach and used a PLATO display terminal as an interactive terminal tied to a laboratory computer system database to develop time-trend plots of physiologically related analytes (22). Reference ranges and information on laboratory variability can be conveyed unobtrusively along with readily used time-trend plots of multiple variables (Fig. 3). By touching options presented on the screen, the user can easily manipulate the form of presentation (e. g., expand time scale, select log-converted display).

Similar approaches have been used to develop monitoring displays by Solberg (14) and Kellerman (23). Enlander (24) and Cimino et al. (25) have described methods for the graphical display of patients' time-trend data for clinical decision-making. Yasaka and colleagues (26) have described graphical methods for summarizing longitudinal studies of individuals participating in a long term health monitoring study. A number of systems for retrospective batch presentation of patient data also have been described (27, 28). Grams et al. (29) have suggested that multivariate time-trend plots pertaining to organ systems might be a useful addition in this area. Combined with pharmacological data, appropriate pharmacokinetic models, and an easy means of user-computer dialogue, the application of time-trend graphs in therapeutic drug monitoring may be another fruitful area for development.

The time-trend display technique promises to be a useful tool because it 1. deals with the more common decision-making process related to patient management rather than diagnosis, 2. builds on data and data systems commonly and readily available, 3. uses a display format that is widely recognized, understood, and accepted by clinical users, and 4. supports the time-consuming and menial tasks of data gathering and assembly.

2.3 Diagnostic displays

The purpose of diagnostic displays is to promote diagnostic inference. A form closely related to that described in the previous section is the display of time-related tests used for diagnosis. Displays have been used for tests of oral glucose tolerance, (12, 30, 31) growth hormone and insulin (12), and cardiac enzymes (32, 33). Usually these displays are accompanied by interpretive statements. Serial displays of organ-specific estimates of lactate dehydrogenase (LDH) isoenzymes may especially be valuable in the diagnosis of recurrent or extending myocardial infarction (34).

In another form of diagnostic display, the magnitude of one or more patient variables is plotted so that similarity of a case to specific diagnostic groups can be visualized and estimated. Diagnostic displays of this type can be classified according to the number of variables represented in the display. Of the many bivariate diagnostic plots that have been developed, the most frequently encountered is the blood-gas map in which pH is plotted against $P(CO_2)$ (35). The bivariate plot is overlaid on a mapping of physiologically significant "regions" (e. g., acute respiratory acidosis) promoting the corresponding diagnostic inference. The bivariate plot also has been used for the display of parathormone level and serum calcium as related to various disease possibilities (36). Stauffer et al. (37) developed a bivariate plot for the interactive interpretation of a pair of thyroid function tests.

When it is necessary to depict more than two variables at once, the constraints of the two-dimensional display screen give rise to an array of methods of presentation. Pain and Duncan (38) have plotted the results of three thyroid function tests independently, within overlapping regions encompassing the hypothyroid, euthyroid, and hyperthyroid states. In another trivariate application regarding the interpretation of liver enzymes, Strandjord et al. (39) present a pseudo-three-dimensional (orthogonal) plot depicting clustering of results of three liver tests in different diseases with an overplot of the individual patient's results. By comparing the patient's position in three-dimensional space with that of a group of patients with known disease, an inference can be made.

The radial plot of Williams et al. (16) becomes a diagnostic display for multivariate data presentation when the shape of the polygon is used to suggest a diagnosis. A related technique for displaying multivariate data is the use of "Chernoff faces" (40). Variables are displayed as effects on cartoon facial features (e.g., breadth of nose proportional to creatinine value, size of mouth related to albumin level) so that a face might emerge that is characteristic for a given disease. The underlying assumption is that people recognize different faces rapidly and automatically enough so that this might aid in diagnostic classification. The utility of the radial plot and Chernoff face in laboratory reporting have not been evaluated extensively. Robertson et al. (41) showed the potential use of the Chernoff face in the recognition of repeated biochemical profiles.

Other techniques for two-dimensional representation of multidimensional laboratory data have been explored. Thompson and Woodbury (42) have demonstrated how multivariate data for patients with suspected myocardial infarct might be conveyed, including the display of multivariate time-trend relationships. Tschope et al. (43) developed a pseudo-three-dimensional plot of multiple variables in which each observation set for one patient falls on the surface of a globe. This was intended for use in differentiating idiopathic hypercalciuria from primary hyperparathyroidism in patients presenting with renal stones. These methods rely on various mathematical techniques for determining the most useful orthogonal separator dimensions in multivariate space. Physicians may not accept such mathematic dimensions for plotting unless the basis appears intuitively sound. Well designed graphic presentations facilitate the necessary intuitive process. Much of this work is limited by the lack of concordance between real data and the idealized assumptions upon which many of these multivariate transformation techniques rest. By using graphical presentation methods, it may be possible to employ more nonparametric methods, which do not rely on strict assumptions regarding data distribution. It may be fruitful to explore multivariate display techniques used in other realms that have not yet been applied to clinical laboratory data (11, 44, 45). Problems of physician acceptance might limit the application of such techniques to laboratory analysis; a laboratory-based consulting physician might then communicate the results and their significance to the clinician. Finally, graphical aids for test selection and interpretation have been introduced (46, 47).

2.4 Laboratory operations and management displays

The final category of clinical uses of graphics in the laboratory consists of displays used to support laboratory operations and management. Computer-based graphic display of quality control data is becoming increasingly common (48, 49). By eliminating the painstaking manual data recording and presentation, more emphasis can be placed on the real-time use of quality control information for error detection and problem solving. Carefully developed displays to allow monitoring of high-speed multichannel instruments may be the only feasible way of ensuring their accuracy in a high-volume laboratory.

Graphic display of variables pertaining to work flow and operations can be very useful in supporting day-to-day laboratory management decisions (50–52). Laboratory forecasting and planning probably can also be aided in this manner.

Employee training might be enhanced through the application of graphical displays. Interactive training sessions could be developed employing computer models of analyzers. Simulated problems could be developed, complete with graphics to represent analyzer meters and controls to provide a lifelike training environment. Computer models with graphical displays have been used in pharmacokinetics training (53). Unfortunately, the use of graphical displays in computer-based training does little to ameliorate the major problems of cost and development time that this form of training presents.

The laboratory medicine researcher could benefit from new approaches to working with masses of experimental data, as have scientists in other areas (54, 55). Some of the interactive methods for exploring data relationships and testing new hypotheses might be especially helpful in research settings (56–58). Graphical display methods will play an important role in a number of investigational analysis systems (59, 60). Image analysis and morphometrics are other forms of computer-based graphics that will play an increasingly important role in research as well as clinical testing (61, 62).

3 The Effectiveness of Graphical Displays in Aiding Decision Making

Although the use of graphical displays to enhance the communication of data is intuitively appealing and has many proponents, the evidence that decision making is improved is conflicting and scattered among many disciplines. Little work with regards to this question has been done in the arena of medical decision making. In this section general studies of the effectiveness of graphics in decision aiding are summarized, methods for enhancing this effectiveness are presented and, finally, some of the few studies related to assessing this effectiveness for reporting and using laboratory results are briefly reviewed.

3.1 Graphical aids for decision makers: studies of effectiveness

While attention to graphical methods for reporting data is a relatively recent concern for computer-oriented information scientists, such methods have long been employed by workers in statistics, cartography, psychology, human factors, education and marketing. In a recent review, DeSanctis summarized and synthesized much of the management science evaluative work to provide a clarified perspective on the actual utility of graphics as decision aids (63). In the relatively small literature base, researchers have used many different dependent variables to assess the decision aiding question. The effects of graphical displays on interpretation accuracy, problem comprehension, task performance, decision quality, speed of comprehension, decision speed, memory and preference have all been employed in various studies making it difficult to develop a coherent summary of research results. For the particular comparison of graphic versus tabular displays, DeSanctis found seven studies that concluded graphs were better decision aids, twelve concluded tables were better and ten found no difference. Rather than regarding these findings as inconclusive, they demonstrate that graphs can be useful in particular decision tasks. The effectiveness of the data display format is largely a function of the particular decision task at hand (64). The problem becomes one of identifying the conditions under which graphs outperform tables. A few studies show graphs are superior in decision tasks where interpolation is required rather than recognition of specific values. A low level of task complexity also favors graphical over tabular methods. For tasks that involve seeing time dependent patterns in a large mass of data, graphs are a good choice (64). Much work remains to be done to clarify the optimum display methodology for particular decision tasks.

Graphical displays may be at a disadvantage to the more common tabular approach because graphical interpretation is a skill that must be learned through experience and can be honed through formal training (65). Unskilled readers will make many common errors in the interpretation of Cartesian graphs (66). Only in the last decade have graphical displays become readily generated by widely accessible computers. Although graphs are becoming more commonly encountered in our daily activities, most individuals aren't as comfortable with using graphs as they are with tabular data display. With repeated exposure and use, the utility of graphical data displays can begin to demonstrate an advantage over tabular displays (67). Thus, the wider availability of graphical displays to provide familiarity may be important for decision effectiveness even when the characteristics of the decision task seem to favor graphics.

3.2 Human perception and graphical display

Fundamental characteristics of human perception influence our ability to use graphical displays. Recent research studies shed important insight for those designing displays to aid decision makers (68). Cleveland (69) has ranked the graphical-perception tasks in order of decreasing human performance accuracy as follows:

1. Position along a common scale
2. Position along identical, nonaligned scales

3. Length
4. Angle, slope
5. Area
6. Volume
7. Color hue, saturation, density

Those developing graphical displays should use methods that keep the graphical perception task as near the top of the list as possible. Thus the popular graphical format of the pie chart should be avoided since it requires assessment of angle and area for proper interpretation. It might be better replaced with a bar chart where the assessment of length, a task in which humans are more accurate, is required. While properly developed time-trend plots can rate at the top of the list, they may not be as effective if it is important to differentiate the rate of change (i. e., slope) of two rapidly changing variables. Cleveland provides many examples of how the knowledge of graphical-perception task accuracy can contribute to more effective displays. A number of other recent books provide a rich source for designing displays that are clear, straightforward and functional (70–72).

In a review of the implications of human perception in the reporting of laboratory results, Politser has provided much needed insight and many useful guidelines (3). The guidelines have been structured around six natural stages of perception: filtering, simplication, coding, grouping, recognition and segregation.

Humans initially edit incoming information by filtering. Laboratory report designers can aid this process by minimizing superfluous data. Unnecessary numbers and heavy grid lines may detract from the value of a graphical display. Tufte refers to such superfluous information on graphs as "chartjunk" (70) and especially decries three forms of it that many computer graphics packages so readily produce. Moire' effects, the vibratory sensation produced by closely placed parallel lines commonly used to shade bar charts, can be very distracting. Heavy grid lines clutter and may hide the data of interest. Finally, self-promoting graphic elements, where the components of graphs take on "decorative forms" and other "computer debris" should be resolutely filtered before they reach the display.

Another step in human perception is simplification. Humans tend to edit data by simplifying them. In tabular displays, five digits of accuracy may impede perception of an important difference in the second most significant digit. Appropriately rounding the data on the report may lead to more effective decision making. Regarding graphical displays, the choice of the more simple display method is to be preferred as is reflected in Cleveland's list of graphical perception tasks described earlier.

In the perception task, coding, data may be dichotomized into normal or abnormal and dealt with more easily. Graphical methods of facilitating this were documented in Section 2.1, referential displays.

Humans perceptually edit data by crudely grouping them by some useful relationship with regards to the task at hand. Variables being graphed might be appropriately grouped by organ system in laboratory reporting to aid in this stage of perception.

Recognition can be fostered with graphical aids by performing transformations of original data to enhance otherwise subtle but important differences. Diagnostic graphs described in Section 2.3 can facilitate this perceptual stage.

Finally, in segregation, humans try to separate the effects of multiple processes. In medicine we try to decide which individual disease process or organ system has caused the observed test abnormality. For instance, the isoenzymes of lactate dehydrogenase (LDH) can be transformed into organ specific components, the time-trend plot of which can be useful in assessing recurrent myocardial injury in a patient already hospitalized for myocardial infarction (34).

3.3 The effectiveness of graphical displays in laboratory reporting

As in other domains, the effects of graphical reporting of laboratory results on the decision maker have been studied little. Graphical display methods have only recently become readily available, and there is little past theoretical framework to draw on in performing such evaluations.

In a study of the use of time trend plots in a university hospital-based general medicine ward (22) the authors studied the acceptability of graphics and its perceived usefulness through survey methods. In this study, a microcomputer with a touch sensitive screen was connected to a laboratory computer and its database. The physician would first select the patient of interest by touching a patient roster screen. Then in quick succession the user could select the laboratory tests and time period of interest by touching the screen. Once this information was entered, the microcomputer requested the relevant tests from the laboratory computer. The microcomputer then received these test results, sorted them with regards to time, then automatically generated time-trend graphical displays of up to six variables per screen. The graphical plots could be printed on a local printer when desired.

A survey of clinical users was conducted after one year of use covering the last third of that year. Letters were sent to housestaff and attending physicians that had been assigned to the medicine ward during the period of interest. Subjects were asked to return the survey form by mail. The returned forms were anonymous. No followup requests of nonresponders could be made. Almost all of the survey questions presented here were of the Likert variety using a five point scale ranging from strongly agree to strongly disagree.

Table 1 summarizes the results of the survey. The response rate of the housestaff was slightly over half while about 75% of the eligible faculty responded. Although the responses were incomplete, the subsequent findings were very consistent with a similar but smaller survey of a different group of physicians done eight weeks after system installation. Twenty of the 23 responding housestaff reported daily usage of the system as compared to only three of the faculty that used the system daily. Most of the responding faculty used the system weekly. Their usage rates and generally favorable view of the system in terms of learnability, usability, value, reliability, speed, and time-saving potential demonstrate that the physicians were generally favorably disposed to the system and reasonably well-experienced with it.

The report itself was almost unanimously regarded highly in terms of use and organization. Similarly 86% of responding housestaff regarded graphical presentation as being helpful in their practice as did 90% of responding faculty physicians. Seventy-four percent of the reporting housestaff reported that often or sometimes

Table 1. Attitudes of physicians toward a graphical laboratory data display system

	Housestaff	Faculty
Eligible Physicians	45	20
Responding Physicians	24	15
Response Rate	53%	75%
Never Used System	1	5
Number of users	23	10
At least weekly usage	22/23*	8/10
THE SYSTEM		
is easy to learn	23/23	8/9
is easy to use	22/22	8/8
is worthwhile	19/21	9/9
is reliable	20/22	7/9
is fast	20/21	6/9
saves time	20/22	8/9
THE REPORT		
is useful	22/22	10/10
is well-organized	21/22	9/9
GRAPHICAL DISPLAYS		
are helpful	19/22	9/10
should be widely available	17/22	6/10
made decisions easier	17/23	5/10

* The first number is the number of subjects who responded in the indicated way. The second number is the number of subjects who responded to that particular statement.

"the graphical report made decisions about laboratory results easier than if only the regular laboratory report had been available". Half of the faculty agreed to this statement. These data indicate that the majority of physicians found that graphical displays were helpful and aided their decision making. Thus this survey of physician attitudes towards graphical methods indicates good acceptance of the reporting method with some indication that it is perceived as favorably affecting clinical care.

Durbridge developed discriminant analysis displays pertaining to common hepatic and biliary disorders (73). These displays were based on eleven biochemical measurements. The clinician-user would indicate diseases under consideration. Then a bivariate plot (i. e., a canonical variate scattergram) was displayed that demonstrated the patient's value relative to values of patients known to have the candidate diseases. Clinicians had no difficulty in understanding the content of the displays. A greater problem was in the choice of diseases to be included in the displays.

The effectiveness of the human decision maker using time-trend plots for the early detection of abnormalities has been compared to that of a time series analysis method (74). Surveillance and early detection of deviation from a homeostatic state are goals common to health care programs for the apparently healthy as well as for groups of patients known to have or have had specific diseases. Automated approaches to detecting deviations have the advantage of being reliably applied, traceable, consistent in outcome, and conserving of professional resources. However, these techniques may be relatively insensitive, require a moderately large number of observations before they become functional, and are not necessarily tuned to the detection of the specific type of deviation of clinical significance. The human observer using graphical trend displays coupled with specific domain knowledge may be a more sensitive and reliable detector of deviations at an earlier time in the course of clinical observations. Time trend graphs representing monthly monitoring of serum carcinoembryonic antigen (CEA) levels in simulated patients with surgically-removed breast cancer were presented to six human observers and a time series analysis routine based on a homeostatic model (75). Based on a small number of simulated cases, the false positive rate of the two approaches appeared comparable. Observers detected deviations earlier than the analytic method in 13 of 39 instances and tied in 15. This preliminary study suggests human observers using time trend graphs can supplement analytic techniques in the early detection of deviations in biologic time series.

But the human observer may be inconsistent, unavailable, distracted or expensive. By embodying the human rules of deviation detection in an expert system, perhaps the advantages of the human expert can be gained while at the same time the disadvantages are eliminated. In a preliminary study of this hypothesis, three of the human observers described their rationale in assigning a level of suspicion regarding the presence of an important deviation as the observation points on the graphs were serially revealed to them. The verbalizations were analyzed to develop rules for an expert system. The expert system tied the observers in 19 of 38 instances and surpassed them seven times. It appears that the human skill of graph reading can at least partially be extracted and embodied in an expert system. If this is so, a more detailed understanding of how people read graphs is in order. This growing understanding will, in turn, guide developers of graphical displays to provide reporting formats that are better matched to graph readers and decision tasks.

4 Summary

Graphical display of laboratory data may ameliorate the effects of human limitations of attention, memory and inference. In terms of patient care, at least three broad classes of clinical graphs can be developed. In referential displays the emphasis is on conveying the existence of an abnormality and to some extent its magnitude. Monitoring displays represent the relationship of analytes over time and are oriented towards revealing important trends and threshold crossings. In diagnostic displays, the patient's measurements are compared to measurements of patient's with known disease to promote classification of the patient into the correct diagnostic group.

Although it is intuitively appealing that graphical displays can facilitate communication and thereby enhance decision making, the evidence for this thesis is limited and equivocal. Few studies involving laboratory analytes have been done in this regard. A growing literature regarding the perception of graphs and the evaluation of the effectiveness of various presentation methods for promoting decision making should serve to guide display developers.

Acknowledgement

Section 2 was adapted in part from the author's article, "Graphical displays in the clinical laboratory" published in the Journal of Clinical Laboratory Automation in 1983 and is included here with permission by the publisher, Appleton-Century-Crofts (see ref. 2).

References

1. Shannon, C. E. & Weaver, W. (1949) The Mathematical Theory of Communication. University of Illinois Press, Urbana.
2. Connelly, D. P. (1983) Graphical displays in the clinical laboratory. J. Clin. Lab. Auto. *3*, 92–100.
3. Politser, P. E. (1986) How to make laboratory information more informative. Clin. Chem. *32*, 1510–1516.
4. Tryer, J. H. & Eadie, M. J. (1976) The Astute Physician. Elsevier, New York.
5. Krieg, A. (1981) Medical information science and the laboratory professional. J. Clin. Lab. Auto. *1*, 133–135.
6. McDonald, C. J. (1976) Computer reminders, the quality of care and the nonperfectibility of man. N. Engl. J. Med. *295*, 1351–1355.
7. Kahneman, D., Slovic, P. and Tversky, A. (eds.) (1982) Judgment Under Uncertainty. Heuristics and Biases. Cambridge University Press, New York.
8. Feinstein, A. R. (1976) Computer malpractice. In: Proceedings of the Sixth Annual Conference of the Society for Computer Medicine (Koss, N., ed.) pp. 1–8, Minneapolis.
9. Myers, W. (1980) Computer graphics: A two-way street. Computer *13*, (7) 49–58.
10. Schmid, C. F. (1954) Handbook of Graphic Presentation. Ronald Press, New York.
11. Fienberg, S. E. (1979) Graphical methods in statistics. Amer. Stat. *33*, 165–178.
12. Altshuler, C. H., Bareta, J., Cafaro, A. F. et al. (1972) The PALI and SLIC systems. CRC Crit. Rev. Clin. Lab. Sci. *3*, 379–402.
13. Rushmer, R. F. (1968) Accentuate the positive. A display system for clinical laboratory data. JAMA *206*, 836–838.
14. Solberg, H. E. & Hansen, E. (1985) CLDS (Clinical laboratory display system): Clinical laboratory data presented by computer colour graphics. In: Computing in Clinical Laboratories Proc. 5th Int. Conf. (Trendelenburg, C., ed.) pp. 61–62, Stuttgart.
15. Van Egmond, J. & Wieme, R. (1977) A new presentation of laboratory data. The patient's evolutive test profile (abstract). In: Medinfo 77 (Shires, D. B., Wolf, H., eds.) p. 1059, North-Holland, New York.
16. Williams, B. T. (ed.) (1982) Computer Aids to Clinical Decisions Vol. II. pp. 170–178, CRC Press, Boca Raton.
17. Zimmer, J. L. (1981) A graphics display for laboratory profiles. In: Proceedings of the Fifth Annual Symposium on Computer Applications in Medical Care (Heffernan, H. G., ed.) p. 1155, IEEE, New York.
18. Love, J. E. (1984) A computer graphic representation of multivariate laboratory data. J. Med. Tech. *1*, 353–358.
19. Gault, M. H., Ackman, D., Morehouse, D. & Dossetor, J. B. (1969) Kidney transplants: Computer display of clinical and laboratory parameters. The first 100 days. Can. Med. Assoc. J. *100*, 359–362.
20. Blum, R. L. (1981) Displaying clinical data from a time-oriented database. Comput. Biol. Med. *11*, 197–210.
21. Osserman, E. F., Katz, L., Sherman, W. H. et al. (1978) Computer-based case tracing (COMTRAC). JAMA *239*, 1772–1776.

22. Connelly, D. P., Lasky, L. C., Keller, R. M. et al. (1982) A system for graphical display of clinical laboratory data. Am. J. Clin. Pathol. *78*, 729–737.
23. Kellerman, W., Peter, K., Jensen, U. et al. (1985) In: Computers in Critical Care and Pulmonary Medicine (Osswald, P. M., ed.) pp. 82–95, Springer, Berlin.
24. Enlander, D. (1981) Computerized graphic representation of clinical chemistry results. J. Clin. Pathol. *34*, 806–808.
25. Cimino, J. J., Munson, P. J., Lewis, T. & Rodbard, D. (1981) Graphical display of clinical and laboratory data using a desktop computer interfaced with a hospital information system. In: Proceedings of the Fifth Annual Symposium on Computer Applications in Medical Care (Heffernan, H. G., ed.) pp. 1085–1088, IEEE, New York.
26. Yasaka, T., Tamura, M. & Kiyoshima, K. (1979) Evaluation of the PL Health Control System after ten years experience. In: Evaluation of Efficiency of Medical Action (Alperovitch, de Dombal, Gremy, eds.) North Holland, Amsterdam.
27. Blum, B. J., Lenhard, R. E., Braine, H. & Kammer, A. (1977) A clinical information system. In: Proceedings of the First Annual Symposium on Computer Applications in Medical Care, pp. 131–138, IEEE, New York.
28. Altshuler, C. H., Bareta, J., Cafaro, A. F. et al. (1975) AIDE (Accessible Information for Diagnosis and Evaluation): An information retrieval system. In: Progress in Clinical Pathology (Stefanini, M., ed.) pp. 307–323, Grune & Stratton, New York.
29. Grams, R. R., Johnson, E. A. & Benson, E. S. (1972) Laboratory data analysis system: Section III – Multivariate normality. Am. J. Clin. Pathol. *58*, 188–200.
30. Saltsman, R. H. W. & Nipper, H. C. (1976) Use of a programmable calculator to display and classify glucose and insulin values in glucose tolerance tests (abstract). Clin. Chem. *22*, 1160.
31. Reece, R. L. (1977) Future of laboratory medicine – Impact of multichannel instruments and computers. Lufkin Medical Laboratory Letter, pp. 359–364, Minneapolis.
32. Talamo, T. S., Losos, F. J. & Mercer, D. W. (1986) Microcomputer-assisted interpretative reporting of sequential cardiac profile data. Am. J. Clin. Pathol. *86*, 336–340.
33. Lupovitch, A. & Kaplan, A. G. (1980) Interpretive reporting of laboratory data. Biochemical profile of an acute myocardial infarct. Am. J. Clin. Pathol. *73*, 767–773.
34. Politser, P. E., Powell, S. H. & Fink, J. (1986) A new method for reporting the sources of abnormal activities of lactate dehydrogenase in serum. Clin. Chem. *32*, 1517–1524.
35. Goldberg, J., Green, S. B., Moss, M. L. et al. (1973) Computer-based instruction and diagnosis of acidbase disorders. JAMA *233*, 269–275.
36. Boyd, J. C., Lewis, J. W., Slatopolsky, E. & Ladenson, J. H. (1981) Parathyrin measured concurrently with free or total calcium in the differential diagnosis of hypercalcemia. Clin. Chem. *27*, 574–579.
37. Stauffer, M., Clayson, K. J., Roby, R. J. & Strandjord, P. E. (1974) A computer-assisted system. Thyroid disease. Am. J. Clin. Pathol. *62*, 766–774.
38. Pain, R. W. & Duncan, B. M. (1976) Pictorial reporting of multiple thyroid function results. Clin. Chem. *22*, 1562–1566.
39. Strandjord, P. E., Clayson, K. J. & Roby, R. J. (1973) Computer assisted pattern recognition and the diagnosis of liver disease. Hum. Pathol. *4*, 67–77.
40. Chernoff, H. (1973) Using faces to represent points in k-dimensional space graphically. J. Am Stat. Assoc. *68*, 361–368.
41. Robertson, E. A., Steirteghem, A. C. V., Byrkit, J. E. & Young, D. S. (1980) Biochemical individuality and the recognition of personal profiles with a computer. Clin. Chem. *26*, 30–36.
42. Thompson, H. K. & Woodbury, M. A. (1970) Clinical data representation in multidimensional space. Comp. Biom. Res. *3*, 58–73.
43. Tschope, W., Schmidt-Gayk, H., Ritz, E. et al. (1979) Analysis of clinical laboratory data by biplot methods using a three-dimensional display. Discrimination of renal-stone patients with idiopathic hypercalciuria and primary hyperparathyroidism. Int. J. Bio-Med. Comp. *10*, 341–351.
44. Wang, P. C. C. (ed.) (1978) Graphical Representation of Multivariate Data. Academic Press, New York.
45. Tukey, J. W. (1977) Exploratory Data Analysis. Addison-Wesley, Reading, Mass.
46. Greenes, R. A. (1983) Interactive microcomputer-based graphical tools for physician decision support. Med. Decis. Making *3*, 15–21.
47. Rossing, R. G. & Hatcher, W. E. (1980) A graphic method for the evaluation of di-

agnostic tests. Meth. Inform. Med. *19*, 149–156.
48. Bold, A. M. (1978) Visual display units as input/output devices for a computer in a clinical chemistry laboratory. In: Computing in Clinical Laboratories (Siemaszko, F., ed.) pp. 174–183. Wiley, New York.
49. Westgard, J. O. (1982) Better quality control through micro-computers. Diag. Med. *5*, 60–74.
50. Bloch, D. M. (1980) Computer-generated management tools for the clinical pathology laboratory. J. Med. Syst. *4*, 367–380.
51. Glick, M. R., Ryder, K. W. and Jackson, S. A. (1986) Graphical comparisons of interferences in clinical chemistry instrumentation. Clin. Chem. *32*, 470–475.
52. Bland, J. M. & Altman, D. G. (1986) Statistical methods for assessing agreement between two methods of clinical measurement. Lancet (8476), 307–310.
53. Florenz, M., Bigger, J. T., Kahn, N. & Hong, S. (1975) An interactive graphic instructional system in pharmacology. Biosci. Commun. *1*, 303–313.
54. Graedel, T. E. & McGill, R. (1982) Graphical presentation of results from scientific computer models. Science *215*, 1191–1198.
55. Jurs, P. C. (1986) Pattern recognition used to investigate multivariate data in analytical chemistry. Science *232*, 1219–1224.
56. Alberts, S. & Meinzer, H. P. (1981) An interpreter for the easy creation of graphic displays of experimental data. Methods Inf. Med. *20*, 28–32.
57. Filliben, J. J. (1981) DATAPLOT – An interactive high-level language for graphics, non-linear fitting, data analysis and mathematics. Comput. Graphics *15*, 199–213.
58. Wainer, H. & Thissen, D. (1981) Graphical data analysis. Ann. Rev. Psychol. *32*, 191–241.
59. Shelly, D. C., Warner, I. M. & Quarles, J. M. (1980) Multiparameter approach to the "fingerprinting" of fluorescent pseudomonads. Clin. Chem. *26*, 1419–1424.
60. Anderson, N. L., Taylor, J., Scandora, A. E. et al. (1981) The TYCHO system for computer analysis of two-dimensional gel electrophoresis patterns. Clin. Chem. *27*, 1807–1820.
61. Goin, J. E., Craven, T. M., Gilbert, J. R. et al. (1986) The "intelligent workstation" for cell-surface phenotyping based on principles of pattern recognition and image analysis. Clin. Chem. *32*, 1655–1659.
62. Simone, J. N., Welling, L. W. & Welling, D. J. (1979) Computer-assisted morphometric analysis for three-dimensional cell shape. Lab. Invest. *41*, 334–339.
63. DeSanctis, G. (1984) Computer graphics as decision aids: Directions for research. Dec. Sci. *15*, 463–483.
64. Dickson, G. W., DeSanctis, G. & McBride, D. J. (1986) Understanding the effectiveness of computer graphics for decision support: A cumulative experimental approach. Comm. ACM *29*, 40–47.
65. McDonald-Ross, M. (1977) Now numbers are shown. AV Comm. Rev. *25*, 359–409.
66. Preece, J. (1983) Graphs are not straightforward. In: The Psychology of Computer Use (Green, T. R. G., Payne, S. J. and vander Veer, G. C., eds.) Academic Press, New York.
67. DeSanctis, G. & Jarvenpaa, S. L. (1985) An investigation of the "tables versus graphs" controversy in a learning environment. Sixth Int. Conf. on Information Systems.
68. Cleveland, W. S. & McGill, R. (1985) Graphical perception and graphical methods for analyzing scientific data. Science *229*, 828–833.
69. Cleveland, W. S. (1985) The Elements of Graphing Data. Wadsworth Advanced Books, Monterey, CA.
70. Tufte, E. R. (1983) The Visual Display of Quantitative Information. Graphics Press, Cheshire, CT.
71. Bertin, J. (1983) Semiology of Graphics. University of Wisconsin Press, Madison, WI.
72. Schmid, C. F. (1983) Statistical Graphics Design Principles and Practices, John Wiley, New York.
73. Durbridge, T. C. (1984) Applying descriptive discriminant analysis as a visual aid for physicians interpreting biochemical test results. Clin. Bioch. *17*, 321–326.
74. Connelly, D. P. & Rich, S. R. (1987) Detection of deviations in time series by the human observer. In: Maintaining a Healthy State Within the Individual (Harris, E. K. and Yasaka, T., eds.) North-Holland, Amsterdam, 211–221.
75. Winkel, P., Bentzon, M. V., Statland, B. E. et al. (1982) Predicting recurrence in patients with breast cancer from cumulative laboratory results: A new technique for the application of time series analysis. Clin. Chem. *78*, 8057–8067.

Chapter 2.4
Digital and Analogous Presentation of Laboratory Results in Dependence of Time

W. Renn, R. Maulbetsch and M. Eggstein

1 Introduction

This volume of Clinical Biochemistry is devoted to the problem of presentation and interpretation of laboratory data. The presentation of these data is usually given in the form of the usual laboratory reports (Chapter 2.2.) or in the form of a graph (Chapter 2.3). Highly sophisticated statistical methods, like discriminant-, factor- or cluster-analysis are used (Part 3 of this volume), for the interpretation of laboratory data. The aim of this chapter is to provide a presentation of time dependent laboratory data in a complete and combined form. This will include digital and analogous presentation of data in a single report, to facilitate the interpretation of the data without additional statistical effort.

Every laboratory result is time dependent and the presentation of laboratory data is incomplete without the exact drawing time of the sample under investigation. However, there are great differences in the importance of the time dependence for the interpretation of laboratory results.

An outpatient with a chronic liver disease for example, showing up at the physician's office at regular time intervals, is checked for his enzyme activities. These values will slowly increase with time, but the exact time intervals are not really important for the diagnosis and therapy control. In contrast, an inpatient with an acute myocardial infarction has to be checked immediately upon admission and thereafter in short time intervals with respect to the typical creatine kinase activities. This gives the physician an impression of the extent of the infarction and allows a prognosis for the patient.

In general, the time dependence of laboratory results taken from patients with acute diseases is of vital interest for the diagnosis and therapy. There are many other fields of medicine where the time dependence is important. The following examples are only a small selection of different usages of time dependenc:

1. The time trend of blood glucose levels which are regularly measured for the therapy control of insulin dependent diabetics.

2. Also regarding diabetes, we have the monitoring of the time dependence of glucose, insulin, and C-peptide after an oral or intravenous glucose tolerance test.

3. The time dependence of hormone levels after endocrine function tests. These laboratory tests give valuable hints for the diagnosis of malfunctions of endocrine hormone releasing- or dependent organs.

4. The time dependence of drug levels in body fluids is the main goal of pharmacokinetics.

In this article, we discuss some methods for the digital and analogous presentation of laboratory results. Therefore, we have to start with a short explanation of what is meant by the technical terms digital and analogous.

In clinical biochemistry, we have two forms of data – analogous and digital data. The digital data are acquired by measurements – the usual laboratory tests. Analogous data are the real values of parameters of clinical interest. However, they are not observable in a strict methological sense. First of all, every measurement and especially the biological ones are disturbed by measurement errors. Secondly, the real values between two sampling times are not observable at all.

However, in practice, a laboratory test at a definite time gives an estimation of the real or analogous value sufficient enough in most cases of clinical interest. The analogous data in a period of time, in cases where the dependence of time is of diagnostical or therapeutical interest, can be estimated from the measurements at the beginning and the end of that period. To get these estimated values, one has to interpolate between the two measurements.

The most common method used is the linear interpolation, because this is the easiest way to get the desired estimation of the unobservable real or analogous data. It can be used without complicated calculations.

Nonlinear regression analysis is the main device applied in pharmacokinetics (1). There, mathematical models and methods are used to interpret data of the time course of drug and metabolite levels in different fluids, tissues, and excreta of the body. Pharmacokinetics offers an approach to improving and perhaps optimising the therapeutic management of individual patients.

The use of nonlinear interpolation methods is also well known from the calculations of calibration curves needed for the evaluation of radio- and enzymeimmuno assays (2). In this case, we have two different approaches. The first one is heuristic, the method of spline functions. It uses polynomials of second or higher degree to interpolate between the measured standard values. The second one has a theoretical background and uses a sound mathematical description of the antigen-antibody binding mechanism.

Nonlinear interpolation by means of systems analysis and mathematical modeling of patients data gathered in the fields of clinical chemistry, endocrinology, and diabetology is applied in scientific work (3). As far as we know, these methods are rarely used with respect to the interpretation of time dependent laboratory tests, as they are taken during the routine work in clinical laboratories or hospitals. This is due to the lack of familiarity on the part of the physician with the mathematical problems involved. However, with the increasing appearance of computers in the hospital, the application of sophisticated mathematical methods is becoming visible.

The presentation of data in clinical biochemistry can also be accomplished in a digital and analogous manner. The digital presentation is most common. It consists, in a tabulated form, of the test results in figures – the well known laboratory reports. Examples are thoroughly discussed in the Chapters 2.2. Analogous presentation of data is accomplished by a graphic device showing the estimated or interpolated values respectively, together with the digital data which are the results of laboratory tests. The application of analogous data presentation using linear interpolation is discussed in Chapter 2.3.

In the present chapter we focus on the combined presentation of digital and analogous data by means of mathematical modeling and nonlinear interpolation. The aim is a complete graphical laboratory report with the goal of assisting the physician on the ward with the interpretation of time dependent laboratory tests.

In Section 2, two examples of analogous presentation by means of the linear interpolation method are introduced. The advantages of this approach along with its drawbacks are discussed. The method of mathematical modeling is outlined in the following sections. In Section 3, the model formulation is discussed, and the model schemes together with the equivalent differential equations and their solutions are presented. The problems of model identification and parameter estimation are shortly outlined in Section 4. The model validation, i.e. the application of the methods to practical examples with respect to clinical chemistry, endocrinology, and diabetology is presented in Section 5. In Section 6, we present the general laboratory report using the methods developed for the digital and analogous presentation of time dependent laboratory data.

2 Linear Interpolation

This section begins with an example of analogous data presentation by means of linear interpolation. This gives an idea of the advantages of the graphic presentation of time dependent laboratory data. In a second example, the pitfalls and drawbacks of the linear interpolation method are also discussed.

The first example comes from diabetology. In our hospital both indoor and outpatients are taught to record their daily blood glucose levels, applied insulin doses, and carbohydrate consumption together with their sporting activities and any additional remarks concerning problems with the therapy. A booklet is used for this diary which consists of forms similar to the one shown in Figure 1. When an outpatient shows up in the doctor's office for a checkup and therapy control, the physician is presented with a tremendous amount of data to find out about the patient's situation. Using the digital presentation of data (Fig. 1), this task is tedious, time consuming, and unsafe, because some data may be overlooked.

To improve the therapy control of diabetic patients, we developed a program package for the analogous presentation of the daily blood glucose levels. The data from the diabetes diary of the patient are typed into a personal computer by a data typist and stored into a data base system. To make these data available to the physician, a dialogue using a screen form has been established, where the physician can specify his request.

This dialogue consists of three steps:

1. The list of the names of patients having data stored in the data base is shown on the screen allowing the patient of interest to be chosen.

Diabetes Therapy Control

Time	Gluc.	CH	Bolus	Basal	Sport	Remarks
3.00	91					
7.00	55	24		12		
9.00	221	12	4			
11.30	195	36	6			
14.00	33	12				
17.00	67	24				
20.00	166	12				
22.00	166			7		
24.00	168					
Date: 25.02.86 Patients Name: Xxxxxx, Xxxxxxxxx						

Fig. 1. Example of the form used for the documentation of the glucose levels, carbohydrate consumption (CH), and insulin doses together with the sporting activities and patient's remarks concerning problems with the therapy of insulin dependent diabetics.

2. The time span of data available for the particular patient is given on the screen, and the desired time span for the graphic report can be selected. The maximum time span for an adequate presentation is seven days.

3. The physician has also the choice between a presentation of the data on the screen or a hardcopy from a matrix printer.

After these selections, a graph of the blood glucose levels in the diary with a linear interpolation between them is drawn. For ease of interpretation of the glucose excursions, the daily means and standard deviations of the blood glucose levels as well as the daily M-values (4) are calculated. The daily M-value is the mean taken over one day of the w-values, which are rather complicate functions of the blood glucose levels:

$$w = \text{abs}\,[(10 \cdot \log(\text{BG}/\text{BGT}))^3] \tag{2.1}$$

BG represents the actual blood glucose level and BGT is the target BG-level being defined as the optimal average BG-value. In the original publication (4) a BGT of 120 mg/dl was adopted. However, in our hospital, we use the intensified insulin therapy, and we found a BGT-value of 100 mg/dl to be more adequate. The symbol abs represents the absolute value, i.e. the w-value is always positive.

In this context it is not necessary to go into the mathematical details of the formula (2.1). Instead, the main properties of the w-value with a target BG-level of 100 mg/dl are discussed:

1. An actual BG-value of 100 mg/dl results in a *w*-value of zero (log(1) = 0).
2. In the lower acceptable range of BG-levels between 60 and 100 mg/dl the *w*-value varies slowly between 11 and 0.
3. In the upper acceptable range of BG-levels between 100 and 180 mg/dl the *w*-value varies slowly between 0 and 17.
4. For BG-levels above 180 mg/dl, the *w*-value increases rapidly above 17.
5. In the hypoglycemid region below 60 mg/dl the *w*-value increases as well, but much more rapidly as in the hyperglycemic region.

Thus, the daily mean of the *w*-values, i.e. the daily *M*-value, gives a good description of the glycemic situation of the patient and emphasizes any hypoglycemid episodes. The calculated values described above are presented together with the graph of the time dependent BG-levels.

In Figure 2 an example of a hardcopy produced with our program is given. In this, the BG-levels over the period of three days of an insulin dependent diabetics are presented. The measured values are symbolized by crosses and the values

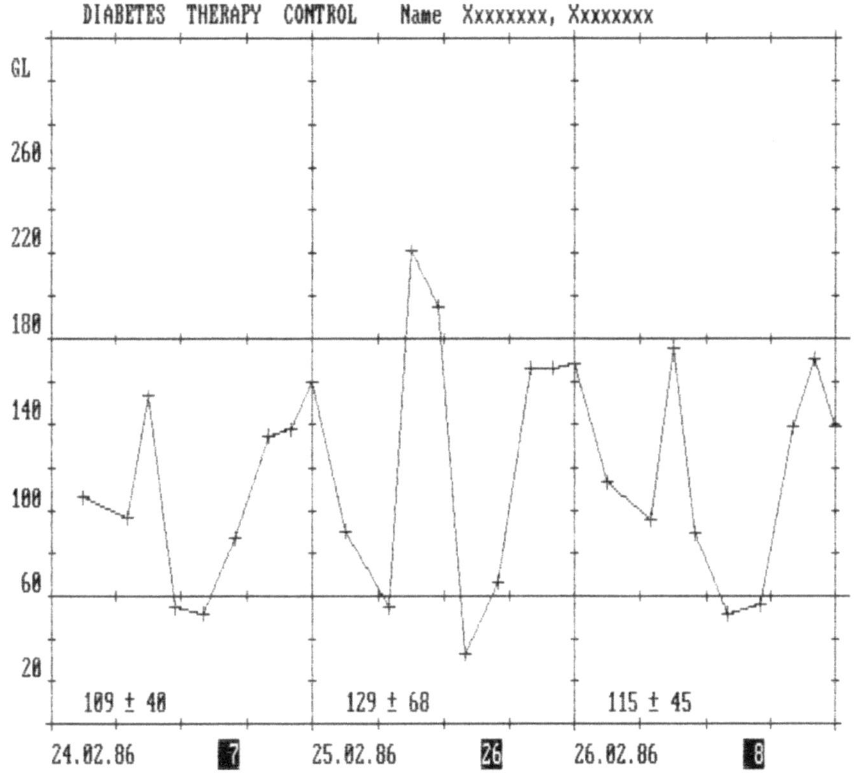

Fig. 2. Glucose levels of an insulin dependent diabetic over three days. The measured values are represented by crosses (+) and the time curves are approximated by the usual linear interpolation. All values are in mg/dl. The euglycaemic range is shown by the straight lines at 60 and 180 mg/dl respectively. The means and standard deviations of the measured glucose levels are printed under the time curve. The date and the calculated *M*-values of the three days are given in the bottom line.

between them are estimated by means of the usual linear interpolation. For ease of interpretation of the glycemic situation, the acceptable range is indicated by horizontal lines at 60 and 180 mg/dl respectively. The calculated daily means and standard deviations of the BG-levels are given at the bottom of the graphic panel. In the bottom line the calculated daily M-values are shown together with the date of the days of interest.

The visual judgement of the patient's glycemic situation is much easier using the analogous presentation of the BG-data compared with the digital one shown in Figure 1. Looking at Figure 2 one is immediately impressed by the hyperglycemic episode in the morning of the second day which is followed by a hypoglycemic excursion in the early afternoon of that day. The daily mean of the BG-levels resembles the differences between these three days, but the M-values emphasize much more expressively the exceptional situation of the second day in our example.

Having developed the program with the output described above, we had a tool for the easy judgement of BG-curves. Using this visual method one can immediately see the strong correlation between BG-levels and HbA1-levels, i.e. the levels of the glycolised haemoglobin. However, we have also found a serious drawback using the data presentation of Figure 2. There, one finds the differences between the various days quite easily, but one never finds the reasons for this differences. There are many cases where the reason for a hyper- or hypoglycemic episode is fairly certain. These are cases where something goes wrong with the insulin therapy or where a mistake with the carbohydrate diet occurs.

The pitfall of the graphic report in the form of Figure 2 is that the data presented are incomplete. For a proper judgement of the glycemic situation of an insulin dependent patient, one needs to know the insulin doses and the consumption of carbohydrates as well. This information is contained in the digital presentation in the form of Figure 1, but it contains the data of only one day. Thus, a comparison between different days is rendered difficult.

To provide the complete information to the physician in a readily conceivable form, we have improved the graphic part of our program by taking into account the carbohydrate intake as well as the doses of short and long acting insulin. In Figure 3, an example is shown with exactly the same data used in Figures 1 and 2. The middle panel is identical to the graph of Figure 2 presenting the BG-curves. The lower panel shows the consumption of carbohydrates in gram. These data are discrete and digital and there is no need for interpolation. For the graphical presentation of these digital data, filled bars are used. In the upper panel, the doses of short acting (open bars) as well as long acting insulin (filled bars) are shown. These data again are discrete and digital and can, therefore, be represented by bars. To emphasize the lowering action of insulin on the BG-levels, the insulin doses are drawn in a inverse top-down representation.

Now, checking the carbohydrate intake of the patient in our example, no differences between the three days occur. Therefore, the outliers discussed above cannot be due to mistakes in the diet. However, if the insulin doses are carefully scanned, one can find out the following history and the reason for the hyper- and hypoglycemic episodes:

On the second day, the early morning bolus of short acting insulin was forgotten by the patient. This mistake in the insulin therapy resulted in the high BG-level of

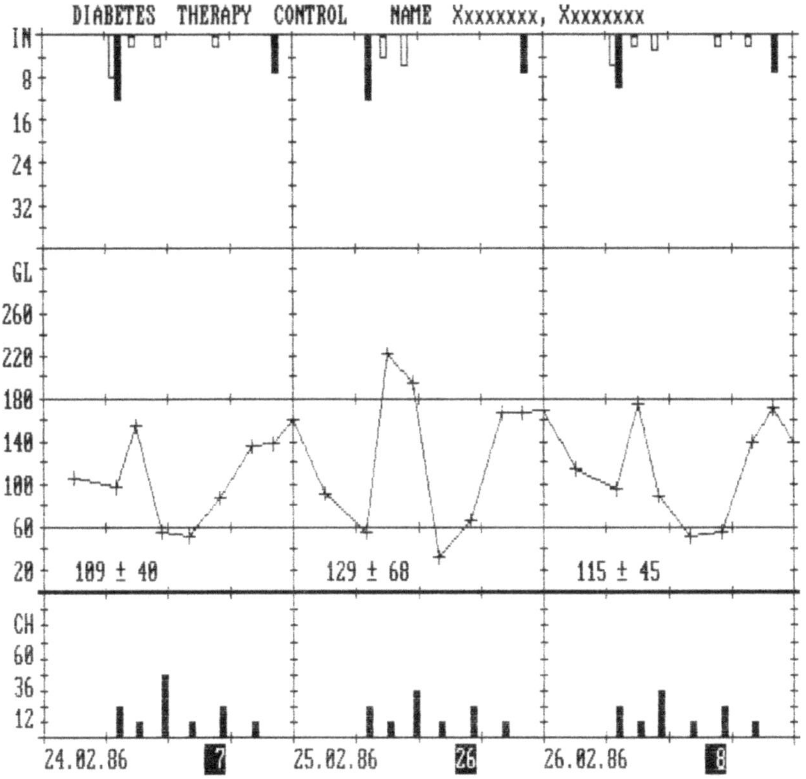

Fig. 3. Diabetes therapy control. The graph presents the glucose levels (GL) together with the carbohydrate consumption (CH), and the insulin doses (IN) of an insulin dependent diabetics over three days. This figure is basically the same as Figure 2, with the exception of the additional upper and lower panel. The filled bars in the later represent the carbohydrate intake during these three days. The open and filled bars in the upper panel represent the doses of short and long acting insulin respectively.

220 mg/dl at 9 a.m. When this value was realized by the patient he compensated with an increased insulin bolus, which was a correct decision. At 11.30 a.m. however, the patient tried to compensate for BG-value of 195 mg/dl with another increased insulin bolus, not realizing that the BG-level was already decreasing. Therefore, he ended with a hypoglycemic BG-value of 30 mg/dl at 2 p.m.

Thus, from the complete presentation of the data influencing the BG-levels, it is possible to find out very subtle details of the patient's insulin therapy and carbohydrate diet.

Inspecting the analogous presentation of diabetes data introduced above more critically, another problem arises. This is the problem of linear interpolation. The question is how closely does the straight line drawn between the measured points approximate the real values? In many cases the exact values are not necessary at all, because they would have no therapeutical consequences if they would be known. However, there are cases, where one would like to get more information out of the data.

If we go back to our example from diabetology and have a look at the BG-levels below 60 mg/dl on the third day, these values between 50 and 55 mg/dl are really borderline, but still acceptable. However, it seems to be very unlikely that there are no even lower values between them. A knowledge of the exact values would probably result in a reduction of the early morning dose of the long acting insulin.

Unfortunately, it is very difficult to find a reliable mathematical model for the BG-homeostasis which can be used for a nonlinear interpolation between the measured BG-levels. The reason for this difficulty is the fact that both glucose as well as insulin are stored or degraded in the liver and the peripheral tissues. In addition, the BG-levels are not only influenced by the insulin doses and carbohydrate consumption, but also by psychological factors like stress and the contra regulatory stress hormones. A mathematical model taking all these interactions into account would be to complicated and could not be fitted to the rather few data available from routine observations. The usage of spline functions does not solve the problem either, because then the results depend heavily on the time intervals between the measured points.

There are, however, problems in the clinical biochemistry which are much more simple than the problems of BG-homeostasis. They can be treated using mathematical modeling and nonlinear regression analysis. Before the mathematical theory is discussed, an example is given as an introduction.

The acute disease of an organ with a high concentration of typical enzymes can be diagnosed by the specific time dependent enzyme pattern in the peripheral blood circulation. As an example, the activities of the creatine kinase CK and the creatine kinase isoenzyme CK-MB are shown in Figure 4. These enzymes, especially CK-MB, are contained specifically in the heart muscle in high concentrations and are released during an acute myocardial infarction (AMI). The data used in our example belong accordingly to a patient suffering from AMI. For the understanding and interpretation of the data presentation used in Figure 4, some technical details and remarks are necessary.

In the work by Connelly presented in Chapter 2.3, a single panel is used for each time dependent laboratory parameter, because the physician is completely free in his choice of parameters he wants to have displayed, regardless whether or not these parameter are correlated. This method has the advantage of having the proper unit on the ordinate of each graph. In contrast, in our program package which will be discussed in Section 6, only laboratory parameters are presented which are related in accordance with the diagnosis or laboratory test. Therefore, a single panel is used to present all the parameters together, allowing one to see the relationship between them. This approach is most useful, if the time dependence of several parameters is needed for a proper judgement of the biochemical reaction. In our example, the two enzymes CK as well as CK-MB are vital for the diagnosis of AMI.

If more than one variable with different dimensions are presented in a single graph, multiple scaling on the ordinate is usually employed. A disadvantage of multiple scaling is that it is very confusing and hampers the interpretation of the data. To circumvent the problem of multiple scaling, all values were normalized by division with the upper limit of the range of reference. In the beginning, this

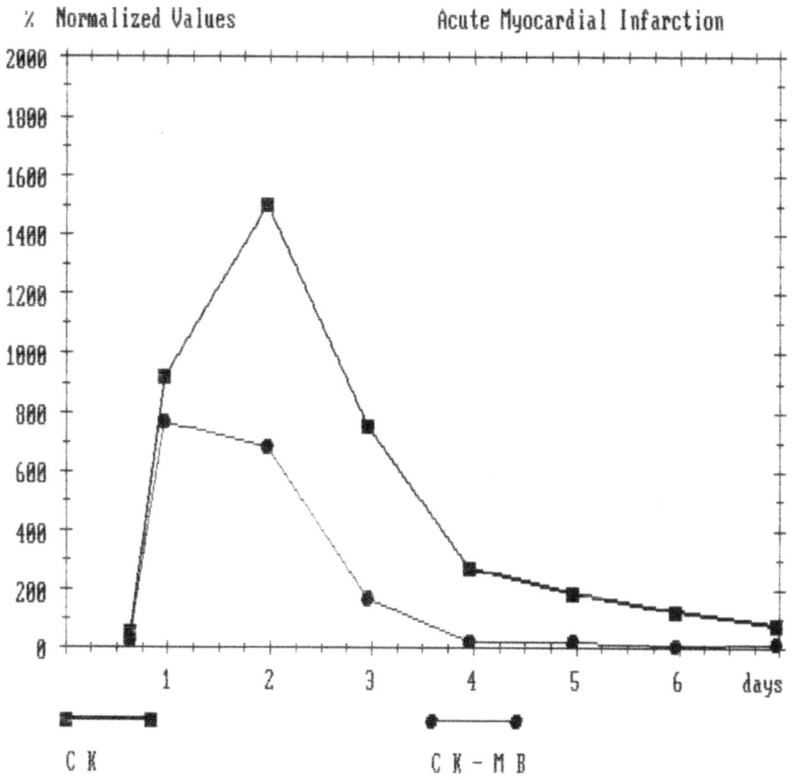

Fig. 4. The time dependent pattern of the enzymes creatinine kinase CK and the isoenzyme CK-MB of a patient suffering from an acute myocardial infarction (AMI). The analogous data between the measurements are approximated by linear interpolation. All data are normalized with the upper bound of the range of reference.

foregoing was a technical device, but afterwards, we determined that the biological and diagnostical importance of a laboratory parameter and its time dependence is impressively emphasized using this normalizes presentation, as will be shown with the examples discussed in Section 5. In addition, the scaling of different parameters in a single graph has a physiological definition, in contrast to the usual multiple scaling method where the choice of the relation between different scales is just a matter of coincidence.

Coming back to the example of AMI, one could use a single scale because both enzymes are measured in international units. However, such a representation would be very unsatisfactory because the CK-MB activities are much lower than CK activities for normal cases of AMI and would be underestimated with respect to its importance for the diagnosis. In the normalized representation of the data, as shown in Figure 4, the time curve of the CK-MB activities is nearly as impressive as the CK curve. The diagnosis of AMI, using this representation, is most stringent.

After a careful inspection of the data presented in Figure 4, one easily finds that the time and the value of the maximum observed enzyme activity are most unlikely

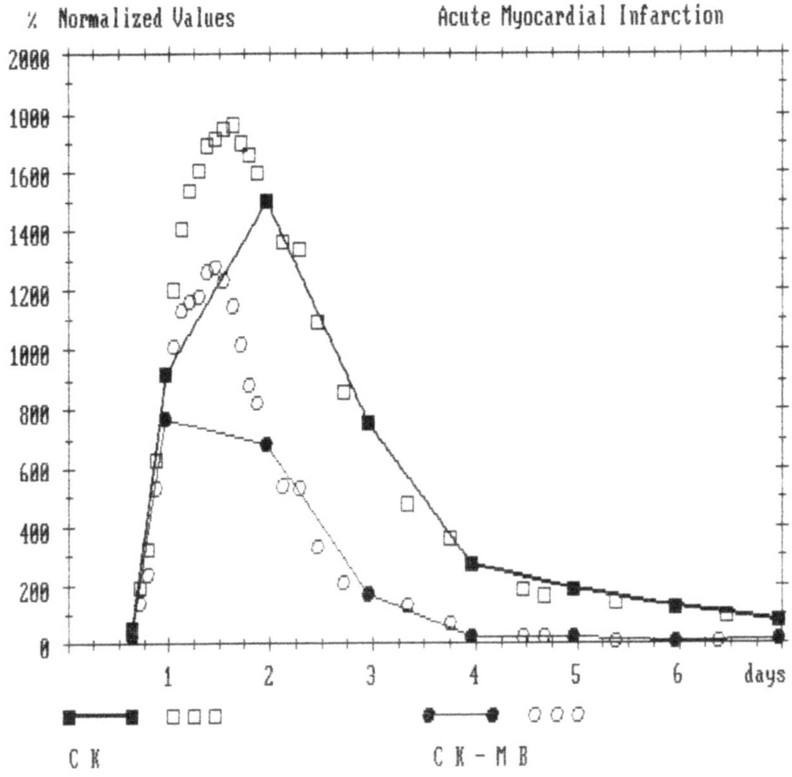

Fig. 5. The discrepancy between linearly interpolated and real data. The enzyme pattern shown is exactly the same as in Figure 4. The filled symbols represent the measurement of samples drawn according to hospital routine. Additional results from samples drawn for scientific purpose are represented by the open symbols.

identical with the true unobserved maximum values. Especially, the time difference between the maxima of the CK-MB and the CK activity of 24 hours seems to be even more unlikely. This problem again brings up the question how close does the linear interpolation approximate the real analogous data? This problem is irrelevant for the diagnosis of AMI, and the data presentation given in Figure 4 suffices. However, for the prognosis of the patients the so-called infarction size, together with the haemodynamic parameters, is most important (5). The infarction size is proportional to the area under the time curve of the enzyme activities (6). If the linear interpolation method is used, the area under the time curves is always underestimated and therefore, the infarction size as well.

To study the differences between linear interpolated and real values, blood samples where drawn every two hours from a patient suffering from AMI, and the CK and CK-MB activities where measured. In Figure 4 the data of this patient, which would have been measured according to hospital routine laboratory tests only are presented. Figure 5 shows all the data of this particular patient. The routine data are represented by the filled symbols, and the data taken for our scientific purpose are given by the open bars.

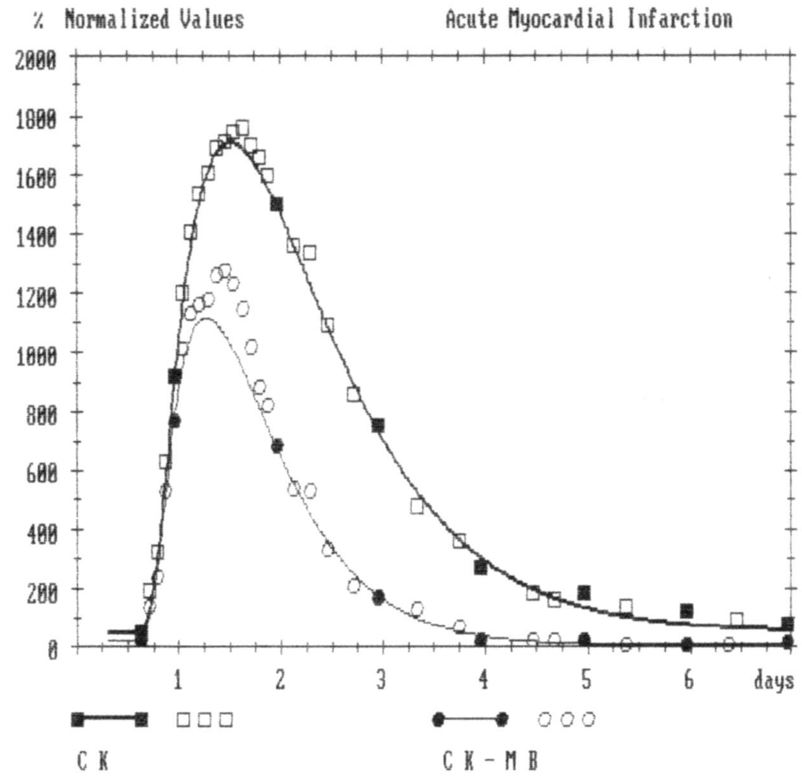

Fig. 6. The discrepancy between nonlinearly interpolated and real data. The enzyme pattern shown is exactly the same as in Figure 4 and 5. The filled symbols represent the measurement of samples drawn according to hospital routine. Additional results from samples drawn for scientific purpose are represented by the open symbols.

Using additional measured points between the routine data, the real values of the enzyme activities can be estimated very closely. Figure 5 clearly shows the differences between the additional data and the linearly interpolated ones. Now the differences between the times of the maxima seem to be plausible. The real maxima lie between the routine measurements. The example given in Figure 5 emphasizes the drawbacks of the linear interpolation method.

Without mathematical proof these drawbacks can be summarized as follows:

1. The times of the maximum values are most likely underestimated (CK-MB) or over-estimated (CK).

2. The maximum value is most likely underestimated.

3. The area under the time curve is always underestimated.

4. The time and value of the maximum can be correct by coincidence between the measured value and the real one.

The area under the time curve is important for many applications because it is proportional to the mass of the substance released during the process being observed. In our example the area under the time curve of the enzyme activity is proportional to the total activity released during AMI. Knowing the enzyme activity contained in one gram of heart muscle (6), it is possible to calculate the size of the heart muscle damaged during AMI in grams.

For this sort of calculation, the theory of mathematical modeling is necessary. This will be discussed in the next sections. In this section the results of the mathematical analysis have been presented, to show the differences between the linear and nonlinear interpolation method. The results of the routine measurements are only taken into account to fit these data to the solution curves of a two-compartment model by means of nonlinear regression. The routine and scientific data together with the nonlinear interpolated time curves are presented in Figure 6.

For the enzyme CK the nonlinear curve coincides closely with the real data and might be identical if one takes measurement errors into account. The calculated CK-MB curve is better than the linear interpolation, but it still underestimates the real data. Thus, Figure 6 shows clearly that the results of a nonlinear regression analysis are not totally exact and do not solve the problem completely. However, they are better than the results of a linear interpolation. The latter is advantageous in cases where no adequate model is available, as discussed above.

3 Model Formulation

The application of the methods of mathematical modeling and dynamic systems analysis is playing an increasingly important role in the study of biological processes, both in physiology and in clinical biochemistry. The proper use of such techniques can provide the clinician, medical or biological research worker, or research student in the life sciences with a greater understanding of the nature and behaviour of the complex processes that occur in physiology. Equally, mathematical models can be used as aids in the diagnosis and treatment of diseases.

In the context of this chapter, we are interested in the application of mathematical modeling to routine data as they are collected during the daily work in a hospital. These data are not as good and complete as data collected during scientific studies. Therefore, for the presentation and interpretation of these data very simple models, i. e. only the so-called minimal models, can be applied. The compartmental models are most appropriate for this purpose.

The most common description of chemical and material transfer processes in biological systems is based on the concept of a compartment. A compartment is fundamentally an idealized store of a substance. If a substance is present in a biological system in several distinguishable forms or locations, then all the substances in a particular form, or all the substances in a particular location, or all the substances in a particular form and location are said to constitute a compartment. Thus, for example, the enzyme activity of CK in the plasma is one compartment and the CK activity in the extra-vascular space a second one. It should be noted that many compartments can coexist in the same location in the biological organism. The function of the compartment as a store can be described by mass balance equations. This leads to the mathematical description of the model. The modeling process is very complicated and tedious. It is still under theoretical investigation (7), but for practical applications the process is straightforward.

The main steps of the modeling process are:

1. Model Formulation
2. Model Identification
3. Model Validation

In this section the model formulation is discussed. The model identification and validation are outlined in the following sections.

Considering a biological system to be modeled as a whole, three distinct components of model formulation can be identified:

1. The formulation of a conceptual model.
2. The mathematical realization of that conceptual model.
3. The solution of the mathematical realization to give the required relations between the variables of interest.

The formulation of a conceptual model is performed by drawing a scheme of the model where all the compartments and the reactions between them are represented.

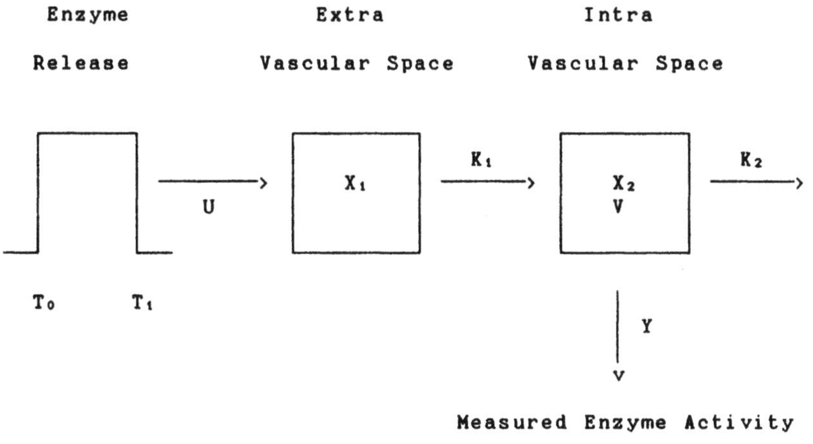

Fig. 7. Model scheme for the AMI. T_0 and T_1 are the times of the beginning and the end of the enzyme release from the heart muscle. This release is described by U the input to the extra vascular space. The diffusion rate from the extra to the intra vascular space is K_1. K_2 is the elimination rate from the intra vascular space. X_1 and X_2 are the enzyme activities in the two compartments. V is the volume of distribution of the plasma compartment, where the enzyme activity concentration Y is observed.

Let us go back to the example of AMI discussed in the previous section. The scheme of the model for this problem is shown in Figure 7. At the time $t = T_0$, the cells in the damaged area of the heart muscle begin to decay and release their content of enzymes. This process comes to an end at $t = T_1$, because the cells are depleted.

The time dependence of the enzyme activity released from the heart muscle is represented by a square wave, i.e. an infusion in medical terms, because it is impossible to get detailed information concerning this process of diffusion. From the heart muscle, the enzymes flow to the surrounding extra-vascular space and enter the intra-vascular space, i.e. the plasma compartment, with the diffusion rate K_1, where they finally are eliminated with the rate K_2. U is the enzyme activity released from the heart muscle and acts as the input to compartment 1. X_1 and X_2 are enzyme activities in the extra- and intra-vascular space respectively and V_1 and V_2 are the volumes of distribution of these compartments. Y is the enzyme activity concentration observed in the plasma.

Having formulated the conceptual model the mathematical realization is achieved by writing down the equations of "mass" balance and the relation between the enzyme activity and its concentration. According to Figure 7, these equations are:

$$\frac{dX_1}{dt} = -K_1 \cdot X_1 + U \tag{3.1 a}$$

$$\frac{dX_2}{dt} = -K_2 \cdot X_2 + K_1 \cdot X_1 \tag{3.1 b}$$

$$Y = \frac{X_2}{V} \tag{3.1 c}$$

where dX/dt is the derivative of X with respect to t, i.e. the change of the quantity X with respect to the time t.

The time dependence of the input to the system, as discussed above, is given by:

$$\begin{aligned} &U = U_0 \text{ for } t \leq T_0 \\ &U = U_0 + \frac{E}{(T_1 - T_0)} \text{ for } T_0 < t \leq T_1 \\ &U = U_0 \qquad \text{for } T_1 < t \end{aligned} \tag{3.2}$$

U_0 is the normal efflux from the heart muscle leading to the normal basal value of the enzyme activity concentration in the plasma. E is the total enzyme activity released during the AMI. The mathematical realization is now defined completely by the equations above.

The final and most difficult step in the formulation of the model is the solution of these differential equations. It would be out of the scope of this article to discuss all the details of this procedure. The interested reader is referred to the literature (1), where the solutions of the most common variations of the two-compartment model are carried out explicitly. Here, the solution of equation (3.1) with the input function (3.2) is presented without the mathematical derivation.

Before the occurrence of the AMI, i.e. for $t < T_0$, the solution of the system of differential equations (3.1) is given by:

$$X_{10} = \frac{U_0}{K_1} \tag{3.3 a}$$

$$X_{20} = \frac{U_0}{K_2} \tag{3.3b}$$

$$Y_0 = \frac{X_{20}}{V} \tag{3.3c}$$

In this state, the patient has a normal basal enzyme activity concentration of Y_0.

After the AMI, i.e. $t > T_0$, the solution is given by:

$$X_1 = X_{10} + A_{11} \cdot f\exp(at) \tag{3.4a}$$

$$X_2 = X_{20} + A_{21} \cdot f\exp(at) + B_{21} \cdot f\exp(bt) \tag{3.4b}$$

$$Y = Y_0 + C \cdot (C_1 \cdot f\exp(at) + (C_2 \cdot f\exp(bt)) \tag{3.4c}$$

The time dependence in the equations above is hidden in the function $f\exp(xt)$. This is a rather complicated function which increases toward a limiting value of $\frac{1}{(T_1 - T_0)}$ during the time of release and decreases afterwards. Their exact definition is:

$$f\exp(xt) = (1 - \exp(xt')) \cdot \frac{\exp(-x \cdot (t - T_0))}{(T_0 - T_1)} \tag{3.5a}$$

with

$$t' = t - T_0 \text{ for } T_0 < t \leq T_1 \tag{3.5b}$$

$$t' = T_1 - T_0 \text{ for } T_1 > t \tag{3.5c}$$

The time behaviour of the solution (3.4) with the definitions (3.5) is governed by the time constants a and b which are simply related to the rates of diffusion and elimination K_1 and K_2:

$$a = K_1 \tag{3.6a}$$

$$b = K_2 \tag{3.6b}$$

The amplitudes in equation (3.4) are defined by:

$$A_{11} = + \frac{E}{K_1} \tag{3.7a}$$

$$A_{21} = - \frac{E}{K_1 - K_2} \tag{3.7b}$$

$$B_{21} = + \frac{E \cdot K_1}{K_2 \cdot (K_1 - K_2)} \tag{3.7c}$$

$$C_1 = - \frac{C \cdot K_2}{K_1 - K_2} \tag{3.7d}$$

$$C_2 = + \frac{C \cdot K_1}{K_1 - K_2} \tag{3.7e}$$

$$C = \frac{(E/V)}{K_2} \tag{3.7f}$$

The area under the time curve representing the enzyme activity concentration and above the basal concentration can now be calculated using the equations above. The result is:

$$\int (Y(t) - Y_0) \cdot dt = C \tag{3.8}$$

Thus far, the model according to Figure 7 has only been developed for the special example of AMI. However, it can be applied to any problem where the following conditions are met:

1. Input is an infusion to compartment 1
2. There is only a flux of material form compartment 1 to compartment 2
3. Elimination takes place from compartment 2 only
4. Measurement is performed only in compartment 2.

If a single condition is changed, for example measurement in compartment 1, the solutions of the differential equations are different.

The case of a bolus injection into compartment 1 and the other conditions remaining unchanged can easily be derived from the formula above. In this case the time points T_0 and T_1 are identical, i.e. $T_1 - T_0 = 0$, and we get from equation (3.5):

$$f\exp(xt) = x \cdot \exp(-x \cdot (t - T_0)) \tag{3.9}$$

Inserting this expression into equation (3.4), the formula holds true for the case of bolus injection. All the other equations remain unchanged.

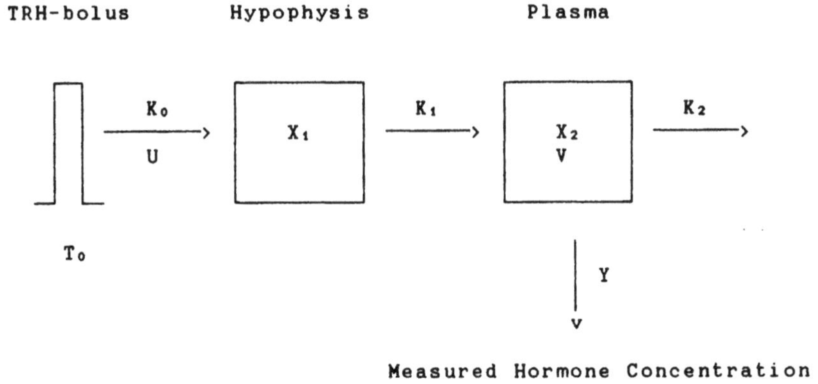

Fig. 8. Model scheme for the TRH-test. T_0 is the time when the stimulating hormone TRH is injected. This bolus is described by U the input to compartment 1 representing the hypophysis. K_0 describes the proportion between the mass of the stimulus TRH and the mass of the hormone which is set free in the hypophysis. The transition rate from the hypophysis to the plasma compartment is K_1. K_2 is the elimination rate from the plasma. X_1 and X_2 are the masses of hormone in the two compartments. V is the volume of distribution of the plasma compartment, where the hormone concentration Y is observed.

The mathematical model of bolus injection derived above can be applied to an other example, the TRH-test used in endocrinology to examine a malfunction of the hypophysis or the existence of a hormone releasing tumour. In Figure 8, the model is shown schematically. At the time $t = T_0$, a bolus of TRH (Thyrotrophic Releasing Hormone), is injected intravenously. This hormone stimulates the hypophysis, i. e. the anterior lobe of the pituitary gland, to release its hormones like TSH (Thyroid Stimulating Hormone), prolactin, growth hormone (GH), and others. The relation between the stimulus and the biological effect of production and release of the hormone in the hypophysis is described by the factor K_0. It is known from the literature (8) that the stimulation of TSH by the releasing hormone TRH is dose dependent. The reason for this is the involvement of receptors in the stimulation process which have a rate limiting action. Therefore, one has to bear in mind that K_0 is some sort of a sigmoidal function of the stimulating dose U. X_1 is the hormone in the pituitary gland and X_2 is the hormone available in the plasma compartment. Y represents the observed hormone concentration in the plasma compartment with V being the volume of distribution. The elimination rates from the two compartments are described by K_1 and K_2 respectively.

This model is equivalent to the model used for the case of AMI, with two exceptions. The input is a bolus injection, therefore one has to use equation (3.9) instead of equation (3.7). In addition, the responsiveness of the hypophysis to the TRH-stimulus K_0 has to be taken into account. Therefore, the input U to compartment 1 has to be replaced by:

$$\begin{aligned} U &= U_0 \cdot K_0 && \text{for } t < T_0 \\ U &= (U_0 + B) \cdot K_0 && \text{for } t = T_0 \\ U &= U_0 \cdot K_0 && \text{for } t > T_0 \end{aligned} \tag{3.10}$$

In this example, U_0 is the constant rate of TRH released from the hypothalamus. B is the TRH-bolus injected at time $t = T_0$.

The solutions of the corresponding differential equations before the stimulation are:

$$X_{10} = \frac{U_0 \cdot K_0}{K_1} \tag{3.11 a}$$

$$X_{20} = \frac{U_0 \cdot K_0}{K_2} \tag{3.11 b}$$

$$Y_0 = \frac{X_{20}}{V} \tag{3.11 c}$$

Y_0 is the basal value of the hormone before the stimulation.

After the stimulation, the solution of equation (3.1) is given by:

$$X_1 = X_{10} + A_1 \cdot \exp(-a \cdot (t - T_0)) \tag{3.12 a}$$

$$X_2 = X_{20} + A_2 \cdot (\exp(-b \cdot (t - T_0)) - \exp(-a \cdot (t - T_0))) \tag{3.12 b}$$

$$Y = Y_0 + D \cdot C_1 \cdot (\exp(-b \cdot (t - T_0)) - \exp(-a \cdot (t - T_0))) \tag{3.12 c}$$

The time behaviour of the solution (3.12) is governed by the time constants a and b which are again identical with K_1 and K_2. The amplitude factors are now given by:

$$A_1 = B \cdot K_0 \tag{3.13 a}$$

$$A_2 = \frac{B \cdot K_0 \cdot K_1}{K_1 - K_2} \tag{3.13 b}$$

$$C_1 = \frac{K_1 \cdot K_2}{K_1 - K_2} \tag{3.13 c}$$

$$D = \frac{B \cdot K_0}{K_2 \cdot V} \tag{3.13 d}$$

The area under the time curve of the hormone concentration and above the basal value, i. e. the total hormone concentration released after the stimulation, is now given by:

$$\int (Y(t) - Y_0) \cdot \mathrm{d}t = D \tag{3.14}$$

After having discussed two simple examples of the two-compartment model, let us have a look at the general two-compartment model. The differences to the models described above are:

1. There are inputs U_1 and U_2 to both compartments.
2. K_{12} and K_{21} represent a flux or signal action respectively between the two compartments in either direction.
3. Elimination from both compartments are possible and described by K_{01} and K_{02}.
4. Two volumes of distribution, V_1 and V_2 appear in the solutions.
5. The concentration of the substances of interest can be observed in the first as well as in the second compartment.

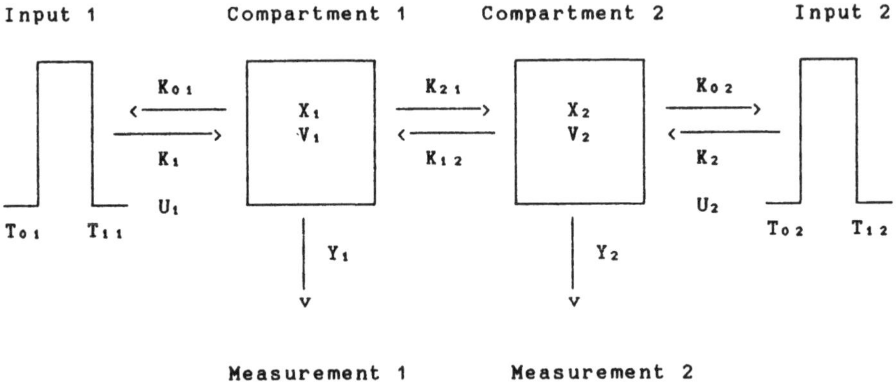

Fig. 9. The genaral two-compartment model with the inputs and outputs U_1, U_2 and Y_1, Y_2 respectively. The beginning and the end of the incoming infusions are given by T_{01}, T_{11} and T_{02}, T_{12}. K_{01} and K_{02} are the elimination rates from compartment 1 and 2. The flux of material or the control signals linking the two compartmental subsystems are represented by K_{21} and K_{12}. X_1, X_2 and V_1, V_2 are the masses of material and the volumes of distribution in the two compartments respectively.

The scheme of the most general two-compartment model is shown in Figure 9. The differential equations equivalent to this scheme are given by:

$$\frac{dX_1}{dt} = -(K_{01} + K_{21}) \cdot X_1 + K_{12} \cdot X_2 + K_1 \cdot U_1 \tag{3.14 a}$$

$$\frac{dX_2}{dt} = +K_{21} \cdot X_1 - (K_{02} + K_{12}) \cdot X_2 + K_2 \cdot U_2 \tag{3.14 b}$$

$$Y_1 = \frac{X_1}{V_1} \tag{3.14 c}$$

$$Y_2 = \frac{X_2}{V_2} \tag{3.14 d}$$

where the input functions are of the infusion type again:

$$\begin{aligned} U &= U_{01} && \text{for} \quad t \leq T_{01} \\ U &= U_{01} + \frac{I_1}{T_{11} - T_{01}} && \text{for } T_{01} < t \leq T_{11} \\ U &= U_{01} && \text{for } T_{11} < t \end{aligned} \tag{3.15 a}$$

and

$$\begin{aligned} U &= U_{02} && \text{for} \quad t \leq T_{02} \\ U &= U_{02} + \frac{I_2}{T_{12} - T_{02}} && \text{for } T_{02} < t \leq T_{12} \\ U &= U_{02} && \text{for } T_{12} < t \end{aligned} \tag{3.15 b}$$

I_1 and I_2 are the total "mass" of substance infused to compartment 1 and compartment 2 respectively. U_{01} and U_{02} are the basal infusion rate to either compartment resulting in the basal levels X_{10} and X_{20} before the perturbation of the system with the additional infusion I_1 and I_2 respectively.

The general solution of equation (3.14) is:

$$\begin{aligned} X_1 = X_{10} &+ A_{11} \cdot f\exp_1(at) + B_{11} \cdot f\exp_1(bt) \\ &+ A_{12} \cdot f\exp_2(at) + B_{12} \cdot f\exp_2(bt) \end{aligned} \tag{3.16 a}$$

$$\begin{aligned} X_2 = X_{20} &+ A_{21} \cdot f\exp_1(at) + B_{21} \cdot f\exp_1(bt) \\ &+ A_{22} \cdot f\exp_2(at) + B_{22} \cdot f\exp_2(bt) \end{aligned} \tag{3.16 b}$$

$$Y_1 = \frac{X_1}{V_1} \tag{3.16 c}$$

$$Y_2 = \frac{X_2}{V_2} \tag{3.16 d}$$

where the time dependence due to input U_1 is given by:

$$f\exp_1(xt) = (1 - \exp(xt')) \cdot \frac{\exp(-x \cdot (t - T_{01}))}{(T_{01} - T_{11})} \tag{3.17 a}$$

with

$$t' = t - T_{01} \text{ for } T_{01} < t \leq T_{11} \tag{3.17 b}$$

$$t' = T_{11} - T_{01} \text{ for } T_{11} > t \tag{3.17 c}$$

and the time dependence due to input U_2 is given by:

$$f\exp_2(xt) = (1 - \exp(xt')) \cdot \frac{\exp(-x \cdot (t - T_{02}))}{(T_{02} - T_{12})} \tag{3.18 a}$$

with

$$t' = \quad t - T_{02} \text{ for } T_{02} < t \leq T_{12} \tag{3.18 b}$$

$$t' = T_{12} - T_{02} \text{ for } T_{12} > t \tag{3.18 c}$$

The time constants a and b govern the time behaviour of the system. Their relation to the rate constants appearing in the model is given by:

$$a + b = K_{21} + K_{01} + K_{12} + K_{02} \tag{3.19 a}$$

$$a \cdot b = K_{21} \cdot K_{02} + K_{12} \cdot K_{01} + K_{01} \cdot K_{02} \tag{3.19 b}$$

The amplitude factors appearing in equation (3.14) are proportional to the quantity of substance infused and the sensitivity parameters K_1 and K_2:

$$A_{11} = K_1 \cdot I_1 \cdot a_{11}; \; B_{11} = K_1 \cdot I_1 \cdot b_{11}; \tag{3.20 a}$$

$$A_{21} = K_1 \cdot I_1 \cdot a_{21}; \; B_{21} = K_1 \cdot I_1 \cdot b_{21}; \tag{3.20 b}$$

$$A_{12} = K_2 \cdot I_2 \cdot a_{12}; \; B_{12} = K_2 \cdot I_2 \cdot b_{12}; \tag{3.20 c}$$

$$A_{22} = K_2 \cdot I_2 \cdot a_{22}; \; B_{22} = K_2 \cdot I_2 \cdot b_{22}; \tag{3.20 d}$$

The constants a_{11}, a_{21}, a_{12}, and a_{22} as well as b_{11}, b_{21}, b_{12}, and b_{22} are complicated functions of the reaction constants K_{12}, K_{21}, K_{01}, and K_{02}. The explicit form of these functions is not given here, because in the practical application of the general model to routinely acquired data all of the amplitude factors (3.20) can be estimated, but the rate constants can not be calculated using these functions, i.e. the rate constants can not be identified. The problem of parameter estimation and identification will be discussed in the next section.

In contrast, the areas under the time curves can always be identified. They are calculated using the integral over the time functions (3.17) and (3.18)

$$\int f\exp_1(xt)\,dt = 1 \tag{3.21 a}$$

and

$$\int f\exp_2(xt)\,dt = 1 \tag{3.21 b}$$

From these relations and equation (3.16) one immediately gets:

$$\int (Y_1(t) - Y_{10})\,dt = (A_{11} + B_{11} + A_{12} + B_{12}) \tag{3.22 a}$$

$$\int (Y_2(t) - Y_{20})\,dt = (A_{21} + B_{21} + A_{22} + B_{22}) \tag{3.22 b}$$

The general two-compartment model will be used for the nonlinear interpolation between data collected during the oral glucose tolerance test (OGTT). As pointed out already in Section 2, the problem of glucose homeostasis is too complicated to be modeled by simple compartmental models. In (10) a model with six compartments is presented for the mathematical modeling of the maximal insulin stimulating test (MIST). For the OGTT this model is not applicable because there are not enough points of measurement.

Therefore, we use a more or less empirical method to get an analogous presentation for the results of the OGTT. This is done by using a two compartmental model of the form given in Figure 9 for each single variable observed during the OGTT. Therefore, the physiological meaning of the two compartments is different for each variable. During the OGTT the variables blood glucose, insulin, and C-peptide levels are measured. This gives us three different cases to differentiate.

1. In the model for the glucose levels, the first compartment describes the glucose levels in the peripheral blood circulation, and the second, the insulin in the body. The input U_1 describes the glucose release of the liver after the oral loading with glucose. U_2 can be used to model the insulin injection applied with an insulin hypoglycemia test. U_1 in the form of a bolus similar to equation (3.10) could be used for the modeling of an intravenous glucose tolerance test (IVGTT).

2. In the model for the insulin levels, the first compartment describes the insulin levels in the portal system, and the second, the blood insulin concentration in the peripheral blood circulation. U_1 describes the insulin release from the liver induced by the oral loading with glucose, and U_2 could be used to model the influence of exogenous insulin applied during the therapy of insulin dependent diabetics.

3. In the model for the C-peptide levels, the first compartment describes the glucose levels in the portal system, and the second, the C-peptide concentration in the body. A differentiation between portal and peripheral system is not necessary for the C-peptide model, because it is assumed that the C-peptide is not influenced during the liver passage. U_1 describes the glucose output from the intestines after the oral loading with glucose.

Using this phenomenological procedure for the modeling of the OGTT, the interactions between the glucose and insulin concentrations are incorporated in the rate constants. Therefore, these quantities are highly dependent on the patient being observed.

4 Model Identification

In general terms, the process of identification consists of the experimental determination of the model, considering both, its true structure and the values of its parameter. Experiments have to be designed whereby the system is perturbed and the resultant response of one or more variables is obtained, thus providing input/output data. Although in some cases static input/output data may be adequate, the dynamic response of one or more of the systems variables to the perturbation is usually required.

The input/output data are compared with the corresponding prediction of the model. Then, the structure and/or parameters of the model are adjusted so as to minimize some function of the error between the experimental data and the model response. Having determined the model experimentally in this manner, the model can be solved, and its validity then further assessed.

In practical terms, applying the theory of dynamic systems to routine clinical data, different types of perturbations of the biological systems under investigation occur:

1. Normal perturbation: An example are the daily meals perturbing the glucose homeostasis of insulin dependent diabetics, discussed in Section 2.

2. Pathological perturbations: After acute diseases the equilibrium of some clinical parameters are heavily perturbed and diverge greatly from the normal state, i. e. the parameters are outside the range of reference. As an example the case AMI was discussed in Section 2 and 3.

3. Diagnostical perturbations: To get some information for the diagnosis of metabolic or endocrine diseases, the natural perturbations are simulated. Examples are the loading and functions tests used in diabetology and endocrinology. The examples of TRH test to examine malfunctions of the anterior lobe of the pituitary gland and the OGTT for the diagnosis of a pathological glucose tolerance were also discussed in the preceding section.

Data from measurements describing such perturbations of biological systems are available daily from the routine laboratory test in the hospital. The theory of model formulation was outlined in the preceding section. The problem of model identification is now to adapt the general solution equivalent to the model to the real data measured for an individual case.

The main steps in this process are:

1. Theoretical (A Priori) Identifiability
2. Parameter Estimation
3. Practical (A Posteriori) Identifiability.

These steps will be shortly outlined below. For full details, the interested reader is referred to the literature (3 and 9).

4.1 Theoretical identifiability

The concept of theoretical, or a priori, identifiability, addresses the question of whether it is at all possible to obtain unique solutions for the unknown parameters of interest in a mathematical model on the basis of data collected in well conceived input/output experiments performed on the dynamic system represented by the model.

Using the solution of the two-compartment model in Figure 7 for the AMI as given in equation (3.4c) together with (3.5), (3.6), (3.7), and (3.8) the diffusion constant K_1, the elimination constant K_2, and the area under the time curve C as well as the time constants T_0 and T_1 can be theoretically determined by fitting this model output function to the data using a least square procedure.

However, the total enzyme activity E, released during the AMI, as well as the volume of distribution of the plasma compartment V cannot be identified, because they only appear in the combination E/V and cannot be separated. Therefore, they are called not identifiable parameters. To circumvent this problem, one can assume a proportionality between the volume of distribution of the plasma and the body weight, to get an estimated total enzyme activity from the result for C (3.7 f.). This method is used for the calculation of the infarction size in the literature (5, 6).

Applying the identification considerations to the case of the TRH-test discussed in Section 3, it is easily shown, that in this case the sensitivity of the hypophysis to the TRH stimulation K_0 and the volume of distribution V are not identifiable because they occur in the combination K_0/V only. However, assuming that the volume of distribution can be estimated from the body weight as in the case of AMI, one gets an estimate of the sensitivity parameter K_0. In contrast, the parameters K_1, K_2, and D can be theoretically identified.

In the case of the models for the OGTT represented by the general two-compartment model shown in Figure 9, with the solution given by (3.16), only the exponents a and b, the amplitudes A_{11} ... B_{22} and the time constants T_0 and T_1 can be identified theoretically. The rate constants K_{01}, K_{21}, K_{02}, and K_{12} as well as the sensitivity constants K_1 and K_2 cannot be identified, because the equations relating these constants with the identifiable constants cannot be solved. This statement is given here without proof and the reader is referred to the text by DiStefano et al. (9) for an exhaustive discussion of the problem of theoretical identifiability.

Finally, the area under the time curve is always identifiable as can be easily seen from the equations (3.8), (3.14), and (3.22). This quantity is also independent of the number of compartments used in the model as proven in (1). Therefore, the area under the time curve or quantities derived thereof are very useful for the interpretation of data collected from dynamic biological systems because they provide a quantitative measure of the perturbation acting on the biological system.

4.2 Parameter estimation

Assuming that the parameters of a postulated model can be identified theoretically, the parameter estimation problem is well posed theoretically, and the data routinely collected in the laboratory can be used for parameter estimation.

Problems in carrying out the estimation process arise due to a number of sources of errors. The most important errors in our applications are:

1. Deviations of the real input from the assumed constant infusion type.
2. Errors due to mistakes during the sample drawing.
3. Errors occurring at the measurement in the laboratory.
4. Model errors, which come about from using a model for identification that does not satisfactorily represent the system being identified.

In the examples discussed so far, all the different types of errors occur. Errors according to 1. and 4. will be neglected in the further proceeding. Errors of the type 2. and 3. have to be taken into account at the parameter estimation process. The solution of the model equations without errors may be given by:

$$Y = Y(t, p) \tag{4.1}$$

i.e. Y is a function of time and some unknown parameters which have to be estimated. In the case of TRH-test, to have a simple example, p represents the identifiable parameters:

$$p = (K_1, K_2, C) \tag{4.2}$$

The real measurement taken at some discrete time points and containing the errors discussed above are symbolized by:

$$Z(t_k) = Y(t_k, p) + e(t_k) \tag{4.3}$$

where e is the error associated with the measurement at time t_k.

If one uses the well known least squares parameter estimation scheme, the cost function C is defined:

$$C = e^T \cdot e \tag{4.4}$$

where e^T represents the transposed of the vector e describing the errors of the different measurements.

Using equation (4.4) it is assumed that the errors taken into account are normally distributed with a constant variance. This assumption is heavily dependent on the process of measurement and if it does not hold the method of maximum likelihood has to be applied for the process of parameter estimation. In (11) these problems are discussed. In the application of parameter estimation to laboratory results, this assumption can be proved by means of the quality control performed in the laboratory.

The most likely parameters are now given by the minimum of the cost function with respect to the parameters, i. e.

$$\frac{\mathrm{d}C}{\mathrm{d}p} = 0 \tag{4.5}$$

Since the model is nonlinear in the parameters, i. e. $Y(t, p)$ is a nonlinear function of the parameters, the cost function (4.4) is not of quadratic form in the parameters. Therefore, an explicit analytical solution of the minimum condition (4.5) is not possible.

To circumvent this problem there is the need to provide initial parameter estimates p^0 of the unknown parameter vector p to be estimated. The choice, which has to rely on independent knowledge or preliminary processing of the data, is particularly crucial. Since, as the cost function is not quadratic, there is the possibility of the iteration process, which has to be carried out to find the global minimum for the cost function, of converging to a local minimum, or even no convergence.

Having provided the initial estimates, there are basically two methods to proceed further:

1. Direct Search Method
2. Gauß – Newton Method.

With the direct search method, new parameters in the vicinity of the initial estimates are used to find the direction in parameter space in which the cost function is decreasing. This search of the minimum can be performed in a deterministic or random manner.

The Gauß–Newton method linearizes the dependence of the solution vector $Y = Y(t, p)$ with respect to changes in the parameter vector, i. e.

$$Y = Y(t, p^0) + S \cdot \delta p \tag{4.6}$$

with

$$\delta p = p - p^0 \tag{4.7}$$

S is the matrix of the partial derivatives of the vector Y with respect to the vector of the parameters p.

Inserting equation (4.6) into the minimum condition (4.5) one gets:

$$\delta p = S \cdot (Z - Y(t, p^0)) \tag{4.7}$$

At this stage a new estimate can be computed for p:

$$p^1 = p^0 + \delta p \tag{4.8}$$

In an iterative manner successive estimates p^2, ..., p^n are obtained, where p^n is the final least square estimate of the parameter vector p.

In contrast to the direct search methods, the Gauß–Newton methods allows an approximate evaluation of the covariance matrix V of the parameter estimates. This is given by:

$$V(p^n) = (S^T \cdot R^{-1} \cdot S)^{-1} \tag{4.9}$$

with R being the covariance matrix of the measurement errors which can be estimated by:

$$R = (Z - Y(t, p^n))^T \cdot (Z - Y(t,p^n) \tag{4.10}$$

A very serious problem with the Gauß–Newton method is the fact, that its success very strongly depends on the proper choice of the initial estimate of the parameters. If there is a local minimum near the initial estimate the iteration process stops there and the true minimum is not found. To avoid this pitfall many tries with different initial parameters have to be performed. In this respect the direct search methods are preferable.

4.3 Practical identifiability

After having outlined the theoretical foundation of the parameter identification and estimation, it is necessary to discuss the practical procedure. In order to have definite conditions, the case of AMI is used as an example. The problems of the TRH-test, OGTT, and similar cases can be treated in the same manner.

The first step of the procedure is the estimation of the initial parameter values. This is achieved by ordering the data with respect to time and searching for the maximum value. After this preprocessing of the data, the initial parameters can estimated as follows:

1. Taking the minimum of the first and last observed value gives the initial value for the basal enzyme activity Y_0.

2. The diffusion constant K_1 is mainly responsible for the increase of the time curve from Y_0 to its maximum value. As a first guess the reciprocal value of the time of the maximum T_{max} is a good approximation for K_1.

3. The elimination constant K_2 is mainly responsible for the exponential decrease of the time curve from T_1 to the end of the enzyme release. Therefore, a first approximation for the elimination constant can be gained by a linear regression of the logarithms of the measured deviations from the basal value $(Y - Y_0)$ against time t for t greater than T_{max}.

4. The area under the time curve can be estimated using the trapezoidal rule as a first approximation.

5. The end of the enzyme infusion T_1 is approximated by the time of the maximum value T_{max}.

6. The onset of the enzyme infusion T_0 is estimated by the time of the first measurement.

The estimation of the initial parameters for the TRH-test and the OGTT can be performed in exactly the same manner using the rules 1. to 4. and 1. to 5. discussed above.

Starting with the initial parameters derived with the methods discussed above, the direct search algorithm by Powell (12) is used to improve the parameter estimation. This algorithm is very robust and comes close to the global minimum avoiding local minima. The final estimation is performed by means of the program NONLIN (13) which uses the Gauß–Newton algorithm.

This combination of initial parameter estimation, direct search, and Gauß–Newton method gives excellent results and enables technicians in the laboratory who are not familiar with parameter estimation problems to get acceptable nonlinear regression curves.

To assess the adequacy of a dynamic model whose parameters have been estimated, three criteria can be adopted. These are the goodness of fit, residual errors, and practical identifiability. We use the program NONLIN in the final parameter estimation step because this program provides all these criteria as can be seen in Table 1. This Table shows the output of NONLIN in a slightly modified version.

4.3.1 Goodness of fit

The goodness of fit resulting from the parameter estimation process can be evaluated by examining the residual sum of squares corresponding to the final parameter estimates p^n. This sum is given by:

$$\Sigma = (Z - Y(t, p^n))^2 \tag{4.11}$$

4.3.2 Examination of residuals

Applying the Gauß–Newton method, a gaussian distribution of the measurement errors has been assumed. After the parameter estimation process this assumption should be tested using the residual errors:

$$e^n = Z - Y(t, p^n) \tag{4.12}$$

However, the data collected during routine laboratory tests are small in number and the distribution of these data cannot be tested. Therefore, a visual method examining the residuals is adopted.

In Table 1 the digital presentation of the residuals is given. A graphic presentation of the residuals is shown in Figure 10.

Table 1. Digital presentation of the parameter estimation results. The program NONLIN performs the final step of the parameter estimation procedure. Its output as presented in this Table provides the information necessary for the practical identification of the parameters. The first block shows the estimates together with the standard deviations and coefficients of variance (*CV*). The absolute and relative deviations between calculated and observed results is given in the second block. In the last line, the coefficients of determination (R^2) and correlation (*COR*) as well as the standard error of the residuals (*S*) are printed.

SUMMARY OF NON-LINEAR ESTIMATION
NAME OF THE PROBLEM: TRH-TEST HORMONE T S H

AFTER 13 ITERATIONS THE ESTIMATES AND THEIR VARIABILITY ARE:

PARAMETER	ESTIMATE	STD. DEV.	CV (%)
K1	.605919E − 01	.195757E − 01	32.3
K2	.226464E − 01	.533841E − 02	23.6
C	.341550E + 03	.275796E + 02	8.1

TIME T	OBS. Y	CALC. Y	OBS-CAL	% DEVIATION
.00000E + 0	.40000E + 1	.40000E + 1	.00000E + 0	.00
.30000E + 2	.13000E + 2	.12440E + 2	.56000E + 0	4.31
.60000E + 2	.10000E + 2	.97300E + 1	.27000E + 0	2.70
.90000E + 2	.70000E + 1	.71400E + 1	−.13000E + 0	−1.86
.12000E + 3	.60000E + 1	.56300E + 1	.36000E + 0	6.00
.18000E + 3	.50000E + 1	.44200E + 1	.57000E + 0	11.40

SUM OF SQUARED OBSERVATIONS = .3950000E + 03
SUM OF SQUARED DEVIATIONS = .8579000E + 00
R-SQUARED = .998 COR = .997 S = .5347585 WITH 3 D.F.

4.3.3 Accuracy of parameter estimates

Using the least square estimation, the approximate computation of the covariance matrix V (4.9) gives a means of evaluating the accuracy with which the parameters can be estimated from the real data. In fact, the diagonal elements, $v(p)$ of the matrix $V(p)$, provide the variances of the parameter estimates.

The accuracy with which the parameter p can be estimated may be expressed in terms of its standard deviation by:

$$p = p^n \pm \sqrt{v} \tag{4.13}$$

The judgement of parameter accuracy may be improved by employing a percentage measure. This is the coefficient of variation (CV), which is given by:

$$\mathrm{CV}(p) = 100 \cdot \frac{\sqrt{v}}{p} \tag{4.14}$$

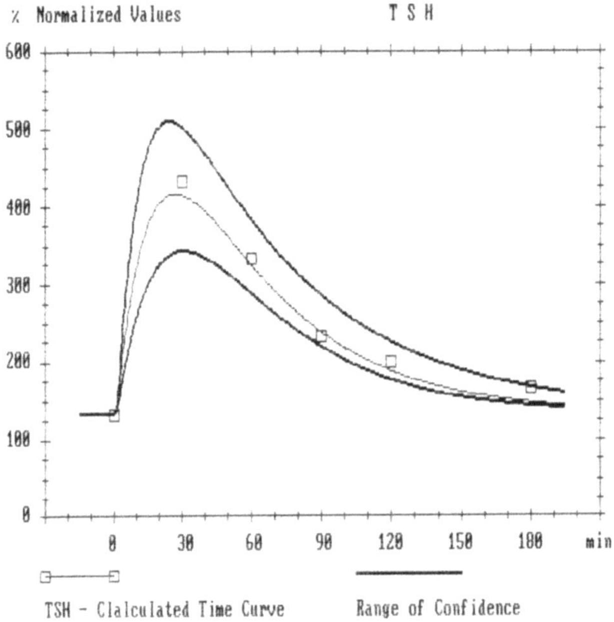

Fig. 10. Analogous presentation of the parameter estimation results. This Figure provides the information necessary for a visual judgement of the practical parameter identification. The residuals, i.e. the differences between the calculated (thin line) and the observed results (open squares), are shown together with the upper and lower bound of the range of confidence (thick lines).

Both the standard deviations and the coefficients of variation of the parameter are provided in the output of the program NONLIN. These quantities are the most important criteria for the practical identifiability of the parameters. If the coefficient of variance becomes greater than 100%, the parameter is called practically not identifiable and the model cannot be used for the interpretation of the data. However, in our application, the analogous presentation, it can still be used, if the residuals are reasonable.

For ease of judgement of the standard deviation connected with the estimated parameters, we use once again a visual method. For each parameter an upper and lower limit can be calculated using the standard deviation:

$$p_u = p^n + \sqrt{v} \tag{4.15a}$$

$$p_l = p^n - \sqrt{v} \tag{4.15b}$$

With these two sets of parameters, two time curves can be drawn, indicating an interval of confidence for the parameter estimation. In Figure 10, this confidence interval is shown for the case of the CK time curve after an AMI. This device is an excellent method to judge the practical identifiability of the model.

5 Nonlinear Interpolation

In this section, we will discuss the model validation. Model validation involves the assessment of the extent to which a model is well founded, is tractable, and fulfills the purpose for which it is formulated.

Although the validation must be tested throughout the process of model formulation, we will confine our discussion to the empirical validation of the models that have been developed to the solution stage. The validation is performed in a pragmatic sense. It will be shown that the models developed above satisfy their objective, i. e. the nonlinear interpolation between the measured laboratory results used for the analogous presentation of time dependent laboratory tests.

The case of AMI has been extensively discussed during the model formulation process. Here, we focus on the examples of TRH-test and OGTT.

In Figure 11 and 12 two examples of a TRH-test are shown, where the hormones TSH, growth hormone (GH), and prolactin were measured after a stimulation of the anterior lobe of the pituitary gland with TRH. For ease of presentation and to get physiologically correct relations between the different hormones, the normalized scale as discussed in Section 2 is again used. The hormone of main interest, TSH, is indicated by the thick line. To differentiate between the time curves of the other two hormones a medium and a thin line are used.

The graph of Figure 11 shows an increased release of prolactin together with a normal TSH and GH release. With these results of the TRH test an adenoma of the hypophysis seems to be very likely.

In contrast, the exagerated release of the growth hormone ocurring after the stimulation with TRH, as presented in Figure 12, proves the diagnosis of acromegaly for the patient to whom this test was applied. The difference between these two diagnoses becomes immediately apparent after the inspection of the two graphs presented in Figure 11 and 12.

An other advantage of the analogous presentation of endocrine function tests by means of the nonlinear interpolation method is the fact that maximum values occuring between the sampling times can be estimated. The best result of a TRH-test can be achieved, if the maximum value can directly be observed with the first sample taken after the stimulation. Because there is an inter-individual variation with respect to the times of the maximum, the real maximum value is observed by coincidence only. Different physicians use different sampling times, but this does not solve the problem. With the nonlinear interpolation, fixed sampling times can be used without loosing information.

Examples of the OGTT with a loading dose of 100 g glucose are presented in Figure 13 and 14 to validate the model for the oral glucose tolerance test. The main variable with this test is, of course, the blood glucose level, which is represented by the thick line. Insulin is the next important parameter and is represented by the medium thick line. The C-peptide time curve, at last, is drawn with the thin line.

Figure 13 shows a normal time curve of the blood glucose levels. The insulin and C-peptide release of this test appears to be optimal. In contrast, Figure 14 shows the result of a pathological disturbed glucose tolerance. The high glucose levels become immediately apparent. The cause of this unbalanced glucose response

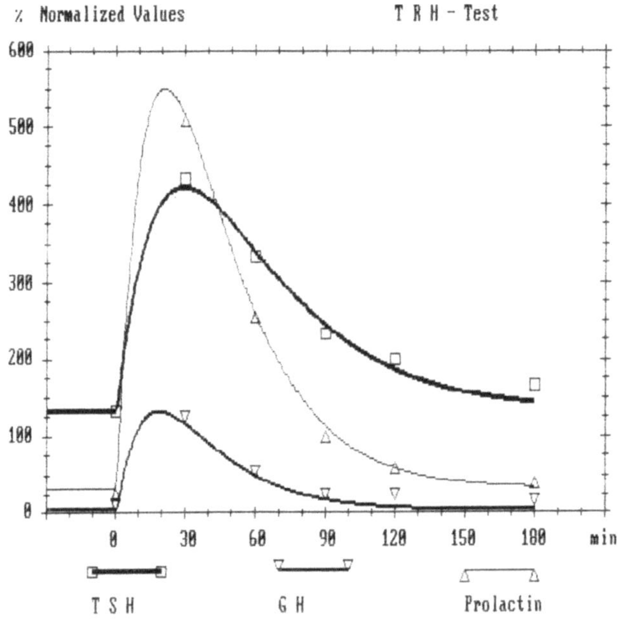

Fig. 11. Analogous presentation of the results of a TRH-test performed on a patient suffering from an adenoma of the pituitary gland. The diagnosis is confirmed by the high release of the hormone Prolactin.

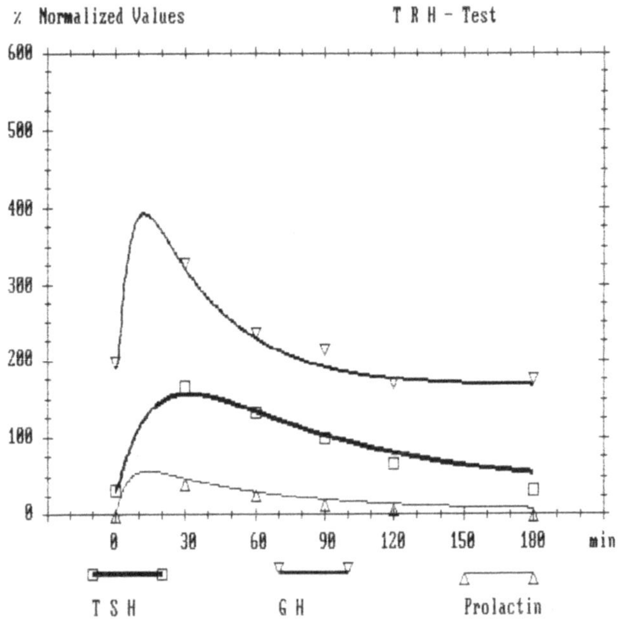

Fig. 12. Analogous presentation of the results of a TRH-test performed on a patient suffering from akromegaly. The diagnosis is confirmed by the high release of the Growth Hormone (GH).

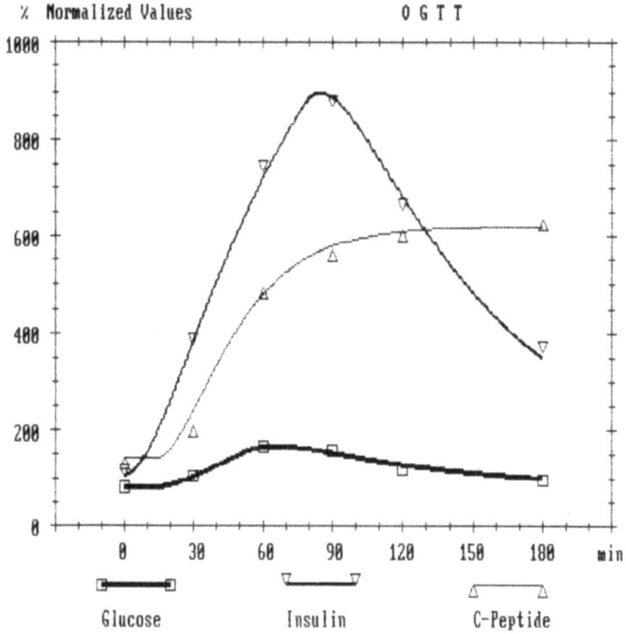

Fig. 13. Analogous presentation of the results of a oral glucose tolerance test (OGTT) performed on a patient with a normal Glucose time curve and normal Insulin and C-peptide responses.

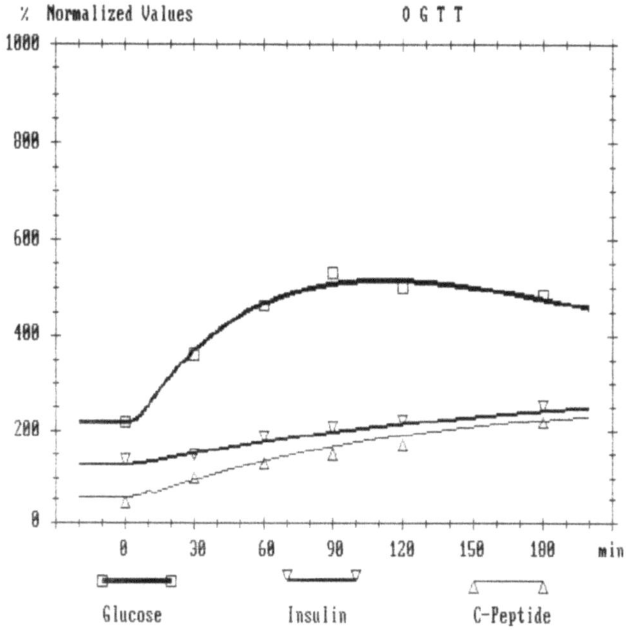

Fig. 14. Analogous presentation of the results of an oral glucose tolerance test (OGTT) performed on a patient suffering from diabetes. The diagnosis is confirmed by the high Blood Glucose Levels and the low response of Insulin and C-peptide.

can be seen as well. It is the insulin release which is not adequate. Both insulin and C-peptide release are to late and the increase of the insulin is too slow. The difference between the normal and pathological OGGT are markedly expressed using the analogous presentation of the data. The relation between the glucose levels and the regulating hormone insulin are correctly represented due to the normalization of these parameters in both cases shown in Figure 13 and 14 respectively.

The four examples presented above demonstrate that the models formulated in Section 3 and identified with the methods explained in Section 4 are of acceptable validity for the purpose of the analogous presentation of time dependent laboratory results.

6 Digital and Analogous Presentation

In the last section, some examples of the analogous presentation of time dependent laboratory results, using the nonlinear interpolation by means of mathematical modeling, were discussed to prove the validity of the models. The advantages of this type of data presentation are as follows:

1. The complete time course can be immediately conceived by visual inspection.
2. The response of the variables due to the perturbation of the dynamic system is apparent.
3. The times of increase and decrease can easily be estimated from the graph.
4. The relations between two variables with respect to their response and the times of increase and decrease become also immediately apparent.
5. The interpretation of the laboratory results is facilitated.

The information gathered from the analogous data presentation is very useful for a quick and exhaustive survey of the time dependent results of a laboratory test, but it also has a serious drawback which must be emphasized. All the information of the graphical representation is more or less quantitative. For example one can see every observed measurement, but the actual quantitative result can only be estimated. This is in contrast to the digital presentation of the laboratory results discussed in Chapter 2.2 of this volume. There, the laboratory results are given in figures which provide the true quantitative information. To combine the advantages of both methods, we developed a program package for the digital and analogous presentation of time dependent laboratory results to assist the physician with the interpretation of the data.

This program package runs on an IBM Personal Computer AT 02 and uses extensively the facilities of the data base management system dBase III (TM). Only the graphs in this article are produced by means of a FORTRAN program, written by W. Renn, and the printouts showing the digital presentation of the laboratory results are produced with a special print program. The FORTRAN program uses the graphic subroutines of the package X-Graph (TM) together with the Hercules

(TM) graphic card. All the other parts of the program package are dBase III (TM) procedures, i. e. the storage and management of the data, the handling of the screen forms, and the control of the program flow are performed using the dBase III (TM) facilities. The FORTRAN program for the graphic presentation and the print program are called from dBase III (TM) procedures and the program flow returns to these procedures after the completion of these special programs. A NEC P3 Pinwriter (TM) is used as a hardcopy and print device.

In detail, the program – called DAP – is used for the evaluation of the following laboratory tests:

1. Time dependent enzyme pattern after acute diseases
2. Metabolic loading tests
3. Endocrine function tests.

After entering the program DAP a menu form, shown in Figure 15, appears on the display and the user is prompted to type in the number of the function he/she wants to be performed and the print indicator depending on whether the presentation shall be on the display or the matrix printer.

The available functions are discussed in the following.

```
Digital and Analogous Presentation of Laboratory Results

1. Entry of               Demographics
2. Entry of               Laboratory Results
3. Nonlinear              Interpolation
4. Digital                Presentation
5. Analogous              Presentation
6. Digital + Analogous    Presentation

[ ]  <——  Enter NUMBER of the desired  FUNCTION
[ ]  <——  Enter PRINT INDICATOR  Y / N ( Yes / No )
```

Fig. 15. Menu form to select the digital and/or analogous presentation of time dependent laboratory results on the display or printer respectively.

The registration of primary data is accomplished by function 1. These include the demographics of the patient, i. e. his name, date of birth, and diagnosis as well as some administrative data like identification number, ward, and date of the test.

The test results are entered with function 2 of the menu. They also have to be typed into the computer, because the personal computer used is a stand alone system which is not connected to the hospital laboratory EDP system. Although the entering of the data on the local level has the disadvantage of possible typing errors, the advantage is, that the program DAP can be used in any laboratory and is completely independent from the existence of a main EDP system.

By means of function 3, the parameter estimation process is carried out by the computer automatically, employing the methods and programs discussed in Section 4. Warnings are given at the display if the convergence of the parameter estimation procedure is not satisfactory. In this case the linear interpolation method is used for the analogous presentation of the data. An example is given in Figure 4. If the parameter estimation converges successfully, the goodness of fit and the accuracy of the parameters can be judged by the digital and analogous presentation of the parameter estimation results as shown in Table 1 and Figure 10 respectively.

Function 4 provides the digital presentation of the time dependent laboratory results. If the print indicator is N (No), the results are shown on the screen in the form presented in Figure 16.

If the print indicator is Y (Yes) the results are printed on the matrix printer in the form shown in Figure 17. The digital presentation of the laboratory results consists of three parts. The first part contains the demographics of the patient. The second part shows the results measured in the laboratory. The final part presents the calculated values which are derived from the parameters of the mathematical model used for the analogous presentation of the laboratory results.

T R H - Test			Date :	**07.01.87**
7/17/00123 **Name, Xxxxxxxxx**		*17.12.24	Ward :	**B2V**
Diagnosis : **Akromegaly**				
Observed Values :	Time (min)	**T S H** (mU/l)	**G H** (pmol/l)	**Prolactin** (μg/l)
	0	1	435	1
	30	5	739	7
	60	4	522	5
	90	3	472	3
	120	2	374	2
	180	1	386	1
Upper Reference Limits :		**3**	**230**	**20**
Calculated Values:	**$T1_{1/2}$** (min)	10	3	3
	$T2_{1/2}$ (min)	51	23	38
	Response (%h)	**146**	**83**	**5**

Fig. 16. Digital presentation of time dependent laboratory results on the screen. The first part containes the demographics of the patient. The second part shows the digital presentation of the laboratory results. The estimated parameters of the model are given in the third part.

T R H - Test Date : **07.01.87**

7/17/00123 **Name, Xxxxxxxx** *17.12.24 **Ward : B2V**

Diagnosis : Akromegaly

Observed Values :	Time (min)	T S H (mU/l)	G H (pmol/l)	Prolactin (µg/l)
	0	1	435	1
	30	5	739	7
	60	4	522	5
	90	3	472	3
	120	2	374	2
	180	1	386	1
Upper Reference Limits :		3	230	20
Calculated Values:	$T1_{1/2}$ (min)	10	3	3
	$T2_{1/2}$ (min)	51	23	38
	Response (%h)	146	83	5

Fig. 17. Digital presentation of time dependent laboratory results on the printer. The first part containes the demographics of the patient. The second part shows the digital presentation of the laboratory results. The estimated parameters of the model are given in the third part.

As an example, the TRH-test is used in Figures 16 and 19 to explain the output of the program DAP. A two-compartment model with bolus injection is used for the TRH-test. The parameters of this model are *K*, *K*, and *D* (the area under the time curve). These describe the influx of hormone to the plasma compartment, the elimination of hormone from the plasma compartment, and the total hormone concentration released after the stimulation respectively.

From the time constants the times of "half-life" are derived using the well known relation:

$$T1_{\frac{1}{2}} = \frac{\ln(2)}{K_1} \tag{6.1}$$

$$T2_{\frac{1}{2}} = \frac{\ln(2)}{K_2} \tag{6.2}$$

$T1_{\frac{1}{2}}$ is a measure of the increase of the hormone after the stimulation. In contrast, $T2_{\frac{1}{2}}$ is a measure of the elimination of the hormone from the plasma compartment.

The area under the time curve is heavily dependent upon the dimensions used. Thus a comparison between the release of different hormones caused by the same amount of the stimulating hormone TRH is strongly hampered. To circumvent this problem and to facilitate the interpretation of the laboratory test, the areas under the time curves are also normalized, using again the upper limit of the reference range (*ULR*). Thus, we have:

$$R = \frac{D}{ULR} \tag{6.3}$$

where R is the so-called percentile hormone response.

The "half-life"-constants as well as the response values calculated with the formulas (6.1) to (6.3) are given in the third part of the digital presentation of the time dependent laboratory results. These quantities can be useful for the interpretation of the data.

With function 5, the user gets the analogous presentation of the time dependent laboratory results. This presentation is the main part of our application and was discussed in the preceding section at length. Examples of the output of DAP were

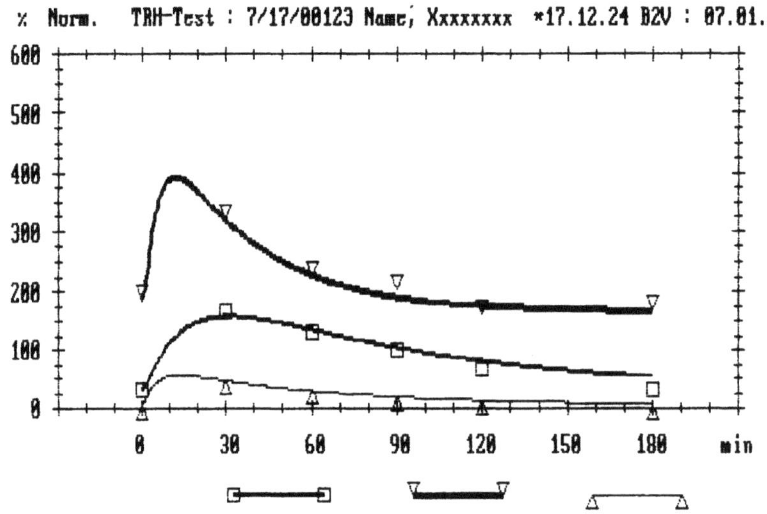

Time (min)	T S H (mU/l)	G H (pmol/l)	Prolactin (µg/l)
0	1	435	1
30	5	739	7
60	4	522	5
90	3	472	3
120	2	374	2
180	1	386	1

Fig. 18. Digital and analogous presentation of time dependent laboratory results on the screen. The demographics of the patient are shown in the first line of the screen. The analogous presentation of the calculated and observed results is shown on the upper half and the digital presentation of the observed values is given on the lower half of the screen.

T R H - Test **Date: 07.01.87**

7/17/00123 **Name, Xxxxxxxx** *17.12.24 **Ward: B2V**

Diagnosis: **Akromegaly**

	Time (min)	T S H (mU/l)	G H (pmol/l)	Prolactin (µg/l)
Observed Values:	0	1	435	1
	30	5	739	7
	60	4	522	5
	90	3	472	3
	120	2	374	2
	180	1	386	1
Upper Reference Limits:		3	230	20
Calculated Values:	$T1_{1/2}$ (min)	10	3	3
	$T2_{1/2}$ (min)	51	23	38
	Response (%h)	146	83	5

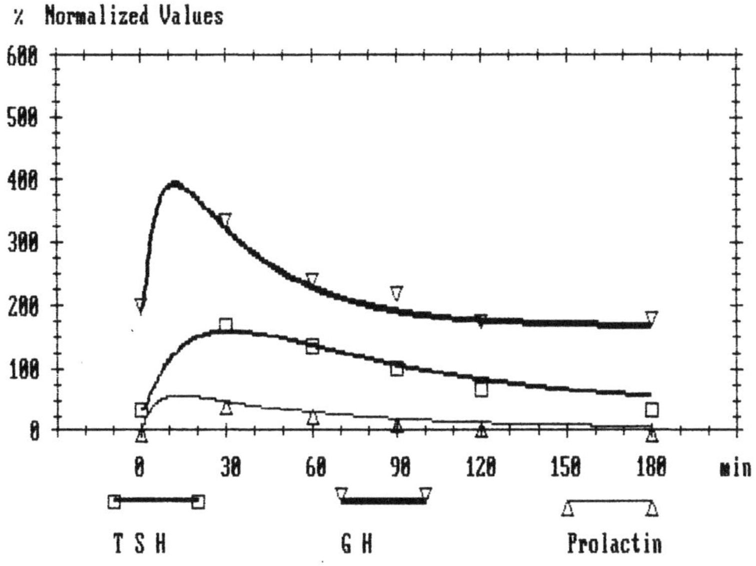

Fig. 19. Digital and analogous presentation of time dependent laboratory results on the printer. The first part containes the demographics of the patient. The second part shows the digital presentation of the laboratory results. The estimated parameters of the model are given in the third part. The analogous presentation of the calculated and observed results is finally shown in the lower half of the laboratory report.

presented in Figures 11 to 14. These graphs are displayed on the screen or by the matrix printer depending on the print indicator chosen by the user.

Function 6 provides the optimum and most complete presentation of time dependent laboratory results with the combination of digital and analogous data. Choosing the print indicator N results in a screen display of the form shown in Figure 18, where the laboratory results are given in a digital presentation under the graph.

Finally, Figure 19 shows the combined digital and analogous presentation of the data which is printed on the matrix printer after having chosen function 6 with print indicator Y. As can be seen, this combined presentation is achieved by merging Figures 17 and 12. The different parts of this combined laboratory report where discussed above. Different character fonts where used to facilitate the differentiation between those parts representing the name and date of the test, the patient's demographics, the observed values, and the calculated values.

As pointed out in Chapter 2.6, different character heights and styles can be used to differentiate the emphasis of the presented information. Since the most important information is the name and date of the test together with the demographics of the patient, we use the large characters for this information. The next important information is the observed laboratory results given in normal character height. Smaller characters are used for the calculated values because the diagnostic importance of these data is not yet established. Finally, the graph is produced using the hardcopy facilities of X-Graph (TM) in connection with the Hercules (TM) graphic card of our personal computer. This gives the contrast between the digital and analogous presentation of the time dependent laboratory results.

7 Conclusion

Modern laboratory diagnostic facilities present the physician on the ward with an ever increasing flood of data. On the one hand the time available for adequate evaluation and application of this information, due to the rationalisation of patient management, is decreasing. This is even more acute with respect to the emergency and intensive care units. On the other hand, these data are the foundation of a timely diagnosis and successful therapy. To assist the physician in this situation, a number of means of data presentation for single laboratory parameters is available. However, multivariate and time dependent laboratory data necessitate a compact and coherent presentation. Since the real information in the digital presentation of data is often lost due to the absolute number of isolated measurements only the analogous presentation of time dependent laboratory results is readily comprehensible.

The program package which we have presented provides a composite information. Due to the availability of the information presented on the screen, the pathologist in the laboratory and the physician on the intensive care unit can make an immediate decision. The physician on the general ward, who gets the printed report in the form shown in Figure 19, has the time dependence of the parameters and their interrelation readily available. Finally, the combined laboratory report with the digital and analogous presentation of the time dependent laboratory results can be included in the patient's files for a permanent documentation.

References

1. Gibaldi, M. & Perrier, D. (1975) Pharmacokinetics, Marcel Dekker, New York, Basel.
2. Rodbard, D. & Frazier, G. R. (1975) Statistical analysis of radioligand assay data, Meth. in Enzymology *37*, 3–22.
3. Carson, E. R. & Cobelli, C. & Finkelstein, L. (1983) The Mathematical modeling of metabolic and endocrine systems, John Wiley & Sons, New York.
4. Schlichtkrull, J. & Munck, O. & Jersild, M. (1965) The M-value, an index of blood-sugar control in diabetics. Acta Med. Scand. *177*, 1, 95–102.
5. Bleifeld, W. et al. (1977) Infarct size estimated from serial serum creatine phosphokinase in relation to left ventricular hemodynamics. Circulation, *55*, 2, 303–311.
6. Hørder, M. et al. (1978) in: Proceedings of the symposium determination of creatine kinase MB activity in the diagnosis of acute myocardial infarction, pp. 63–84, Diagnostica Merck.
7. Schneider, B. (1984) in: Simulationstechnik 1984 (Breitenecker, F., Kleinert, W., eds.) Informatik-Fachberichte *85*, 333–338, Springer, Berlin, Heidelberg, New York, Tokyo.
8. Saratchandran, P. & Carson, E. R. & Reeve, J. (1976) An improved mathematical model of human thyroid hormone regulation. Clin. Endocrinol. *5*, 473–483.
9. Di Stefano, J. J. & Cobelli, C. (1980) On parameter and structural identifiability: nonunique observability, reconstructibility for identifiable systems. IEEE Trans. Automat. Contr. *AC-25*, 830–833.
10. Renn, W. et al. (1984) in: Simulationstechnik (Breitenecker, F., Kleinert, W., eds.) Informatik-Fachberichte *85*, 419–426, Springer, Berlin, Heidelberg, New York, Tokyo.
11. Renn, W. et al. (1985) in: Simulationstechnik (Möller, D. P. F., ed.) Informatik-Fachberichte *109*, 297–303, Springer, Berlin, Heidelberg, New York, Tokyo.
12. Powell, M. J. D. (1965) An optimization program using a direct search algorithm. Comput. Journal *7*, 87–91.
13. Metzler, C. M. et al. (1974) A package of computer programs for pharmacokinetic modeling. Biometrics *30*, 562–563.

Chapter 2.5
Computer Graphics for Data Presentation

I. O. Angell

Before considering the practicalities of the graphical display of data it is well worth spending time discussing the selection of a graphics terminal. You should start by asking whether you need to purchase a graphics device at all! As an alternative you can beg, steal, borrow (or even pay for time on) someone else's graphics equipment. The need for graphical display of data goes without saying, all experimental results benefit from an attractive presentation; but the cost-effectiveness, in time and money, must be a major factor in any decision. The answer to this dilemma lies in the complexity of your particular graphical needs.

If you need relatively few very specialised images then it makes far more sense to subcontract the work, and so there is only one question:

1. Where do I get the work done?

Should you decide that you really need your own equipment, then a further three questions must be addressed:

2. What type of graphics device?
3. Which manufacturer?
4. What software?

These latter three questions are so closely interconnected that no one question should not be treated in isolation.

It is these questions that I will attempt to answer in this chapter. Most of what I have to say is common sense; but it still has to be said: I had to learn by trial and error. Also I hope my illustrations will give you some ideas for creating eye-catching presentations of your data. The figures I will use are not related to clinical medicine, but the methods they represent are totally independent of particular applications, and so very relevant lessons can be learned from them.

I will first deal with question 1, so that I can concentrate on the more relevant latter three questions in the remainder of the chapter. There are a growing number of graphical artists who now advertise a computer graphics service. Their names can be found in trade magazines, e. g. Computer Images, or from national and international 'Creative Arts' and 'Computer Graphics' trade exhibitions where they display their wares. The stands at such shows are usually manned by the artists themselves and so this gives newcomers the opportunity to get 'free' advice on 'state-of-the-art' graphics and more importantly how much it will cost. Even if you have your own graphics equipment it may be more cost-effective to approach such companies for particular jobs. In fact some of the diagrams given in this article were produced by such a company, Digital Arts, London, and used with their permission.

Now the answer of "What Manufacturer?". Unless you have a very specialist application with hardware/software requirements that automatically limit choice, then if possible *always* buy from a major manufacturer. They may prove slightly more expensive than 'look alikes', but the peace of mind created by guaranteed delivery, follow-up facilities, reliability, large service and spares departments, standardisation, good documentation, education facilities, the possibility of expanding or upgrading, a large range of software, and most important the existence of a large international user community, make the extra cost well worthwhile.

As for "What type of device?", the answer really depends on your particular applications, expectations and finance. The best way I can answer this question is to give examples of the different devices I have used and, in the discussion of each type, make relevant references to the other questions as well as give a rough order of magnitude for the costs involved. It is essential that you know both the potential and limitations of these devices in order to make sensible decisions and to avoid (unreasonable) disappointments.

First the flat-bed plotter, which can use microcomputers, minis or main-frames as host. This is still a very popular hard copy device: the CalComp, Hewlett-Packard and Tektronix multi-coloured pen plotters give excellent quality line drawings, in particular for overhead projector foils, which may be produced quickly and are of enormous value for lectures, demonstrations and presentations. The brand-name consumables (i. e. pens and transparent foils) tend to be a little expensive, but good value alternatives are readily available. You will find that the quality of the pen line drawing does degrade rapidly on certain cheaper types of plotter, so it is essential to use new pens for important drawings. These are mechanical devices so, before buying, check on the size and accuracy of the incremental step: the 'resolution'. Test the device with a random-walk program that creates a sequence of lines that should start and end at the same co-ordinate point and repeat the process several times: such repetition of the same diagram will show up any inaccuracies in the motor system by the lines getting successively broader and the vertices of the line-diagram spreading over an unacceptable area. Make sure that any software you buy for your plotter (or any other graphics device for that matter) is transportable. This usually means that the software generates calls to routines of a particular graphics standard (e. g. the rapidly emerging international standard: the Graphical Kernel System, G.K.S.), and that the necessary 'driver routines' for your plotter are included. Should you subsequently change graphics device, (to another plotter or to an alternative terminal) then, given drivers for the new device, your software should transfer directly with a minimum of fuss. (That's the theory anyway!)

Most of the plotters have a built-in digitising facility of reasonable quality: you may find this a very convenient alternative to 'mouse or light-pen input'. Some form of interactive input is essential if you intend to use any large scale geographic data, such as the map data shown in Figure 1. The plotter equipment will cost anything from around a few hundred dollars to tens of thousands for really high quality plotters such as the Benson range or microfilm plotters.

Then of course we must look at the very important area of raster graphics on microcomputers. There are a wide range available at the bottom end of the market (hundreds of dollars) but these tend to appear and disappear very quickly. A better

Fig. 1. The map of the earth.

long term strategy is to buy the slightly more expensive systems like the Apple Macintosh and IBM PC (a few thousands of dollars), which although they may become outmoded, are made by established companies who will ensure that new models will be upwardly compatible with older models in order to maintain the market need for software stability.

You must check the resolution of the device i.e. the number of pixels – horizontal and vertical – on the device screen. *Note* this is not necessarily the number of addressable points on the display that are referenced by software, for example you may find that a pixel is equivalent to a four by four block of addressable points. Find out how many colours are available at any one time, and the total number in the palette. Check on the built-in properties – area fill, pattern fill, pies, soft characters, pixel blocks or sprites, text-only modes – as well as the speed of drawing.

The cost of the cheaper microcomputers is so low that time and effort are the only considerations that need concern you. Unfortunately these low cost systems do tend to have inadequate documentation, and specialist knowledge is required to find your way round all the PEEKs and POKEs. If you have a particular scientific application, and know of an available package (inexpensive; preferably free) that runs on such a system, then you cannot lose, but you must be prepared to discard the software after the demise of the microcomputer. As a general rule don't write your own software, leave it to those professionals who spend a great deal of time working with these systems. It may be worthwhile, however, spending some time on the better micros, in particular on an IBM PC (or PS/2). This machine sells in such large numbers that the graphics commands on the PC have become a de facto standard. The documentation is very good, there is an enormous amount of quality professional software and hardware add-ons available and there is compatibility throughout the range; from the Standard to the XT, AT and beyond. Beware of 'look alikes' though: they claim to be 98% compatible (say), but that problem 2% is more than likely to be in the graphics display! The IBM PC is very good for the production of 'spreadsheets' and medium resolution data diagrams, so-called, 'business graphics' (pie charts, bar charts and graphs). A few hundred dollars will give you packages that mix word-processing, data-base management, mixed text/graphics display and desk top publishing (laser printing). 'Public domain' business graphics software is readily available (including some by the author, see 'Advanced Graphics with the IBM PC' published by Macmillan). It is a simple matter to create your own 'soft character sets' and 'pixel blocks' in conjunction with a graphics editor to add extra quality and text clarity to the data displays. For the best pictures though it is best to use add-on frame stores: see Figures 2, 3 and 4. The text only mode of 8 multi-coloured pages is an ideal text information display system. Animation, using both multiple-screen buffering and colour redefinition (also see the above mentioned book) will revolutionise hitherto static presentations. Many future conferences will be using microcomputers linked to television projection systems to illustrate the talks. The days when audio-visual aids meant solely hand-written overhead projection foils and diapositive slides of photographs is over!

If required, hard copy can be produced by simply photographing the screen with a SLR camera. To get good results I use Kodak Ektrachrome and set the

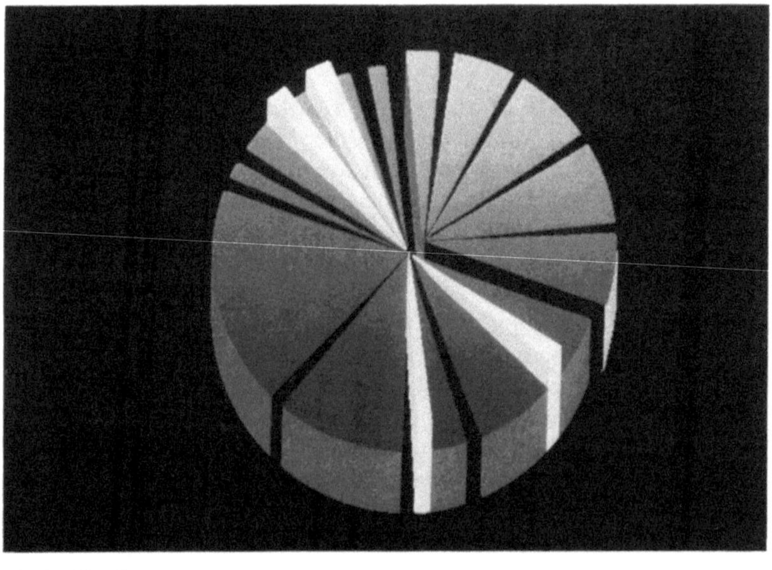

Fig. 2. Pie chart.

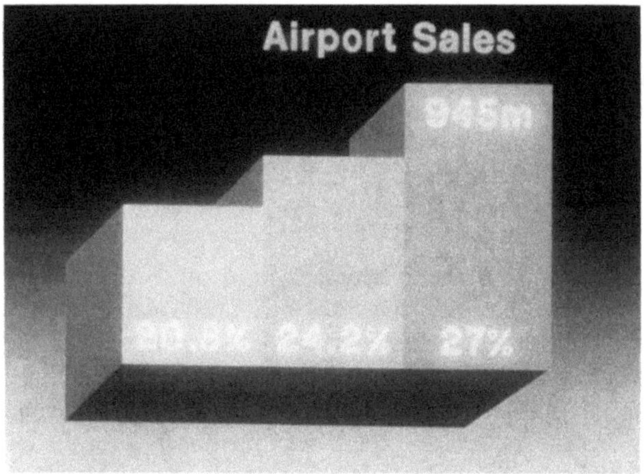

Fig. 3. Bar chart (labelled: airport sales).

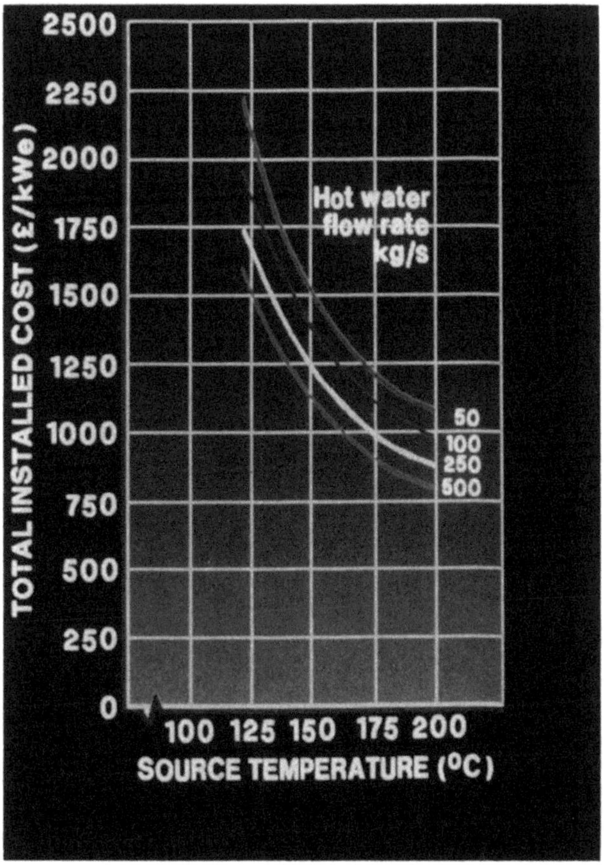

Fig. 4. Graph of cost against temperature.

shutter speed to between a quarter and a half second. You will have to experiment with the aperture value – it will depend on the intensitiy of the major background colour. Figures 2, 3 and 4 were produced in this way. You can of course buy a graphics printer for 'screen dumps', or a special purpose photographic peripheral or video recording and editing systems: these of course increase the expenditure to the order of thousands to tens of thousand of dollars, but you should not forget that much of the expensive equipment is available on short-term hire (a day at a time).

Special graphics boards and chips are available to add on to the micros which vastly increase their power and potential: the IBM CGA, EGA and VGA boards or the Pluto boards give increased resolution, a much larger palette of colours and many more built-in graphics options. Floating-point accelerator chips enable micros to service the 3-dimensional graphics systems which until recently have been the province of specialised research groups. Figure 5 was produced with an IBM XT and Pluto 2 board. Software is available to run graphics editors and paint-box programs, which use a mouse or light pen to draw interactively on the screen. A good system (hardware and software) will cost around five thousand dollars. Such programs are sophisticated graphics editors and will include many advanced options for the production of high quality data displays, in particular anti-aliased character sets which do not have the jagged edges of the basic microcomputer character set. For example Figure 2 was shows a typical pre-labelled pie-chart constructed by a business-graphics program library, while Figures 3 and 4 show respectively a bar-chart and a graph produced by the same library, but after they have been labelled using a graphics editor. Another example of such a paint-box generated display is given in Figure 6. This time it was constructed solely by mouse input into the paint-

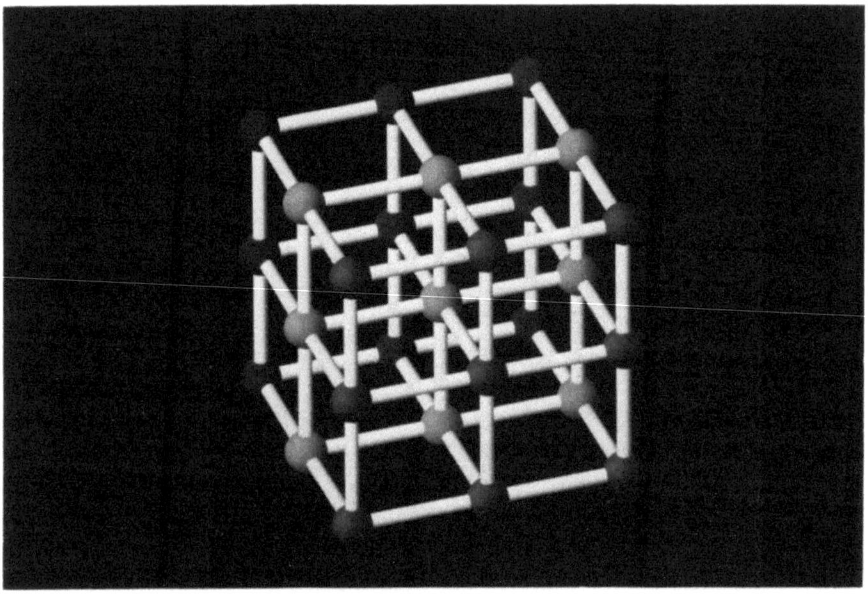

Fig. 5. Ball and spoke model.

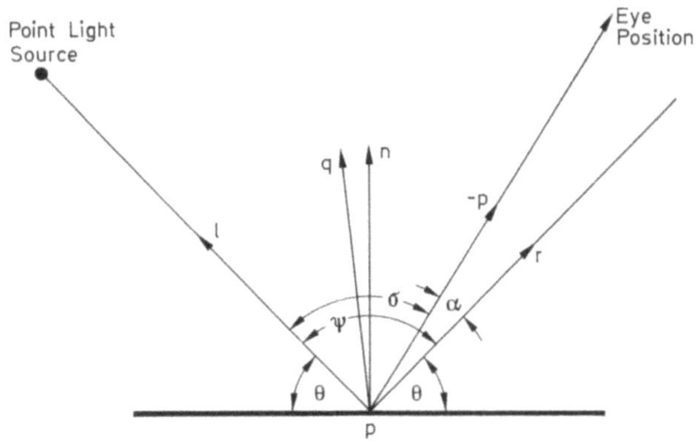

Fig. 6. Point light source and light position.

box program, and shows a geometrical figure describing the theory of 'smooth shading' used in the production of Figure 5. Furthermore it is possible to extend these systems so that pictures can be digitised via a video camera and then manipulated on the screen ('frame grabbing'): Figure 7. Now of course the cost often exceeds ten thousand dollars, and some very specialised video equipment may even exceed a hundred thousand dollars. The artistic eye and the practical experience to use a paint-box program effectively usually means getting outside help or investing a great deal of time. For the more artistic displays it is often cheaper, certainly quicker, and the end result is far superior if you employ a professional computer graphics artist.

Next comes the stand-alone workstation. Machines like the SUN, APOLLO, PERQ and the VAX station give you multi-user, high resolution interactive graphics, with window manager and menu/icon based software, and mouse/tablet input. Ideal for Human Computer Interaction (H. C. I.) methods for graphical communications between the user and computer. It is essential that you get the full range of software and include a networking interface: now the price is above twenty thousand dollars. Workstations are an excellent high resolution preview system for constructing complex diagrams before networking them to expensive shared facilities such as microfilm plotters, photo-typesetting equipment, or laser printers. At this level of sophistication you must ensure that all software (whether written by you, or purchased) can be integrated into a distributed multi-terminal type environment. This leads on to state-of-the-art research level for multi-media documents and computer conferencing, combining text, computer generated graphics, bit-map graphics, video and sound: very expensive and to be avoided initially until standards have been agreed on.

Finally we have the dedicated graphics terminal, linked to a host mini-computer or main frame, or with built-in processor(s). Machines like the expanded microcomputers mentioned above, or the standard Tektronix 4010 line drawing display costing around a thousand dollars (Figure 8 was produced on a 4010 before being transferred to microfilm), or the colour Tektronix 4100 series of around ten thousand dollars, or high quality displays such as Sigma, Silicon Graphics, AED, Megatek, Ramtek (tens of thousands of dollars) and the ultra expensive (and

Fig. 7. Mona Lisa.

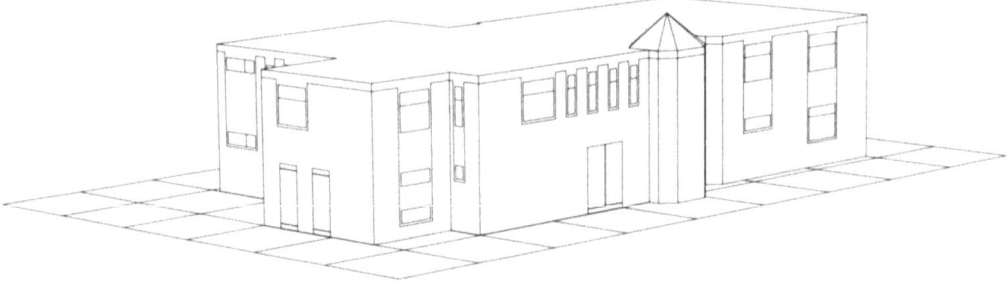

Fig. 8. Architectural building.

superb) machines like the Evans and Sutherland (hundreds of thousands). These machines come with a large range of backup facilities: accurate digitising tablets, hardware anti-aliasing, hardware smooth shading, hardware three dimensional transformations etc. You must be very sure that you really need equipment of such sophistication before spending such large sums of money! All of the equipment in this price range is very good and very reliable, however, make sure there is adequate

Fig. 9. Chess pieces.

local support. There is nothing worse than having work delayed while waiting for a spare part from the U.S.A., Japan or other parts of Europe, just because your local supplier has inadequate stock.

Such equipment is ideal for state-of-the-art geometric modelling (using polygonal mesh, oct-tree and ray tracing methods): see Figures 5 and 9. Now we are in the region of capital intensive graphics: enormous software and manpower backup, expensive CAD/CAM packages, film and video animation, graphics data-bases: all running on dedicated computers like VAXs or even Crays (millions of dollars), and the next generation of parallel computers. If you have access to high-resolution hard copy, then there is no need for top-of-the-range terminals. You can use low-cost (?) terminals to give you a reasonable impression of a complex picture, and then network it to the hard-copy device. Pictures which could eventually contain thousands of different shades of colour, include smooth shading, shadows, reflections, transparency etc. can be tested first on a machine which allows you only (say) 8 colours from a palette of sixty four colours. Expanded micros can give you 256 or more colours with the processing power to go with it. In fact Figure 9 was constructed on an 8-colour machine before being transferred to a high resolution display!

Having taken this overview of different types of graphics device, let us now consider the decision processes for choosing a graphics terminal. Firstly you must have a clear view of your present (and future) requirements and realise that cost will probably be the major consideration in your decision. Take into account the time and effort needed to run a graphics facility: maintenance, manpower and software costs. Remember that many graphics requirements do not need state-of-the-art (hence expensive) technology. You must decide on the configuration of your graphics system within this cost limitation: always keep open the option of sharing facilities and hiring equipment and/or expertise. Use the experience of other graphics users in your locality, many of the general problems with computer devices can be attributed to local aberrations in service and spares. Enlist the advice of these users when choosing machines – but do check that they don't have a commercial vested interest in these machines. Be aware of other groups working in similar research fields and/or using the same equipment: a short telephone call could save you hours, if not days, of wasted effort. Check with local universities and join a Special Interest Group or User Group and share your experiences: SIG's are a major source of reliable information regarding common problems, communication peculiarities, software and hardware difficulties and, most importantly, where to find help. If you intend networking into a large local, national or international facility, then check with these centres regarding compatibility of software and hardware. Furthermore, should you intend to configure a number of graphics devices together then, if possible use the equipment of a single manufacturer.

Finally we must consider a software strategy. In general don't write your own software – it is far cheaper and more reliable to buy it. Obviously this is a vast generalisation: scientific software is very specialised and you may be forced to write your own, or indeed your institution may already employ the experienced manpower that is needed! In this case use a major programming language: most graphics professional programs are written in Fortran 77, Basic, C or Pascal. Ensure that all software is transportable by linking your programs into a graphics standard, thus making them device independent. When buying software get a list of users of the package and take advantage of their experiences, and most important have a predetermined 'benchmark' before setting out on your quest.

Because of future trends in the use of computers, make sure that any software you use is modular in design and fits into a distributed computer system. Do not limit your outlook to a single processor! With a proper modular design, your main processor can deal with major file handling and 'number-crunching' tasks, while graphics can be relegated to secondary processors (microcomputers may be suitable), perhaps using pre-programmed utilities such as unusual character sets and precompiled graphical document layout. As for the future, don't think of computer graphics as just a quick way of producing book-like documents with a 'cut-and-paste' approach. The interactive facility, multi-media aspects and inexpensive mass storage (Compact Disc ROMs, and hopefully soon video disc writable memory) and the reality of animation will imply a total re-think of what we mean by a document. 'Books' and articles will no longer be 'carved in stone', they will be multi-user, interactive and infinitely flexible, containing animated pictures and sound: in fact what is generally described as a computer-conferencing system. All pictorial examples can be self-generated under the control of an expert system, so that the book itself will change to suit the experience of the readers.

I hope that my list of warnings hasn't put you off computer graphics. It is going to be central to developments in computer conferencing and really is a most satisfying field of study, and the future is very exciting. Most medical and scientific establishments will be setting up Computerised Audio-Visual departments: the prices mentioned in this chapter will drop and as costs tumble video animation and real-time animation in addition to static displays will become commonplace. Dedicated graphics processors and international graphics, networking, conferencing, document and data-base standards will make the next generation of terminals much easier to use and the transfer of graphics programs and data, not only between machines but also between continents, will be a simple matter of pressing a button. And the future starts now! I hope you will all join in the exiting new developments, and that you make the right choices of equipment. I wish you good luck with all your graphics projects.

References

1. Angell, I. O. (1985), Advanced Graphics on the IBM Personal Computer. MacMillan, London.
2. Angell, I. O. & Griffith, G. (1987) High Resolution Computer Graphics Using FORTRAN 77. MacMillan, London.
3. Foley, J. D. & Van Dam, A. (1981) Fundamentals of Interactive Computer Graphics. Addison-Wesley, Reading, Massachusetts.
4. Harrington, S. (1983) Computer Graphics: a Programming Approach. McGraw-Hill, London.
5. Hopgood, F. R. A., Duce, D. A. et al. (1983) Introduction to the Graphical Kernel System (GKS). Academic Press, London.
6. Newman, W. M. & Sproull, R. F. (1973) Principles of Interactive Computer Graphics. McGraw-Hill, London.

Chapter 2.6
Use of Colour in Tables and Graphic Displays

L. Reynolds

1 Introduction

Clinical data are of crucial importance. It is therefore essential that they should be clearly presented, and designed for maximum ease of use. While it might be argued that those using such data are highly motivated and will do their best with whatever displays are available, there is still no good reason for producing displays that are anything less than the best that can be achieved within the limitations of the medium. However highly motivated the user is, it is easy to make mistakes in reading off and interpreting data when working under stress. When dealing with clinical data, such mistakes do not merely waste time and cause frustration. They can be dangerous. The importance of clear and unambiguous presentation cannot, therefore, be overemphasised.

Over the years, skilled typographers and graphic designers have developed very effective techniques for the presentation of data in conventionally printed tables, charts and graphs. There is also a substantial body of research literature relating to the presentation of data of various kinds in various forms (1, 2, 3). Much of this wisdom has to do with basic perceptual principles and it can be applied directly to the presentation of information in other media too. Other aspects of this established body of knowledge cannot be transferred so readily however. The constraints presented by computer-generated displays of various kinds are not the same as those associated with conventionally printed images. This can have far-reaching effects on the way in which the data must be organised, as well as on details of layout, typography and graphic representation.

The standard 80-column alphanumeric VDU and the line-printer printout impose very severe constraints on the information designer, but with the advent of affordable computer graphics systems the possibilities for effective data display via the computer are immeasurably improved. Many computer systems offer colour too, a facility that has always been too costly for extensive use in printed materials. With these new freedoms we might expect a dramatic improvement in the presentation of computer-generated data, but it is unfortunately true that a very sophisticated tool in the hands of an inexperienced designer can often produce infinitely worse results than a much simpler device. And whereas typographers and designers working in print will usually have years of training in the effective presentation of information, most scientists designing computer-produced data displays in the course of their work do not have this advantage.

The purpose of this Chapter, therefore, is to summarise some basic principles in the presentation of data, to consider the relative merits of tables and graphic displays, and to examine the implications of using colour.

2 Visual Coding

2.1 The concept of visual coding

There are three essential requirements for the effective presentation of information in any medium. Some visual means must be found for *emphasising* important items of information, for *relating* items that are linked in some way, and for *separating* or dividing items that are unrelated.

When information is printed by conventional means, these distinctions can be made relatively easily using the rich variety of typographic variations available and the subtle variations in spacing that can be achieved. These two methods are often referred to as 'typographic coding' and 'spatial coding' of information. Typographic coding includes the use of changes in type size, style, or weight, and the use of additional characters such as parentheses. Spatial coding includes the use of devices such as variations in line spacing, indentation, the arrangement of information in columns and so on.

Typographic coding can be used very effectively to create hierarchies of emphasis. For example, bold type and/or larger sizes of type are often used to visually indicate that a hierarchical system of headings exists in a document. More subtle variations, such as the use of italics, can be used to distinguish or 'separate' information that is different in kind. The use of italics to indicate words in a foreign language is an example of this.

Typographic coding alone is not enough however. For clarity it is essential that information should be appropriately coded spatially. Hierarchies of emphasis can be created using space alone if necessary. For example, the more space there is around a heading, the more important that heading will appear to be. Space is also a powerful means of separating items such as paragraphs, or columns in a table. By the same token, the items within the paragraph or column are thereby strongly linked.

Appropriate visual coding is essential for the design of any kind of information. This is because human beings have an inbuilt tendency to try to form visual groupings within whatever they are looking at. There are various ways in which we try to group things, as described by the Gestalt psychologists (4). Grouping by similarity is the basis of typographic coding. Headings in the same size and weight will be seen as being of the same importance, and different from those in other sizes or weights. The principle of grouping by proximity is the basis of spatial coding. We tend to 'read' the relative amounts of space between items as a indication of the closeness of their logical relationship.

Because of the fundamental nature of these grouping activities, inadequate or inappropriate typographic or spatial coding is likely to result in great confusion for the user of the information. For example, it is important to establish a system of spacing within a set of data and to use it consistently throughout. It is not good practice to change this spacing in part of the display only, in order to fit in all the data. If this kind of adjustment seems unavoidable, then the initial layout is at fault and should be redesigned.

2.2 Visual coding in computer-generated information

In designing information which is to be output on a computer-driven device of some kind, the range of possibilities for typographic and spatial coding of information is usually much more limited than is the case with conventionally printed materials.

Many systems still provide only one character set, so variations in letter form are not possible. If more than one character size is available, different sizes cannot necessarily be used on the same display at the same time. Larger characters can of course be very useful for headings. An italicised version of the font is sometimes provided, and an overprinted bold or a brighter screen image are often available for emphasis, but these are not usually as effective as their conventionally printed counterparts. Many VDUs offer reverse video for emphasis on the screen, but it can interfere with legibility. This is especially true where the data consist of capital letters or numerals, and the top edge of the reversed panel is close to the tops of the characters.

All-capital lettering can of course be used for emphasis, but it is only suitable for headings or very short statements. All-capital printing has been shown to reduce the legibility of continuous printed text (5). This is largely because words in capitals have a much less distinctive shape than words in lowercase letters with their ascenders and descenders. There is every reason to think that the same will be true for all-capital text on a computer printout or VDU.

The possibilities for spatial coding are often very restricted too. There is typically a fixed number of character positions, with inflexible letter and word spacing and a line feed increment in units of whole lines. Subtle variations in spacing are therefore not possible. This has considerable implications for the design of tabular matter and of charts and graphs. There is also the problem of fitting the information into the space available. On VDUs in particular, the relatively small number of lines available per screen means that space is at a premium and it is not always possible to use as much space between items as would ideally be required for maximum clarity of the information.

It seems likely though that many of these constraints will gradually be removed. The advent of desktop publishing systems capable of producing hard-copy output and screen displays in a variety of styles and sizes of lettering is a very encouraging development. It must surely be only a matter of time before some of these facilities are made available on other kinds of system too.

In the meanwhile, colour is becoming widely available for both alphanumeric and graphic applications. To a certain extent it can help to compensate for the restricted typographic and spatial coding possibilities but, as we shall see, it must be used with care.

3 General Principles in the Use of Colour

3.1 Colour and legibility

Before examining the ways in which colour can be used to help convey the structure of information, it is important to consider its implications for legibility.

In the case of output from a colour printer or plotter, the effect of colour on legibility is fairly easily predicted. The greater the contrast of the ink with the paper, the more legible the display will be. A contrast of 70 per cent or more is recommended for good legibility (5). Yellow and other pale colours are therefore not recommended for use on white paper. With colour on VDUs, however, the situation is rather more complex.

VDU colours are generated not by the use of differently coloured pigments but by the additive mixing of different wavelengths of light. A typical seven-colour system will include the three additive primaries, red, blue and green, the three corresponding complementaries, cyan, magenta and yellow, plus white. The white signal is obtained by mixing green, red and blue signals. If we give white a nominal luminance of 100 per cent, then green has a luminance of 59 per cent, red 30 per cent, and blue only 11 per cent. The complementaries are each obtained by adding two primary signals. Thus yellow has a luminance of 89 per cent (green plus red), cyan 70 per cent (green plus blue), and magenta 41 per cent (red plus blue). The contrast that these colours make with the dark background therefore varies considerably. In terms of luminance alone, white, yellow, cyan and green, in that order, are likely to be the most legible colours, and magenta, red and blue the least legible. The three darkest colours are therefore best avoided for important data. (In some systems, though, it is possible to boost the luminance of the darker colours, thus making them more legible.)

The relative legibility of the colours is also affected by the fact that the human eye is not equally sensitive to all wavelengths of light. When the eye is light-adapted, as it would be when looking at a VDU in normal room lighting, it is most sensitive to wavelengths in the yellow/green area of the spectrum (6). Yellow is only 90 per cent as efficient a stimulus as green, orange 80 per cent and red only 6 per cent. Blue in particular tends to have a lower subjective brightness than other colours, especially when viewed against a dark background (7).

Visual acuity is another factor which must be considered. Acuity is highest for green and white light. The human eye's resolving power for red is only about one third of that for white, and for blue it is only about one fifth of that for white (6).

Yet another significant factor is the phenomenon of chromatic aberration. Different wavelengths of light are refracted by different amounts as they pass through the eye. This means that only one colour at a time can be sharply in focus on the retina. If yellow/green or white are in focus, then blue will be focussed in front of the retina and red behind it. This not only results in blurred images for the out-of-focus colours, but it can also cause differently coloured images to appear to be in different planes. This effect can be seen very clearly when red and blue are used in close proximity on a VDU, and it is visually disturbing.

Thus, bearing in mind the differences in the luminances of the colours, their influence on sensitivity and acuity, and the effects of chromatic aberration, it would seem that white, yellow, cyan and green are the most suitable colours for the display of important data on a dark background. Blue, red and magenta should be avoided for this purpose.

Many of the newer colour-graphics systems offer background colour as an option. Background colour is an invaluable aid in the creation of graphs and charts, and it can also be useful as a means of highlighting headings or individual items of information which have some special significance. The use of coloured backgrounds for large quantities of alphanumeric information is not recommended however. The resultant glare from the screen is likely to be very tiring, particularly if there is perceptible flicker in the image.

Where lettering is superimposed on a coloured background, the important factor for legibility is that the lettering and background should differ as much as possible in luminance as well as in hue. Combinations such as yellow and cyan are very difficult to read. Red and green would of course be a disastrous combination for those with a red/green colour vision deficiency. The best light-on-dark combinations are probably white or yellow on blue, white on red, or white on magenta. Among the most legible and pleasing dark-on-light combinations will be blue, red or magenta (and possibly green) on white, or blue on yellow (red and magenta are too close to yellow in hue).

3.2 Colour as a means of coding information structure

3.2.1 **Colour as a grouping cue**

Colour is a very strong grouping cue, operating on the principle of similarity. Display items having the same colour will automatically be perceived as being related in some way, and different from items in other colours. It is therefore essential that colour should be used logically and consistently, or non-existent relationships or similarities between items of information may be inferred by the user.

For greatest effect, colour must be used sparingly. Research has shown that if three or four colours are used in a target detection task on a VDU, items of the same colour tend to form a 'gestalt' and are easily distinguished from the rest of the display. If more colours are used, however, the information becomes too fragmented and the gestalt is lost (8).

It is important to realise that colour should not be used to compensate for a poor layout. A monochrome version of the same information should make perfect sense. The colour should merely reinforce the existing visual coding.

3.2.2 **Colour for emphasis**

Colour can also be used very effectively as a means of indicating different levels of emphasis. The luminance hierarchy described above seems to result in a corresponding perceptual hierarchy. The brightest colours appear to dominate the display. It

Fig. 1. An illogical use of colour and a fragmented layout.

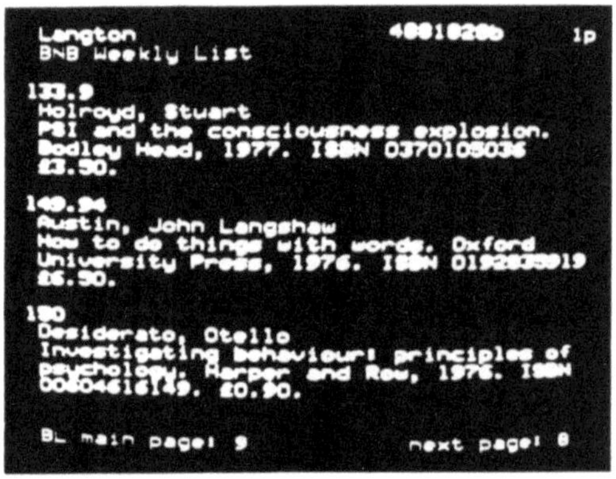

Fig. 2. Use of colour according to the luminance hierarchy and a simpler layout give the information a visual structure that is immediately obvious.

would therefore seem logical to use the bright colours for information requiring the greatest emphasis, in the same way as bold or larger type size might be used in print. The perceived order of importance of the four most legible colours is likely to be white first, followed by yellow, cyan and green. Thus white or yellow might be used for headings and cyan or green for the main body of data.

The sense of order that comes from using colour in this hierarchical way is very noticeable. Figure 1 shows an example of an experimental bibliographic display with the colours assigned somewhat randomly. Figure 2 shows the same information but with the classmark in white, the author in yellow and the remainder of the information in cyan. This reflects the relative importance of these items for a user scanning the data.

If, on the other hand, the intention is to make qualitative distinctions between items without implying an order of importance, then it is best to choose colours that are close together in the hierarchy, such as yellow and cyan, or cyan and green.

3.2.3 **Colour coding**

With alphanumeric data, colour is best used to code the structure of the information rather than its content. That is to say, it should be used, as suggested above, as a means of creating hierarchies of emphasis and of grouping related information. The use of colour to code content (for example, by using different colours for data from different clinical tests) is generally less successful for several reasons. The different luminances of the colours will suggest the existence of a hierarchy, thus giving prominence to some information at the expense of the remainder. If this hierarchy is not intended, confusion will result.

Coding by association is often particularly unwise. The data producer may be tempted to use less legible colours (e. g. red for data relating to arterial blood, blue for venous blood), and his or her perceived colour associations may not correspond with those of the data user. A much better solution would be to label each column clearly in the same dominant colour, with all the data in a single less dominant colour.

The associative use of colour may occasionally be justified however. The use of red to indicate danger is an obvious example. It must be remembered though that red is not a very legible colour on many displays. In this case it may be unwise to use red for the actual data, but it may be worth experimenting with a red flag of some kind (an arrow for example).

If, in spite of the above, there are strong reasons for colour coding alphanumeric information on the basis of its content, it is important to keep the number of colours to a minimum or there may be problems with colour identification and discrimination. Identification can be a problem if the complete set of colours is not available for comparison on any given display. There will probably be little difficulty with the seven colours described here, but with systems offering a wider range of colours, users may easily become confused. Even with a small number of easily identifiable colours, there is still the risk that users may forget the significance of a given colour.

An additional problem is that of simultaneous colour contrast. A large area of one colour may distort perception of small areas of other colours nearby. The distortion is always towards the complementary of the colour occupying the larger area. White in particular may be affected. Next to blue it may take on a yellowish hue, and if there is no yellow on the screen for comparison, confusion can result. Next to magenta, white may look greenish; next to red, blue-greenish; next to green, pinkish.

Ideally, therefore, colour coding should always be redundant. It should merely reinforce a message given elsewhere in words. This will also avoid problems for users with defective colour vision.

3.2.4 **Flashing**

The use of flashing characters is a very effective way of attracting the user's attention. So much so that it can be extremely distracting and irritating if used frequently. If this device is employed, then users should have the option of stopping the flashing once their attention has been attracted. This will allow them to read the message more easily. Ideally the flashing character should be a flag of some kind next to the data to which attention must be drawn, rather than the data themselves. A flashing red arrow, for example, might be used to draw attention to a dangerously high value. Red conventionally symbolises danger, but is not very prominent on a dark screen. The flashing will compensate for the low luminance.

4 Presentation of Tabular Information

4.1 The structure of tables

The ease and speed with which tables can be understood depends very much on the way they are structured. Wright (9) suggests that those producing tables must ask themselves what information readers will already have when they consult a table, and what information they will be seeking from it. The row and column headings should relate to the information users already have, thus leading them to the information they seek in the body of the table.

It is also important to present the information in a sequence that will be compatible with the user's likely reading strategy (10). This usually operates from left to right and from top to bottom.

Ideally, any items that need to be compared with one another should be listed vertically, particularly in the case of numerical values. Thus, if several variables are being monitored over time, the variables should form the column headings and the times the row headings. In this way the values for any given variable over time can easily be compared. This must be borne in mind if there is any question of swapping the rows and columns to make the table fit the available space more easily.

If tables are transferred from a line printer output to an 80-column screen, or from an 80-column screen to a 40-column screen, the structure is likely to need re-thinking. Complex tables may well need to be split into several sub-tables which the user can then access via an initial index frame. Norrish (11) points out that this exercise requires a clear understanding of the structure of the information, together with a knowledge of the kinds of information that users may want to combine at any one time, and of the strategies that they are likely to adopt in searching for the information they want. The information provider or designer must also have come to terms with the visual potential and limitations of the new display. If the transfer is well done, users may find that several small tables are easier and quicker to use than one large one.

4.2 Layout in tables

A vertical, columnar structure is usually the most striking visual characteristic of tables, but it is often only the first column of data that is read vertically. Having selected an entry in that column, users will then wish to scan across the table horizontally. Vertical scanning is carried out merely to check the column heading.

In this situation, data within each vertical column are related in kind, but items in a given row have a functional relationship with the item in the first column. It is this relationship that interests the reader and that must be emphasised in this kind of table.

In two-way matrices, where the reader must select a row heading and a column heading and follow both to locate an item in the body of the table, both horizontal and vertical scanning are of course important. Vertical scanning rarely presents any difficulty however, so once again it is horizontal emphasis that is required.

To help with reading across, columns should have just enough space between them to separate them clearly, but no more. They should never be spread out to fill the space available. This can result in a very real danger of users making

	Group one	Group two	Group three	sex
Class A	1234	14	986	♂
	4567	6	431	♀
Class B	9849	84	18	♂
	8321	100	1000	♀
Class C	4679	8	864	♂
	2131	2	9	♀

no!

		GROUP		
Class		I	II	III
A	♂	1234	14	986
	♀	4567	6	431
B	♂	9849	84	18
	♀	8321	100	1000
C	♂	4679	8	864
	♀	2131	2	9

?

		GROUP		
Class		I	II	III
A	♂	1234	14	986
	♀	4567	6	431
B	♂	9849	84	18
	♀	8321	100	1000
C	♂	4679	8	864
	♀	2131	2	9

yes

Fig. 3. Three versions of the same data using different systems of ruled lines. (Reproduced from Reynolds and Simmonds (19) with the permission of the publisher.)

mistakes in reading across between columns. Research on indexes presented on microfilm has shown that wide column spacing reduces the speed and accuracy with which they can be used (12).

Positive assistance can be given to the user by allowing relatively generous space between rows. Thin horizontal rules are often helpful, particularly where the amount of vertical space available for a table is limited or where space alone does not give sufficient emphasis. Strong vertical rules should be avoided as these will merely draw the eyes down the columns instead of across. Similarly, heavy boxes should also be avoided as they will interfere with scanning. Figure 3 shows how much clearer and less cluttered a table can be when only light horizontal rules are used.

In some situations, large gaps between columns are unavoidable because of variations in the length of entries. It is especially important here to allow adequate space between rows. In long tables it can be very helpful to leave additional space after, say, every fifth row. This is not recommended for small tables, however, as the reader may try to attach some significance to the resultant grouping of rows. Leader dots can also be very helpful. In experiments on microfilm indexes they improved speed and accuracy of searching with wide column spacings, even though subjects said they disliked the look of the dots (12). In some cases it may be worth considering changing the order of the columns so that the very variable length information occurs on the right and the less variable information on the left. Hartley (13) has shown that left-ranging tables are certainly no less easy to read than centred tables, and they are simpler to produce (Figure 4).

Row and column headings should be carefully positioned so that it is immediately clear which data they refer to. Long column headings are often responsible for

Distribution of places of meeting of spouses

Place of meeting	%
Dance or dance hall	27.3
Private house	17.6
Work or Forces	14.6
Street or public transport	9.7
Cafe or pub	6.1

Distribution of places of meeting of spouses

%	Place of meeting
27.3	Dance or dance hall
17.6	Private house
14.6	Work or Forces
9.7	Street or public transport
6.1	Cafe or pub

Fig. 4. In the upper version of this table the information has been arranged about a central axis in an attempt to produce a 'balanced' appearance. In the lower version the information has been arranged for maximum ease of use. (After Hartley (13)).

unnecessarily wide spacing between columns. If a long heading can be abbreviated without too much loss of meaning, this may be the better option. A key should be provided if there is any possibility of users not understanding the abbreviation. In large tables it may also be worth considering repeating the column and row headings at the bottom and right-hand side of the table respectively.

4.3 Colour in tables

If carefully used, colour can be very helpful in improving the presentation of clinical data in tables. If it is misused, however, it can be a positive hindrance. Legibility and comprehensibility may both suffer.

Fig. 5

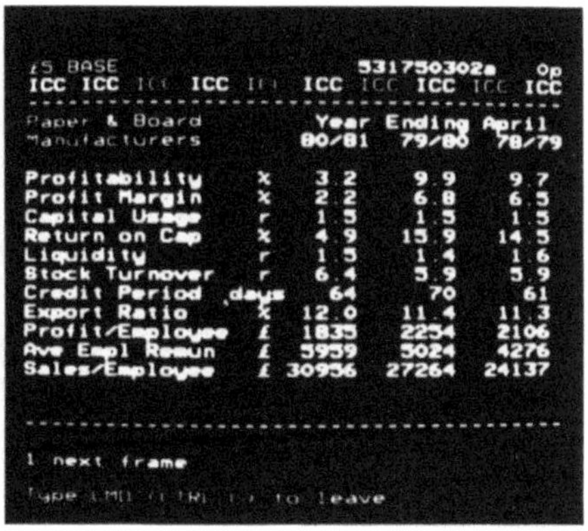

Fig. 6

Fig. 5. The use of a different colour for each column of information draws the eye down the table instead of across.

Fig. 6. The use of colour to reflect the structure of the information (see text).

It is rarely helpful to present different colums of information in different colours, simply because they contain different kinds of information. This will merely draw the eye down the table in the same way as vertical rules. In Figure 5, the three columns of page numbers are in different colours. The reader wishes to link the page numbers horizontally with the subjects, not to compare the page numbers in any given column. The three colours (red, white and cyan) are not helpful, therefore, and their different luminances suggest differences in importance. It is also worth noting that the columns are much farther apart than is necessary, hence the need for the leader dots. The arrangement shown in Figure 6 is much more satisfactory. Here white is used for the row and column headings and cyan for the data in the body of the table. The colour therefore relates to the structure of the table rather than to its content.

Figure 7 is another interesting example of the vertical use of colour. Here there are two sets of columns; the company names and changes in price are in yellow and the share prices themselves in cyan. Although cyan is slightly lower in the

Fig. 7 Fig. 8

Fig. 7. If colour is not used carefully, figure/ground separation may result in inappropriate emphasis (see text).

Fig. 8. The use of two different colours on alternate lines obscures the structure of this table.

luminance hierarchy than yellow, it is the share prices rather than the company names that attract attention first. This is because of a perceptual phenomenon known as figure/ground separation. If we are to make sense of our surroundings we must decide which areas in our visual field we should regard as objects (or 'figures'), and which areas we should regard as background (or 'ground'). We have a tendency to perceive small enclosed areas as objects and the remainder as background. In this example, there is less cyan than yellow, and the two columns of cyan are surrounded by yellow. The prices are therefore seen as 'figure', and the remainder of the table as 'ground'. It would be more in keeping with the structure of the table if the two columns of company names were in one colour (preferably the brighter colour), and the remainder of the table in the other colour. The company names would then become the 'figure' and attention would be drawn to them first. Thus, with colours close together in the luminance hierarchy it is likely to be the figure/ground phenomenon that determines which information is seen as being most important. With greater differences in luminance, however, the brighter colour may still dominate even if it does occupy a larger area.

Figure 8 shows another inappropriate use of colour, this time in a horizontal direction. Here white and yellow are used on alternate lines. The intention was to use colour to link each place name with its one-word description and temperature value, but there is in any case no problem in reading across between the columns because they are so close together. The result is very confusing, because the structure of the table is obscured. Once again the best solution would be to give the two columns of place names in one colour and the remaining information in another.

Tables with alternating lines of colour have been compared with single-colour versions in experiments at the University of Reading (14). Green and yellow striped tables, such as that shown in Figure 9, were tested against yellow-only and green-only versions. A slight advantage was found in favour of the plain versions. Once again, because the alternating colour did not relate to the structure of the table, it seemed to destroy its visual coherence.

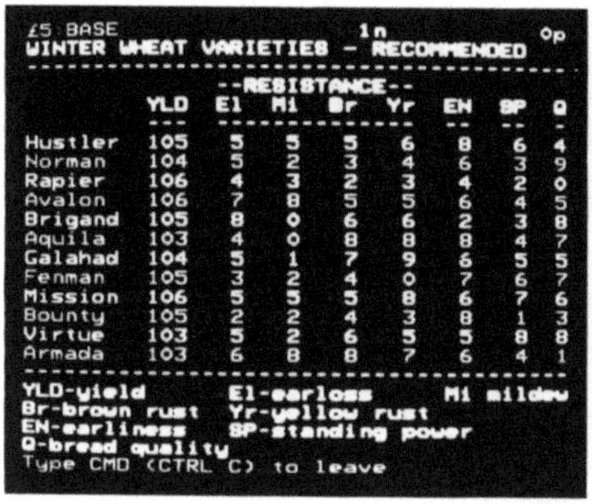

£5 BASE 1n 0p

WINTER WHEAT VARIETIES - RECOMMENDED

	YLD	El	Mi	Br	Yr	EN	SP	Q
		--RESISTANCE--						
Hustler	105	5	5	5	6	8	6	4
Norman	104	5	2	3	4	6	3	9
Rapier	106	4	3	2	3	4	2	0
Avalon	106	7	8	5	5	6	4	5
Brigand	105	8	0	6	6	2	3	8
Aquila	103	4	0	8	8	8	4	7
Galahad	104	5	1	7	9	6	5	5
Fenman	105	3	2	4	0	7	6	7
Mission	106	5	5	5	8	6	7	6
Bounty	105	2	2	4	3	8	1	3
Virtue	103	5	2	6	5	5	8	8
Armada	103	6	8	8	7	6	4	1

YLD-yield El-earloss Mi mildew
Br-brown rust Yr-yellow rust
EN-earliness SP-standing power
Q-bread quality
Type CMD (CTRL C) to leave

Fig. 9

224 ORACLE 224 Thu21 May ITV 1104 44
LATEST PRICES
3/4

	WEDS CLOSE PRICE	PRICE AT 10.00		WEDS CLOSE PRICE	PRICE AT 10.00
ICI	284	288	NatWest	368	368
Imp Gas	215	213	Ocean T	131½	131½
Imperial	71½	71½	Pearson	222	222
IMI	65	65½	Phoenix	254	254
Inchcape	458	453	Pilkntn	296	296
Land Sec	403	405	Plessey	276	279
L & Gen	218	218	P & O	132	132
Lloyds	340	340	Pru	218	218
Lonhro	92	92	Racal	364	364
Lucas	213	213	Rank	184	185
Mk & Sp	135	135	Rk Hovis	51	51
MEPC	225	225	Reckitt	250	250
Metal Box	200	200	Redland	170	173
Midland	310	310	Reed Int	243	244
Mthrcare	236	234	RTZ	538	540

More follow shortly

Fig. 10

Fig. 9. A table with alternate lines in green and yellow, as tested by Norrish (14).

Fig. 10. Inappropriate use of both space and colour disguises the structure of this information.

In Section 4.2 it was suggested that although it can sometimes be helpful to leave a space after every fifth line in a long table, this is not advisable with short tables. Figure 10 illustrates the problem. Here we have groups of six, five and four lines, to which we immediately try to attach some significance. The grouping is strongly reinforced by the use of green for the upper and lower blocks and yellow for the central block. There is, of course, no significance whatever in the grouping, and on first inspection the strong horizontal emphasis completely obscures the fact that there are two columns of company names and not just one. The sensible solution, once again, would be to use a brighter colour for the two columns of company names and a second colour for the lists of prices. The line spaces serve no useful purpose and should be omitted.

In Figure 11, there is only one set of columns and not two, so the use of yellow and green for alternate blocks of three lines is not so confusing. The equal size of the blocks means that users are less likely to try to attribute some non-existent

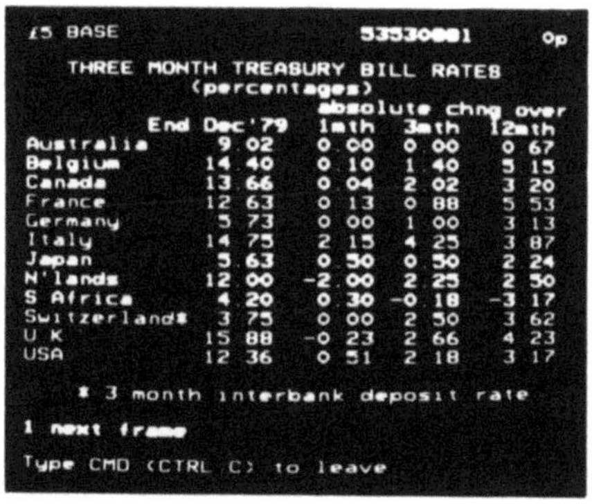

£5 BASE 53530001 0p

THREE MONTH TREASURY BILL RATES
(percentages)

	End Dec'79	absolute chng over 1mth	3mth	12mth
Australia	9.02	0.00	0.00	0.67
Belgium	14.40	0.10	1.40	5.15
Canada	13.66	0.04	2.02	3.20
France	12.63	0.13	0.88	5.53
Germany	5.73	0.00	1.00	3.13
Italy	14.75	2.15	4.25	3.87
Japan	5.63	0.50	0.50	2.24
N'lands	12.00	-2.00	2.25	2.50
S Africa	4.20	0.30	-0.18	-3.17
Switzerland*	3.75	0.00	2.50	3.62
U K	15.88	-0.23	2.66	4.23
USA	12.36	0.51	2.18	3.17

* 3 month interbank deposit rate

1 next frame

Type CMD (CTRL C) to leave

Fig. 11. Here yellow and green are used for alternate blocks of three lines. This probably neither confuses nor assists the user.

significance to the grouping. The use of colour in this way has the same effect as the use of space, while leaving more room for the data themselves. One could still question, however, whether this use of colour is necessary as there is no real problem in reading across between columns. Yet again, the use of a bright colour for row and column headings and a second colour for the data would possibly be preferable.

Colour in tables, then, is generally best used to emphasise the structure of the table. In most cases this means using colour to distinguish the column and/or row headings from the body of the table. Wherever possible, the colour should merely reinforce what is already a satisfactory layout. It should not be used to compensate for an unsuitable layout that has not been properly thought through.

The use of colour to emphasise particular pieces of information is an exception to the principle of using it to convey structure rather than content. For example, a second colour might be used to emphasise all values above some critical level, or to draw attention to values that have changed since some previous tabulation. Red is often assumed to be a good colour to use in this kind of situation because of its association with danger and the need to pay attention. Red works well on paper but, as we have seen, on screen it is unlikely to be as succesful. If the red is of relatively low luminance, any data in red will be less conspicuous and less legible than data in brighter colours. The best answer then may be to present the bulk of the data in, say, green, and any critical values in white. In some systems, however, it is possible to adjust the colours and a brighter and more acceptable red can be created.

5 Graphic Displays

5.1 A comparison with tabular formats

On gaining access to computer graphics facilities it is tempting to use them at every opportunity, simply because they are there. This is a trap that must be avoided. If tabulated data have been adequate in the past, it is important to consider not only what is to be gained by presenting these data in graphical form, but also what might be lost.

Most authorities agree that charts and graphs are particularly useful for revealing broad trends over time, and for showing other qualitative aspects of data. Here they have an advantage over tabular formats. Charts and graphs are not so useful for precise quantitative information because it is difficult to read off values accurately. For precision, tables are better.

Tufte (15) makes the point that tables are usually superior to graphics in reporting on small data sets of twenty numbers or less. Graphics, on the other hand, are especially powerful with large data sets where trends would be difficult to visualise from numerical data alone. Ehrenberg (16) suggests that while charts and graphs work well with relatively simple information, they are not so successful with complex information where several variables are involved. For a chart or graph to communicate succesfully, the message must be simple. If it is too simple it runs the risk of becoming trivial, so it must communicate something new, Ehrenberg argues. He suggests that graphs are:

- helpful for getting the feel of new data – is the relationship a straight line or a curve, is there much scatter?
- useful for presenting findings to others who are not familiar with the data.

In effect he is saying that charts and graphs are most useful for communicating trends to the 'ignorant' – ourselves or others. For detailed analysis of the data, tables are more effective. There is, of course, no contradiction in this argument. If the facilities are available it may be helpful to have graphic displays in addition to tables, but for different purposes.

5.2 The impact of computer graphics

Charts and graphs are very difficult to produce satisfactorily on a computer line printer because of the lack of precision available for the positioning of data points, scale calibrations and the various labels necessary; lines linking data points are rarely possible on graphs. With graphics facilities, however, the situation is very different. Acceptable graphs and charts can be produced very rapidly. Unfortunately the standard chart and graph displays produced by some software are not as well designed as they might be, but with the more sophisticated packages the user can change any undesirable features. As a general rule, the simpler the presentation the better. As Tufte (15) points out, the emphasis should always be on the data, not on showing off the drawing ability of the computer.

5.3 Charts versus graphs

Wright (9) suggests that bar charts may be preferable to line graphs where the data can be subdivided in a number of different ways and where the reader is required to make static comparisons. Line graphs would seem to be superior where dynamic comparisons of data plotted against time are required. Schutz (17) found that where extrapolation of a trend was required, line graphs were better then vertical bar charts and vertical bar charts were better than horizontal bar charts. Wright (9) cites a study which suggests that a pie chart may be superior to a bar chart where three or four comparisons are being made, but for accuracy a table would be superior to both.

5.4 Pie charts

Most graphics programs include pie charts in their repertoire. Their main application is for static comparisons, especially where a total amount can be broken down in various different ways. The main disadvantage of pie charts is that they rely on proportional areas to denote relative quantities. As Petterson (18) points out, we are much better able to make comparisons of length than comparisons of area. It is therefore important always to give the percentage value for each segment on a pie chart, and this figure should ideally be given in the segment itself rather than in a separate key. Pie charts are therefore unsuitable if a number of very small divisions of the whole is required. They are also unsuitable for comparisons over time. For example, it would be unreasonable to present readers with a series of pie charts, perhaps one each for a series of years, and to expect them to be able to make accurate comparisons of the size of corresponding segments across the years. Bar charts or graphs would be much more suitable for this purpose.

5.5 Bar charts

5.5.1 **Structure**

The first consideration in designing a bar chart should be to ensure that the data are structured appropriately. As Wright (9) has pointed out, where there are several variables involved it will be possible to plot the data in several different ways. This is illustrated in Figure 12. The best way of presenting the data will depend on the users' information requirements. In this example the first version is best for comparing the total annual expenditures, the second version is best for comparing expenditure on the same item in different years, and the third version is best for comparing expenditure on different items in the same year.

5.5.2 **Detail**

Once the overall structure of the chart has been decided upon, the details of the design can be considered. The keyword here must be simplicity. Tufte (15), in his fascinating and very well illustrated book on charts and graphs, describes the concepts of 'data ink' and 'chartjunk'. Of the total ink used in printing a chart or graph, data ink is the proportion that actually carries information. Non-data ink, or redundant data ink, he says, often takes the form of chartjunk. Chartjunk is Tufte's word for 'graphical decoration' or unnecessary embellishment. The higher the proportion of data ink and the less chartjunk there is, the more effective the graph or chart will be. These concepts are obviously directly applicable to computer graphics as well as to printed materials.

Following on from this, it is not a good idea to embellish bar charts by depicting the bars as three-dimensional 'towers' drawn in perspective. The third dimension usually conveys no additional information and is therefore redundant. It can also make an accurate comparison of the lengths of the bars quite difficult.

The way in which scales are shown should also be carefully considered. For accuracy in reading off values it is often helpful to extend the vertical scale markers as shown in Figure 13. As well as increasing accuracy, the arrangement shown in version B reduces the clutter on the vertical axis and emphasises the white data bars. Any such grid lines should always be thinner or otherwise less prominent than the lines delimiting the data bars. Where the data bars are shown as solids, an even better solution is to dispense with the grid lines drawn in on version B and to indicate them merely as thin breaks in the solid data bars.

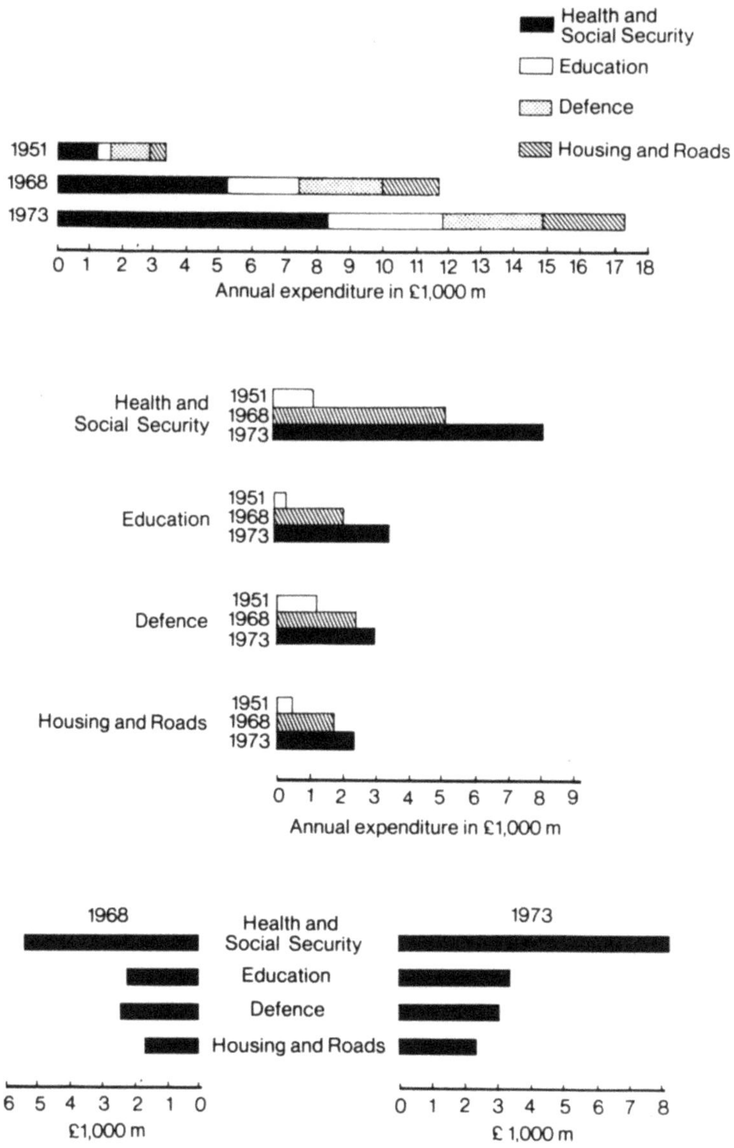

Fig. 12. The best way of organising a bar chart will depend on the users' requirements. (After Wright (9), reproduced from Reynolds and Simmonds (19) with the permission of the publisher.)

Many graphics systems offer a wide range of shadings, but more than one shading should not be used on a bar chart without good reason. If the different bars represent the same thing varying over time, they should be shown in the same way. Only if the bars, or sections of the bars, represent different things should they be differently shaded. In choosing a set of shadings to be used on adjacent bars, it

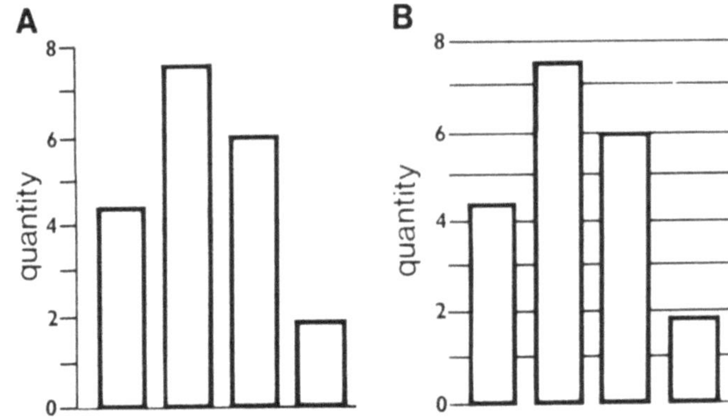

Fig. 13. The height of each bar can be more easily assessed in B. This version is also more aesthetically pleasing because the scale provides a background against which the white bars stand out more easily than in A. (Reproduced from Reynolds and Simmonds (19) with the permission of the publisher.)

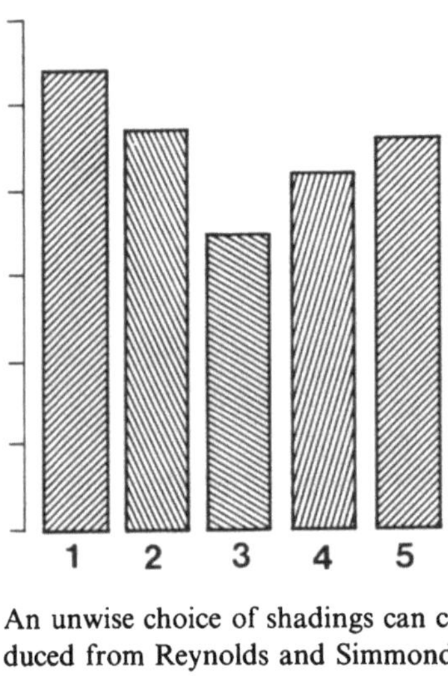

Fig. 14. An unwise choice of shadings can create marked optical distortions in the bars. (Reproduced from Reynolds and Simmonds (19) with the permission of the publisher.)

is important to avoid the kind of optical distortion that is created by the shadings in Figure 14. If space permits, direct labelling of the bars is always preferable to the use of a key.

5.5.3 **Use of colour**

Colour can be very useful as a means of coding bars or bar segments, though if the colours are not fairly close together in the luminance hierarchy, differences in importance may be implied. With small luminance differences, however, the colour coding may be very confusing for users with colour vision defects. If there is space to label the bars, reinforcing the colour coding, then there will be no problem. Figure 15 shows an example of colour coded bars (in red and yellow) comparing the performance of two unit trusts over time. As this example shows, fairly satisfactory bar charts can be prepared even on relatively low-resolution systems such as Prestel (British Telecom's viewdata system). There would of course be a problem in reading off the heights of the bars with any precision, but in this case a tabular format was also available. An additional point of interest here is that with this particular display the user can choose any two sets of data from a menu offering alternatives. He or she can then go on to change one or other of them at a time. This kind of flexibility is of course one of the great advantages that the computer has to offer in presenting data, especially where there are too many variables to be shown together on a chart at one time.

Colour is also valuable as a means of drawing attention to data points that exceed a certain value. It has been used in this way in Figure 16, where yellow is used below the critical value and cyan above it. Red, a relatively dark colour, has sensibly been used for the axis lines so that the scale values and data bars stand out clearly.

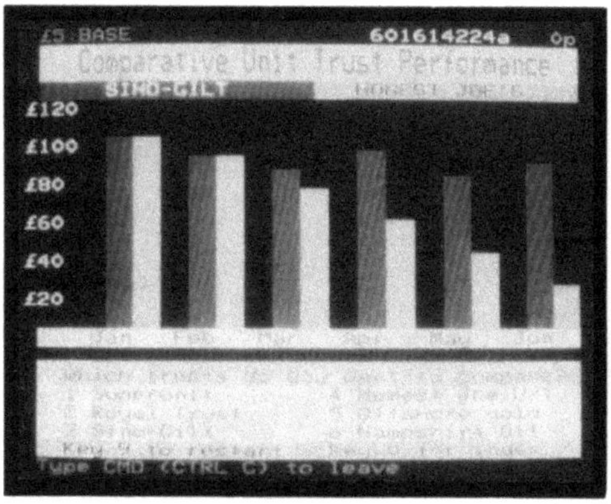

Fig. 15

Fig. 16

Fig. 15. An interactive bar chart on Prestel (see text).

Fig. 16. Use of colour to draw attention to data points exceeding a certain value.

5.6 Graphs

5.6.1 Structure and detail

For a discussion of the general principles involved in constructing graphs that are easily comprehensible and not misleading in any way, the reader is referred to Reynolds and Simmonds (19) and to Tufte (15).

The data points are the most important feature on any graph since any lines linking them are nearly always a matter of conjecture. Data points should therefore be made to stand out well. Data points and any linking lines should both have more visual emphasis than the axes or any grid lines. As suggested for bar charts,

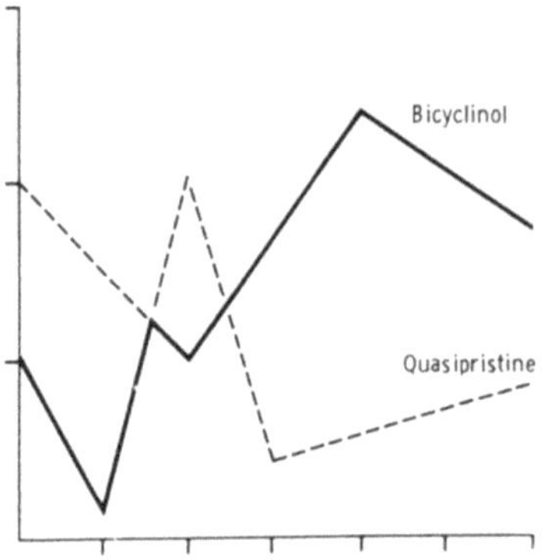

Direct labelling leads to fast comprehension

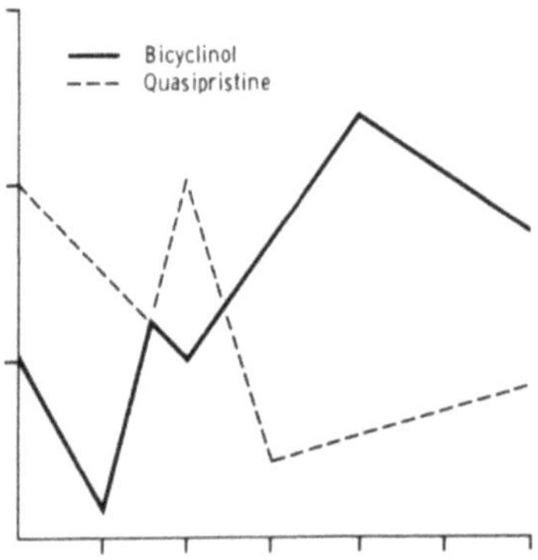

Indirect labelling requires double scanning, and comprehension is therefore slower. If, for some reason, a key is needed, then it is best shown within the graph area, as in this example.

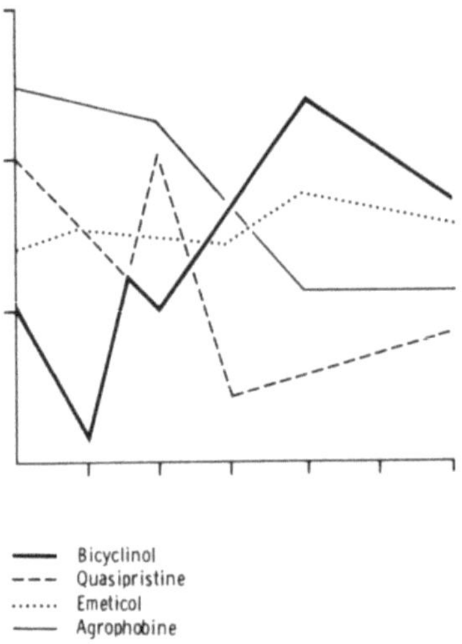

An undesirable situation may not be easy to avoid. In a complex graph, where space is limited, a separate key may become necessary.

Fig. 17. Direct and indirect labelling of graph lines. (Reproduced from Reynolds and Simmonds (19) with the permission of the publisher.)

with some graphs it may be helpful to fill in the area under the curve as a solid and show the grid lines on the screen as thin black lines running through the solid area.

If possible, all labels should read horizontally, particularly if users are working from a screen display. Unfortunately, lack of space may sometimes mean that a vertical label is unavoidable on the vertical axis. Direct labelling of graph lines should be used wherever possible. Keys make more work for the reader. With multiple-line graphs, however, the use of a key may be unavoidable, as shown in Figure 17.

To display line graphs adequately, a relatively high resolution system will be needed. Figure 18 shows a graph on Prestel, which has a resolution of only 80 × 144 pixels in the graphics mode. A graph such as this can show data only in a relatively crude form. Tabulated data were available as a back-up in this case.

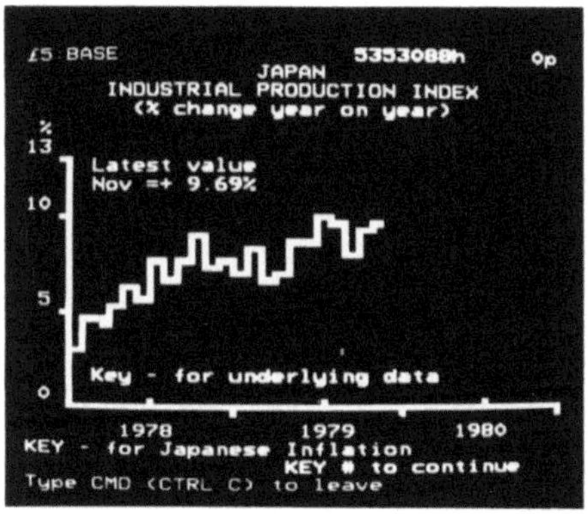

Fig. 18

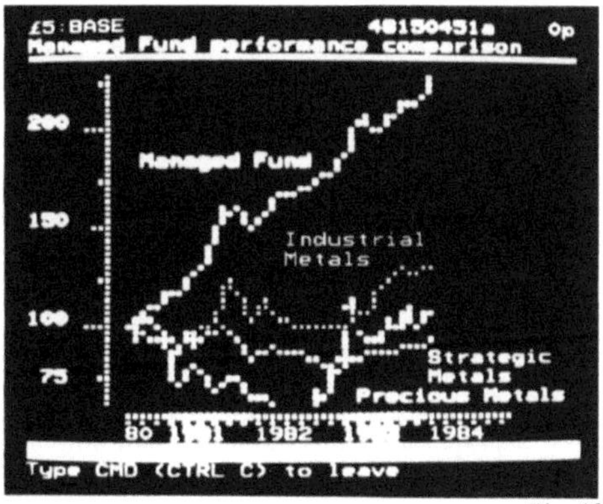

Fig. 19

Fig. 18. A simple line graph on Prestel.

Fig. 19. A multi-line graph on Prestel. The lines and their associated labels are colour coded.

5.6.2 **Use of colour**

Colour can be very helpful in displaying information in graphs, though the problems of those with defective colour vision must once again be borne in mind.

Colour is likely to be particularly useful with multi-line graphs. Unless lines are adequately coded and crossing points clearly indicated, monochrome multi-line graphs can be very ambiguous. Even on Prestel, colour makes a multi-line graph feasible, as shown in Figure 19. Here the lines are directly labelled, with the colour of the label matching the colour of the line. There is then no ambiguity if the positioning of the labels is difficult through lack of space. With adequate space available, however, the lines would have more impact if the axes, scale calibrations, and all labels were all in the same less dominant colour. With better resolution one would of course expect to see the actual data points clearly distinguished and joined by smooth lines (preferably stopping just short of the data points to give them more emphasis). Line breaks at crossing points would also be helpful, especially

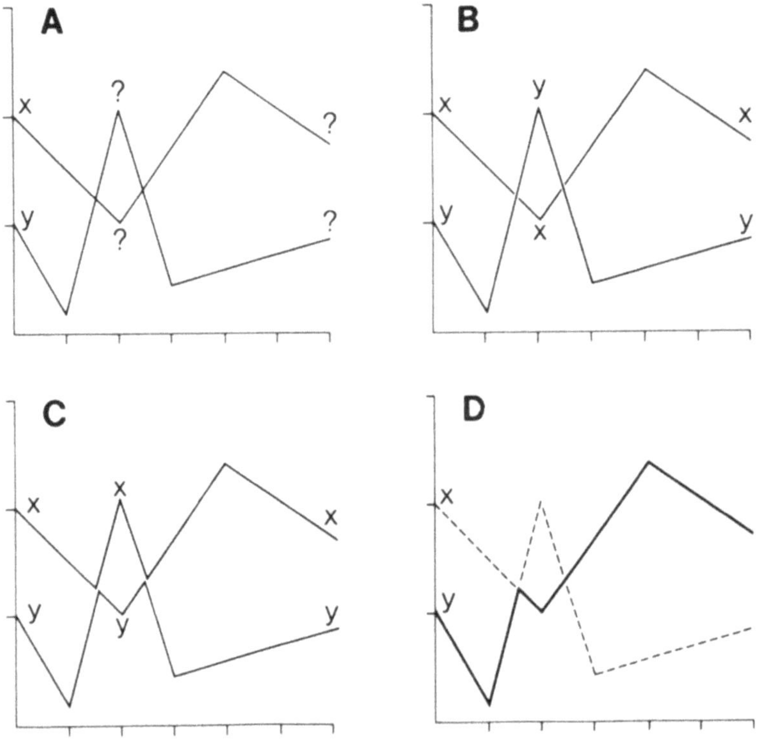

Fig. 20. Lines without a hierarchical order cause confusion. Graph A could be interpreted as in graphs B or C. Graph D is not ambiguous because the lines are coded and they have not been allowed to touch one another. (Reproduced from Reynolds and Simmonds (19) with the permission of the publishers.)

for those with colour vision problems. The difficulty that can occur in interpreting cross-over points is illustrated in Figure 20.

While colour can be helpful for multi-line graphs, it should not be used to create graphics of the kind shown in Figure 21. It is unscientific to try to plot differently measured variables on the same axes. The best solution in print would be as shown in Figure 22. In situations where the depth of the screen is not sufficient for this kind of display, there is always the possibility of using the system interactively and comparing variables two at a time.

To sum up, the ideal use of colour will be in a graphics system with good resolution where it can be used merely to reinforce a graphic display that would be unambiguous if displayed in monochrome. The use of colour to help compensate for low resolution or lack of space is usually much less satisfactory. If the display is used in monochrome, printed out in one colour only, or used by someone with colour vision problems, then any ambiguities will be immediately obvious. Finally, it is important to keep the number of colours used to a minimum and to use them in a logical way. Using colour for the sake of it will obscure the truth rather than helping to reveal it.

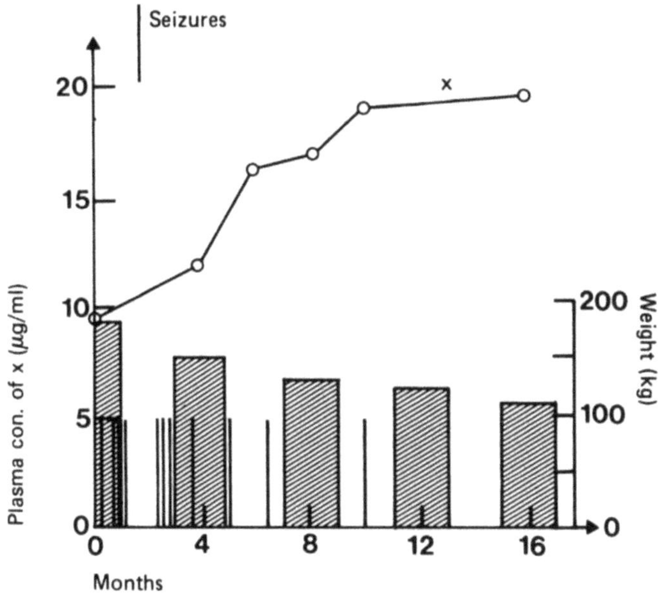

Fig. 21. A badly designed graph, based on examples from the scientific press. Faults: inward facing scale calibrations; the label on the right hand scale reads downwards; it is not clear from the x-axis label what is being measured in months; the key for 'seizures' is too far from the area where seizures occur; 'con' on the y-axis is not a standard abbreviation; arrows at the ends of axes are unnecessary, as is the double zero at the intersection of the two axes. (Reproduced from Reynolds and Simmonds (19) with the permission of the publishers.)

6 Conclusions

Until relatively recently, most computer users expected and accepted that computer-produced data displays, whether tables or graphs, would be visually crude. With the lower cost and greater variety of colour-graphics systems now available, this need no longer be the case. Many graphics systems are capable of producing images that approach the quality of a hand-drawn and conventionally printed chart or graph. The possibility of using colour and interactive techniques gives additional advantages that cannot be matched by the printed image. In theory, then, we have the tools for presenting data more clearly and comprehensibly than ever before. But the existence of these tools is not in itself sufficient. It is essential that we learn how to use them properly.

As with the presentation of any kind of information in any medium, the first requirements are a knowledge of:

- the kind of information that users need
- their preferred ways of accessing it
- the limitations and potential of the display medium or system.

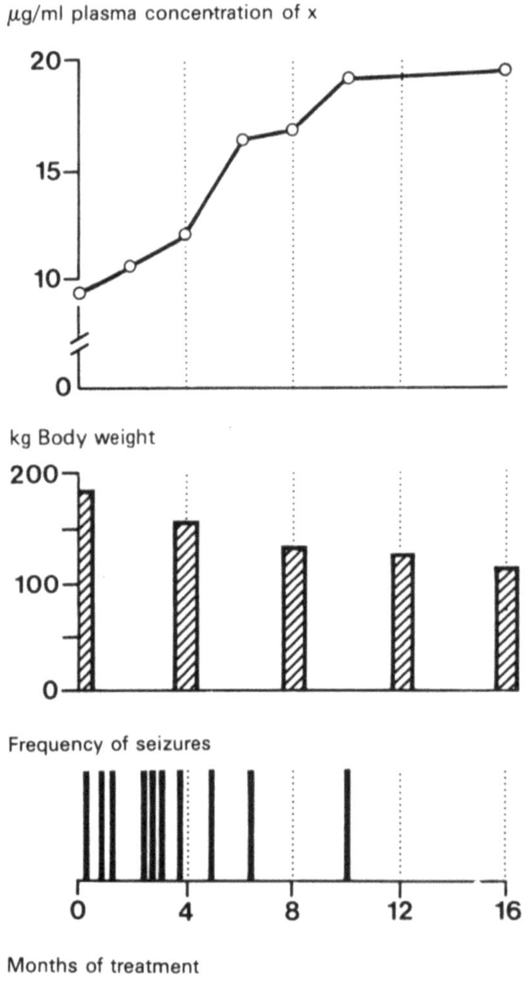

Fig. 22. Here each of the three sets of data in Figure 21 has been provided with its own vertical scale. The three graphs are linked through the common horizontal scale. (Reproduced from Reynolds and Simmonds (19) with the permission of the publishers.)

The answers to these questions will determine the overall structure of the display. Within any given display, every mark should have a purpose and there should be a good reason why it appears where it does and in the form it does. With complex data there is no room in the display for unnecessary embellishment. It will only distract and confuse. Designing information of this kind requires rigorous analytical thinking. The aesthetic qualities of the result will be achieved through the design of a display that fulfils its function cleanly and simply, not through the use of elaborate detail or colour for no reason other than because they are available.

References

1. Spencer, H. (1969) The Visible Word: Problems of Legibility. Lund Humphries, London.
2. Reynolds, L. (1979) Legibility studies: their relevance to present-day documentation methods. Journal of Documentation *35*, 307–340.
3. Easterby, R. & Zwaga, H. (eds.) (1984) Information Design. John Wiley and Sons Ltd, Chichester, New York.
4. Scrivener, S. (1983) Perceptual factors in the design of information displays. In: Computer Graphics '83, pp. 85–100. Online International, Pinner, UK.
5. Tinker, M. (1963) Legibility of Print. Iowa State University Press, Ames.
6. Graham, C. H. (1965) Vision and Visual Perception. John Wiley and Sons Ltd, New York.
7. Ton, W. H. (1969) Optimal visual characteristics for large screen displays. Information Display *6*, 48–52.
8. Cahill, M. C. & Carter, R. C. (1976) Colour code size for searching displays of different density. Human Factors *18*, 273–280.
9. Wright, P. (1977) Presenting technical information: a survey of research findings. Instructional Science *6*, 93–134.
10. Frase, L. T. (1969) Tabular and diagrammatic presentation of verbal materials. Perceptual and Motor Skills *29*, 320–322.
11. Norrish, P. (1984) Moving tables from paper to screen. Visible Language *18*, 154–170.
12. Reynolds, L. & Spencer, H. (1979) Two Experiments on the Layout of Information on Computer Output Microfilm. Graphic Information Research Unit, Royal College of Art, London.
13. Hartley, J. (1978) Designing Instructional Text. Kogan Page, London.
14. Norrish, P. (1987) The Graphic Translatability of Text. British Library Research and Development Report 5854. Department of Typography and Graphic Communication, University of Reading.
15. Tufte, E. R. (1983) The Visual Display of Quantitative Information. Graphics Press, Cheshire, Connecticut.
16. Ehrenberg, A. S. C. (1975) Data Reduction: Analysing and Interpreting Statistical Data. John Wiley and Sons, London, New York.
17. Schutz, H. G. (1961) An evaluation of formats for graphic trend displays. Human Factors *3*, 99–107.
18. Petterson, R. (1983) Visuals for Instruction. Institutionen för ADB, Stockholms Universitet, Stockholm.
19. Reynolds, L. & Simmonds, D. (1982) Presentation of Data in Science. Martinus Nijhoff Publishers, The Hague, Boston, London.

3 Interpretation of Laboratory Results

Chapter 3.1
The Bayes Approach: Systematic Graphic Evaluation of Diagnostic Tests

W. Gerhardt

1 Terminology

The regrettable lack of an international nomenclature in the field of test evaluation has led to a bewildering number of different concepts, symbols, and names. In order to understand current test reports a dictionary of synonyms and definitions may be required (Appendix, tables 1 – 3).

Table 1. List of current names (incomplete). Terminology and definitions used in this paper are specified in the first two columns.

Working terminology	Definition	Current names	Ref.
2-class test:	test with only two. classes of test results: neg and pos or 0 and 1.	Qualitative test, dichotomous test, binary test	1 2 3
3-class test:	Test with three test-classes: neg, inconclusive, pos or 0, 1, and 2.	None	–
Multi-class test:	Test with n classes of test results	Quantitative test	1,2,3
Discriminator:	Boundary between two adjoining test classes	Discrimination value, Cut-off value, Reference value, Operating point	4 5 6 7
DIS	Test persons classified as having the disease to be detected by the test	Diseased persons	–
REF	Test persons classified as not having the disease to be detected by the test	Reference persons	–
Pos	Test persons allocated to the positive test class	Test positive persons	–
Neg	Test persons allocated to the negative test class	Test negative persons	–

In contrast, the following descriptions will preferentially be based on simple, mnemonic abbreviations and some of the most commonly used conventional names (Tables 1 and 2, table 3 in Appendix).

Test evaluations are frequently reported in probability notation like: $P(D|Si)$ = probability of disease given the test result, Si (1, 3, 5, 7, 8). However, the term "test evaluation" implies a description of test performance *after* the test has been investigated in a clinical study under specified conditions. "It is meaningless to predicate the probability of a completed event" (ref. 8, p. 223). When a clinical study has been carried out these quantities are no longer probabilities but number fractions. They may be calculated as X/Y, i.e. the number of X divided by Y. Like prevalence they vary between zero and one (8, 9).

In this chapter, therefore, the name "prior probability" is replaced by "fraction diseased persons in the study group", or "study fraction DIS". Likewise, "predictive value of a positive result (PVpos, refs. 1, 2, 4) is the same as "the fraction test-positive diseased among all persons allocated to the positive test class (4, 9).

The concept of probability only becomes relevant when discussing the probable performance of the test at a future application under, hopefully, similar conditions as in the evaluation study.

2 Function of a Diagnostic Test

2.1 Study groups

A diagnostic test functions as a sorting device. It separates – more or less efficiently – a number of diseased persons, *DIS*, classified as having a certain disease, from a number of reference persons, *REF* (Table 1), classified as not having that disease. A proper clinical study requires that these classifications are based on other criteria than the test in question (selection of study groups see sections 5 and 6, ref. 9).

2.2 Anterior prevalence of disease

The fraction DIS in the mixture of DIS and REF in the study group before the test is called *anterior or prior prevalence of disease* (1, 4, 6, 9) dictionary of concepts and synonyms in Table 3 in the appendix).

Figure 1 illustrates that a test allocates a number of test persons from a study group with one anterior prevalence of disease to a number of test classes, each with another posterior prevalence (= class fraction) of disease. The fraction DIS within each test class is highly dependent on the anterior prevalence of DIS.

Figures 1 a and b show one and the same 3-class test applied to two different prevalences of DIS, 0.50 and 0.10, with unchanged discriminator positions. The fractions DIS within each test class obviously decreased with decreased anterior prevalence of DIS.

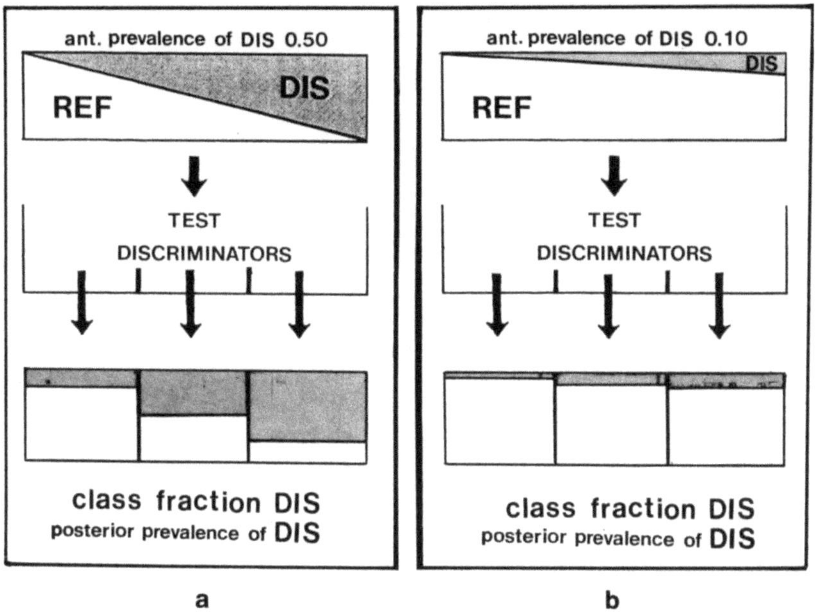

Fig. 1. Schematic presentation of one and the same three-class test applied at two different anterior prevalences of disease: 0.50 (a) and 0.10 (b). The test allocated all persons in the study group to three test classes. Each test class contained a new test-class fraction (or posterior prevalence) of diseased and a complementary fraction reference persons.
(b) Compared to a, the lower anterior prevalence of DIS, 0.10, has decreased the class fractions DIS, and increased the class fractions REF (from Scand. J. Clin. Lab. Invest. Suppl. 181, 1986 (ref. 9), with kind permission).

The test classes are separated by discriminators (ref. 4, Table 1). In the process of evaluating a quantitative test the evaluator may select any number of discriminator positions. These need not be equidistantly spaced, cf. the three-class test in Figure 5a. Dependent on whether one, two, or several discriminators are selected the test will be evaluated as a two-class, three-class, or multi-class test (ref. 9, Table 1).

2.3 2 × 2 tables

Routine tests are still generally evaluated and used as two-class tests allocating diseased and reference persons to a negative and a positive test class, respectively. The four basic test quantities may be named *DIS*, *REF*, *NEG* and *POS* (Tables 1 and 2, Fig. 2). The four possible classifications by the test (see below) are most rationally inserted into a 2 × 2 table (Fig. 2).

The lay-out of such tables is analogous to that of the double histograms presented later in this chapter (9).

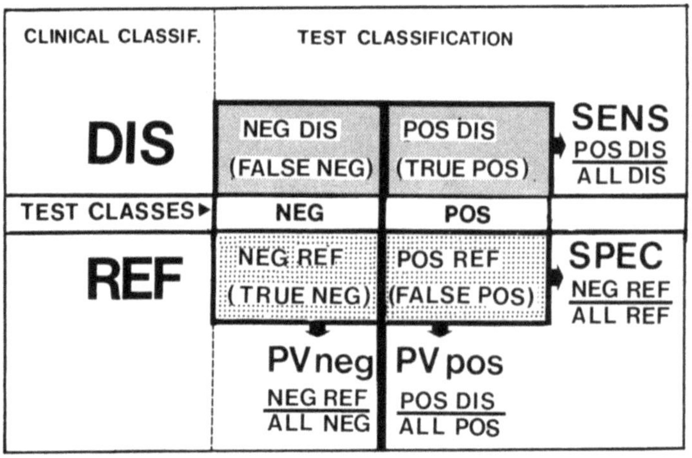

Fig. 2. Schematic presentation of a two-class test evaluation matrix. The lay-out is the same as that of the following test histograms.
Upper field: all persons classified as having the disease in question (DIS).
Lower field: all persons classified as not having that disease (REF).
The test has allocated all DIS and REF to either the NEG or the POS test class. The four basic test quantities are called: NEG DIS, POS DIS, NEG REF, and POS REF. The four derived test evaluation quantities are:
Horizontally, upper field: fraction POS DIS among ALL DIS = sensitivity.
Horizontally, lower field: fraction NEG REF among ALL REF = specificity.
Vertically, NEG test class: fraction NEG REF among ALL NEG = PVneg.
Vertically, POS test class: fraction POS DIS among ALL POS = PVpos.
When the study has been carried and a discriminator position selected, the four basic test quantities are known and Bayes equation is not required for calculation of the derived test evaluation quantities (from Scand. J. Clin. Lab. Invest. Suppl. 181, 1986 (ref. 9), with kind permission).

Upper field: Contains all persons in the study who have been classified by independent means as having the disease in question at the relevant stage and prevalence of that disease. These are designated by the abbreviation DIS.

Lower field: Contains all reference persons in the study who have been classified by independent means as not having the disease in question. They frequently have other diseases and symptoms leading to clinical suspicion of the disease in question, being the differential diagnostically relevant group for a realistic study of test performance.

2.4 Discriminations

The test sorts diseased and reference persons into a number of test classes separated by *discriminators* (definitions in Table 1). In Figure 2 the vertical bar represents one discriminator.

Left vertical column: Contains all persons allocated to the negative test class.

Right vertical column: Contains all persons allocated to the positive test class.

The resulting four cells in the 2 × 2 table may now be unambiguously defined as *NEG DIS*, *POS DIS*, (upper field) and *NEG REF*, and *POS REF* (lower field, Fig. 2).

Test performance may now be described by four derived quantities with the conventional names: *sensitivity*, *specificity*, and *predictive values of a negative or a positive result* (1, 2, 4, 6). All of these are *number fractions* and vary between zero and one (8, 9) (synonyms: Tables 2 and 3, Appendix).

In Figure 2,

Horizontally: *Sensitivity* = the fraction POS DIS among ALL DIS (upper field). *Specificity* = the fraction NEG REF among ALL REF (lower field).
Sensitivity and specificity indicate the fractions of DIS and REF, respectively, that are correctly classified by the test. It is obvious that both sensitivity and specificity are *independent* of the anterior prevalence of disease.

Vertically: The test has allocated all test persons to two test classes, a negative one and a positive one. Each class contains a fraction DIS. This may be called the *class fraction DIS* or *Class Prevalence of Disease*, abbreviated *CPD* (9). The complementary class fraction of REF = class prevalence of REFpersons, *CPR* (9). In a two-class test with only one negative and one positive test class CPD is the same as PVpos, and CPR is the same as PVneg (ref. 9, table 3).

Being number fractions anterior and posterior prevalences vary between zero and one (1, 2, 4, 9).

The *Predictive value of a positive result*, *PVpos* or *CPD*, (9, 10) is the fraction POS DIS within the positive test class.

The *Predictive value of a negative result*, *PVneg* or *CPR* (ref. 9, Table 2) is the fraction NEG REF within the negative test class. Within one and the same test class CPD and CPR are complementary values: CPD = 1 − CPR.

Figure 2 demonstrates that the traditional Bayesian equations (Appendix) are not required for these calculations.

PVneg and PVpos indicate the risks of disease (DIS) and not-disease (REF) within the positive and negative test classes, respectively. PVneg and PVpos are highly dependent on the anterior prevalence of disease (cf. also Fig. 1).

The choices of number of test classes and of discriminator positions should preferable be made in a dialogue with the clinicians involved and be guided by the wanted clinical information. This again depends on the intended future application of the test to a defined clinical situation or within a diagnostic strategy (see further sections 5–8). Accordingly, test data for evaluation of a clinical study should be presented in such a way that the test evaluator and clinician together may follow all changes of the dependent quantities such as *sensitivity*, *specificity*, CPD, *predictive values* and (not advised), *efficiency* (see Appendix) as functions of discriminator positions; the latter two concepts even as functions of anterior prevalence of disease. This may be accomplished by double histograms with superimposed curves of these quantities as shown in the following.

A graphic survey of these working concepts is given in Table 2.

Table 2. Quantities in test evaluations:
Survey over working concepts, definitions, and names. (Modified from Scand. J. Clin. Lab. Invest. Suppl. 181, 1986 (rel. 9) with kind permission).

Schematic representation	Kind of quantity number fraction	Trivial name
	Fraction diseased in the study F DIS(ALL) = n DIS / ALL	Prevalence of diseased persons in the study "fraction DIS in study"
	Fraction (test-)positives among the diseased F POS DIS(ALL DIS) = n POS DIS/ALL DIS	Sensitivity "fraction POS among ALL DIS"
	Fraction (test-)negatives among the REF persons F NEG REF(ALL REF) = n NEG REF/ALL REF	Specificity "fraction NEG among ALL REF"
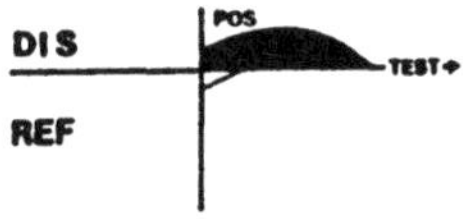	Fraction diseased among all (test-)positives F POS DIS(ALL POS) = n POS DIS/ALL POS	PVpos "fraction DIS among ALL POS"
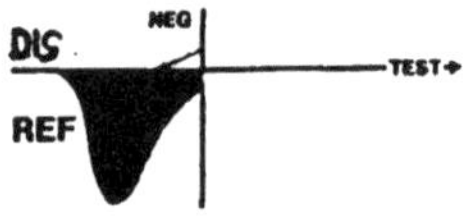	Fraction REF persons among all (test-)negatives F NEG REF(ALL NEG) = n NEG REF/ALL NEG	PVneg "fraction REF among ALL NEG"
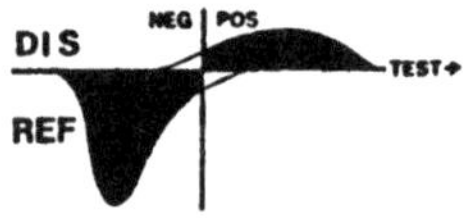	Fraction correct classi- fications among all (test-)classifications F correct. classif. (ALL) = (n POS DIS + n NEG REF)/ALL	Efficiency "fraction correct classifications among all classifications"

3 Systematic Graphic Evaluation of a Diagnostic Test: S-Myoglobin at Suspected AMI

3.1 Samples drawn 8 – 12 hours after symptoms

The double histograms Figures 3 – 5 present stepwise evaluation of test data from a clinical study. As an example, S-myoglobin was determined by a RIA-method in 291 men, all admitted to a coronary care unit on suspicion of an acute myocardial infarction (AMI). The patients were classified into 130 AMI and 161 NOT-AMI (REF) by WHO criteria, not using S-myoglobin data for the classification. The prevalence of AMI in the study was 0.45 (10).

The following histograms were all designed in the same way:

Upper field: distribution of AMI patients. The numbers of AMI, shown on top of each test class, are plotted as a function of test value (abscissa).

Lower field: distribution of REF-persons. All were suspected of an AMI. Most had chest pain. Consequently, the distribution shows a tail of relatively high S-myoglobin values from skeletal muscle damage.

In contrast to a traditional reference group of e. g. healthy blood donors, this reference material may be used in a realistic estimate of the function of the test under the conditions it is intended for (9).

Superimposed curves: The values of the dependent variables: sensitivity, specificity, PVpos, PVneg, (and efficiency) may be plotted as functions of discriminator position. The fraction AMI within each test class, CPD, has been plotted and the points connected to a continuous CPD curve. All these variables vary between zero and one. Consequently, any intersect between one of these and any discriminator value may be read directly off one and the same scale on the right ordinate, cf. Figure 4 a.

The plotting program has placed the abscissas in the individual Figures 3 – 5 at different heights for optimal use of paper space. This does not interfere with the reading of the value of any point on the curves from an intersect with the right ordinate.

On a graphic computer such histograms may be used as working tools. Let our clinician request an evaluation of S-myoglobin in discriminating between AMI and NOT-AMI sampled in the time interval 8 – 12 hours after onset of acute symptoms at the AMI prevalence in the coronary care unit. Let a high sensitivity of at least 0.95 be wanted.

Figures 3 – 5 demonstrate a stepwise, systematic evaluation of the test. Observe, that in the following discriminator selection and evaluation of test performance is based on the empirical distributions of actual test data. No assumptions of mathematically defined distributions need to be made.

Figure 3 a: First a survey plot was made of the distributions of AMI and REF at a class-width of 200 µg/l and 40 test classes within the range 0 – 8000 µg/l. This showed both distributions to be skewed. There were only few AMI cases above 2000 µg/l and few REFpersons above 600 µg/l. Consequently the next step was to study the region that appeared relevant for possible discriminator positions.

Figure 3 b: This is a closer study of the distributions at a smaller class width of 50 µg/l in the range 0 – 2000 µg/l. Sensitivity, specificity, and CPD curves were plotted in order to serve as guides for discriminator selection. The curves demonstrate the inverse relationship between sensitivity and specificity. *The choice of a discriminator position is thus always a compromise*. In this case a high sensitivity of about 0.95 would be obtained by selecting a discriminator position between 100 and 150 µg/l. The corresponding specificity is about 0.70.

The large fluctuations of the CPD curve were due to the relatively small test class width selected, cf. the plot at class width 125 µg/l in Figure 4 b, below.

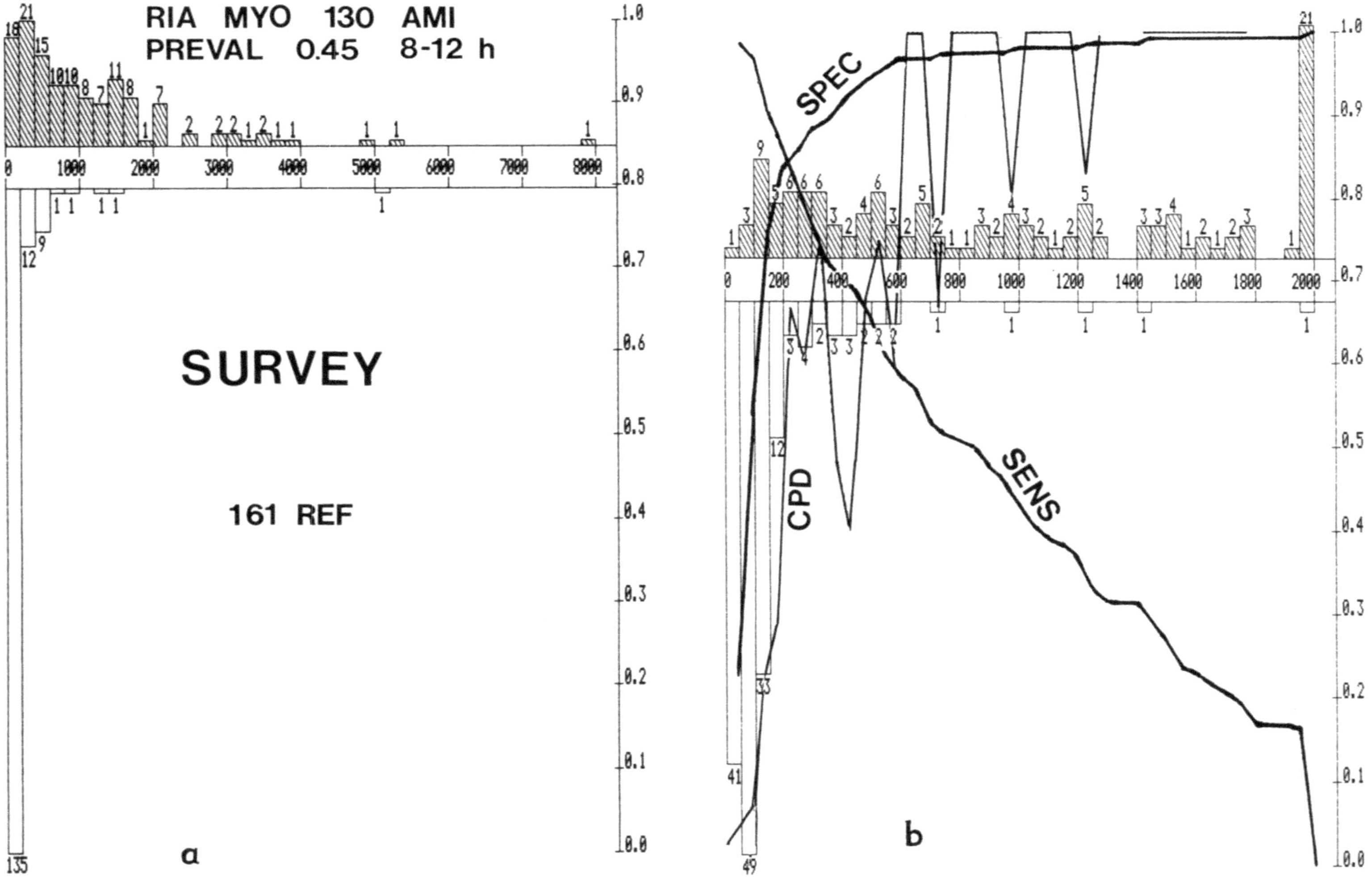

RIA MYO 130 AMI
PREVAL 0.45 8-12 h
SURVEY
161 REF
a
SPEC
SENS
CPD
b

Some preliminary conclusions could be drawn already at this stage; the CPD curve indicated that the risk of an AMI in this study group increased with increasing S-myoglobin values as follows:

1. Values below about 150 µg/l excluded an AMI with a class prevalence of reference persons (CPR) about 0.90.

2. S-myoglobin values in the range 150 – 500 µg/l showed CPD curves with great fluctuations around 0.5 indicating the existence of an intermediate, inconclusive test class, cf. Figure 4 b.

3. A possible third test class with S-myoglobin values above about 500 µg/l contained a mixture of about 90% AMI and 10% REFpersons.

Consequently, the next step was to study these informations in detail.

Figure 4 a: Presents an expanded view of the relevant region of possible discriminators plotted at a still smaller test class width of 25 µg/l. A sensitivity of 0.95 required a discriminator position at 125 µg/l (follow the arrows). The corresponding specificity was about 0.70 (arrows).

Evaluated as a *two-class test* with this discriminator S-myoglobin detected 95% of all AMI at a PVpos around 0.70, and excluded about 70% of all NOT-AMI at a PVneg around 0.95, cf. Figure 5 b.

However, a closer study of the plot indicated that the distributions in the intermediate range 125 µg/l to about 500 µg/l corresponded to a CPD (= PVpos) of about 0.50 instead of 0.70.

Figure 4 b: The study was now replotted for a survey at multiples of the discriminator value, 125 µg/l. At this class width the fluctuations of the CPD curve were considerably reduced and confirmed that a two-class evaluation would actually represent an *oversimplification* of test performance. The courses of the CPD curves indicated that a three-class test evaluation would be more correct. Note, that the PV curves cannot provide such information, cf. Figure 7.

The study was replotted accordingly (Fig. 5 a).

Figure 5 a: The study plotted as a *three-class test* with discriminator 1 at 125 µg/l and discriminator 2 at 500 µg/l. Test performance may now be summarized as follows:

Fig. 3. (a) Double histogram presenting the distributions of 291 men and women suspected of having an acute myocardial infarction (AMI) plotted as a function of S-myoglobin values. Prevalence of AMI 0.45. Samples were drawn within 8 – 12 hours after onset of acute symptoms.
Abscissa: S-myoglobin, 40 test classes, class-width 200 µg/l. The numbers of AMI and REFpersons, respectively, are printed above and below each test class.
Upper field: Distribution of 130 AMI patients.
Lower field: Distribution of 161 REFpersons, all suspected of an AMI.
(b) Same study as in (a) replotted at 40 test classes, class-width 50 µg/l. The derived test quantities: sensitivity and specificity have been plotted as functions of discriminator position. The fraction AMI within each test class = class prevalence of disease (CPD) has been calculated and the points connected to a CPD curve. The values of any of these quantities may be read for any discriminator position by extending a horizontal line from the intersect of that discriminator and the curve to an intersect with the right ordinate, cf. Figure 4a.

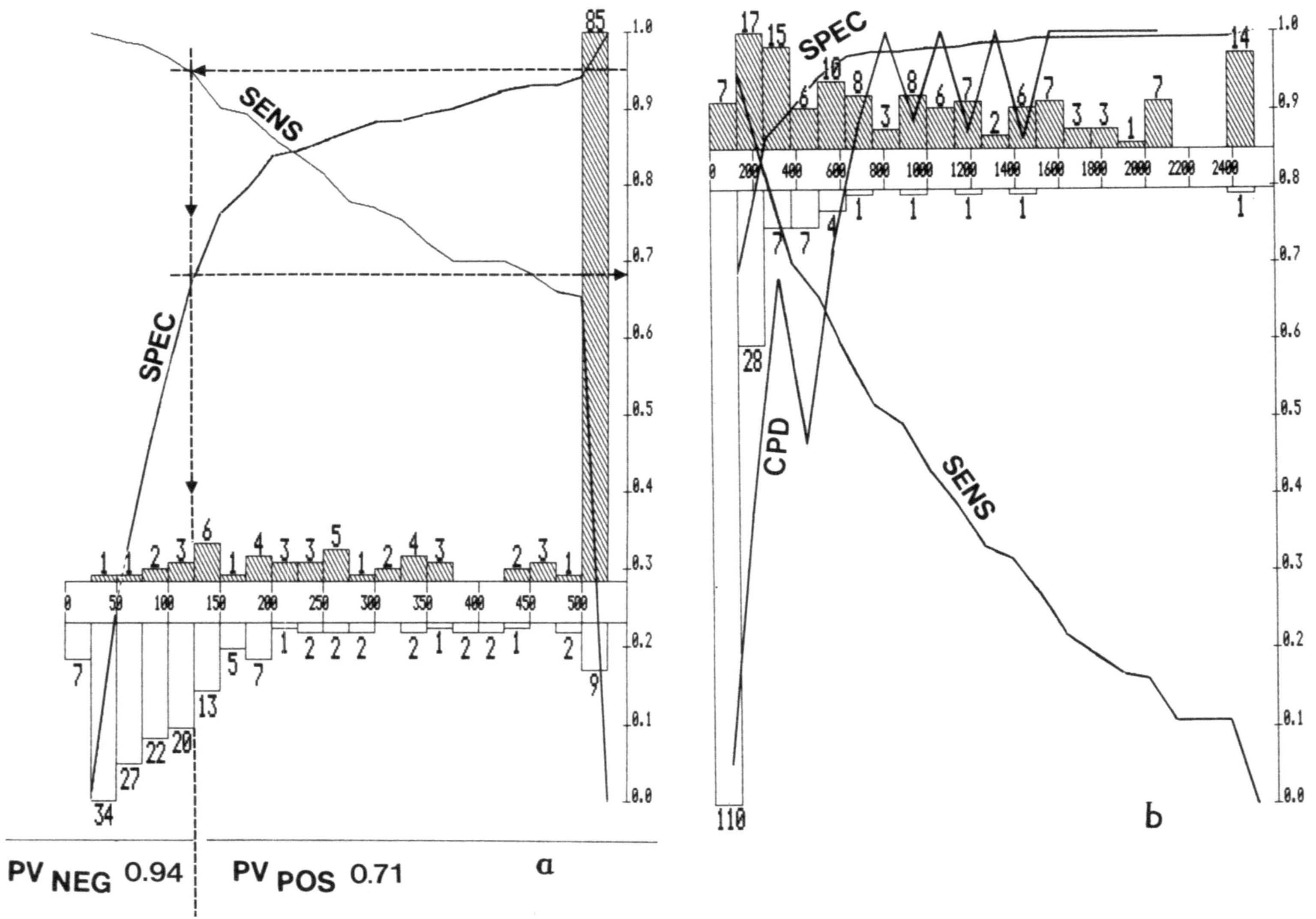
SENS
SPEC
CPD
PV NEG 0.94
PV POS 0.71
a
b

1. Test class one contained 5% of all AMI, 68% of all REF, and excluded an AMI with a CPR (=PVneg) around 0.94.

2. Test class two contained 29% of all AMI, 26% of all REF, and was inconclusive with a CPD around 0.48.

3. Test class three verified AMI in about 65% of all AMI cases with a CPD (=PVpos) around 0.90.

Figure 5 b: For comparison, the test was plotted as a conventional *two-class test*. The data confirmed the conclusions from Figure 2 a, and may be summarized:

1. As a two-class test with one discriminator at 125 μg/l the positive test class of 8–12 hour S-myoglobin detected about 95% of all AMI with a PVpos around 0.71.

2. Values below 125 μg/l excluded about 70% of all NOT-AMI with a PVneg around 0.94.

Both the two-class and the three-class test evaluations confirmed the preliminary conclusions from the previous plots that S-myoglobin sampled 8–12 hours after symptoms was a better tool for *exclusion* than for *verification* of an AMI. The uncertainty implied in such estimates may, of course, be calculated by conventional statistical methods. This is beyond the scope of this chapter.

The test evaluation above was based on a primary clinical request for about 95% sensitivity – at the cost of specificity.

Obviously, a similar evaluation might have been carried out based on a primary clinical request for a high specificity instead. A discriminator position around 600 μg/l would give a 97% specificity (which is the same as the conventional 97th percentile) and a corresponding sensitivity of about 42% (Fig. 4 b).

3.2 Samples drawn 4–8 hours after symptoms

S-myoglobin has been reported to increase already within about 4 hours after an AMI (11). Let the clinician request an evaluation of S-myoglobin for *very early* confirmation or exclusion of an AMI.

Fig. 4. (a) Same study as in Figure 3 replotted at 21 test classes, class-width 25 μg/l. All data over 500 μg/l have been cumulated in the last test class.
The discriminator that gives a sensitivity of 0.95 is found by extending a horizontal line from 0.95 on the right ordinate to an intersect with the sensitivity curve and another from this point vertically downwards to an intersect with the abscissa and reading the value (125 μg/l).
The corresponding specificity is found by extending a horizontal line from the intersect between the vertical discriminator and the specificity curve toward the right ordinate and reading the value at that intersect (0.68).
PVneg and PVpos have been calculated for the discriminator position 125 μg/l.
(b) Same study replotted at 20 classes, class-width = the discriminator value 125 μg/l. All data over 2375 μg/l have been cumulated in the last test class. Sensitivity, specificity, and CPD curves have been superimposed on the histogram. Note, that the fluctuations of the CPD curve have been greatly reduced by the larger class-width as compared to Figure 3 b. Details, see text.

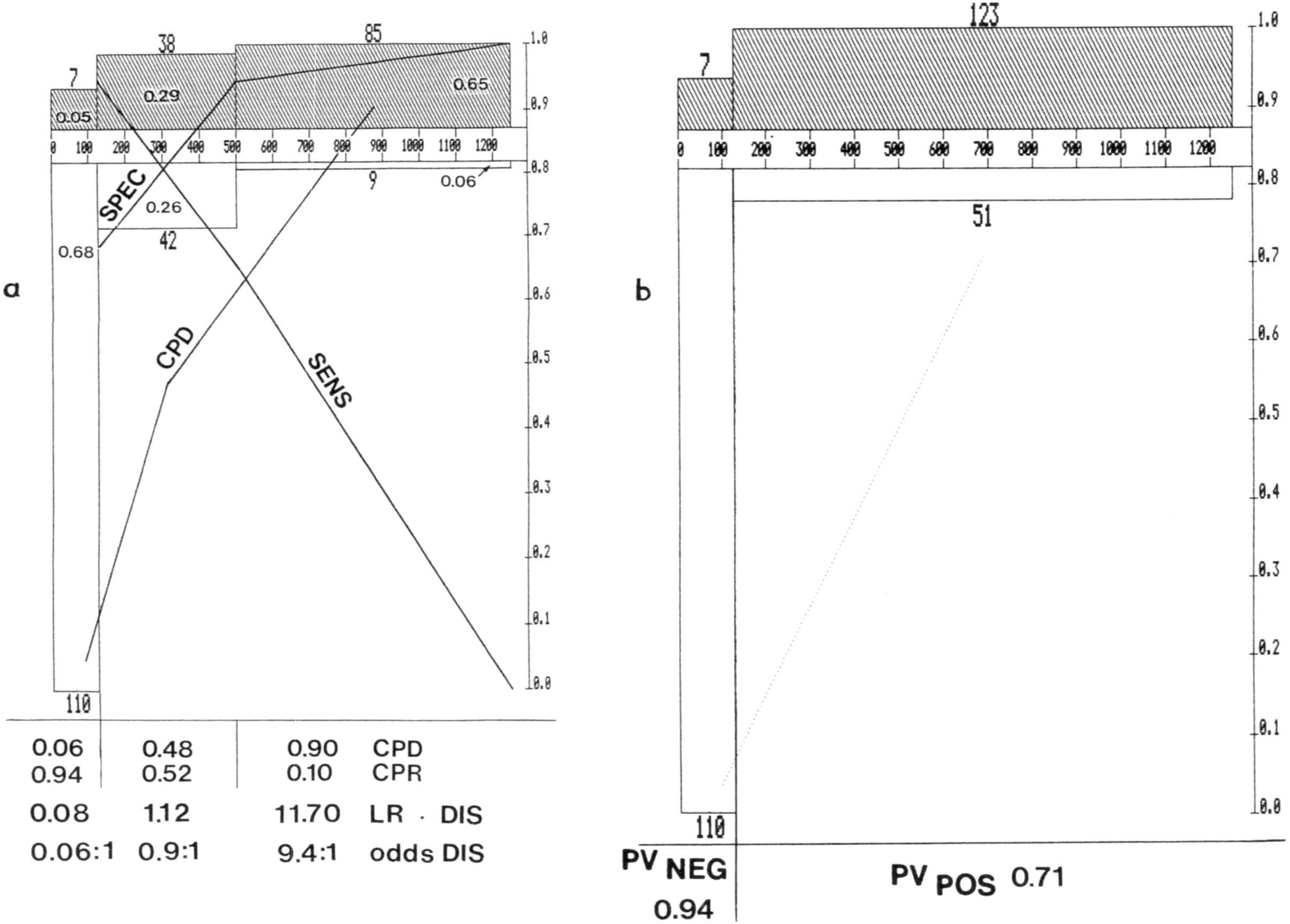

0.06	0.48	0.90	CPD
0.94	0.52	0.10	CPR
0.08	1.12	11.70	LR · DIS
0.06:1	0.9:1	9.4:1	odds DIS

PV_{NEG} 0.94 PV_{POS} 0.71

Figure 6 presents a study of the same patient material but with sampling 4–8 hours after onset of acute symptoms. To obtain a similar sensitivity as in the 8–12 hour study the discriminator value must be lowered to 75 µg/l with a corresponding loss in specificity to about 0.40, and a decrease of PVpos to about 0.60. The necessity for decreasing the discriminator value to order to retain a high sensitivity may be immediately understood by comparing the two plots, Figure 4a and Figure 6.

Comparing Figure 4a to Figure 6 it is apparent that S-myoglobin in the AMI patients 4–8 hours after onset of acute symptoms is still remaining at relatively low values. Consequently, a lower discriminator at 4–8 hours is required to obtain the same sensitivity as with the 8–12 hour samples. The histograms with the superimposed test quantities have been used as tools both for selecting clinicallly relevant discriminator positions and a comparison of the performances of one and the same test in two study groups. Below is discussed the predictive values, PVpos and PVneg, at different anterior prevalences of AMI. The reader may then draw his own conclusions regarding performance of 4–8 hour and 8–12 hour S-myoglobin in the coronary unit.

4 Anterior Prevalence, CPD, and Predictive Values

Hitherto, the problem of selecting a clinically meaningful discriminator position has been discussed mainly from the viewpoints of sensitivity and specificity. However, for routine use of the test, the clinician also wants an estimate of the risks of disease and not-disease given by the results of the test. These are conventionally expressed by "predictive values", PVpos and PVneg, which are restricted to two-class tests (2, 9). In contrast, test class fractions (or class prevalence) of DIS and

Fig. 5. (a) Same study as in previous figures, replotted as a three-class test evaluation with discriminators at 125 µg/l and 500 µg/l. As before, the figures above and below each test class show the numbers of AMI and REF within these classes.
The fraction AMI (CPD) and the fraction REF (CPR) among ALL within each test class are shown at the bottom of the figure.
(For the discussion of likelihood and odds concepts see appendix). The figures within each test class in the upper field show the fraction AMI of ALL AMI that have been allocated to that class. This is the same as the: (test class) – likelihood of AMI.
Likewise, the figures within each test class in the lower field show the fraction REF of ALL REF that were allocated to that class: (test class) – likelihood of REF.
Likelihood ratio (LR) of AMI for test class two, for example, then becomes 0.29/0.26 = 1.12. (LR) is an odds multiplication factor:
ant. odds DIS 0.81 : 1 × LR 1.12 = post. odds DIS 0.91 : 1).
This can be calculated directly from the graph: 38 DIS : 42 REF, reducible to 0.91 : 1. (All figures rounded for didactic purposes.)
(b) Same study replotted as a two-class test evaluation with one discriminator at 125 µg/l. The CPD line, PVneg and PVpos are shown.
CPR in test class 1 = PVneg, CPD in test class two = PVpos.

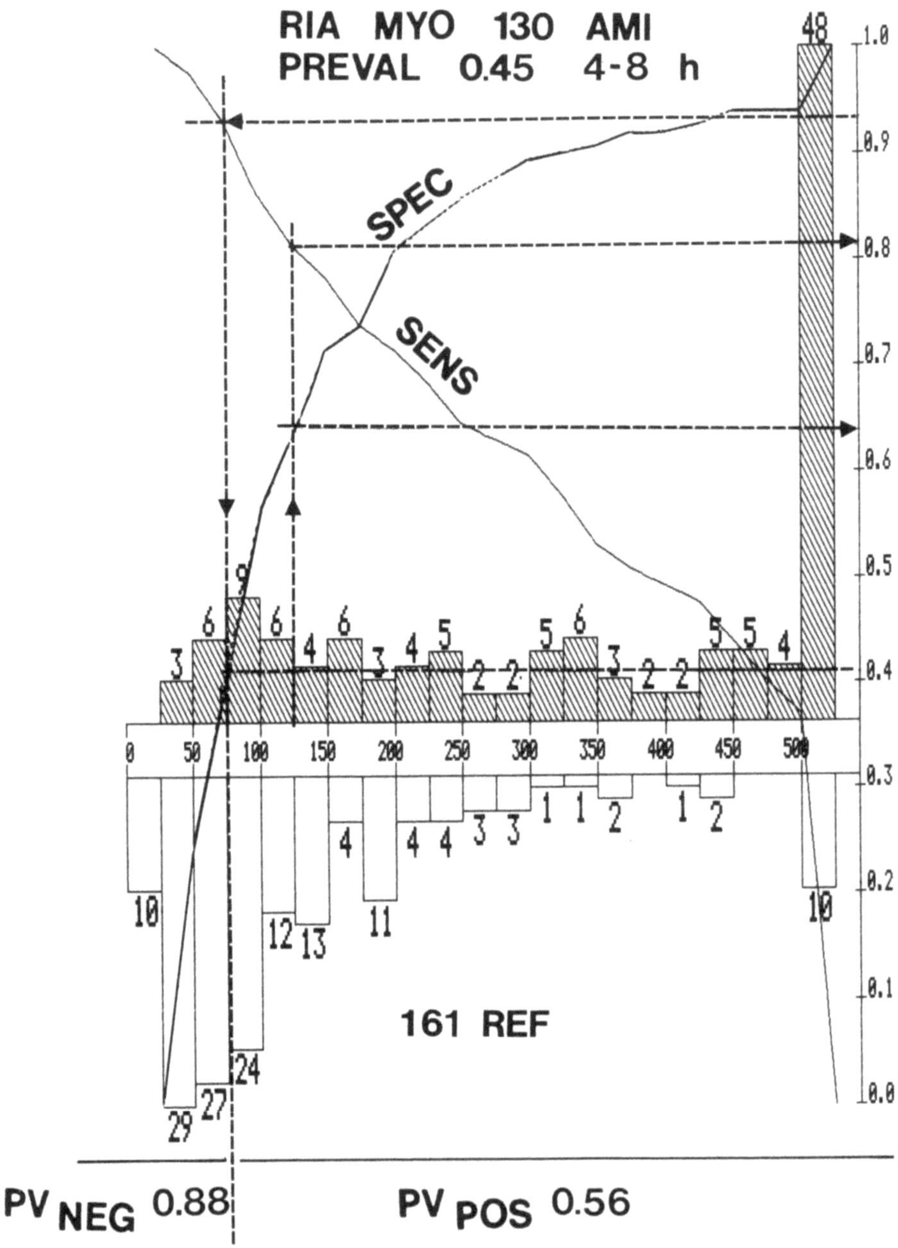

Fig. 6. Distributions of S-myoglobin in the same patients as in Figures 3–5, but measured in samples drawn 4–8 hours after onset of acute symptoms. The discriminator value required for a sensitivity of 0.95 lies at 75 µg/l.
The corresponding specificity has decreased to about 0.40.
PVneg and PVpos for a discriminator value of 75 µg/l are shown below.
For comparison with the 8–12 hour values, the discriminator at 125 µg/l is also shown. The corresponding sensitivity and specificity lie around 0.82 and 0.63, respectively.

REF apply to any test class of an n-class test evaluation (9, 13). The CPD curves may indicate the clinically meaningful number and range of test classes.

In contrast to sensitivity and specificity, CPD and predictive values are highly dependent on the anterior prevalence of disease in the clinical situation for which the test is intended. Ideally, a pilot study should be carried out at this prevalence. This may not always be possible. For instance, let us consider evaluation of a diagnostic test intended for screening for a certain disease at a prevalence of 0.01 (1% DIS). If we assume that the minimum number of DIS required for a study is 100, this corresponds to nearly 10 000 REFpersons! Obviously, testing of this number of persons is frequently unrealistic, and by the time this feat has been accomplished the test may be obsolete.

However, provided the appropriate DIS and REF groups have been selected it is possible to carry out and evaluate a pilot study with a smaller number of test persons at a different prevalence and subsequently recalculate the data to estimate test performance at the clinically relevant anterior prevalence of disease (1, 9, 12, 13). This is accomplished by keeping the diseased group constant and multiplying each test class in the reference group by a prevalence factor (1, 9, 12, 13). The unchanged DIS group will thus be "diluted" by an enlarged REF group with unchanged distribution. This is equivalent to multiplying the scale of the plot of the reference group in the histogram by the appropriate prevalence factor. This calculation may be carried out for evaluation at any number of test classes.

The prevalence- or scale factor is calculated as follows (9, 13):

$$F = \left(\frac{1}{\text{wanted new fraction DIS}} - 1\right) \times \frac{\text{ALL DIS}}{\text{ALL REF}}$$

In Figure 7 a, PVpos and PVneg have been plotted as a function of discriminator position at eleven test classes, class width 75 µg/l. The curves demonstrate the inverse relationship between the two variables. Figures 3–6 demonstrated that moving the discriminator from higher toward lower values increased sensitivity and PVneg but also decreased specificity and PVpos (1). Again, the selection of discriminator position(s) is a compromise.

Let the clinician want an evaluation of the performance of S-myoglobin as a two-class test, sampled within 4–8 hours after symptoms in the medical emergency room instead of in the coronary care unit with its AMI prevalence of 0.50. As the clinical criteria for "AMI-suspect" may be wider in this clinical situation he estimates an AMI prevalence among the medical emergencies to 0.20. (In actual practice it may be even lower.)

The histogram in Figure 7 b shows the data from the 4–8 hour S-myoglobin study recalculated from the empirical prevalence of 0.44 to 0.20 as described above. The number of REFpersons has been increased by a factor 3.224 from 161 to 519 without changing the distribution on the test classes. The discriminator position and consequently sensitivity and specificity remain constant.

The PVpos curve has been displaced downwards toward *lower* values, and the PVneg curve has been displaced toward *higher* values.

Again, the reader may now draw his own conclusions as to the proposed use of the test.

a

b

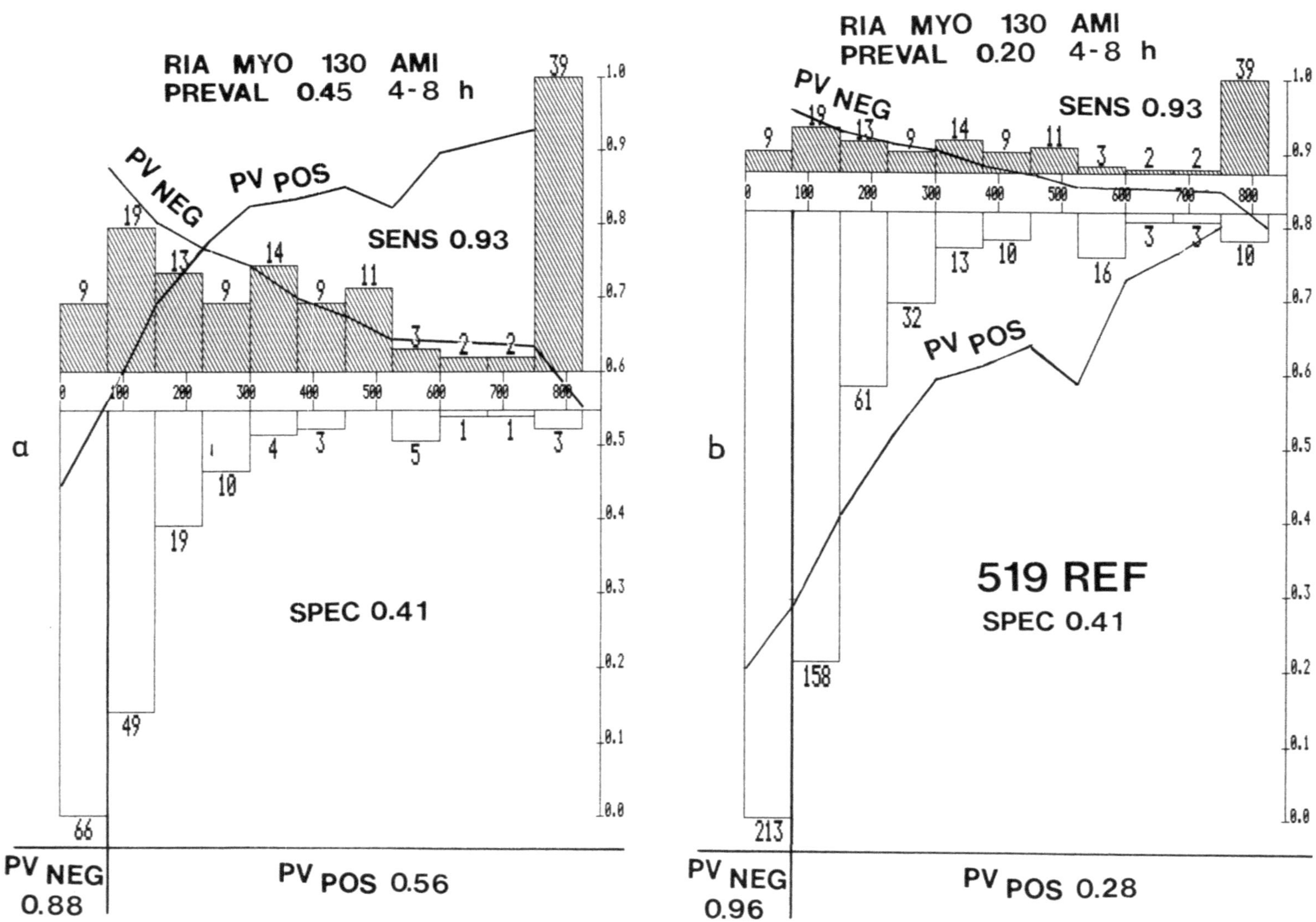

RIA MYO 130 AMI
PREVAL 0.45 4-8 h
PV NEG
PV POS
SENS 0.93
SPEC 0.41
PV NEG 0.88
PV POS 0.56
RIA MYO 130 AMI
PREVAL 0.20 4-8 h
SENS 0.93
519 REF
SPEC 0.41
PV NEG 0.96
PV POS 0.28

5 Selection of Sample Groups for Clinical Studies

Classification. The sample groups of a clinical study should be classified by criteria other than the test that is to be evaluated.

Diseased persons. The stage of disease should correspond to that which may be expected in the clinical situation for which the test is intended (9).

Reference persons. This group should be the clinically relevant one frequently consisting of persons with symptoms that makes the clinician suspect the disease in question, but in whom this disease has been excluded by other criteria than the test. Only *rarely*, e.g. in screening, will the relevant reference group consist of *healthy* persons, such as the traditional blood spenders, laboratory technicians etc. (9).

Prevalence of disease. The prevalence of the evaluation study should preferentially correspond to that in the clinical situation for which the test is intended. A minimum requirement is that CPD and predictive values are recalculated for estimates of the clinically relevant prevalence, (9).

Sample size. Statistical calculations may indicate the minimum size required for reliable estimates of the performance of the test under the specified conditions (8, 14). However, it should be considered that observations in biological materials are subject to variations. The chance of getting *identical* data by repeating the study or in the subsequent clinical application is small. Consequently, the data should be considered *estimates* with a given uncertainty depending on selection and size of the sample groups and analytical imprecision. Reporting of test evaluation data with three decimal accuracy is foreign to the real world.

It may well be more advantageous to evaluate the performance of a given test by several smaller studies in different clinics than attempting to achieve a high statistical accuracy by one very large study. By making certain assumptions of the distributions of DIS and REF confidence intervals may be calculated for the quantities sensitivity, specificity, predictive values etc. (14). A discussion of such procedures is beyond the scope of this chapter.

Fig. 7. Histograms showing the dependence of predictive values on the anterior prevalence of disease.

(a) Same study as in Figure 6, plotted at 11 test classes, class-width 75 μg/l. All data above 750 μg/l have been cumulated in the last test class. The curves for PVpos and PVneg at the prevalence of 0.45 AMI have been superimposed on the graph.

PVneg and PVpos have been calculated for a discriminator position of 75 μg/l.

(b) Same study replotted at an anterior prevalence of 0.20 AMI. The unchanged group of 130 AMI has been "diluted" by an increased group of 519 REFpersons with unchanged distribution. The displacement of the PVneg curve upwards, and of the PVpos curve downwards are evident.

6 "Comment Corriger la Fortune"

Breaking any of the rules for selection of study materials may lead to unrealistic evaluations of test performance. Selection of unfelicituous DIS or REF groups (or both) may "improve" or "impair" apparent test function. Four different traps are presented in the following with illustrative references. For didactic purposes this is done in a somewhat unconventional way – with an expression from the world of gambling: "comment corriger la fortune" ...

Let Figure 8 show the distributions after a *correctly* done hypothetical study for comparisons with the following four error type situations. DIS and REFgroups have been selected according to the guidelines described above. Evaluated as a two-

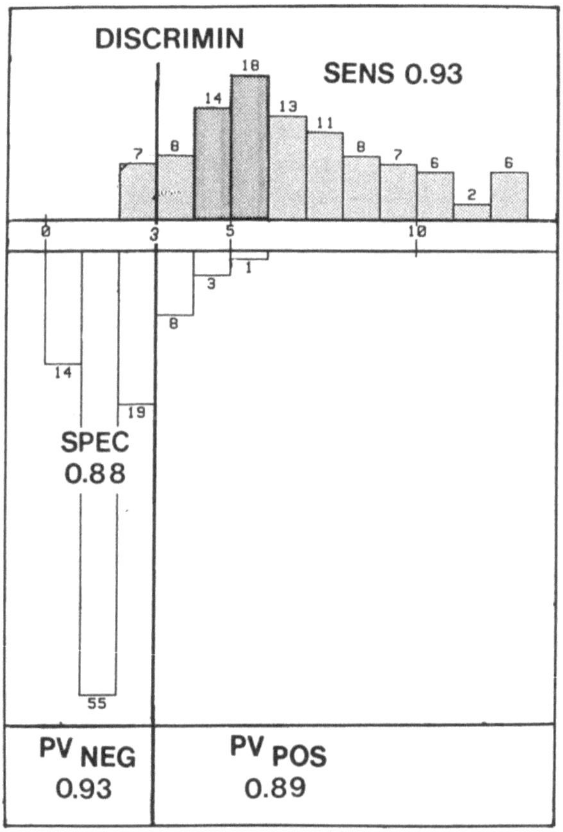

Fig. 8. "Correctly planned clinical study" with a hypothetical test. With a discriminator at position 3 on the abscissa the positive test class detected 93% of all DIS with a PVpos of 0.89. The negative test class excluded 88% of all REF with a PVneg of 0.93.
This study is used for comparison with the following graphs illustrating how to "corriger la fortune" (from Scand. J. Clin. Lab. Invest. Suppl. 181, 1986 (ref. 9), with kind permission).

class test with the discriminator placed at position 3 on the abscissa the test detected 93% of all DIS with a PVpos of 89% and excluded 88% of all REF with a PVneg of 93%.

6.1 How to improve test results

1. Select a DIS group with advanced stages of the disease (call them typical cases). This is symbolized in Figure 9a. Some examples:
   S-gammaglutamyltransferase (S-GT) for the detection of "hidden alcoholics": select chronic alcoholics in a clinic for alcoholics. At a specificity of about 0.95, the sensitivity for light consumers may be around 0.08–0.15, and for advanced alcoholics 0.40–0.88 (15–19).
   S-creatine kinase (S-CK) and *S-CK MB* at suspected AMI: select only clinically- and ECG-confirmed large infarctions.
   S-prostatic acid phosphatase (S-PAP) for detection of early prostatic carcinoma: select only patients with advanced carcinomas with metastases. This will increase test sensitivity (20–24), cf. the "syndrome of early enthusiasm and subsequent disillusionment (22).
   Carcino-embryonal antigen (S-CEA) as an early tumormarker: Select patients with advanced colo-rectal carcinomas (25).
   S-thyroid stimulating hormone (S-TSH) with increased analytical detectability: select clinically well-confirmed advanced hyper and hypothyroids (26).
   Figure 9a: The distributions of the DIS group has been shifted toward higher test values. Accordingly, the discriminator may be moved to position 5 on the abscissa increasing specificity with no loss of sensitivity. The test now detects all DIS with a PVpos of 99% and excludes 99% of all REF with a PVneg of 100%.
2. Select a REF group consisting of only completely healthy persons (Fig. 9b). For the same examples as above:
   S-GT: drug-free vegetarians screened for liver-biliary disease.
   S-CK and *S-CK MB*: young healthy individuals with restricted physical activity and no signs of coronary disease.
   S-PAP: select a reference group of young, healthy males, or women, or newborns.
   S-TSH: select healthy persons with no indication of thyroid disease (26).
   Figure 9b: With the discriminator remaining at position 3 as in Figure 8 the specificity has actually increased from 88% to 99%. The test now detects 93% of all DIS with a PVpos of 99% and excludes 99% of all REF with a PVneg of 93%.

6.2. How to impair test results

1. Select a DIS group that has been poorly classified and only contains very low stages of the disease (Fig. 9c). Same examples as previously:
   S-GT: alcoholics defined as persons who drink more than one bottle of light beer per day.
   S-CK, CK MB: border line AMI cases, unstable angina pectoris.
   S-TSH: "subclinical hyperthyroids" and "compensated hypothyroids".

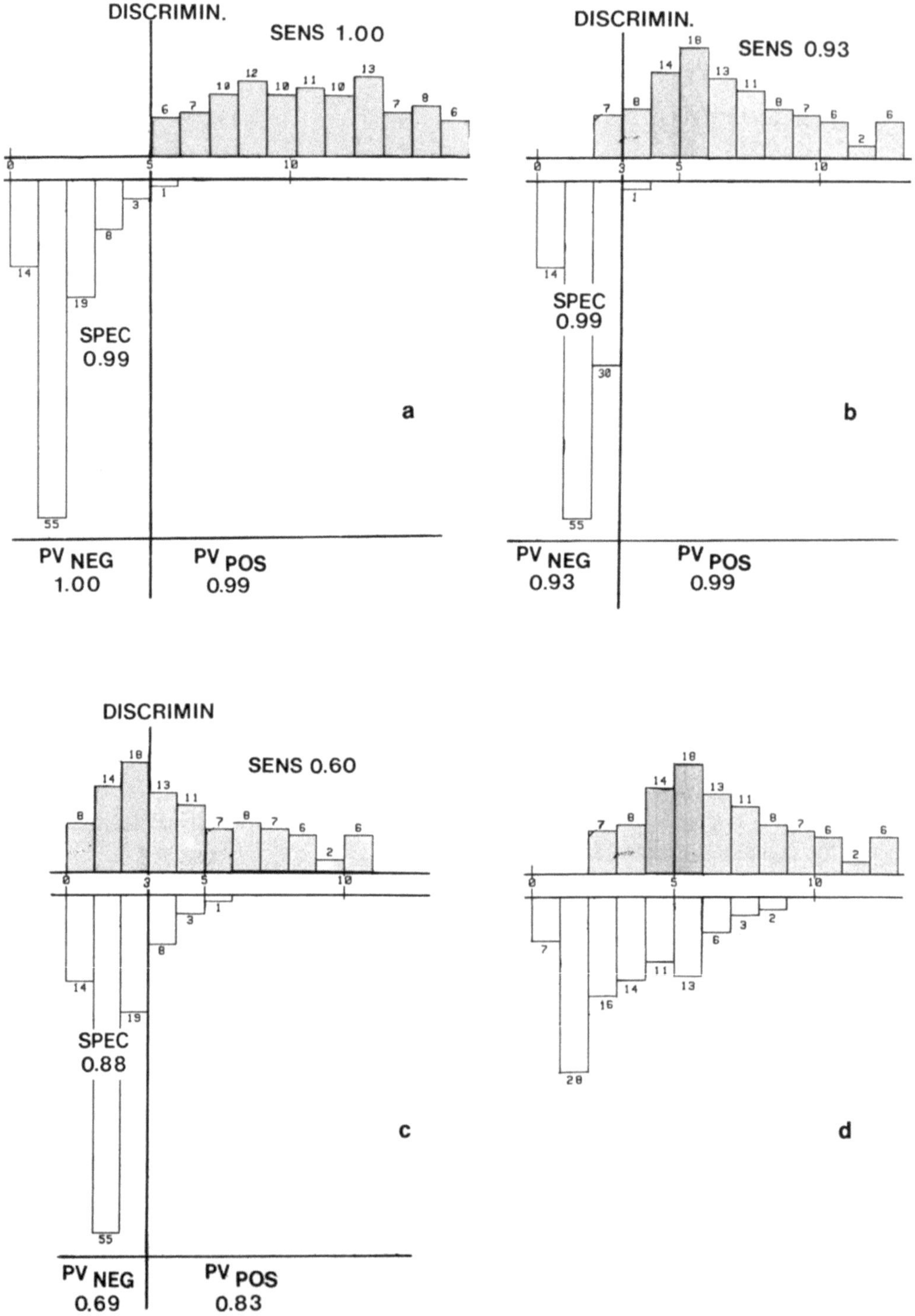
DISCRIMIN.
SENS 1.00
SPEC
0.99
a
PV NEG
1.00
PV POS
0.99
DISCRIMIN.
SENS 0.93
SPEC
0.99
b
PV NEG
0.93
PV POS
0.99
DISCRIMIN
SENS 0.60
SPEC
0.88
c
PV NEG
0.69
PV POS
0.83
d

Figure 9c: The distribution of the diseased group has shifted toward lower test values. With the discriminator remaining at position 3 as in Figure 8, the test now detects only 60% of the DIS with a PVneg of 83% and excludes 88% of all REF with a PVneg of 69%.

2. Select a REF group predominated by conditions known to cause false positive results with the test (Fig. 9d). Some examples:
S-GT for detection of alcoholism: liver-biliary diseases and drug-addicts (27).
S-CK and *S-CK MB* at suspected AMI: select extensive skeletal muscle damages (28, 29) marathon runners (30), and Duchenne muscle dystrophies.
S-lactate dehydrogenase isoenzyme 1 (S-LDH H4) at suspected AMI: select a REF group of severe liver diseases with increases of all 5 LDH isoenzymes (31) instead of AMI-suspected NOT-AMI patients (32).
Figure 9d: The distribution of the reference group has now been shifted so far toward high test values that selection of a meaningful discriminator is no longer possible.

7 Proper Reporting of Test Evaluations

As mentioned in the introduction unfortunately no internationally standardized format for test evaluation reports exists. It would obviously be a great advantage to all if test data were always presented in such a way that direct comparison of distributions of test data, and of test evaluation quantities between different studies with the same or similar tests were possible. Regrettably, several types of statistical data reduction and certain types of graphs make such comparison difficult or impossible.

In contrast, presentation of test data in double histograms with numbers of DIS and REF within each test class, and superimposed curves of the dependent test

Fig. 9. (a) "Improved study": The DIS group has been selected among advanced stages of the disease such as large AMI, carcinomas with metastases, chronic alcoholics, advanced hyperthyroidism etc. Consequently, the discriminator could be moved to position 5. The positive test class now detected all DIS with PVpos of 0.99. The negative test class excluded 99% of all REF with a PVneg of 1.00.
(b) "Improved study": The REF group has been selected among completely healthy persons instead of patients suspected of having the disease in question. Retaining the discriminator at the initial position 3, the positive test class detected 93% of all DIS with a PVpos of 0.99. The negative test class excluded 99% of all REF with a PVneg of 0.93.
(c) "Impaired study": The DIS group consisted of borderline cases. With the discriminator remaining at position 3 the positive test class detected 60% of all DIS with a PVpos of 0.83. The negative test class excluded 88% of all REF at a PVneg of 0.69.
(d) "Impaired study": The REF group has been selected exclusively among patients with other diseases that cause high test values. Test performance is now so poor that a meaningful discriminator position can hardly be found (from Scand. J. Clin. Lab. Invest. Suppl. 181, 1986 (ref. 9), with kind permission).

quantities plotted as functions of discriminator positions permit comparison of data distributions and clinical information value. The plots may be recalculated by the reader himself to any other prevalence as described above.

7.1 2 × 2 tables for presentation and reconstruction of test data

A minimum requirement is presentation of test evaluations in a complete 2 × 2 table as shown in Figure 1. It has been amply documented by frustrated editors, referees, and readers that even this modest requirement is not accomplished in a large fraction of test reports. Test evaluation data are often described verbally in complicated texts which mainly serve to confuse the reader, hide the essential data, and make it difficult to detect omissions of results necessary for a meaningful evaluation. Presentation in a 2 × 2 table prevents omissions and makes errors in calculation of the derived test evaluation quantities immediately obvious.

As an example, a screening study for early prostatic adenocarcinoma (PAC) with S-PAP was described as follows (23): "Cooper has screened 6320 men over the age of 45 years ... using a radioimmunoassay for prostatic acid phosphatase. Of the 6320 men, 444 had an elevated serum PAP level (7% of the total). Of the 444 cases of elevated serum PAP levels, 62 were found to be associated with patients having adenocarcinoma of the prostate, 15 with localized disease, and 47 with advanced disease. Only 14% of the positive results were due to patients with adenocarcinoma of the prostate, a very low predictive value. Furthermore, 67% of the 444 men with elevated levels of PAP had benign prostatic hyperplasia".

Although this kind of description may be less than crystal clear to the reader such texts are nonetheless rather frequently encountered e. g. in abstracts. After studying the text the reader may want to ask the following question: How large a fraction of the localized or advanced prostatic adenocarcinomas were detected at which PVpos by the test and how large a fraction of the reference persons were excluded at which PVneg?

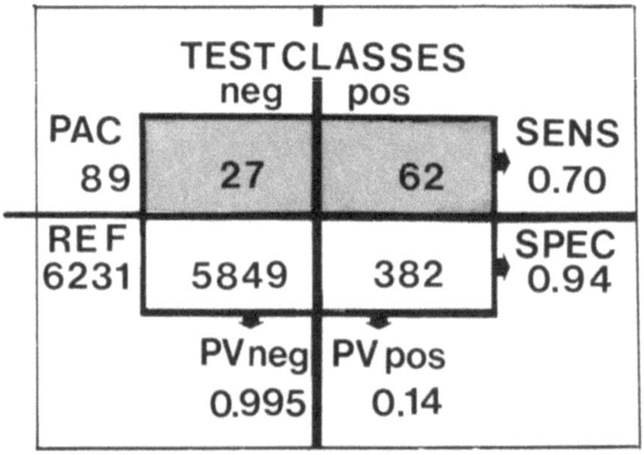

Fig. 10. 2 × 2 table. Reconstruction of S-PAP screening study for prostatic adenocarcinomas.

To answer these questions the reader is advised to draw the frame of a 2 × 2 table as shown in Figure 2 and successively insert the data from the text above into the appropriate cells. From the description we can extract the numbers of ALL persons, ALL POS, POS PAC, and PVpos. The remaining data must be reconstructed. For this we need either sensitivity or specificity which are not given in the quoted text. Fortunately, the referent and others (21, 24) have reported an overall sensitivity for RIA-PAP assays for PAC of about 0.70. Tentatively using this figure, a complete 2 × 2 table may be successively reconstructed (Fig. 10) as follows:

First, the number of PAC, REF, and prevalence of PAC in the study:

ALL test persons:	6320
ALL PAC: 62 POS PAC × 1.43 (=1/SENS) =	89
ALL REF: 6320−89 PAC =	6231
Prevalence of PAC in the study: 89 PAC/6320 =	0.014

We may now fill in the appropriate cells in the 2 × 2 table (Fig. 10):

Positive test class:	
ALL POS:	444
POS PAC:	62
POS REF: 444 POS−62 POS PAC =	382
PVpos:...62 POS PAC/444 POS =	0.14
Negative test class:	
NEG PAC: 89 PAC−62 POS PAC =	27
NEG REF: 6231 REF − 382 POS REF =	5849
ALL NEG: 27 NEG PAC + 6231 NEG REF =	5876
PVneg: 5849 NEG REF/5876 REF =	0.995
Specificity: 5849 NEG REF/6231 REF =	0.939

The reconstructed data (Fig. 9) may now be compared to those of the original report (33) stating an over-all sensitivity in screening for all stages of PAC of 0.73 with a corresponding specificity of 0.94.

From the quoted text the reader may calculate that test sensitivity for detecting *low-stage* PAC with this test was 0.17 with PVpos of 0.04. This confirms that despite several enthusiastic reports S-PAP remains a poor tool for detection of early stages of prostatic carcinomas (20−22).

It might be argued that at least the test is efficient in excluding PAC with a PVneg of 0.995 (as an exception calculated with three decimals; otherwise it would be rounded off to 1.00!) However, this may be compared to the PVneg *without any test at all* being 0.986, namely the *anterior prevalence of reference persons* in the study.

8 Diagnostic Strategies

The clinical information value of tests may be enhanced when the tests are part of a diagnostic strategy. The design and performance of strategies with two-class tests may be described by 2 × 2 tables. In principle, tests may be combined in two ways: in sequence or in parallel.

Below, we shall deal with the following situations:

1. A sequence of tests in which the results of a preceding test selects the samples for the following test.

2. A combination of parallel tests, such as are obtained from e. g. multi-channel analyzers or "test profiles", with various degrees of overlapping among the tests in the combination.

8.1 Two-class tests in sequence

A diagnostic strategy may consist of several sequential steps. Frequently the clinician functions as the first test in the sequence. An example may be the detection of diabetes mellitus (DM):

Step 1: clinical suspicion of DM leads to:
Step 2: determination of one or more fasting B-glucose. Values below 5 mmol/l exclude manifest DM with a high degree of certainty. Values above 5 mmol/l lead to:
Step 3: an Oral Glucose Tolerance Test (OGTT). OGTT 2-hour values above 11 mmol/l confirm the diagnosis of DM (WHO, ref. 34).

The prevalence of DM increases from one step to the next: from about 2% in the population, to a higher value among those who seek the clinician, to perhaps 20% among those with both clinical suspicion *and* fasting B-Glucose above 5 mmol/l to 100% among those who then *also* show a positive OGTT.

The reason for the increase in prevalence through the sequence is the exclusion of patients by negative results at each step.

Figure 11 demonstrates these principles by means of a sequence of four hypothetical tests. The tests 1–4 represent *different* tests. For didactic reasons, however, all four have the same, high sensitivity (to minimize loss of patients through the sequence, and give a high PVneg) and the same, low specificity (to demonstrate the increase of specificity through such a sequence).

The sequence functions according to the following principles:
A number of test-negative patients are excluded at each step.
All test-positive patients continue to the following step.

Figure 11 now illustrates three rules of function of a sequence of two-class tests:

1. The over-all *sensitivity* decreases through the sequence. The final sensitivity is the product of the sensitivities of all tests (8, 9). In the figure, the resulting sensitivity is $0.99^4 = 0.96$.
2. The over-all *specificity* increases through the sequence. It may be calculated as the sum of all reference persons being excluded through the sequence divided by the number of ALL REF in the study. In Figure 10, the specificity of the sequence 844 NEG REF/900 REF = 0.94.
3. The *prevalence of DIS* increases through the sequence. The number of DIS decreases by only one per cent per step, whereas 50% of the remaining REF are excluded at each step. The PVpos of a preceding step becomes the anterior prevalence of DIS for the following step. The anterior prevalence of DIS in the example was 0.10. The posterior prevalence of DIS in the positive test class of

the last step was 0.63. This may also be described as an increase of the concentration of DIS through the sequence. In Figure 11 the "concentration factor was 0.63/0.10 = 6.3.

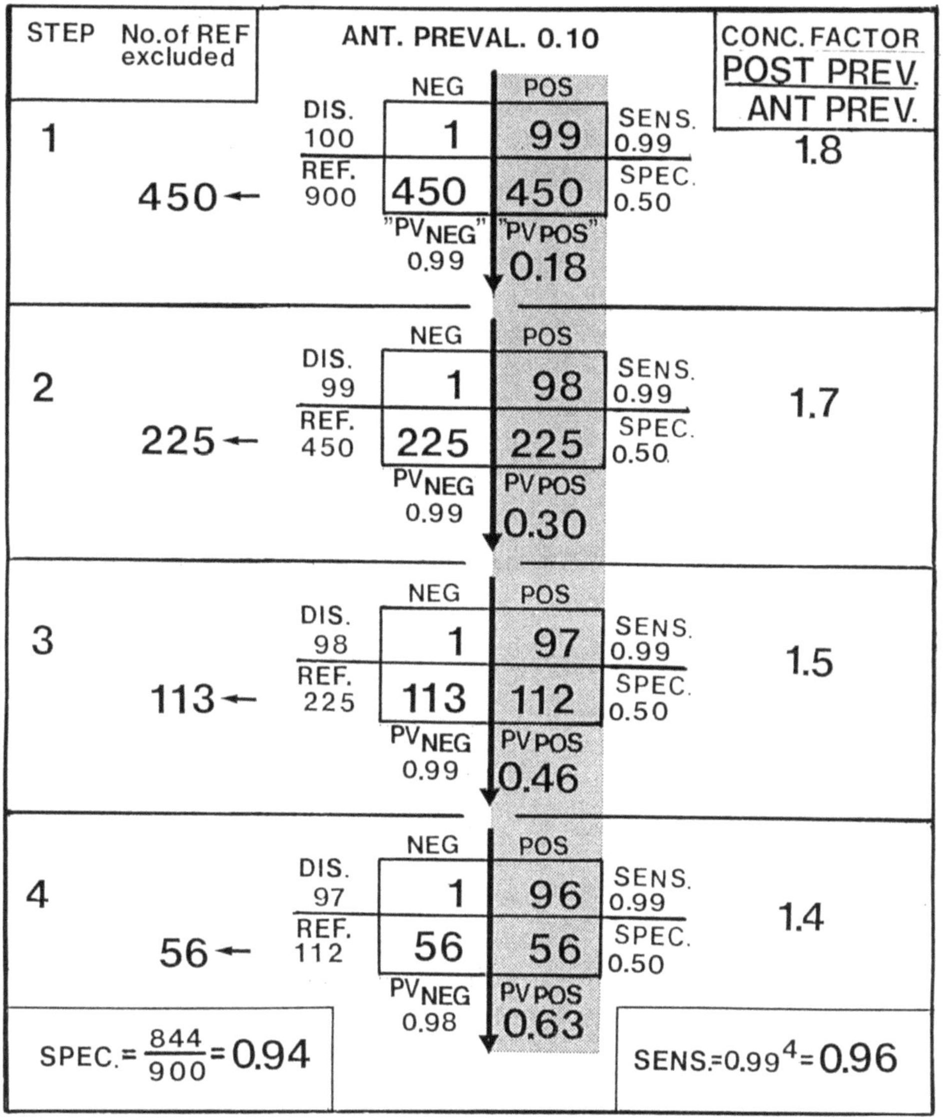

Fig. 11. Sequence of four different hypothetical tests, each with a sensitivity of 0.99 and a specificity of 0.50. Each subsequent test is applied to the samples in the positive test class of the preceding test. PVpos of a preceding test becomes the anterior prevalence of disease for the following test. Sensitivity decreases, Specificity and the prevalence of disease increase through the sequence. See text for details (modified from Scand. J. Clin. Lab. Invest. Suppl. 181, 1986 (ref. 9), with kind permission).

The performance of this diagnostic strategy may be summarized as follows:
Over-all sensitivity: 96 POS DIS/100 DIS = ... 0.96
PVpos: increased through tests 1–4 to 96 POS DIS/152 POS = ... 0.63
Over-all specificity: 844 NEG REF/900 REF = ... 0.94
PVneg: decreases through tests 1–4 to step 4: 56 NEG REF/57 NEG = ... 0.98
The performance of the diagnostic strategy may be verbally expressed as: The sequence detected 96% of all DIS at a PVpos of 0.63 and excluded 94% of all REF at an overall PVneg of 0.99 (844 NEG REF/848 ALL NEG).

Figure 11 has been designed to show the principles of function of a sequential test strategy. In real life each of the tests may he assumed to be a different method at a sensitivity and specificity different from those of the others. Each preceding test allocates certain fractions of the DIS and REF to the negative and positive test classes, respectively.

Sequences may be designed so that also the *test-negative* persons, or, for three-class tests, the patients in the inconclusive test class continue to another test or series of tests. In any case, allocation to one class in a preceding test selects the sample group for the following test. The sensitivity and specificity of the following test depends on this selection by the preceding test. Consequently, the performance of sequential tests must be determined through *actual clinical studies of the complete sequence* (9).

Studies of the performance of each individual test on the *initial sample group* are not sufficient and may lead to the erroneous conclusion (9) that the resulting PVneg and PVpos are *independent* of the order of the tests in the sequence as has been claimed by Galen and Gambino (6).

8.2 The clinician as the first test in a diagnostic strategy

In the description above the word "test" is used in its widest sense. It may be a chemical test, an X-ray or even a clinician. In fact, a diagnostic strategy is frequently initiated by a clinician. The clinician thus becomes the first test in the sequence as discussed above. An actual example is provided by the "Scandinavian (SCE) diagnostic strategy at suspected acute myocardial infarction" (35, 37). This consists of three sequential steps:

1. Clinical suspicion of an AMI.
2. Determination of S-creatine kinase (S-CK) on samples drawn within 10–20 hours after onset of acute symptoms.
3. Determination of S-creatine kinase B-subunit activity on all S-CK positive samples.

Let the prevalence of AMI among all patients admitted to the medical emergency unit be 0.05. In contrast, the prevalence of AMI in the group of patients considered clinically as AMI suspects is much higher and depends on the clinician's decision criteria. If the clinician is inexperienced or overly cautious the prevalence may be relatively low. If the clinician is highly experienced it may be higher. In most coronary care units the prevalence of AMI lies around 0.50 (36). In other words, the clinicians' average PVpos must lie around 0.50 (9).

One frequently encounters the statement that a PVpos of 0.50 is equivalent to the flipping of a coin. Like most generalizations this statement is not always true. In the example of the medical emergency unit given above the 5% AMI were "diluted" by 95% patients with other diseases. Through the decisions of the clinician the

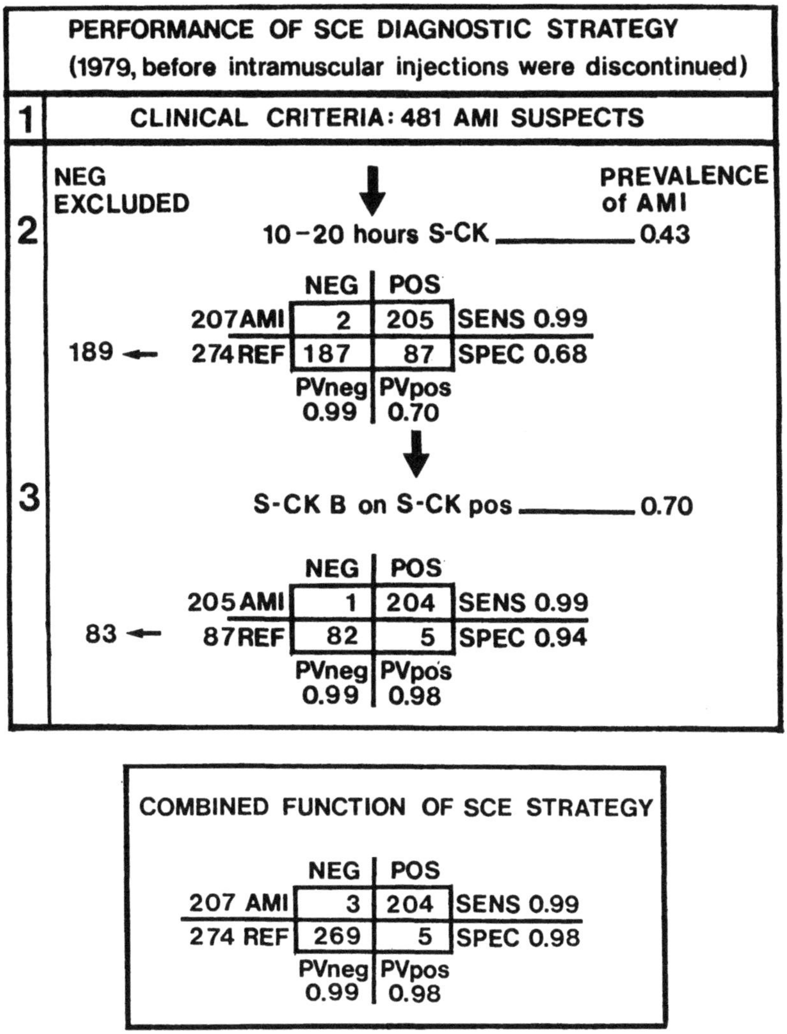

Fig. 12. Performance of the SCE 3-step diagnostic strategy at suspected AMI: (1) clinical suspicion, (2) two 10–20 hour S-CK, and (3) S-CK B on all S-CK positive samples. Data from ref. 35. The anterior prevalence of AMI was 0.43. The study included patients receiving i. m. injections of analgetica and anti-arrhytmica, etc. and represented a "worst case" situation. Sensitivity decreased, specificity and prevalence of AMI increased through the sequence. The combined performance is summarized in the box below:
The positive test class detected 99% of all AMI with a PVpos of 0.98.
The negative test class excluded 98% of all REF with a PVneg of 0.99 (from Scand. J. Clin. Lab. Invest. Suppl. 181, 1986 (ref. 9), with kind permission).

prevalence of AMI in the AMI-suspected group has been increased by a factor 10 – from 0.05 to 0.50. Consequently, the next test in the sequence, S-CK, was applied at this AMI prevalence (cf. the increases of prevalence through Fig. 11).

The performance of the SCE diagnostic strategy has been described in detail elsewhere with presentation of the distributions in double-histograms (35, 37). Figure 12 summarizes the performance as evaluated in a SCE study. This material comprised 481 patients consecutively admitted to a coronary care unit during two periods 1977 – 1979 (35). During this period intramuscular injections of analgetica, anti-arrhytmica etc. were still routinely administered and some patients with skeletal muscle damage happened to be included. The material was thus comparable to the "worst case" sample group encountered in European coronary care units in terms of potentially false positive values as reflected in the specificity data in Figure 12.

The prevalence of AMI in the AMI-suspected group was 0.43. S-CK was determined in two samples drawn within the diagnostic time window 10 – 20 hours after onset of the acute symptoms. The discriminator values for total S-CK were different for men and women and selected to give a sensitivity of at least 0.99. As a consequence, specificity for S-CK in this study was as low as 0.68 (Fig. 12). The consequences of falsely positive results were simply a number of additional S-CK B determinations. A new evaluation was later carried out in a coronary care unit material after all intramuscular injections had been discontinued. This showed an increase in specificity of S-CK to around 0.85 (37).

The performance of the diagnostic strategy in Figure 12 may be described as follows:

Two 10 – 20 hour S-CK determinations were carried out at an AMI prevalence of 0.43. Both S-CK being negative excluded an AMI in 68% of all REFpersons at a PVneg of 0.99 within 24 hours. These patients could be moved to a less expensive bed or be discharged.

A positive S-CK indicated an AMI in 99% of all the AMI cases at a PVpos of 0.70. This consequently, was the prevalence of AMI in the group in which S-CK B was subsequently determined. Negative S-CK B results excluded an AMI in 94% of the falsely S-CK positive REFpersons at a PVneg of 0.99. A positive S-CK B result confirmed an AMI in 99% of the S-CK positive AMI at a PVpos of 0.98. The box in Figure 12 summarizes the performance of the whole sequence:
99% of all the AMI were confirmed at a PVpos of 0.98. AMI was excluded in 98% of all the REFpersons at a PVneg of 0.99.

8.3 Two-class tests in parallel

Obviously, test sensitivity will be increased when several tests are applied simultaneously, and a positive test result is defined as a positive test result with *any single test* in the combination.

A negative test result in such combinations then requires *all tests in combination* being negative (6, 9, 38).

Test evaluations in the literature sometimes deal with the claimed advantages of parallel testing. An example is screening for "hidden alcoholism" showing considerable increases in diagnostic sensitivity as a function of the number of parallel

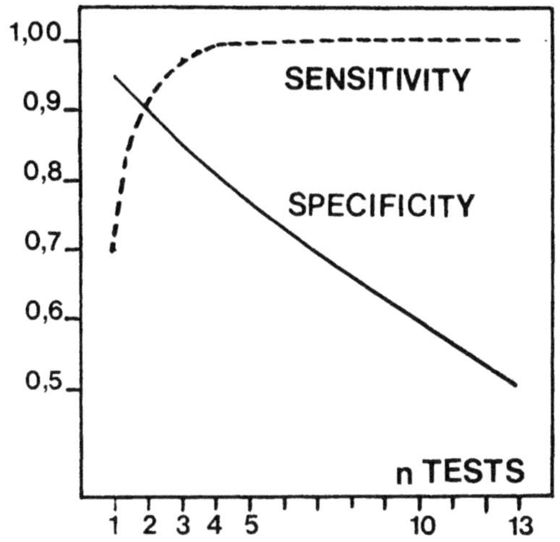

Fig. 13. Combined sensitivity and specificity (ordinate) plotted as function of the number of hypothetical tests applied in parallel (abscissa).
Sensitivity (stippled curve) increases according to one – or several – possible curves depending on the degree of overlapping among the sensitivities of the tests in the combination.
Specificity (full drawn curve) decreases and has here been calculated as the product of the specificities of the individual tests: 0.95^n.
Provided that the tests are pathophysiologically independent of each other (a most unlikely occurrence) the combined specificity should decrease to 0.50 when the number of parallel tests has reached 13 (from Scand. J. Clin. Lab. Invest. Suppl. 181, 1986 (ref. 9), with kind permission).

tests included. A Swedish alcohol research group reported this increase as follows: S-gammaglutamyltransferase (S-GT): 42%, S-GT + Ery-MCV: 67%, S-GT + Ery MCV + S-ASAT: 68%, and finally S-GT + Ery-MCV + S-ASAT + S-albumin 74% (39).
The corresponding changes in specificity were not studied at all (!). With parallel testing sensitivity increases and specificity decreases as a function of the number of tests. This is symbolized in Figure 13. If the tests in this combination had independent specificities of 0.95 the over-all specificity would decrease as the product of the single specificities, 0.95^n, and should theoretically reach 0.50 for 13 tests (1, 8, 9, 40). In reality, however, neither over-all sensitivity nor specificity can be accurately predicted but must be established in *actual clinical studies.*

Theoretically, at least three possibilities may be considered:

1. *Multiple determinations with one and the same test.* Sensitivity and specificity should remain unchanged within the errors of analytical imprecision.
2. *Wholly independent tests.* The over-all sensitivity may approach the sum of the individual sensitivities and increase asymptotically toward 1.00. The over-all specificity may approach the product of the individual specificities. Such conditions are rarely encountered.

3. *The tests in the parallel combination have partially overlapping sensitivities and specificities.* The over-all performance can be determined only through actual studies. As an example, parallel determinations of S-barbiturate were carried out simultaneously by thin-layer chromatography (TLC) and by an EMIT technique. The referee method was gas chromatography (GC) (38). The individual sensitivities of TLC and GC were 0.67 and 0.57, respectively. The over-all sensitivity in the parallel combination had increased to 0.90.
   The individual specificities were 0.50 and 0.83, respectively. The over-all specificity had decreased to 0.40 (38).
   Summarizing, the increased sensitivity by a parallel combination was obtained at the cost of a dereased specificity (6, 9, 38).

8.4 Tests for screening

Tests applied at very low prevalences of disease as e. g. unconditional screening must fulfill particularly high requirements regarding both sensitivity and specificity. As it may be assumed that the purpose of such screenings is the detection of a disease that can be treated, sensitivity should be high in order that as few cases as possible are missed.

As the prevalence of disease in the sample group is very low, specificity must also be very high to avoid an unacceptably high number of false positives.

ANT. PREVAL. 0.0003

S-TSH

	NEG	POS	
DIS. 100	1	99	SENS. 0.9900
REF. 300000	299940	60	SPEC. 0.9998
	PV_{NEG} >0.99	PV POS 0.62	

$S\text{-}T_4$

	NEG	POS	
DIS. 100	1	99	SENS. 0.990
REF. 300000	297000	3000	SPEC. 0.990
	PV_{NEG} >0.99	PV POS 0.03	

Fig. 14. Screening for congenital hypothyroidism with S-TSH and S-T4. At the very low anterior prevalence of hypothyroidism 0.0003 a difference in specificity of only 0.0098 changed the PVpos from 0.03 to 0.62 (from Scand. J. Clin. Lab. Invest. Suppl. 181, 1986 (ref. 9), with kind permission).

Test performance at such situations may be illustrated by the example of screening for congenital hypothyroidism with S-TSH and S-T4, respectively, at an anterior prevalence of disease of about 0.0003. In the 2 × 2 tables in Figure 14 (modified after Keller (41)) both sensitivity and specificity have – for demonstration purposes – for once been given with three- and four-decimal accuracy.

Let the sensitivity of both tests be as high as 0.9900. Let the specificities of S-TSH and S-T4 be 0.9998 and 0.9900, respectively. Let both tests be applied to the same sample groups, prevalence of hypothyroidism 0.0003.

The difference in test performance induced by the small difference in specificity is striking: PVpos for S-TSH is 0.62, for S-T4 0.03! The number of S-T4 false positives is 3000 as compared to 60 for S-TSH. The concentration factors, posterior prevalence/anterior prevalence are 2066 for S-TSH and 100 for S-T4.

Appendix

Bayes theorem

The alert reader may have observed that the preceding sections have dealt with Bayesian concepts without using the Bayesian equations at all. The famous theorem was first presented by the Reverend Bayes in: "Toward Solving a Problem in the Doctrine of Chances" published posthumously by the Royal Philosophical Society in 1763 (42). This theorem is a method of inverting the order of the conditional probability:
P(*Si*|*D*) to *P*(*D*|Si), see ref. 8 and below. For a two-class test this is the same as calculating PVpos from the known values of sensitivity, specificity, and anterior prevalence of disease.

The formula is presented below in traditional probability notation [1.], in terms of sensitivity and specificity [2.], and in the working terminology used in the preceding sections of this chapter [3.].

1. Probability notation (8):

$$P(D \mid Si) = \frac{P(D) \qquad\qquad \times \boxed{P(Si \mid D)}}{P(D) \times P(Si \mid D) + P(\bar{D}) \times \boxed{P(Si \mid \bar{D})}}$$

D = diseased
$\bar{D}$ = not-diseased = REF
$P(D \mid Si)$ = probability of disease given test result *Si*
$P(D)$ = prior (ant.) prevalence of disease
$P(Si \mid D)$ = probability (likelihood) of test result *Si* given disease
$P(Si \mid \bar{D})$ = probability (likelihood) of test result *Si* given not-disease

The ratio within the box = likelihood ratio (see below).

Probability algebra is the study of predictive statements about the outcome of an experiment from the specifications of the appropriate mathematical model (8). Probabilities may be expressed as number fractions varying between zero and one (8).

When a study of a test has *actually been carried out* the above probabilities have become *empirical facts* (ref. 8, p. 223) and may be described and calculated as simple fractions which are easier to handle than probability notation.

This may be illustrated by the traditional example of throwing dice.

Before the throw, calculate the probability of getting a "6". Obviously, this is one out of a total of six possibilities, i.e., the probability of this event occurring is $1/6 = 0.1666 \ldots$

Throw the dice. Now the anterior *probability* has become converted into a posterior *fact* – it no longer has any meaning to talk about probability. It either is a "6" or one of the five other possibilities. All you have to do is look and count.

2. In terms of predictive values, sensitivity, and specificity:

$$\text{PVpos} = \frac{\text{preval. of DIS} \times \boxed{\text{sensitivity}}}{\text{preval. of DIS} \times \text{sensitivity} + \text{preval. of REF} \times \boxed{(1 - \text{specificity})}}$$

3. In the terminology of this chapter:

$$\text{PVpos} = \frac{\dfrac{\text{ALL DIS}}{\text{ALL DIS} + \text{ALL REF}} \times \boxed{\dfrac{\text{POS DIS}}{\text{ALL DIS}}}}{\dfrac{\text{ALL DIS}}{\text{ALL DIS} + \text{ALL REF}} \times \dfrac{\text{POS DIS}}{\text{ALL DIS}} + \dfrac{\text{ALL REF}}{\text{ALL DIS} + \text{ALL REF}} \times \boxed{\dfrac{\text{POS REF}}{\text{ALL REF}}}}$$

$$\text{PVpos} = \frac{\dfrac{\text{POS DIS}}{\text{ALL DIS} + \text{ALL REF}}}{\dfrac{\text{POS DIS}}{\text{ALL DIS} + \text{ALL REF}} + \dfrac{\text{POS REF}}{\text{ALL DIS} + \text{ALL REF}}}$$

$$\text{PVpos} = \frac{\text{POS DIS}}{\text{POS DIS} + \text{POS REF}}$$

Once the test data have been arranged in the four cells of a 2 × 2 table as shown in Figure 1 PVpos may be calculated simply as the fraction DIS among ALL within the positive test class. This is the same as the last expression in 3. In other words, the long formulas above are not necessary for calculating the "predictive values".

Likelihood ratios, class prevalence of disease (CPD), predictive values, posterior odds of disease, and risk quotients (RQ)

The ratio within the boxes in the equations in 1., 2., and 3. is traditionally called a likelihood ratio (3, 9, 43–47).

LR is calculated as the ratio of two conditional probabilities:

$$\text{LR} = \frac{P(X \mid \text{DIS}) = (\text{likelihood of test result } x \text{ given DIS})}{P(X \mid \text{REF}) = (\text{likelihood of test result } x \text{ given REF})}$$

This is the same as:

$$\text{LR} = \frac{\text{the probability of DIS being allocated to test class X}}{\text{the probability of REF being allocated to test class X}}$$

$$\text{LR} = \frac{\text{the fraction DIS (of all DIS) allocated to test class X}}{\text{the fraction REF (of all REF) allocated to test class X}}$$

and even simpler:

$$\text{LR} = \frac{n\ \text{DIS in test class X/ALL DIS}}{n\ \text{REF in test class X/ALL REF}}$$

The term LR must be defined. It may be LR DIS, i.e. the ratio above, or LR REF, i.e. the inverted ratio; further, we may talk of LR in the whole study or of LR within test class X. Conventionally, however, it is taken for granted that LR means LR DIS within a specified test class.

In the special case of a two-class test LR generally indicates the LR DIS within the positive test class, and LR = sensitivity/(1-specificity).

These are also the data points for ROC curves, see below.

The function of LR may be explained through the concepts of *odds*.

No standardized format exists for this concept. As a matter of fact, many statistical handbooks do not even mention odds. Still, they have become a popular test evaluation quantity (48).

Odds DIS are expressed as a ratio, e.g. 4 DIS : 1 REF. Traditionally odds are written in the format X : 1 (48). In the case of the dice, odds *for* throwing a "six" are 1 : 5 which reduces to 0.2 : 1, and odds *against* it are 5 : 1. One has the choice either to retain a consistent format like "odds for DIS" which unavoidably includes odds smaller than 1 : 1, or switching from "odds for DIS" to "odds for REF" substituting "odds for DIS 0.2 : 1 by "odds for REF" 5 : 1.

The use of odds in algebraic expressions can easily lead to confusion because there is a misleading similarity beween the decimal value of this ratio and conventional fractions such as probability, prevalence etc. Odds may be converted into fractions: odds/(1 + odds).

As an example, odds for "six", 0.2 : 1 = 0.2/(1 + 0.2) = 0.2/1.2 = 0.1666.

Odds for DIS 4 : 1 do not reduce to 4 but to 4/5 = 0.80.

anterior odds DIS = DIS : REF in the study group before the test.

posterior odds DIS = DIS : REF within test class X- after the test.

The relationship between posterior odds DIS and Class prevalence (fraction) DIS, CPD, or PVpos is:

$$\text{CPD} = \frac{\text{POS DIS}}{\text{POS DIS} + \text{POS REF}} = \frac{\dfrac{\text{POS DIS}}{\text{POS REF}}}{1 + \dfrac{\text{POS DIS}}{\text{POS REF}}} = \frac{\text{post odds DIS}}{1 + \text{post odds DIS}}$$

Let post odds DIS = 4 DIS : 1 REF, then CPD = 4/1 + 4 = 0.80.

The relationship between anterior odds DIS and anterior prevalence of DIS is as follows:

$$\text{anterior odds DIS} = \frac{\text{anterior prevalence of DIS}}{1 - \text{anterior prevalence of DIS}}$$

The relationship between LR DIS, anterior odds DIS, and posterior odds DIS in test class X is:

$$\text{LR} = \frac{\text{DIS in X/ALL DIS}}{\text{REF in X/ALL REF}} = \frac{\text{DIS in X} \times \text{ALL REF}}{\text{REF in X} \times \text{ALL DIS}}$$

$$= \frac{\text{DIS in X/REF in X}}{\text{ALL DIS/ALL REF}} = \frac{\text{post odds DIS in X}}{\text{ant. odds DIS}}$$

LR functions as a multiplication rule:

anterior odds DIS × LR = posterior odds DIS.

In other words, LR indicates the factor by which posterior odds are changed within test class X for a specified test. It is apparent from the formulas above that LR is independent of the anterior prevalence (and anterior odds) of disease (9, 49).

Thus, an LR DIS = 4 within test class X means that allocation to test class X increases posterior odds DIS over anterior odds by a factor 4 at any ant. prevalence of DIS.

The above relationships are illustrated in Figure 5 a. In the following calculation examples all figures are given with only two decimals and have been rounded off to demonstrate the *principles* only. Taking test class 3
Anterior odds DIS is 130 DIS : 161 REF = 0.81 : 1 (ant. preval. 0.45)
LR DIS in test class 3 = 0.654/0.056 = 11.68.
Posterior odds DIS = 85 DIS : 9 REF which reduces to 9.44 : 1.

If we multiply ant. odds DIS with LR DIS we also get post. odds DIS : ant. odds DIS 0.81 : 1 × LR DIS 11.68 = post. odds DIS 9.46 : 1.

We can calculate the Class Prevalence of DIS (CPD) from post. odds DIS:
CPD = 9.44/10.44 = 0.90, which, of course, could also have been calculated completely without LR or odds DIS, simply as:
(test class three) 85 DIS/85 DIS + 9 REF = 0.90 (9, 49).

In contrast to LR DIS, posterior odds DIS and CPD change with anterior prevalence and anterior odds DIS. Posterior odds DIS are also called *risk quotients* (RQ). Provided that the distributions of test data are approximately Gaussian, log RQ may be plotted as a function of test value for the region of overlap as an approximately straight line, usually indicating increasing risk of disease with increasing test values (47).

However, as is obvious from the conversion equations given above, the information given by risk quotients and by CPD is very similar and actually identical in the special case of anterior prevalence DIS = 0.5 = (anterior odds DIS = 1).

$$\text{LR} = \frac{\text{post. odds DIS}}{\text{ant. odds DIS}} = \frac{\text{post. odds DIS}}{1} = \text{post. odds DIS}$$

At an anterior prevalence of DIS of 0.50:
CPD = post. odds DIS/(1 + post. odds DIS) (above). Plots of LR DIS, posterior odds DIS (= RQ), and CPD as functions of test value (test class) will present the same information.

Therefore, it is actually a practical choice whether one wants to use the relatively complicated calculations of posterior odds and RQ through likelihood ratios.

Likelihood reduces information to one single number and gives no information on sensitivity and specificity. Further, LR and RQ are difficult to evaluate, as they vary from zero to infinity.

In contrast, CPD is simple to calculate, varies from zero to one, and may thus easily be superimposed on histograms together with curves for the other test evaluation quantities as sensitivity, specificity, predictive values etc. (9).

Receiver operating characteristic (ROC)

ROC curves show pairs of corresponding values of sensitivity and 1-specificity (or of their complementary values) plotted as a function of discriminator position. At each data point the multi-class test is reduced to, and evaluated as a two-class test. The basic principles of ROC plots may easily be understood through a double histogram (Fig. 15).

Figure 15 a presents the distributions of 100 DIS and 100 REF after a clinical study with a hypothetical test. We may now describe 4 variants of ROCs.

ROC plots are used to indicate the "optimal" discriminator position for reducing a multi-class test to a two-class test. Current literature has provided a number of rather complex formulas for calculating this discriminator position (1, 5, 7, 9, 12, 50 – 58). Greatly simplified the four ROC variants may be described as follows:

ROC 1: seek the discriminator position that gives the maximal sum of sensitivity + specificity. This is also the definition of "efficiency II", see below.
ROC 2: seek the minimum sum of 1-sensitivity and 1-specificity.
ROC 3: seek the maximal difference between sensitivity and 1-specificity.
ROC 4: seek the maximal difference between specificity and 1-sensitivity.

Figure 15 b shows the corresponding ROC plots, calculated from the distributions shown in histogram 15 a. The four axes radiate from the value 0.5 in the center. Sensitivity and specificity increase periferically. Accordingly, their complementary values increase in the opposite direction. For example, ROC 3 in the upper left quadrant: sensitivity increases upwards, and 1-specificity toward the right.

Simply explained, a perfect test would have a discriminator position that would give *both* a sensitivity *and* a specificity of 1.00, i. e. the ROC 3 curve in Figure 15 b would reach this point in the upper left corner. A less perfect test would produce a ROC curving more or less toward the upper left corner, frequently of an irregular shape making the selection of an "optimal" point difficult (12, 50 – 56).

The histogram in Figure 15 b may also be used for a "popular" explanation in terms of areas. For a perfect test with *both* sensitivity *and* specificity = 1.00, the plot of e. g. ROC 1 would cover the area:
sensitivity 1.00 × specificity 1.00 = area 1.00.

With less than such a perfect test the area below the ROC would be some fraction of 1.00. All four ROC variants give essentially the same information. This can be demonstrated by superimposing "relative ROCs" directly on the double histogram (9).

ROC plots are based on sensitivity and specificity which are *independent* of the prevalence of disease (1, 9, 12, 50 – 58).

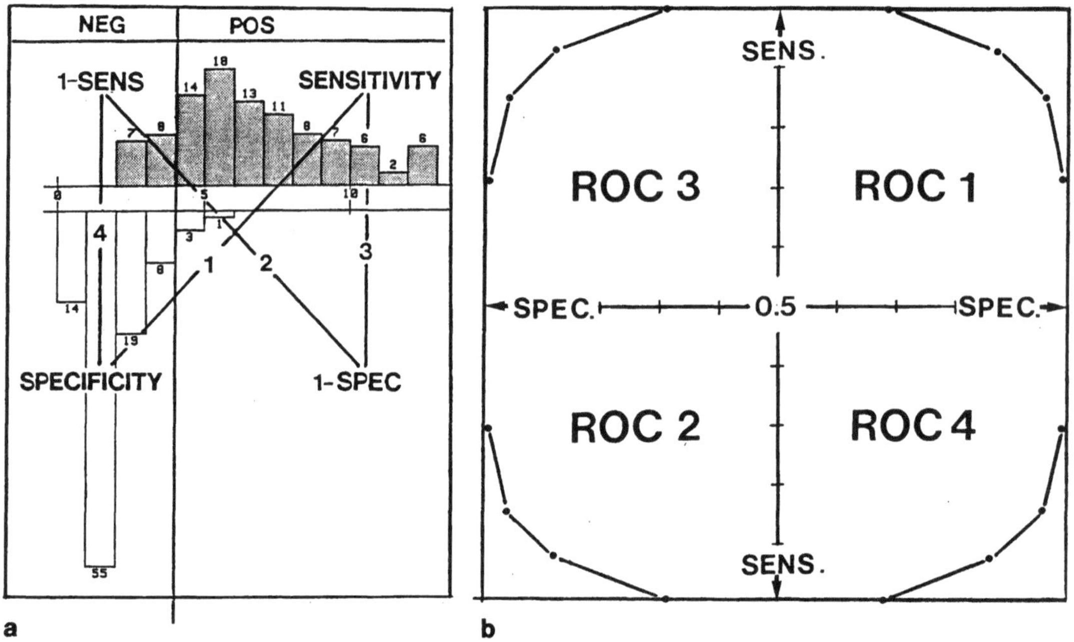

a b

Pairs of corresponding sensitivity, specificity, and complementary values

Discriminator position	sensitivity	specificity	1-sensitivity	1-specificity
0	1.00	0.00	0.00	1.00
1	1.00	0.14	0.00	0.86
2	1.00	0.69	0.00	0.31
3	0.93	0.88	0.07	0.12
4	0.85	0.96	0.15	0.04
5	0.71	0.99	0.29	0.01
6	0.53	1.00	0.47	0.00
7	0.40	1.00	0.60	0.00
8	0.29	1.00	0.71	0.00
9	0.21	1.00	0.79	0.00
10	0.14	1.00	0.86	0.00
11	0.08	1.00	0.92	0.00
12	0.06	1.00	0.94	0.00

Fig. 15. The four types of ROC curves in (b) have been calculated from the test distribution shown in (a): Pairs of corresponding sensitivity and specificity or their complementary values have been plotted as a function of discriminator positions 2 through 5 (Table). Note, that the four types of ROCs are mirror images of each other with respect to the horizontal and vertical axes. Details are given in the text (from Scand. J. Clin. Lab. Invest. Suppl. 181, 1986 (ref. 9), with kind permission).

It is apparent from the calculation of ROCs that these provide direct information only on sensitivity and specificity, but *not* on the fraction DIS or REF (predictive values) within the test classes (9, 50).

ROC 3 is based on the data points: y, x = sensitivity, 1-specificity. Thus, likelihood ratio of disease may be calculated directly from each of these points as: sensitivity/1-specificity. As described above, post. odds DIS for each data-pair may then be calculated through multiplication of ant. odds DIS with LR DIS.

Efficiency

The concept of efficiency of effectiveness has been introduced to select an "optimal" discriminator position by a simple calculation. Two types of efficiency are in common use. "Efficiency I" may be defined as the fraction correct classifications among all classifications by the test (1, 6, 59–60) (Fig. 16). It varies between zero and one, and changes with prevalence in a complex way, depending on the relationship between sensitivity and specificity (9, 59).

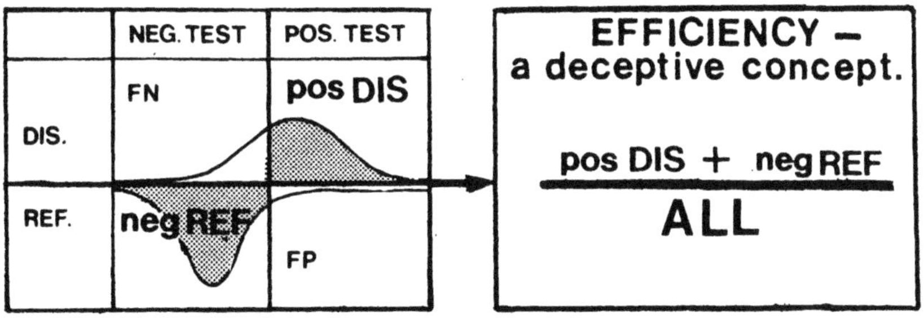

Fig. 16. "Efficiency I": Illustrates the data reduction from the left: distribution in histograms... through the 8 test evaluation quantities: NEG DIS, POS DIS, NEG REF, POS REF, Sensitivity, specificity, PVneg, and PVpos to one single fraction.

"Efficiency II" is simply the sum of sensitivity and specificity (61), varies between zero and two, and is consequently independent of prevalence. It corresponds to ROC 1 (9). Both concepts are based on the oversimplification that POS DIS, NEG DIS, POS REF and NEG REF all are of the same weight. In most clinical situations this is not the case.

Comparison of likelihood ratio, post. odds DIS, ROCs, and efficiency with double histograms with plots of test quantities

Current reports frequently describe the functions of test-class likelihood ratios of DIS (3, 45) and ROCs (50–58) in so advanced mathematical language that they are next to incomprehensible to the not-specialist. It is doubtful whether such statistical efforts are in proportion to the repeatibility of clinical studies. An attempt to summarize the basic functions and limitations of these three evaluation concepts is given below.

It should be emphasized, that this discussion is restricted to the *direct information* provided by each quantity. Obviously, there are several possibilities of calculating the predictive values by inserting the appropriate values of sensitivity, specificity, and anterior prevalence of disease into the complete Bayes' equation.

Test class likelihood ratio of disease

1. indicates the factor by which posterior odds within a test class are increased relative to the anterior odds of disease.
2. is independent of anterior prevalence (or odds) of disease.
3. can be calculated for *n* test classes.
4. may vary from zero to infinity.
5. equals posterior odds DIS at anterior prevalence 0.50.

Test class posterior odds DIS = risk quotient

1. can be calculated for *n* test classes.
2. indicates the ratio DIS:REF within any of *n* test classes.
3. may vary from zero to infinity.
4. changes with anterior prevalence (or odds) of disease.

ROCs

1. are based on weighted or unweighted plots of sensitivity versus specificity (or their complementary values) as functions of discriminator positions. At each discriminator position they deal with a *2-class test* evaluation.
2. sensitivity and specificity, and accordingly ROCs, are independent of the anterior prevalence (or odds) of disease.
3. provide no direct information on predictive values. These must be separately calculated through LR DIS $\times$ ant. odds = post. odds and further post odds DIS/(1 + post odds DIS) = PVpos. Comparison of tests by ROCs thus becomes comparisons of sensitivities and specificities only, and not of predictive values at the relevant anterior prevalence of disease. About half the information required for a meaningful test evaluation or comparison is missing.
4. do not provide the reader with information of the underlying distributions of DIS and REF (9, 62).

Double Histograms with plots of test quantities

1. present the distributions of test data of DIS and REF at any number of test classes, class width, and optional discriminator positions based on the empirical results in the actual study. They provide immediate and intuitive understanding

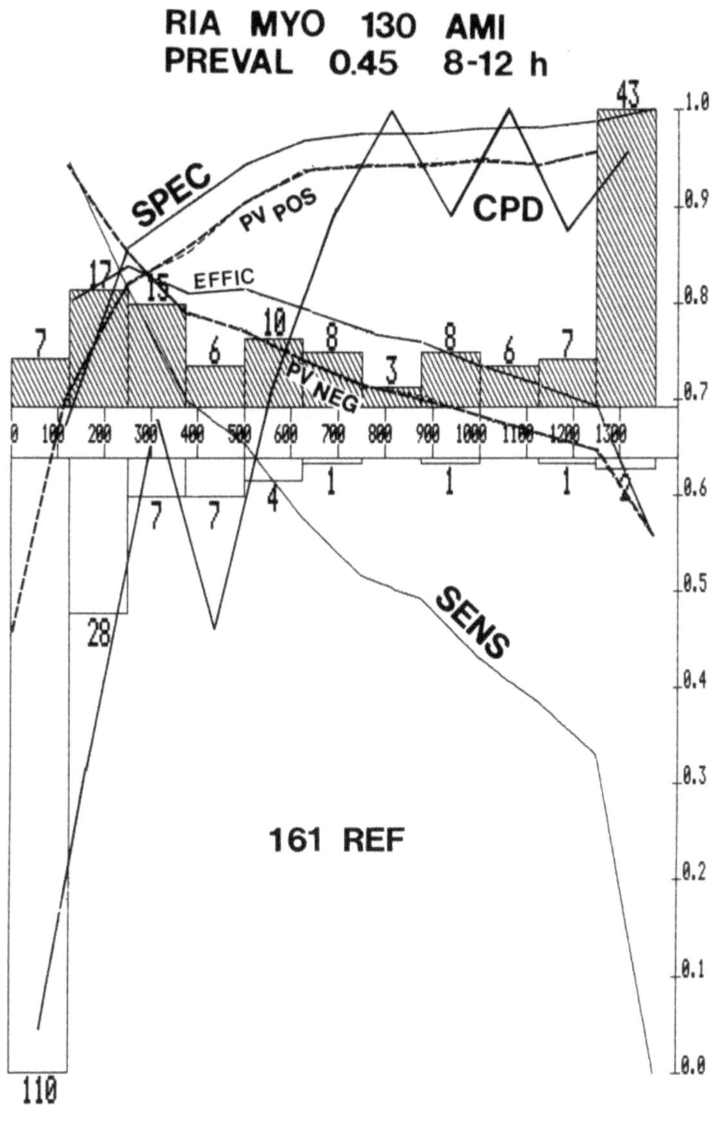

Discr.	PosDis	NegRef	Sens	Spec	CPD	PVpos	PVneg	Effic
0	130	0	1.0000	0.0000	0.0000	0.4467	------	0.4467
125	123	110	0.9462	0.6832	0.0598	0.7069	0.9402	0.8007
250	106	138	0.8154	0.8571	0.3778	0.8217	0.8519	0.8385
375	91	145	0.7000	0.9006	0.6818	0.8505	0.7880	0.8110
500	85	152	0.6538	0.9441	0.4615	0.9043	0.7716	0.8144
625	75	156	0.5769	0.9689	0.7143	0.9375	0.7393	0.7938
750	67	157	0.5154	0.9752	0.8889	0.9437	0.7136	0.7698
875	64	157	0.4923	0.9752	1.0000	0.9412	0.7040	0.7595
1000	56	158	0.4308	0.9814	0.8889	0.9492	0.6810	0.7354
1125	50	158	0.3846	0.9814	1.0000	0.9434	0.6639	0.7148
1250	43	159	0.3308	0.9876	0.8750	0.9556	0.6463	0.6942
1375	0	161	0.0000	1.0000	0.9556	------	0.5533	0.5533

Fig. 17. Example of double histogram with superimposed curves of the test evaluation quantities plotted as function of the discriminator position: sensitivity, specificity, PVpos, PVneg, efficiency and CPD. Values for any of these quantities at any discriminator position may be read off the corresponding intersect with the right ordinate, cf. Figures 4 and 6.

of test performance without requiring any advanced mathematical qualifications and are thus useful in the dialogue e. g. between clinical chemist and the clinician.
2. present plots of the test quantities sensitivity, specificity, PVpos, PVneg or class fraction of DIS (class prevalence of DIS, CPD) at optional test class values and classwidth, at the clinically relevant anterior prevalence of disease.
3. Using a graphic computer the histograms may be used as working tools estimating test performance at any discriminator positions and anterior prevalence of disease (diskettes available from the author).
4. Comparison of different studies of similar tests include all test quantities and may be recalculated to the relevant prevalence.

Figure 17 presents the distributions of 8 – 12 hour RIA myoglobin data discussed previously in a comprehensive graph at an anterior prevalence of AMI of 0.45, class width 125 μg/l, range 0 – 1325 μg/l, with superimposed curves of the five test evaluation quantities: sensitivity, specificity, PVpos, PVneg, efficiency I, and the corresponding numerical data tabulated below.

Synonym Dictionary of some current concepts in test evaluation

Table 3. Some current concepts in test evaluation

Working terminology: name and abbreviation	Definition	Current names	Ref.
For a 2-class test only:			
POS DIS	POS DIS	true positives	1 – 9
Sensitivity	$\frac{\text{POS DIS}}{\text{ALL DIS}}$	diagnostic sensitivity	1
		clinical sensitivity	4
		technical sensitivity	59
		nosological sensitivity	69
		prior probability	8, 59
		positivity in disease	6
		true positive ratio	5
		1-β error	67
		1-α error	7
NEG REF	NEG REF	true negatives	1 – 9
Specificity	$\frac{\text{NEG REF}}{\text{ALL REF}}$	diagnostic specificity	1
		clinical specificity	4
		technical specificity	59
		nosological specificity	66
		negativity in disease	6
		true negative ratio	5
		1-α error	67
		1-β error	7

Continued Table 3: see page 201

Continued Table 3

Working terminology: name and abbreviation	Definition	Current names	Ref.
NEG DIS	NEG DIS	false negatives	1–9
		alpha error	6, 7
		beta error	67
		error of the first kind	7
		error of omission	6
POS REF	POS REF	false positives	1–9
		alpha error	67
		beta error	7
		error of the second kind	6, 7
		error of commision	6
PVpos	$\frac{\text{POS DIS}}{\text{ALL POS}}$	predictive value of a positive result	1–9
		diagnostic sensitivity	59
		diagnostic specificity	66
		predictor for disease	65
PVneg	$\frac{\text{NEG REF}}{\text{ALL NEG}}$	predictive value of a negative result	1–9
		diagnostic specificity	59
		diagnostic sensitivity	66
		indicator for health	65
generally applicable to n-class test evaluations:			
CFD, CPD	$\frac{\text{class X DIS}}{\text{class X ALL}}$	class X fraction (prevalence) of DIS	9, 13
CFR, CPR	$\frac{\text{class X REF}}{\text{class X ALL}}$	class X fraction (prevalence) of REF	9, 13

"we frequently hear decision makers, including clinicians, discussing alpha and beta errors type I and type II errors, errors of omission and commision, and false negatives and false positives. What do they mean?" (Galen and Gambino, ref. 6, p. 127)

Acknowledgements

The author wants to give his warmest thanks for numerous valuable suggestions, encouragement, and positive criticism to the following persons: professors H. Keller, St. Gallen, Switzerland, O. Siggaard-Andersen, Herlev University Hospital, Denmark, and M. Hörder, Odense University Hospital, Denmark, who also supported the development of a graphic test evaluation programme which may be run on any IBM compatible PC with MS DOS 2.2 x or higher, and at least 640 kb RAM (diskettes available from the author); to J. F. Bergseng, public health statistician; and to K. Nörregaard-Hansen, Odense, for permission to use his S-myoglobin data as examples.

References

1. Büttner, J. (1977) Die Beurteilung des diagnostischen Wertes klinisch-chemischer Untersuchungen. J. Clin. Chem. Clin. Biochem. *15*, 1–12.
2. Vecchio, T. J. (1966) Predictive value of a single test in unselected populations. N. Engl. J. Med. *274*, 1171–1173.
3. van der Helm, H. J. & Hische, E. A. H. (1979) Application of Bayes Theorem to results of quantitative clinical chemical determinations. Clin. Chem. *25*, 985–988.
4. Sunderman, F. W. (1975) Current concepts of "Normal Values", "Reference Values", and "Discrimination Values" in Clinical Chemistry. Clin. Chem. *21*, 1873–1877.
5. McNeil, B. J., Keeler, E. & Adelstein, S. J. (1975) Primer on certain elements of medical decision making. N. Engl. J. Med. *293*, 211–215.
6. Galen, R. S. & Gambino, S. R. (1975) Beyond normality: The predictive value and efficiency of medical diagnosis. Wiley J. and Sons, New York.
7. Lusted, L. B. (1968) Introduction to Medical Decision making. C. C. Thomas, Springfield, Illinois.
8. Murphy, E. A. (1970) Probability in Medicine. John Hopkins, Baltimore.
9. Gerhardt, W. & Keller, H. (1986) evaluation of test data from clinical studies. I. Terminology, graphic interpretation, diagnostic strategies, and selection of sample groups. II. Critical review of the concepts of efficiency, receiver operating characteristics (ROC), and likelihood ratios. Scand. J. Clin. Lab. Invest. Suppl. 181, pp. 0–74.
10. Nörregård-Hansen, K., Gerhardt, W., Olsson, V. & Pedersen, T. S. (1986) Serum Myoglobin as a three-class test in the diagnosis of acute myocardial infarction. (abstract). Scand. J. Clin. Lab. Invest. Suppl. 185, 163.
11. McComb, J. M., McMaster, E. A., McKenzie, G. & Adghey, A. A. J. (1984) Myoglobin and creatine kinase in acute myocardial infarction. Brit. H. J. *51*, 189–194.
12. Metz, C. E. (1978) Basic principles of ROC analysis. Seminars in Nuclear Med. *8*, 283–298.
13. Gerhardt, W., Berning, H., Olsen, J. S. & Bagger, P. (1983) Evaluation of Quantitative Tests by Class Prevalence of Disease. in: Proc. 5th Colloque Int. de Pont a Mousson (Siest, G., ed.) pp. 869–874, Masson et cie. Paris.
14. Linnet, K. (1985) Precision of sensitivity estimations in diagnostic test evaluations. Power functions for comparisons of sensitivities of two tests. Clin. Chem. *31*, 574–580.
15. Whitfield, I. B., Hensley, W. J., Bryder, D. & Gallagher, H. (1978) Some laboratory correlates of drinking habits. Ann. Clin. Biochem. *15*, 297–303.
16. Whitfield, I. B., Hensley, W. J., Bryden, D. & Gallagher, H. (1978) Estimation of alcohol intake from laboratory results. Ann. Clin. Biochem. *15*, 304–336.
17. Wadstein, J. (1980) Biochemical Changes in hospitalized chronic male alcoholics. Thesis, Malmö.
18. Eckardt, M. J., Rawlings, R. R., Ryback, R. S., Martin, P. R. & Gottschalk, L. A. (1984) Effects of abstinence on the ability of clinical laboratory tests to identify male alcoholics. Am. J. Clin. Pathol. *82*, 305–310.
19. Gluud, C., Andersen, I., Dietrichson, O., Gluud, B., Jaconsen, A. et al. (1981) Gamma Glutamyltransferase, aspartate aminotransferase, and alkaline phosphatase as markers of alcohol consumption in out-patient alcoholics. Europ. J. Clin. Invest. 11, 171–176.
20. Gerhardt, W. (1985) Planning and evaluation of clinical studies of diagnostic tests. Failure of S-prostatic acid phosphatase in screening for prostatic carcinoma. Scand. J. Clin. Lab. Invest. *179*, 31–41.
21. Watson, R. A. & Tang, D. B. (1980) The predictive value of prostatic acid phosphatase as a screening test for prostatic cancer. N. Engl. J. Med. *303*, 497–499.

22. Henderson, A. R. & Nealon, D. A. (1982) Clinical enzymology. in: Clinical Biochemistry Reviews, vol. 3 (Goldberg, D. M., ed.), pp. 187–197. Wiley, New York.
23. Bruce, A. W. & Mahan, D. E. (1982) The role of prostatic acid phosphatase in the investigation and treatment of adenocarcionoma of the prostate. Ann. N. Y. Acad. Sci. *390*, 110–121.
24. Shaw, L. M., Yang, N., Neal, M. & Croop, W. (1982) Immunological and clinical specificity of the immunochemical determination of prostatic acid phosphatase. Ann. N. Y. Acad. Sci. *390*, 73–85.
25. Wagener, C. (1984) Diagnostic sensitivity, diagnostic specificity, and predictive values of the determination of tumour markers. J. Clin. Chem. Clin. Biochem. *22*, 969–979.
26. Paterson, N., Biggart, E. M., Chapman, R. S. & Beastall, G. H. (1985) Evaluation of a time-resolved immunofluorometric assay for serum thyroid stimulating hormone. Ann. Clin. Biochem. *22*, 606–611.
27. Ellis, G., Worthy, E. & Goldberg, D. M. (1979) Lack of value of serum glutamyltransferase in the diagnosis of hepatobiliary diseases. Clin. Biochem. *12*, 142–145.
28. Prellwitz, W. (1981) Creatine kinase isoenzymes in direct skeletal muscle damage. in: Creatine Kinase Isoenzymes (Lang, H., ed.), pp. 170–175. Springer, Berlin.
29. Shahangian, S., Ash, K. O., Wahlström, N. O. et al. (1984) Creatine kinase and lactate dehydrogenase isoenzymes in serum of patients suffering burns, Blunt Trauma, or myocardial infarction. Clin. Chem. *30*, 1332–1338.
30. Apple, F. S., Rogers, M. A. & Sherman, W. M. (1984) Profile of creatine kinase isoenzymes in skeletal muscle of Marathon runners. Clin. Chem. *30*, 413–416.
31. Fuentes, J., Martinez, M., Fuste, M. & Biosca, C. (1984) Isoenzyme 1 of lactate dehydrogenase is increased in hepatic diseases. Clin. Chem. *30*, 1424–1425.
32. Gerhardt, W., Hofvendahl, S., Ljungdahl, L., Waldenström, J., Tryding, N. et al. (1983) Serum LD-1 activity in suspected acute myocardial infarction. Clin. Chem. *29*, 1057–1060.
33. Cooper, J. F. (1980) The radioimmunochemical measurement of prostatic acid phosphatase: Current state of the art. Urolog. Clin. N. America *7*, 653–665.
34. WHO Diabetis mellitus. Report of a WHO Study Group. WHO Techn. Report Series 727 WHO, Geneva 1985.
35. Gerhardt, W., Waldenström, J., Hörder, M., Hofvendahl, S., Billström, R. et al. (1982) Creatine kinase and creatine kinase B-subunit activity in serum in cases of suspected myocardial infarction. Clin. Chem. *28*, 227–283.
36. Mulley, A. G., Thibault, G. E., Hughes, R. A., Barnett, G. O., Reder, V. A. & Sherman, E. L. (1980) The course of patients with suspected myocardial infarction. N. Engl. J. Med. *302*, 943–948.
37. Gerhardt, W., Ljungdahl, L., & Hofvendahl, S. Serum Enzymes at Suspected Acute Myocardial Infarction. Pictorial Atlas With Case Histories. GIT Verlag 1988..
38. Oellerich, M. & Haeckel, R. (1978) Diagnostischer Wert klinisch-chemischer Untersuchungen. Med. Welt *29*, 866–872.
39. Dahlgren, L. & Hollstedt, C. (1983) Biokemiske markörer behjälpelige vid tidigdiagnos av kvinnliga alkoholister. Läkartidningen *50*, 4890–4893.
40. Durbridge, T. C. (1983) Clinical acceptance of a multi-test reference region for biochemical panel results. Clin. Chem. *29*, 1724–1726.
41. Keller, H. & Gessner, U. (1980) Auswahl und Interpretation diagnostischer Parameter. Internist *21*, 173–180.
42. Bayes, T. H. (1763) An essay towards solving a problem in the doctrine of chances. Philosoph. Transact. R. Soc. London. *53*, 370–418.
43. Hall, G. H. (1967) The clinical application of Bayes theorem. Lancet, 555–557.
44. Hische, E. A. H., van der Helm, H. J., van Walbeek, H. K. (1982) The cerebrospinal fluid immunoglobulin G index as a diagnostic aid in multiple sclerosis: A Bayesian approach. Clin. Chem. *28*, 354–355.
45. van der Helm, H. J., Hische, E. A. H. & Bolhuis, P. A. (1982) Bayes Theorem and the estimation of the likelihood ratio. Clin. Chem. *28*, 1250–1251.
46. Albert, A. A. (1982) On the use and computation of likelihood ratios in Clinical Chemistry. Clin. Chem. *28*, 1113–1119.
47. Keller, H. (1981) The likelihood quotient line, a tool for assessment of diagnostic validity of clinical chemical tests. J. Clin. Chem. Clin. Biochem. *19*, 1211–1212.
48. Simel, D. L. (1985) Playing the Odds. Lancet, 329–330.
49. Gerhardt, W. (1987). Likelihood ratios, odds, and class prevalence of disease compared. Scand. J. Clin. Lab. Invest. *47*, 635–636.

50. National Committee for Clinical Laboratory Standards. Assessment of clinical sensitivity and specificity of laboratory tests; Proposed Guideline. NCCLS publication GP 10–P. Villanova, PA.: NCCLS; 1987.
51. Egan, J. P. (1975) Signal Detection Theory and ROC Analysis. Academic Press, New York.
52. Haeckel, R., Riedel, H. & Büttner, J. (1981) Estimation of decision criteria for the uric acid concentration for the early diagnosis of gestosis. J. Clin. Chem. Clin. Biochem. *19*, 173–176.
53. Jacobsson, G., Groth, T. & deVerdier, C. H. (1980) Pancreatic iso-amylase in serum as a diagnostic test in different clinical situations. A simulation study. Scand. J. Clin. Lab. Invest. *40*, suppl. 155, 75–77.
54. Robertson, E. A. & Zweig, M. H. (1981) Use of receiver operating characteristic curves to evaluate the clinical performance of analytical systems. Clin. Chem. *27*, 1569–1574.
55. Hanley, J. A. & McNeil, B. J. (1982) The meaning and use of the area under a receiver operating characteristic (ROC) curve. Diagnostic Radiology *143*, 29–36.
56. Herman, G. A., Sugiura, H. T. & Krumm, R. P. (1986) Comparison of thyrotropin assays by relative operating characteristics. Arch. Pathol. Lab. Med. *110*, 21–25.
57. Noe, D. A. (1983) Selecting a diagnostic studys cutoff value by using its receiver operating characteristic curve. Clin. Chem. *29*, 571.
58. Beck, J. R. & Schultz, E. K. (1986) The use of relative operating characteristic (ROC) curves in test performance evaluation. Arch. Pathol. Lab. Med. *110*, 13–20.
59. Werner, M., Brooks, S. H. & Wette, D. (1973) Strategy for cost-effective laboratory testing. Human Pathol. *4*, 17–23.
60. Rossing, R. G. & Hatcher, W. E. (1980) A graphic method for the evaluation of diagnostic tests. Meth. Inform. Med. *19*, 149–156.
61. Stamm, D., Hansert, E. & Feuerlein, W. (1984) Detection and exclusion of alcoholism in men on the basis of clinical laboratory findings. J. Clin. Chem. Clin. Biochem. *22*, 79–96.
62. Fuentes, J. & Martinez, M. (1983) An alternative to receiver-operating characteristic curves. Clin. Chem. *29*, 1445.
63. Kark, S. L. (1974) Epidemiology and Community Medicine. Appleton-Century, New York.
64. Spicer, C. C. (1980) Test reduction: II-Bayes's theorem and the evaluation of tests. Brit. Med. J. *283*, 592–594.
65. Sebag, J. & Hall, P. (1975) Decision-making in medical research and clinical practice: Theory and methodology of laboratory data. Evaluation by predictors, indicators, and indices. Methods Inform. Med. *14*, 113–117.
66. Wulff, H. (1976) Rational Diagnosis and Treatment. Blackwell, London.
67. Feinstein, A. R. (1977) Clinical Biostatistics. C. V. Mosby, St. Louis.

Chapter 3.2
Deterministic Methods for Computerized Interpretation of Laboratory Tests

M. D. D. McNeely

1 Introduction

The clinical laboratory's duty to interpret the results it produces can be greatly enhanced by recent developments in computer technology. Compared to hand-written or even typed reports, interpretations printed by a computer have a better format, can be prepared with greater efficiency, enjoy consistency of the report content and may be refined and improved over time.

Software is the key to interpretive reporting and several different approaches are possible (1).

Probability models were the first to be studied intensively (2). In specific settings they have been very successful. As a general approach they are impractical because they demand precise a priori as well as post-test probability data for every disorder under consideration. They become extremely complex when numerous tests are considered.

Multivariate analysis is an intriguing mathematical technique able to produce "clusters" from the results of several tests. This method has been used diagnostically but is most successful in studying the diagnostic power of various test combinations. As a routine interpretive approach it has several problems. A high level of mathematical sophistication is required to work with the technique. Exact probability data is needed for every disorder under consideration, and the results for every test employed in the diagnostic model must always be provided. A recent book applies a multivariate analysis to a variety of clinical conditions (3). Both mathematical and graphical techniques are demonstrated.

Mathematical modelling techniques are particularly useful for pharmacologic manipulations (4), and blood-gas and acid-base interpretation (5). They will find increasing application in endocrinology and metabolism. Such methods cannot be applied to a wide variety of problems.

Artificial Intelligence is developing at a rapid rate and methods evolving from this area of endeavor (such as expert systems) will become widely used (6, 7). However, the field is still in its infancy, many technical problems have not yet been solved, and a high degree of sophistication is required for its application.

For these reasons, Deterministic Methods are currently the most practical for developing computer generated interpretive reports. Deterministic Methods are computer programs which allow the computer to locate and print a specific interpretation which has been designed for each possible combination of test results. The advantage of this approach lies in its simplicity and general applicability. Its disadvantage lies in its tendency to become exponentially complex after three or four test results are considered together. However, this is not a serious problem since many of the interpretations made in the clinical laboratory involve one or two test combinations, or unique patterns of larger test panels.

2 Hardware and Software Considerations

2.1 Microcomputer vs. minicomputer

The type of computer employed in interpretive reporting depends upon the software system to be used.

Microcomputers are particularly useful for interpretive reporting because they have sufficient power to perform most tasks, are easily programmed and will usually be accessible so that modifications can be made.

Developing an interpretive reporting system on a Laboratory Information System (LIS) is fraught with problems at the present time. The reporting options in these systems constrain the textual nature of the interpretive reports. As a rule, software modifications are not permitted in commercial LIS nor is there provision for user developed features. The need to assure 24 hour operation of the LIS may compromise the ability to develop new programs and modify existing ones.

At the present time microcomputer offer the most practical tool for the development of automated interpretive reporting systems. It is gradually becoming possible to develop such systems on LIS.

2.2 Language

Computer programs for Deterministic Interpretive reporting are not complex and unlike AI methods, can be run with very little computer power. Any higher level computer language is suitable for Deterministic Interpretive reporting. Speed is not crucial and therefore the language may be compiled or interpreted. Because of the need to make changes in the logic structure as the program is being developed and as new facts become known, the computer code should be convenient to change. Because the interpretations are text intensive, it is important to use a language with convenient string-handling capability.

For all these reasons, BASIC is a very suitable language choice.

Regardless of the computer language used it is crucial to document the programs in detail. It is important to keep all the source documents which provided the medical facts used to create interpretations. It is vital to maintain careful flow diagrams of the logic upon which the interpretation is based.

2.3 Commercial programming aids

A number of commercial programs are available to assist in the development of interpretive reporting systems (8).

Texteditors are a valuable adjunct which allow the rapid composition of the interpretive comments.

Word processors are a standard utility on most computers and, if they have a good document retrieval system can be used directly in report generation.

Special Form Generator programs can be purchased. These are particularly suited to the preparation of specialized reports.

Programs, designed specifically to aid in the creation of interpretive reports are now becoming available. One of these, developed by Dr. Edward Hobart, Jr. will be described below.

2.4 Commercial deterministic interpretive reporting systems

Several Deterministic Interpretive reporting programs are available commercially. The Pribor System (9) was the first of these to become available and evolved from a system employed routinely by a pathology group.

Another program for the interpretation of protein electrophoresis patterns was developed using an AI Expert system shell (EXPERT, Rutgers University) (10). When judged to be functioning in an optimum fashion its logic was converted to assembly language code for a microprocessor and used as a component of a commercial electrophoretic system (11). This is an example of AI used to produce a deterministic interpretive system.

2.5 Essential features

Whether the system used is developed in-house or obtained from a commercial vendor several important points prevail:

1. The system must be documented so that all medical facts can be re-evaluated at any time.
2. The system must be amenable to change since interpretations will differ depending upon the clinical setting in which it is used (out-patients, pediatric hospital, chronic care institution, etc.) and because new information will constantly become available and must be added.
3. The system must be under a physician's control so that medical responsibility for the interpretations is always present.

2.6 Composing the comment

Composing the interpretive comment is the key to creating a useful reporting system. It is important to craft the interpretive comment with care. Not only must the medical information be correct but it must also be presented in a clear and unambiguous fashion. Simply generating a list of possible disorders has limited usefulness but is the approach most often taken. This "memory jogging" approach can be justified on the basis that a diagnosis never considered is one that will never be made.

It is more helpful to the practitioner receiving the report for the list of disorders to be priorized. For example, a fictitious interpretation might be phrased: "Results of this magnitude are highly suggestive of pancreatitis, but have also been reported in association with parotitis, bowel obstruction and ectopic pregnancy. The drugs morphine and acetylcholine are known to cause false positive results. Rare causes are bronchogenic carcinoma, abdominal trauma and renal failure."

It is often possible to list the possible diagnoses with an indication of how often the test abnormality is a feature of the condition. For example: "Values of this magnitude are seen in 90% of persons suffering from rheumatoid arthritis, in 40% of those with systemic lupus, and in less than 10% of those with mixed connective tissue disease."

Physiologic explanation of the results is very useful when it can be done. For example: "Primary hypothyroidism is highly likely. The TSH is raised in an attempt to compensate for the low T_4 (Thyroxine)..." Suggestions regarding the diagnostic approach that should follow are also helpful. For example: "This cortisol result is above the upper limit of normal. To clarify the diagnosis a Dexamethasone Suppression Test is suggested and can be arranged by phoning 387-3333."

2.7 Information sources

Conventional textbooks and journal articles provide little information which can be easily incorporated into an interpretive reporting system. Several sources which are particularly helpful are: Tietz (12), Young (13), Speicher (14), and Friedman (15).

3 "Canned" Comments

3.1 Definition

The simplest form of Deterministic Interpretation is the "canned comment". Here, the computer system stores a selection of pre-written comments which can be appended to the laboratory report. This type of interpretation is a common feature of LIS and is usually invoked by requiring each result to be reviewed and an appropriate comment selected by the viewer from a menu of options displayed on a video terminal.

3.2 Word processor applications

Conventional word processors may be used to develop very sophisticated systems which do not rely on computer programming. The word processor is used to create as many interpretive comments as necessary. A word processor with a sophisticated indexing capability is recommended because this will allow all interpretations related to a particular test to be grouped together and retrieved quickly using key words. For example, a word processing program may allow all the interpretations for the Schilling Test to be catalogued together. The reviewing physician may be able to call the appropriate interpretation using a code form of the results. For example, the code S1L2N stands for "S" = Schilling, "1L" = Part 1 is low, and "2N" = Part 2 is normal.

Results		*Code*
Part I	Part II	
2%	15%	S1L2N
10%	not done	S1L2X
2%	3%	S1L2L

Christie has developed a complete system of computer generated clinical pathology consultations.

The system is described in detail in a forthcoming book (16). His system of interpretations is designed to provide assistance to family practitioners in community practise. Consultations have been created for cardiac risk, tumor markers, enzymes, electrolyte abnormalities, hormones, acid-base, hematology, blood banking, therapeutic drug monitoring, clinical microscopy and microbiology. For each test (or category) there are several options available. For example, for the CEA (carcinoembryonic antigen) there are seven consultation letters.

Code	*Title*
CEAABNC	CEA Abnormal, No change
CEAAADEC	CEA Abnormal, Decreasing
CEAABINC	CEA Abnormal, Increasing
CEAABMET	CEA Abnormal, Metastases
CEA2.5 – 5	CEA 2.5 – 5 ng/mL range
CEAAB > 5	CEA Abnormal, > 5
CEANORM	CEA Normal

Each consultation letter provides demographic information, the test result(s), detailed reference ranges, a clear and detailed scientific explanation of the results and the clinical significance. Literature references are also listed when applicable.

3.3 Menu driven and canned comments

A similar approach uses a menu of options which can be selected by the reviewing physician.

A hematology interpretation program developed by Brigden and Morgan is an example. This system has been designed to assist a hematologist in his review of abnormal blood films. The hematologist first reviews the film under the microscope. He then turns to the computer where 13 categories of abnormality are displayed. They are:

(1) normocytic anemias, (2) microcytic anemias, (3) macrocytic anemias, (4) mixed anemias, (5) hemolytic anemias, (6) erythrocytosis, (7) red cell morphology, (8) red cell inclusions, (9) quantitative white cell changes, (10) qualitative white cell changes, (11) platelet abnormalities, (12) rouleau, (13) non-diagnostic.

The appropriate number is typed in and a new list of possibilities is displayed. For example, a microcytic picture is present, number 2 is selected. This causes 4

ISLAND MEDICAL LABORATORIES
HEMATOLOGY BONE MARROW REPORT

[redacted]	[redacted]	2-501 [redacted]	Y-156	Nov04/86rp
SURNAME	FIRST NAMES	ADDRESS	LAB No.	DATE
03[redacted]	381[redacted]		DR. L. [redacted]	
BIRTH DATE		PHONE	DOCTOR	

BONE MARROW REPORT: SPECIMEN - ASPIRATE PUNCTURE BY: M.L. BRIGDEN, M.D.

The aspirate site is P.I. spine. The bone density of the puncture is normal.

The marrow aspirate is moderately hypercellular. The E:G ratio is 1:4. Erythropoiesis is quantitatively normal. Erythroid maturation is normoblastic with a few megaloblastoid features and occasional bi-nucleated late normoblasts. Hemoglobinization is slightly decreased. Granulocytes are quantitatively increased. There is normal maturation with increased numbers of myelocytes. There are giant metamyelocytes and staff cells. There are increased numbers of eosinophils. Lymphocytes are quantitatively normal although a few particle smears display extensive infiltration of mature lymphocytes. The estimate is 20%. Plasma cells are quantitatively increased. The estimate is 3%. An occasional bi-nucleated cell is noted. Megakaryocytes are quantitatively normal and are budding platelets. Foreign cells are not present but numerous storage cells are noted and mast cells, both fixed and free, are increased. Iron stores are moderately increased. Ringed sideroblasts are absent.

IMPRESSION:

Nonspecific changes most likely reactive in nature. Await biospy.

M.L. BRIGDEN, M.D.
F.R.C.P.(C) HEME

Fig. 1.

possibilities to be presented: (1) microcytic hypochromic anemia, (2) microcytic anemia with normal MCHC, (3) microcytosis (low MCV), and (4) microcytic hypochromic with fragmentation.

The appropriate option may be selected. A pre-determined, canned comment is then generated. The opportunity for additional free text commentary is also available. This system has a total of 61 comments which are capable of providing over 70% of all abnormal out-patient blood films to be given an interpretive comment without additional free text. Less than 10% of all blood films require a complete free text commentary.

At the next level of sophistication are systems which compose the text of an interpretive comment following the entry of various descriptive parameters. For example, a system used for generating bone marrow reports has also been developed by Brigden and Morgan. This program displays a series of questions on the VDT of the computer. Each question has a menu of possible responses. The hematologist selects the appropriate response to each question. The question to be presented depends upon the response to the previous question. When all questions have been answered a text commentary is automatically composed and printed (Fig. 1). Systems like this have the advantage of instantly providing a finished report, allowing electronic transfer and efficient storage of the interpretation. The interpretation can be compressed into a series of numbers (menu responses), segregated by commas. This not only allows for memory efficient storage but permits database storage and subsequent retrieval of reports by specific characteristics (microcytosis, hypocellular, etc.). Free text entry with standardized numerical coding is an important additional feature.

The most elaborate example of this type of program is one developed by Gantner for recording a standard or a forensic autopsy (17). It asks sufficient questions to allow the generation of a complete autopsy report. The system performs automatic coding of the causes of death.

Such programs can be extremely useful provided they are not used to reduce secreratial time while forcing increased expensive professional time.

Many pathologists in the U.S.A. prefer "canned" interpretive comments which require professional selection over interpretive comments selected by computer logic. This is because payment for the interpretation may depend upon the active selection of the comment by a laboratory physician.

4 Logic Techniques

4.1 Branching logic

The standard approach employed in deterministic schemes is branching logic. Here, the logic is represented by a tree, each branch of which represents a new decision with two (or more) possible options. The point at which a branch must be selected is called a decision node. Figure 2 shows an example of branching logic applied to serum gastrin interpretation.

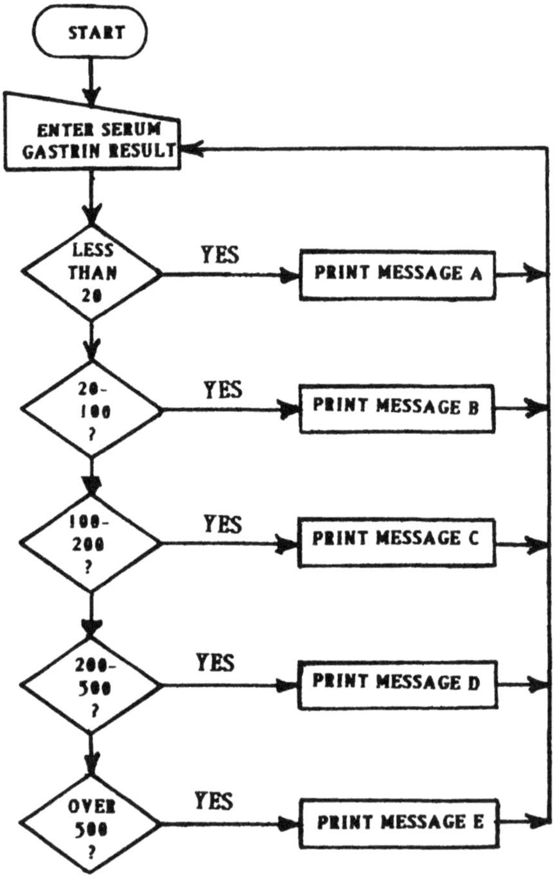

Fig. 2.

Any algorithmic decision tree which can be drawn on paper can be encoded into a straight-forward computer program. Each decision node is encoded using an IF/THEN program statement.

The combination of two or more tests can obviously be handled in the same way. However, when the number of clinically significant test value ranges is 6 (low, borderline low, normal, borderline high, high and very high) and three tests are employed in a group, the number of combinations is 216. This number is difficult to portray in a logic tree. When four tests are considered the number becomes an unmanageable 1296.

A method which assists in this process and allows a large tree to be created is a commercial program developed by Hobart, Jr. (18). In Hobart's system, an algorithm generator allows the algorithm path, the comments and the decision levels to be modified by a series of higher level commands rather than by programming. In this system, the decision tree can be balanced or unbalanced. Each decision level can include Boolean logic, numeric decision points, table look up, or calculations. Each algorithm can have over 50 decision levels and each level can have multiple branches.

4.2 Octal coding

An early paper by Dito (19) provided a logic system for interpretation using an octal coding technique. In this method the results of any test were classified as :

Value	*Test Result*	*Compressed Format*
1	Decreased	1
2	Normal	2
3	Normal to decreased	1,2
4	Increased	4
5	Never normal but may be increased or decreased	1,4
6	Normal to increased	2,4
7	Decreased, normal, or increased	1,2,4
0	Do not know result	0

All test possibilities could be defined by using a compressed format consisting of the digits 0,1,2,4.

A panel of four test results could therefore be represented in octal code as a four digit number in which only the numbers 1, 2 or 4 are present. Each disorder with an effect on the test panel could be represented by several code numbers (one for each possible combination seen in association with that disorder). A library of codes was easily created (and modified) against which a patient's results could be compared in order to produce a "list" of possible disorders. This system was particularly suitable for use on earlier computers with limited memory.

4.3 Binary branching tree

If the tests employed in the testing panel produce binary results (POSITIVE vs. NEGATIVE; REACTIVE vs. NON-REACTIVE) the creation of a binary branching tree is possible and a logic table can also be used. This is demonstrated in Figure 3 which shows a Hepatitis B serology scheme.

It can be seen that the logic is depicted by both methods. The branching graph is somewhat more complex than the table but has the advantage of being a clearer representation of the sequence required for computer coding.

4.4 Matrix logic

A more complex matrix logic method was developed in 1979 by Gordon-Findlay (20). This method can be used for tests with numerous clinically significant value ranges and can theoretically be adapted for N-tests. In fact, the practical limit is three or four analytical results. This program allows a non-programmer to create an interpretive system. From a menu of options, the author may select a "CREATE MATRIX" option. This asks for the name of the new matrix, the limits of the matrix, the name of the test represented on the horizontal axis, the incremental values into which the horizontal axis is divided, the name of the test represented by the vertical axis and the incremental divisions of the vertical axis.

Decision Tree Formulation of the Logic

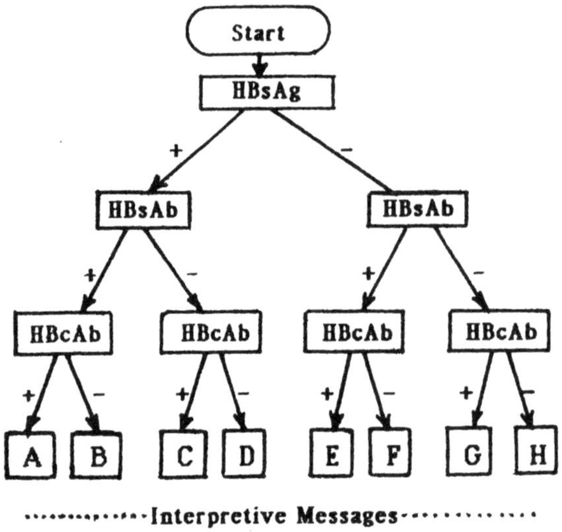

Decision Matrix Formulation of the Logic

HBsAg (Hepatitis B surface Antigen)	HBsAb (Hepatitis B surface Antibody)	HBcAb (Hepatitis B core Antibody)	Interpretive Message
+	+	+	A
+	+	-	B
+	-	+	C
+	-	-	D
-	+	+	E
-	+	-	F
-	-	+	G
-	-	-	H

Fig. 3.

A PRINT MATRIX OPTION allows the matrix to be printed out. This empty matrix can then be used to manually define the logic. For illustrative purposes a trivial example has been given (Fig. 4). For every combination of the two tests a specific interpretation can be defined. A positive whole number representing a unique interpretive comment can be entered into each matrix cell. It is clear that an unlimited number of unique comments can be accommodated by this system. It is also obvious that several test combinations (cells) will share a common

MATRIX LOGIC

T4 (Thyroxine) in µg/dL vs. TSH (Thyrotropin) IU/dL

T4	0-.9	1-1.9	2-2.9	3-3.9	4-4.9	5-5.9	6-6.9	7-7.9	8-8.9	9-9.9	>= 10
TSH											
0 - .9	-1	-1	-1	-1	1	1	1	1	1	1	0
1 - 1.9	-1	-1	-1	-1	1	1	1	1	1	1	0
2 - 2.9	-1	-1	-1	-1	1	1	1	1	1	1	0
3 - 3.9	-1	-1	-1	-1	1	1	1	1	1	1	0
4 - 4.9	-1	-1	-1	-1	1	1	1	1	1	1	0
5 - 5.9	-1	-1	-1	-1	8	5	5	5	5	5	0
6 - 6.9	-1	-1	-1	-1	8	5	5	5	5	5	0
7 - 7.9	-1	-1	-1	-1	8	5	5	5	5	5	0
8 - 8.9	4	4	4	4	4	5	5	5	5	5	0
9 - 9.9	4	4	4	4	4	5	5	5	5	5	0
10 - 10.9	4	4	4	4	4	5	5	5	5	5	0
11 - 11.9	2	2	2	2	2	3	3	3	3	3	0
>= 12	2	2	2	2	2	3	3	3	3	3	0

If -1 is inserted in the primary matrix (above) the matrix below is entered

FTI	0-.9	1-1.9	2-2.9	3-3.9	4-4.9	5-5.9	6-6.9	7-7.9	8-8.9	9-9.9	>= 10
TSH											
0 - .9	7	7	7	8	1	1	1	1	1	1	0
1 - 1.9	7	7	7	8	1	1	1	1	1	1	0
2 - 2.9	7	7	7	8	1	1	1	1	1	1	0
3 - 3.9	7	7	7	8	1	1	1	1	1	1	0
4 - 4.9	7	7	7	8	1	1	1	1	1	1	0
5 - 5.9	9	9	9	9	6	6	6	6	6	6	0
6 - 6.9	9	9	9	9	6	6	6	6	6	6	0
7 - 7.9	9	9	9	9	6	6	6	6	6	6	0

Each positive number in the matrix indicates a different interpretive message.

Fig. 4.

interpretive comment. This results in zones in the matrix with identical comment codes. For absurd results (or result combinations for which a comment has not yet been prepared, a zero (0) left in the matrix will cause the computer to reject the report and force manual intervention. If necessary, a negative whole number may be entered into the matrix cell. When this is encountered the program seeks out a new matrix which requires the result of a new test. Obviously, many additional matrices can be linked in this fashion.

Myhre has employed a three-dimensional matrix to structure the logic of a program for paternity testing (21). In this clinical dilemma is to determine the statistical chance that a man is the actual father when the blood types of a mother and child are known. Myhre's approach was to visualize all the possible combinations as a three-dimensional matrix with each individual's blood group located on the X, Y and Z axes. The matrix cell, identified by the intersection of the three blood types, provides the required statistical information.

4.5 Other approaches

Commercially available data base programs (LOTUS 1–2–3, Lotus Development Corporation, 55 Wheeler Street, Cambridge, MA 02188 and dBase III, Ashton-Tate, 10150 W. Jefferson, Culver City, CA 90230) have been used on an experimental basis to establish a matrix of representative cases. The database search capability can be used to locate cases with matching test result patterns. This approach has been employed by several workers (personal communication) but no formal publications have been prepared.

The BASIC programming language can be used to create the production rules and an inference engine like those used in expert systems. This allows an expert system of very limited scope to be created. This approach has been used successfully for teaching purposes. Although it is feasible to produce interpretive logic in this way, branching methods are equally effective for small projects.

5 Conclusion

There is little doubt that the future of computerized interpretive reporting lies in the further development of mathematical models of pharmacokinetic and metabolic physiology and in AI techniques. At the present time however, deterministic methods which rely on the selection (by human or machine logic) of an interpretive message specific for every test result (or combination of results) represents the state-of-the-art and can be employed in a wide variety of test interpretations.

References

1. McNeely, M. D. D. (1983) Computerized Interpretation of Laboratory Tests: an Overview of Systems, Basic Principles and Techniques. Clin. Biochem. *16*, 141–146.
2. Zlotnick, H. (1972) Bayes' Theorem for Intelligence Analysis. In: Computer diagnosis and diagnostic methods (Jaquez, J. A., ed.), pp. 180–190. Charles C. Thomas, Springfield, Illinois.
3. Wolf, P. L., Durbridge, T. & Lott, J. A. (1986) Interpretation of Biochemical Panels Utilizing Computer Analysis, pp. 7–19. Field, Rich and Associates, Inc., New York.
4. Tallarida, R. J. & Murray, R. B. (1981) Manual of Pharmacologic Calculations with Computer Programs. Springer, New York.

5. Patil, R. S. (1979) Design of a Program for Expert Diagnosis of Acid Base and Electrolyte Disturbances. Report no. TM-132, Laboratory for Computer Science, Massachusetts Institute of Technology.
6. Buchanan, B. G. & Shortliffe, E. H. (1984) Rule-Based expert systems, p. 748. Addison-Wesley Publishing Company, Reading, Massachusetts.
7. Clancy, W. J. & Shortliffe, E. H. (eds.) (1984) Readings in Artificial Intelligence, the First Decade, pp. 339–360. Addison-Wesley Publishing Company, Reading, Massachusetts.
8. Sher, P. P. (1987) General Purpose Commercial Software. In: Microcomputer Applications in the Clinical Laboratory (McNeely, M. D. D., ed.), ASCP Press, Chicago.
9. Product Literature: The Pribor System™, Health Development Inc., 1165 West Third Avenue, Columbus, Ohio 43212.
10. Weiss, S. M. & Kulikowski, C. A. (1979) EXPERT: A system for developing consultation models. In: Proceedings of the Sixth International Joint Conference on Artificial Intelligence, pp. 942–947. Stanford University, Department of Computer Science, Stanford, California.
11. Weiss, S. M., Kulikowski, C. A. & Galen, R. S. (1984) Developing Microprocessor-based Expert Models for Instrument Interpretation. In: Readings in Artificial Intelligence, the First Decade (Clancy, W. J. & Shortliffe, E. H., eds.), pp. 339–360. Addison-Wesley Publishing Company, Reading, Massachusetts.
12. Tietz, N. W. (ed.) (1983) Clinical Guide to Laboratory Tests, p. 695. W. B. Saunders Company, Philadelphia.
13. Young, D. S., Pestaner, L. C. & Gibberman, V. (1975) Effects of drugs on clinical laboratory tests. Clin. Chem. *21*, (5): ID-432D.
14. Speicher, C. E. & Smith, J. W. (1983) Choosing Effective Laboratory Tests, p. 389. W. B. Saunders Company, Philadelphia.
15. Friedman, R. B., Anderson, R. E., Entine, S. M. & Hirschberg, S. (1980) Effects of diseases on clinical laboratory tests. Clin. Chem. *26*, (4): ID-476D.
16. Christie, R. W. Actions and Decisions. ASCP Press, Chicago (In Press).
17. Gantner, G. & Gantner, T. (1985) Coding of Forensic Factors and Circumstances Surrounding Disease, Injury and Death. In: Health Data Coding and Classification Systems (Cote, R. A., Protti, D. J. and Scherrer, J. R., eds.) pp. 107–117. Elsevier Science Publishers B. V.
18. Product Literature: Partners in Medical Computing. 735 Stonegate, Libertyville, Illinois 60048.
19. Dito, W. R. (1977) An octal algorithm for pattern coding and computer-assisted interpretive reporting. Am. J. Clin. Pathol. 575–583.
20. Gordon-Findlay, L. (1980) A Computer Algorithm for Thyroid Test Interpretation. Work-term Report, University of Victoria, Department of Mathematics.
21. Myhre, B. (1982) The Use of the Microcomputer as an Aid in Parentage Testing. In: Probability of Inclusion in Paternity Testing: a Technical Workshop. American Association of Blood Banks, p. 83–91.

Chapter 3.3
Information Theoretical Model of a Clinical Chemical Test

J. Büttner

1 Introduction

Clinical chemical examinations are "purpose oriented, largely planned processes for obtaining relevant medical information" (1). Below such investigation will be referred to simply as "test". The last part of a test, which takes place in several stages, is the evaluation of the examination results obtained with regard to the medical question put and is referred to as "interpretation". For a correct and thorough interpretation a *theoretical model* is necessary which describes the test not only on the analytical chemical or pathophysiological level. It should be possible using such a model to recognize and quantitatively determine the *efficacy* or *validity* of the test (i. e. the conformity of the test result with the matter which it was intended to investigate (2)). In addition, the model might help in estimating the risk of erroneous decisions, the choice of the test suitable for a given question, optimization of the so-called test parameters (for example the decision limits), the choice of suitable test combinations in planning diagnosis strategies and in cost-benefit analyses. Finally, the model could be used to assess the position of a test within the framework of a complex diagnostic process or a therapeutical treatment.

Theoretical models of clinical chemical tests have been developed in various ways on the basis of the theory of probability (3). The model explained below, which is based on the information theory founded by Claude E. Shannon (4) in 1948, has various advantages over earlier models (5, 6, 7).

2 Information Theoretical Model of a Test

The definition of a test cited at the beginning is directed to obtaining medically relevant information. What does "information" mean? How can it be defined? The concept in common usage is vague and not suitable for our purposes. In the information theory the term "information" is used in a restricted sense. Information is considered as a message on the occurrence of an event. Such a message is selected from a supply of possible messages and at the receiver an existing uncertainty is eliminated. The definition used is as follows: "Information is the elimination of uncertainty on the occurrence of possible events". This concept disregards the meaning or value of a message, i. e. the semantic aspect.

A simple information theoretical model of a diagnostic test could be formulated as follows: Prior to execution of the test there is uncertainty for example about the presence of a certain disease, i. e. several possibilities exist. After execution of the test in the ideal case a decision has been made for one of the possibilities and the uncertainty has been completely eliminated. The test is thus a message that a specific event has occurred.

3 Quantification of the Uncertainty

Shannon (4) has shown a possibility of quantifying uncertainty. Uncertainty is related to the occurrence of possible events. Therefore its magnitude will depend on the amount of possible events and on the probability with which the individual events occur. For various reasons Shannon chose a logarithmic measure for the uncertainty which is referred to as (information) *entropy* H and defined as:

$$H = \sum_{i=1}^{n} P_i \cdot \mathrm{ld}\left(\frac{1}{P_i}\right) \tag{1}$$

wherein $P_i = P_1, P_2, P_3, \ldots, P_n$ denote probabilities for the occurrence of the events $E_1, E_2, E_3, \ldots, E_n$. "ld" (logarithmus dualis) means logarithm to the base 2[1]. The unit of entropy H is "binary digit" ("bit"). Due to the logarithmic definition the entropies H_i of the individual components give additively the entropy H.

With entropy as a quantitative measure of uncertainty we can evaluate the *information gain* of a diagnostic test by calculating the entropy difference before and after execution of the test.

A simple example can be used to clarify this. Let us assume that in a diagnostic process a decision must be made between 10 equally probable disease entities (each of them having a probability of $P_i = 1/10 = 0.1$). In the case of complete uncertainty, i. e. in the absence of any additional information whatever, the calculation according to eq. (1) gives an entropy of $H = 10 \cdot (0.1 \cdot \mathrm{ld}(1/0.1)) = 3.32$ bit. We assume that after execution of an "ideal" diagnostic test a decision for one out of the 10 possibilities is possible. In this case the entropy becomes $H = 0$. The difference $H_{\text{before test}} - H_{\text{after test}} = 3.32$ bit is a measure of the information gained by the test.

[1] Natural logarithms (ln) are converted to ld as follows: $\ln(x) \cdot 1/\ln(2) = \ln(x) \cdot 1.4427 = \mathrm{ld}(x)$

4 Sophistication of the Information Theoretical Model of a Clinical Chemical Test

Our very simple model of a clinical chemical test has an important shortcoming. In reality our tests are not "ideal" (or "pathognomonic"). Due to inherent errors of the test system a test will produce to a certain degree false positive and false negative results. Furthermore many tests have several "outcomes". In the simplest case only two ("negative" and "positive" (so-called "binary test")), in most cases many (a quantitative test produces results which are given on a continuous scale). Therefore the test result can be considered as one out of a set of possible events. In other words there is some uncertainty or entropy in the test result, the test can not completely eliminate the uncertainty originally existing in the diagnostic problem. A refined model of a clinical chemical test should take account of this.

From an information theoretical point of view a *test system* consisting of a "medical question" and a "test" has two sets of possible events which will be denoted by $D_i = D_1, D_2, D_3, \ldots, D_m$ (for the diseases in the case of diagnosis) and $T_i = T_1, T_2, T_3, \ldots, T_n$ (for the test outcomes). The probabilities for the occurrence of the D's are called *input probabilities* and will be denoted by $P(D_i)$, the probabilities of the T's are called *output probabilities* and denoted by $P(T_i)$ (Table 1).

Table 1. Characteristic probabilities used in test models

Symbolic notation for a binary test	Model Information theoretical	Probability theoretical
$P(D)$, $P(-D)$	input probability	apriori probability, prevalence
$P(T\|D)$	transition probabilities	sensitivity (SENS)
$P(-T\|-D)$		specificity (SPEZ)
$P(T\|-D)$		(1-specificity)
$P(-T\|D)$		(1-sensitivity)
$P(D\|T)$	backward probabilities	predictive value (pos.)
$P(-D\|-T)$		predictive value (neg.)
$P(T)$, $P(-T)$	output probability	aposteriori probability (see Figures 7 and 8)

A test result T will provide any kind of information on D only if there is a connection between T and D. In the ideal test we will get the result $T = T_y$ only if $D = D_x$. Without any dependence between T and D the test would be worthless.

Using eq. (1) we can calculate two entropies, $H(D)$, the entropy due to the medical question, and $H(T)$, the entropy resulting from the test outcomes. As a combined measure of the uncertainty of the test system we define the *joint entropy* $H(D, T)$ which will assume different values in accordance with the mutual dependence of D and T. In the case of complete independence the joint entropy $H(D, T)$ is obtained additively from the entropies $H(D)$ and $H(T)$:

$$H(D, T) = H(D) + H(T) \tag{2}$$

If there is a dependence between D and T the uncertainty of the system is reduced because part of the information on D is also contained in T. The joint entropy becomes

$$H(D, T) < H(D) + H(T) \tag{3}$$

The exact calculation of $H(D, T)$ requires the introduction of conditional probabilities which are defined by

$$P(T \mid D) = \frac{P(D, T)}{P(T)} \tag{4}$$

and

$$P(D \mid T) = \frac{P(D, T)}{P(D)} \tag{5}$$

In information theory these conditional probabilities are called *transition probability* (probability that a transmitted sign D is received as T) and *backward probability* (probability that a received sign T was sent as D) (Table 1). Using Shannon's logarithmic measure (eq. (1)) we obtain from eqs. (4) and (5)

$$H(D, T) = H(D) + H(T|D) = H(T) + H(D|T) \tag{6}$$

In information theory special terms are used for the conditional entropies. $H(D|T)$ is called *equivocation* and $H(T|D)$ *ambiguity*. Equivocation is the uncertainty about D when T is given. It is that part of $H(D)$ which can not be reduced by the test due to disturbances of the test. Ambiguity is the uncertainty on T when D is given. It is part of $H(T)$ and has no relation to the medical question, i. e. the "source". It is due to "noise".

The *information gain* (denoted as I) mentioned above, i. e. the reduction of the uncertainty of the system on execution of the test, then results from the difference between the sum of the entropies before the test and the joint entropy

$$I(D; T) = \{H(D) + H(T)\} - H(D, T) \tag{7}$$

or with the conditional entropies according to eq. (6)

$$I(D; T) = H(D) - H(D|T) = H(T) - H(T|D) \tag{8}$$

In information theory the entropy difference $I(D; T)$ is designated *transinformation* (instead of "information gain"). To explain the relationship between the different entropies the so-called *Berger diagram* is helpful (Fig. 1).

For the case of independence of D and T eqs. (2) and (7) give

$$I(D; T) = 0 \tag{9}$$

i. e. the test does not reduce the uncertainty at all. For total dependence of D and T we get

$$H(D, T) = H(D) + 0 = H(D) \tag{10}$$

and with eq. (7)

$$I(D; T) = H(T) \tag{11}$$

This is the case of the "ideal" or "pathognomonic" test which unfortunately is exceedingly rare in reality.

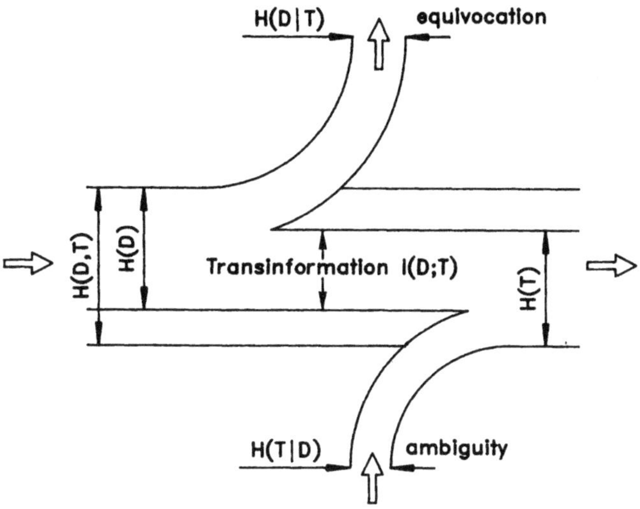

Fig. 1. Berger diagram.

A further important quantity is the maximum transinformation which can be achieved with a test under optimum conditions. It is called *test capacity* C (or "channel capacity" with reference to the "communication channel" (see below)) and is obtained by maximizing eq. (7)

$$C = \max\{H(D) + H(T) - H(D, T)\} \tag{12}$$

Whether or not this test capacity or maximum transinformation C is reached in a real case depends, using the language of communications engineering, on whether the channel is adapted in optimum manner to the message source. Applied to our model this means the following: the maximum value of the transinformation for a given test is reached only for certain input probabilities. This *optimum input probability* $P(D)_{opt}$ can be calculated in simple cases. The transinformation I thus designates the efficacy of the combination of message source (i. e. question put) and test. We are referring to this combination as "test system". The transinformation I is governed by the *input probabilities* $P(D)$ and the *transition probabilities* $P(T|D)$. In contrast the test capacity (= maximum transinformation) is a measure of the information characteristics of the test without taking account of the adaptation of the message source (question put). The test capacity depends solely on the transition probabilities $P(T|D)$.

The relationships between the various entropy quantities can be illustrated very well by symbols of the set theory. In Figure 2 the sets of the entropies are symbolically represented by circles (the diameter corresponds roughly with the amount of H). The transinformation is then obtained as intersection $H(D) \cap H(T)$. The residue sets $H(D) \setminus H(T) = H(D|T)$ and $H(T) \setminus H(D) = H(T|D)$ are equivocation and ambiguity (see above).

Figure 3 illustrates the different cases of dependency between the entropies $H(D)$ and $H(T)$ as described in eqs. (7), (9), and (11).

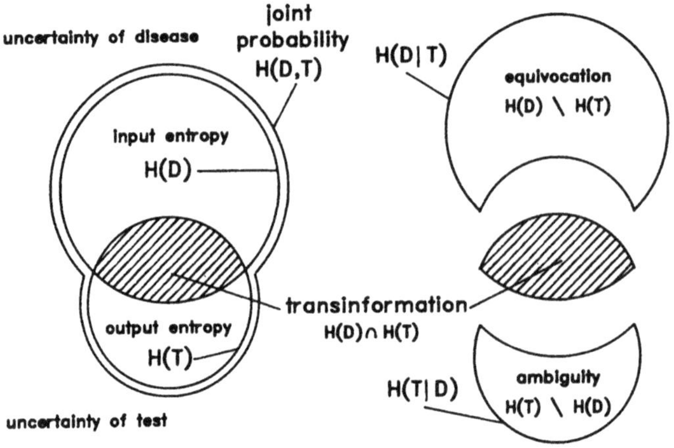

Fig. 2. Information theoretical model of a test symbolized by set theory.

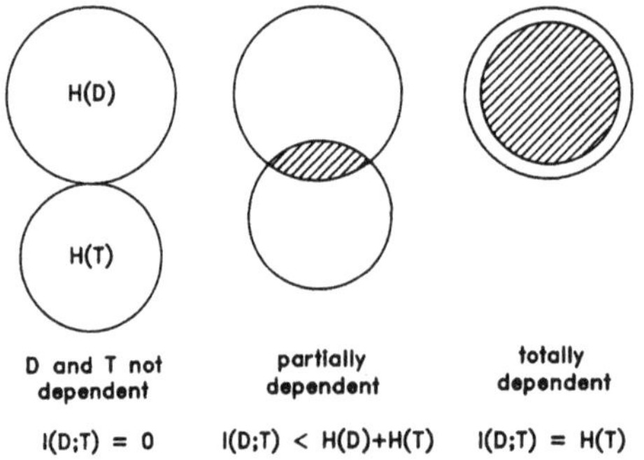

Fig. 3. Representation of transinformation in different cases of dependency of entropies $H(D)$ and $H(T)$. See equations (7), (9), (11).

Let us summarize our ideas so far: our model describes a clinical chemical test by linking the question put (for example in the case of diagnostic application the number of disease entities to be recognized, expressed as entropy) to the test result (e. g. the number of test outcomes, likewise expressed as entropy). The information gain or transinformation is calculated from the entropy difference as a measure of the efficacy of the test.

5 Calculation of the Information Theoretical Quantities (8, 9, 10, 11)

The calculation of the information theoretical quantities of the test model varies according to whether we are dealing with a simple qualitative test, a quantitative test or the combination of several tests. In this Section the calculation will first be

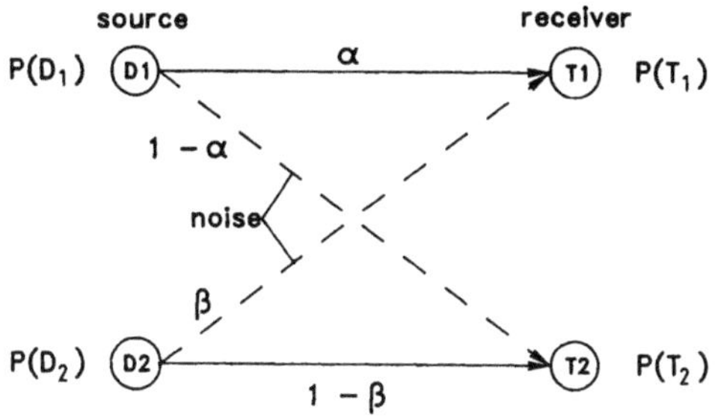

Fig. 4. Qualitative test described as discrete, binary communication channel. Abbreviations: $a = P(T_1|D_1)$, $1 - a = P(T_2|D_1)$, $\beta = P(T_1|D_2)$, $1 - \beta = P(T_2|D_2)$.

represented for a qualitative test. To make the calculations as clear as possible to understand it has been found useful in information theory, adapted to actual circumstances in communications engineering, to use the scheme of a *communication channel.* This consists of the message source or *sender* and the *receiver* and indicates the various possibilities used as a basis for the information theoretical calculations. Such a communication channel can easily be applied to our test model. The sender corresponds to the question put, the receiver to the test outcome. Figure 4 shows the scheme for the simple case of a qualitative test with two inputs (two diagnosis alternatives which mutually exclude each other, e. g. "disease present" and "disease not present"). For the input probabilities, since the alternatives mutually exclude each other, we have

$$\sum_{i=1}^{2} P(D_i) = 1 \tag{13}$$

For our example we shall further assume that the test has two outputs which likewise mutually exclude each other ("positive", "negative"). The following relationship then holds for the "output probabilities"

$$\sum_{j=1}^{2} P(T_j) = 1 \tag{14}$$

The connecting lines of our scheme correspond to the various "true" (or correct) and "false" transitions. They are characterized by conditional probabilities introduced above (transition probabilities).

The data base taken for the calculation is a frequency table ("fourfold table") as made up in the clinical investigation of a test at defined patient collectives (e. g. "ill" and "not ill"). It contains frequencies for the various combinations of the diagnosis alternatives D_i and the test outputs T_j. From the frequencies the required conditional probabilities $P(T_j|D_i)$ can be estimated as relative frequencies. In addition, estimates of the input probabilities are necessary which result for example from the prevalences ($P(D)$) of the disease to be detected.

Using eq. (1) we can calculate the entropies in the following manner:

$$H(D) = \sum_{i=1}^{m} P(D_i) \cdot \mathrm{ld} \left[\frac{1}{P(D_i)}\right] \tag{15}$$

$$H(T) = \sum_{j=1}^{n} P(T_j) \cdot \mathrm{ld} \left[\frac{1}{P(T_j)}\right] \tag{16}$$

$$H(D, T) = \sum_{i=1}^{m} \sum_{j=1}^{n} P(D_i, T_j) \cdot \mathrm{ld} \left[\frac{1}{P(D_i, T_j)} \right] \tag{17}$$

It is easier to arrange the transition probabilities $P(T_j \mid D_i)$ and the input probabilities $P(D_i)$ in matrices and to calculate the remaining probabilities by simple matrix operations. A detailed calculation is given in my former paper (7).

The transinformation $I(D;\ T)$ can be calculated by eqs. (7) or (8) or by the equation

$$I(D; T) = \sum_{i=1}^{m} \sum_{j=1}^{n} P(D_i, T_j) \cdot \mathrm{ld} \left[\frac{P(D_i, T_j)}{P(D_i) \cdot P(T_j)} \right] \tag{18}$$

The calculation of the entropies can also be carried out directly on the basis of the frequencies without previous estimate of the probabilities. In this case in the Shannon formula eq. (1) P_i is replaced by n_i/N (N = total number). This gives

$$H = \mathrm{ld}(N) - \frac{1}{N} \cdot \sum_{i=1}^{n} n_i \cdot \mathrm{ld}(n_i) \tag{19}$$

The frequencies for a simple qualitative test are given as a $(m \cdot n)$-table where m denotes the possibilities at the input, n at the output. We use the following total and marginal sums:

total number $$N = \sum_{i=1}^{m} \sum_{j=1}^{n} n_{ij} \tag{20}$$

marginal sum input $$n(D)_i = \sum_{i=1}^{m} n_{ij} = n_{i\cdot} \tag{21}$$

marginal sum output $$n(T)_j = \sum_{j=1}^{n} n_{ij} = n_{\cdot j} \tag{22}$$

Using this notation we get:

$$H(D) = \mathrm{ld}(N) - \frac{1}{N} \cdot \sum_{i=1}^{m} n_{i.} \cdot \mathrm{ld}(n_{i.}) \tag{23}$$

$$H(T) = \mathrm{ld}(N) - \frac{1}{N} \cdot \sum_{j=1}^{n} n_{.j} \cdot \mathrm{ld}(n_{.j}) \tag{24}$$

$$H(D, T) = \mathrm{ld}(N) - \frac{1}{N} \cdot \sum_{i=1}^{m} \sum_{j=1}^{n} n_{ij} \cdot \mathrm{ld}(n_{ij}) \tag{25}$$

The transinformation $I(D, T)$ is calculated in simple manner from the calculated entropies using eq. (7). The calculation of the test capacity C is more involved, at least when the number of possibilities in the test input (m) differs from the number of possibilities in the test output (n). A few important methods are explained in the Appendix.

To illustrate the information theoretical quantities in Table 2 a simple numerical example is worked out. The example shows that the information gain (transinformation) on execution of this test is 0.208 bit. Under optimum conditions, i. e. when the input probability (e. g. the prevalence of the assumed disease) lies at 0.47 and not as in the example at 0.20, the same test would give a maximum transinformation (test capacity C) of 0.297 bit.

Table 2. Numeric example for a simple qualitative test

Given: frequency table (−D denotes "disease absent", −T "test negative")

		Test output $j = 1\ (T)$	$j = 2\ (-T)$	Row sum
Test input	$i = 1$ (D)	35	15	50
	$i = 2$ (−D)	20	180	200
Column sum		55	195	250

Calculated (with eqs. (23, 24, 25, 6):

input entropy	$H(D)$	= 0.7219 bit
output entropy	$H(T)$	= 0.7602 bit
joint entropy	$H(D, T)$	= 1.2734 bit
equivocation	$H(D\|T)$	= 0.5132 bit
ambiguity	$H(T\|D)$	= 0.5515 bit
transinformation	$I(D; T)$	= 0.2078 bit
test capacity	C	= 0.2967 bit
optimum input probability	$P(D)_{opt}$	= 0.4719

Conventional Measures:

sensitivity	= 0.7
specificity	= 0.9
apriori probability $P(D)$	= 0.2

6 Extension of the Model to Tests with Quantitative Results

The majority of tests used in Clinical Chemistry give quantitative results, i. e. results imaged on a continuous scale. In contrast to the qualitative test hitherto considered the outputs of quantitative tests cannot be represented by discrete probabilities but only by continuous probability density functions. If these are well defined distribution functions (e. g. the Gaussian distribution) they can be used for calculating the entropies, although this is a more complicated mathematical procedure.

Eq. (18) was used for calculation of transinformation in the qualitative (discrete) case. Replacing the discrete probabilities $P(X)$ by probability density functions $w(X)$ we get in the quantitative (continuous) case:

$$I(D;\ T) = \int_{-\infty}^{\infty} \int_{-\infty}^{\infty} w_2(D,\ T) \cdot \mathrm{ld} \left[\frac{w_2(D,\ T)}{w_D(D) \cdot w_T(T)}\ \mathrm{d}D\,\mathrm{d}T \right] \tag{26}$$

In practical use however probability density functions are unknown and empirical distributions will have to be taken as a basis. The simplest thing to do in this case is to order the test results into classes, i. e. to convert results existing on a continuous

Table 3. Calculation of the information theoretical quantities of a quantitative test

Example: GGT in the diagnosis of alcoholism (Data from Stamm et al. (23, 24))
Population groups: *D:* alcoholism confirmed by outer criteria
−D: alcoholism excluded by outer criteria
Frequency table: lower limit of classes (in U/l) = (class no. · 20) − 19

	Class 1	2	3	4	5	6	7	8	9	10	Row sum
D	7	21	11	8	3	5	4	2	1	19	81
−D	33	26	4	4	1	0	1	0	0	1	70
Column sum	40	47	15	12	4	5	5	2	1	20	151

Calculated entropies:

input entropy	$H(D)$	= 0.9962 bit
output entropy	$H(T)$	= 2.6343 bit
joint entropy	$H(D, T)$	= 3.3596 bit
equivocation	$H(D\|T)$	= 0.7254 bit
ambiguity	$H(T\|D)$	= 2.3635 bit
transinformation	$I(D; T)$	= 0.2708 bit
test capacity	C	= 0.2750 bit
optimum input probability	$P(D)_{opt}$	= 0.4681

scale into discrete frequencies (in information theory this procedure is referred to as "quantization"). In Table 3 an example is given of a quantitative test in which a division into 10 classes has been made. The calculation of the entropies is identical to that in the example in Table 2 only now there are $m = 10$ alternatives at the test output.

It should be pointed out here that in Clinical Chemistry in the practical evaluation of the results of a quantitative test usually a decision criterion ("cut-off point") is used. Results lying on one side of the decision criterion are allocated to the one collective, e. g. "diseased". Results on the other side are allocated to the other collective, e. g. "non diseased". A "decision scheme", an "interpreter" is used, or expressed in the language of communications engineering a "decoder" (8). Considered from the point of view of information theory this procedure reduces the transinformation obtained; part of the possible information from the test is dispensed with.

7 Treatment of Test Combinations (Multivariate Model) (12, 13)

It has been known in clinical diagnostics for a long time that the combination of various tests in many cases leads to an improvement in the diagnostic sharpness. As examples the combination of enzyme activity determinations in the diagnosis

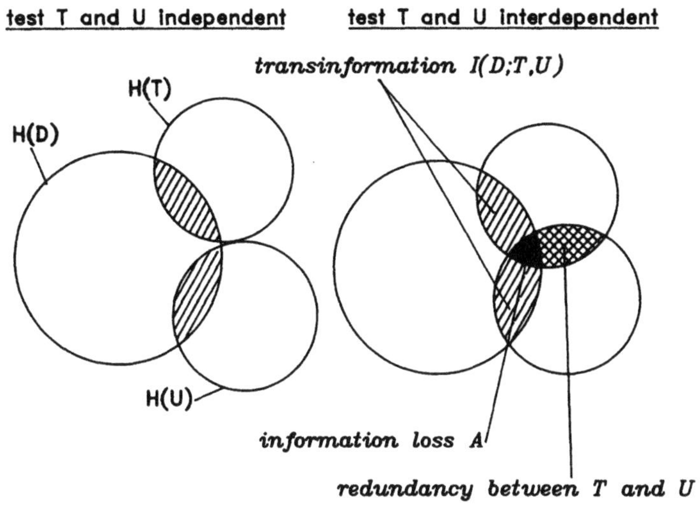

Fig. 5. Combination of two tests symbolized by set theory.

of liver diseases can be mentioned. An exact analysis of such test combinations is often made difficult by the individual quantities to a large extent not being independent of each other. Thus, erythrocyte number and hemoglobin content in the blood or calcium concentration and albumin concentration in serum are very much interdependent. Account must also be taken of whether the combined tests are performed in "parallel", i. e. simultaneously or "sequentially", i. e. one after the other.

To enable such questions to be dealt with it is desirable to extend the information theoretical test model also to multiple tests dependent on each other (so-called test combinations). To illustrate this clearly, firstly the combination of two tests will be discussed. The model can however be extended without difficulties to n-dimensional problems. In Figure 5 the representation according to the theory of sets, which we have utilized to describe the model for the single test, is applied to the combination of two tests. If the tests, which will be denoted below by the symbols T and U, are independent of each other then the transinformation $I(D; T, U)$ is simply the sum of the intersections

$$I(D; T, U) = \{H(D) \cap H(T)\} + \{H(D) \cap H(U)\} \tag{27}$$

The important case to be considered now is that of the mutual dependence of test T and U. This case is apparent on the right side of Figure 5 from the overlapping of the entropies $H(T)$ and $H(U)$. The intersection $H(T) \cap H(U)$ obviously represents the part of the information which is contained in both tests. In communications engineering this is referred to as "redundancy". Figure 5 shows a further important change compared with the case of mutual independence of T and U: The transinformation $I(D; T, U)$ is no longer the sum of the intersections (eq. (27)) but reduced by an amount which can be interpreted as "information decrease by interaction". This results from part of the transinformation which the test T

Table 4. Calculation of entropies for the parallel combination of two qualitative tests

Frequencies: $(m \cdot n \cdot p)$-table, notation of frequencies similar to eq. (20, 21, 22)
Indices used: input D, index $i = 1, 2, 3 \ldots m$
output T, index $j = 1, 2, 3 \ldots n$
U, index $k = 1, 2, 3 \ldots p$

Entropies:

(4.1) $$H(D) = \mathrm{ld}(N) \cdot \frac{1}{N} \cdot \sum_{i=1}^{m} n_{i..} \cdot \mathrm{ld}(n_{i..})$$

(4.2) $$H(T) = \mathrm{ld}(N) \cdot \frac{1}{N} \cdot \sum_{j=1}^{n} n_{.j.} \cdot \mathrm{ld}(n_{.j.})$$

(4.3) $$H(U) = \mathrm{ld}(N) \cdot \frac{1}{N} \cdot \sum_{k=1}^{p} n_{..k} \cdot \mathrm{ld}(n_{..k})$$

(4.4) $$H(D, T) = \mathrm{ld}(N) \cdot \frac{1}{N} \cdot \sum_{i=1}^{m} \sum_{j=1}^{n} n_{ij.} \cdot \mathrm{ld}(n_{ij.})$$

(4.5) $$H(D, U) = \mathrm{ld}(N) \cdot \frac{1}{N} \cdot \sum_{i=1}^{m} \sum_{k=1}^{p} n_{i.k} \cdot \mathrm{ld}(n_{i.k})$$

(4.6) $$H(T, U) = \mathrm{ld}(N) \cdot \frac{1}{N} \cdot \sum_{j=1}^{n} \sum_{k=1}^{p} n_{.jk} \cdot \mathrm{ld}(n_{.jk})$$

(4.7) $$H(D, T, U) = \mathrm{ld}(N) \cdot \frac{1}{N} \cdot \sum_{i=1}^{m} \sum_{j=1}^{n} \sum_{p=1}^{p} n_{ijk} \cdot \mathrm{ld}(n_{ijk})$$

furnishes on D also being furnished by test U. An exact calculation shows that in some cases a negative value occurs for the information decrease. This case is referred to as "information increase by interaction".

The calculation of the information theoretical quantities will now be made below for the simple case of the parallel combination of two tests T and U. Since both tests are to be assumed as qualitative, i.e. each with binary output, for the combination the four test outputs shown in Figure 6 result. The basis of the calculation is a frequency table which, corresponding to the two outputs, has $2 \cdot 4 = 8$ fields. The entropies can easily be calculated in a manner similar to that adopted in the simple qualitative test (eqs. (23) to (25)). The equations are summarized in Table 4. The various transinformations are calculated in the following manner, as can be derived from Figure 5:

$$I(D; T) = H(D) + H(T) - H(D, T) \tag{28}$$

$$I(D; U) = H(D) + H(U) - H(D, U) \tag{29}$$

$$I(T; U) = H(T) + H(U) - H(T, U) \tag{30}$$

$$I(D; T, U) = H(D) + H(T, U) - H(D, T, U) \tag{31}$$

$$I(D; T; U) = H(D) + H(T) + H(U) - H(D, T, U) \tag{32}$$

The quantity $I(T; U)$ calculated with eq. (30) corresponds to the redundancy between the two tests T and U. Finally, the transinformation amount must be calculated which with dependent tests is defined by the intersection $H(T, U) \cap H(D)$. This quantity, which will be designated A, is the result of

Table 5. Numeric example for the parallel combination of two qualitative tests

Given: Frequency table

		Output TU $j=1$ $k=1$	$T-U$ $j=1$ $k=2$	$-TU$ $j=2$ $k=1$	$-T-U$ $j=2$ $k=2$
Input	$D\,i=1$	435	44	22	5
	$-D\,i=2$	2450	1562	68	5448

Calculated:

$H(D)$	= 0.2882 bit	
$H(T)$	= 0.9921 bit	
$H(U)$	= 0.8770 bit	
$H(D, T)$	= 1.2358 bit	
$H(D, U)$	= 1.1025 bit	
$H(T, U)$	= 1.4792 bit	
$H(D, T, U)$	= 1.6971 bit	
$I(T; U)$	= 0.3898 bit	redundancy of tests T and U
$I(D; T)$	= 0.0445 bit	transinformation test T
$I(D; U)$	= 0.0627 bit	transinformation test U
$I(D; T, U)$	= 0.0704 bit	transinformation of combination T and U (in case of independency of T and U the transinformation becomes 0.1072 bit (acc. to eq. (27))
$I(D; T; U)$	= 0.4602 bit	information gain of the total system
A	= 0.0368 bit	information loss due to mutual dependency of T and U
C	= 0.4020 bit	test capacity of the combination
$P(D)_{opt}$	= 0.5570	optimum input probability

interaction and may be positive ("information decrease", "information loss") or negative ("information increase") (see above). It can be calculated from

$$A = \{I(D; T) + I(D; U)\} - I(D; T, U) \tag{33}$$

In Table 5 a numerical example is calculated. The results show that the two tests have a considerable redundancy. As a result the transinformation on executing both tests is reduced to about 2/3. Nevertheless, the combination gives a larger transinformation than either of the individual tests. The transinformation is relatively small in this example compared with the test capacity, this being due to the low input probability $P(D)$.

In the same manner with our model the combination of more than two tests can also be dealt with. The necessary calculations can be carried out easiest using matrices. In the combination of several tests the already mentioned decision scheme at the test output ("decoder") becomes of particular significance because in clinical practice as far as possible a "yes/no" answer is desired to a question put in order to derive from this answer an immediate direction for necessary action. The scheme for the parallel combination of two qualitative tests (Fig. 6) shows that at first there are 4 test outputs. By suitable decoders these 4 outputs can be reduced. Thus, the decoder could be "positive if test T *and* U are positive" ($T \cap U$). The decoder "positive if test T *or* U are positive" ($T \cup U$) would also be conceivable. Finally,

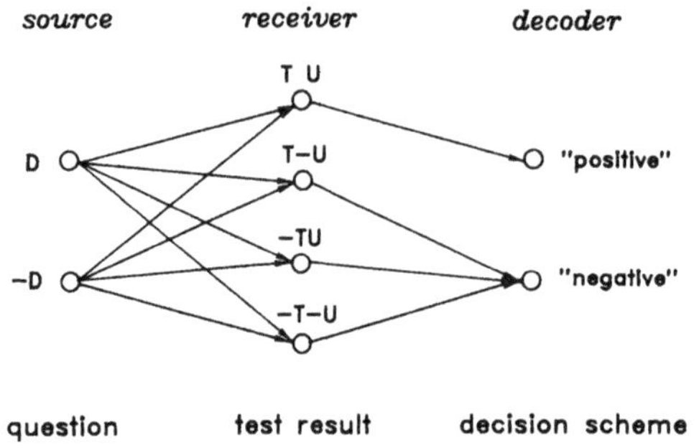

Fig. 6. Parallel combination of two qualitative tests described as communication channel with "ideal observer" decoder.

the cases $T, -U$ and $-T, U$ can be combined as "doubtful" with the direction that a control of some sort is to follow. A rational method of determining a decoder is based on the principle of minimizing the probability of erroneous decisions. The decoder in this case is construed so that the result at the test output is interpreted always as the most probable message. The decision scheme is referred to as "ideal observer" (14). It will be explained using the example of Table 5. For simplification the four test outputs will be designated as follows: $T, U = y_1$, $T, -U = y_2$, $-T, -U = y_3$, $-T, U = y_4$ (Table 6). For the calculation the output probabilities $P(y_j)$ and the backward probabilities $P(D_i|y_j)$ are required and they can be calculated as follows:

$$P(y_j) = \sum_{i=1}^{2} P(D_i) \cdot P(y_j \mid D_i) \tag{34}$$

$$P(D_i \mid y_j) = \frac{P(D_i) \cdot P(y_j \mid D_i)}{P(y_j)} \tag{35}$$

The calculation is carried out easiest using matrices. A calculation scheme will be found in my former paper (7). On page 489 of that paper due to a printing error the indices of the "backward matrix" are interchanged. It should read:

$$E = \begin{bmatrix} \dfrac{P(T_1, D_1)}{P(T_1)} & \dfrac{P(T_2, D_1)}{P(T_2)} \\ \dfrac{P(T_1, D_2)}{P(T_1)} & \dfrac{P(T_2, D_2)}{P(T_2)} \end{bmatrix} = \begin{bmatrix} P(D_1 \mid T_1) & P(D_1 \mid T_2) \\ P(D_2 \mid T_1) & P(D_2 \mid T_2) \end{bmatrix}$$

By means of the table thus calculated of the backward probabilities (Table 6) for each of the four test outputs the input symbol corresponding to the higher backward probability is now selected. We thus denote y_1 as D_1, y_2 as D_2, y_3 as D_1 and y_4 as D_2.

The error probability $P(e)$ which can be calculated (8) to check the decoder devised is:

$$P(e) = \sum_{j=1}^{4} P(y_j) \cdot P(e \mid y_j) \tag{36}$$

Table 6. Numeric example for the construction of an "ideal observer" decision scheme

Given: Frequency table from Table 5
assuming input probabilities of $P(D) = 0.5$ and $P(-D) = 0.5$ the transition, output and backward probabilities were calculated

Transition probabilities $P(y_i|D)$:

		Output $TU = y_1$	$T-U = y_2$	$-TU = y_3$	$-T-U = y_4$
Input	D	0.8595	0.0872	0.0435	0.0098
	$-D$	0.2572	0.1640	0.0072	0.5716
Output probabilities $P(y_j)$:					
		0.5584	0.1256	0.0254	0.2907
Backward probabilities $P(D_i|y_j)$:					
Input	D	*0.7697*	0.3471	*0.8580*	0.0169
	$-D$	0.2303	*0.6529*	0.1420	*0.9831*

Decision scheme:

"denote y_1 and y_3 as D, y_2 and y_4 as $-D$"

Error probability $P(E)$ *for the decision scheme:*

$P(e) = (0.5584 \cdot 0.2303) + (0.1256 \cdot 0.3471) + (0.0254 \cdot 0.1420) + (0.2907 \cdot 0.0169) = 0.1807$

using the decision scheme "$y_1 = D$, $y_2, y_3, y_4 = -D$" the error increases to $P(e) = 0.1988$

wherein $P(e \mid y_j)$ are the backward probabilities for the outputs which do not correspond to the selected assignment. For the decision scheme mentioned above "positive if *T and U* are positive" which does not correspond to the "ideal observer" in the example calculated in Table 6 a higher error probability results.

8 Relationship between the Information Theoretical and the Probability Theoretical Model of a Test

Between the information theoretical model of a clinical chemical test introduced and the usual probability theoretical model there are close relationships. Both models proceed on the basis of frequencies or probabilities (Table 1). Since the transinformation measures a relationship between "input-side" and "output-side" probabilities it is to be expected that relationships will also exist to the coherence measures of statistics. This is indeed the case. For example it can be derived from eq. (26) that in the case of two partially dependent Gaussian distributions $g_A(A)$ and $g_B(B)$ the transinformation depends on the correlation coefficient only:

$$I(A; B) = \frac{1}{2} \cdot \mathrm{ld}(1 - \varrho^2) \qquad (37)$$

Kullback (15) has examined the relationships between information theoretical measures and statistics in detail. In his terminology the transinformation is called "discrimination information". Miller and Madow (16) showed – starting from likelihood ratio – that the quantity

$$\chi_T^2 = 2 \cdot \ln(2) \cdot N \cdot I = 1.3863 \cdot N \cdot I \tag{38}$$

N = total number of observations
I = transinformation

is approximately χ^2-distributed. This gives the possibility for testing the significance of transinformation measures.

Example:

We chose the numerical example of a qualitative test given in Table 2. To test the hypothesis of a dependency between output and input we calculate χ_T^2 according to eq. (35). With $N = 250$ and $I = 0.2078$ bit we get $\chi_T^2 = 72.02$ with $(2-1)(2-1) = 1$ degree of freedom. This means that the transinformation I is significantly different from 0, i. e. output and input of the test are dependent.

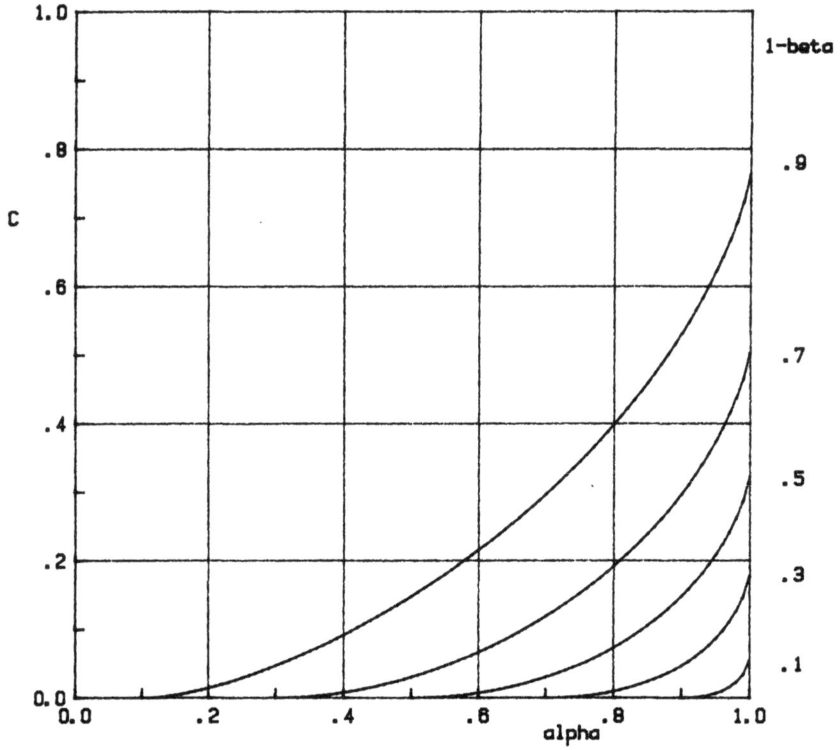

Fig. 7. Relationship between test capacity C and sensitivity/specificity of a qualitative, binary test. Abbreviations: sensitivity = α, specificity = $1 - \beta$.

Using the statistic $2 \cdot \ln(2) \cdot N \cdot I$ Kullback and coworkers (17) proposed the "2 I-Test" for examination of contingency and dependency in frequency tables (this test is identical with the "log likelihood ratio test ("G-test")" described by Woolf (18).

Table 7. Relationship between test capacity and specificity/sensitivity in case of a qualitative (binary) test

Abbreviations used: SENS = sensitivity = $P(T|D)$
SPEZ = specificity = $P(-T|-D)$

Condition	Communication channel	Test capacity C
SENS = (1 − SPEZ)	totally disturbed	$C = 0$
SENS = SPEZ	symmetric	$C = \mathrm{ld}(2) + \mathrm{SENS} \cdot \mathrm{ld}(\mathrm{SENS}) + (1 - \mathrm{SENS}) \cdot \mathrm{ld}(1 - \mathrm{SENS})$
SENS ≠ SPEZ	asymmetric	$C = \mathrm{f}\{\mathrm{SENS}, (1 - \mathrm{SPEZ})\}$ (see Figs. 7 and 8)

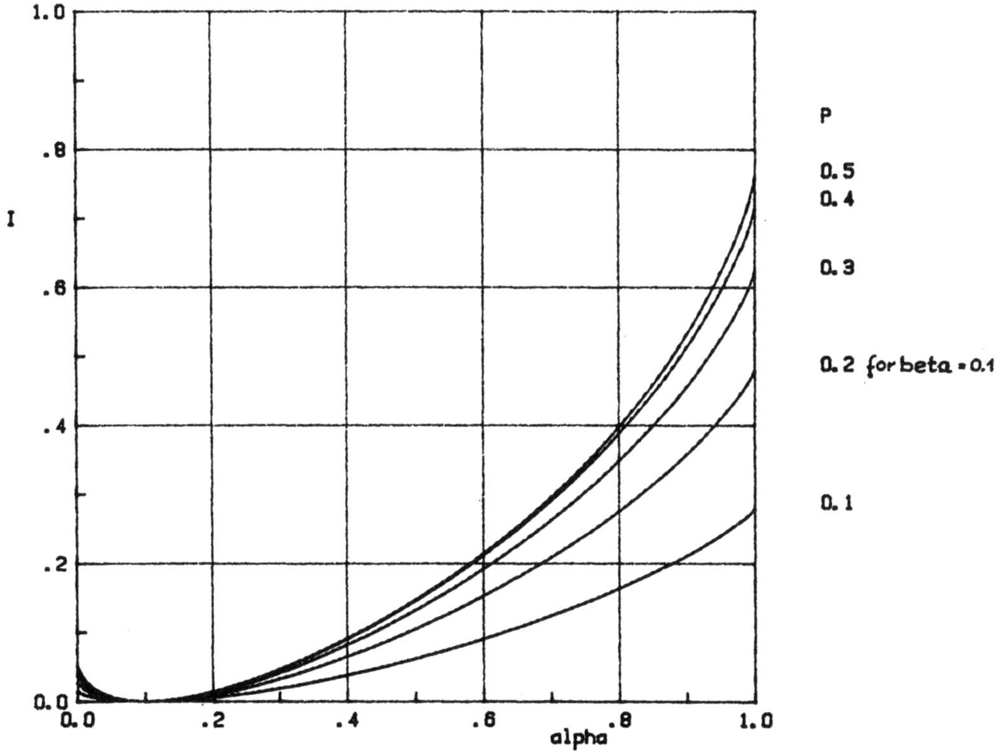

Fig. 8. Dependency of transinformation I from sensitivity, specificity and input probability of a qualitative, binary test. The graph is construed for $\beta = (1 - \text{specificity}) = 0.1$. Abbreviations: a = sensitivity, P = input probability $P(D)$.

For evaluating clinical chemical tests with the probability theoretical model preferably specificity and sensitivity (2, 19, 20, 21) are used. Hansert et al. (22) employ the efficiency (16), which is the sum of sensitivity and specificity. As already pointed out the test capacity C of a test depends only on the sensitivity and specificity but the transinformation I depends additionally on the input probability. Table 7 illustrates the quantitative relationships for simple cases of the binary test. In Figure 7 the relationship between test capacity C, sensitivity and specificity is shown graphically and in Figure 8 this is done accordingly for the transinformation. For the numerical example in Table 2 test capacity C and transinformation I can also be derived graphically from Figures 7 and 8.

9 Practical Applications

A great number of applications are conceivable for the universal model of a test illustrated. Below a few possibilities will be pointed out with the aid of examples. As in Table 3 as well for the examples the data material from an extensive study by Stamm and Hansert (23, 24) on clinical chemical diagnostics of alcoholism is employed.

9.1 Quality criterion of a test

To enable a test to be assessed as regards its diagnostic efficacy and compared with other tests which are possible choices for answering the same question, a quality criterion is desirable. It should be suitable both for qualitative and quantitative tests and independent from assumptions on the distribution of the data. The numerical example in Table 8 shows that of the four tests GGT, AST, MCV and urea when using the *conventional* decision limits (upper or lower limit of the reference interval) AST supplies most information on the presence of alcoholism, followed by GGT. As explained above, the test capacity designates the test without taking account of the apriori probability, i. e. the given disease prevalence for the practical application. If account is taken of this as well the transinformation should be use as quality criterion (Section 9.3).

9.2 Optimizing the decision limit of a test

It has been explained in Section 6 that in Clinical Chemistry quantitative tests are usually reduced to a binary result, i. e. a "yes/no" decision. This is done by comparison of the test result with the reference interval (so-called transversal evaluation). Experience has shown that for many medical questions flexible decision limits provide better results. Hansert et al. (22) have recently discussed this from a theoretical point of view. Now, the information theoretical test model is very well

Table 8. Use of test capacity and transinformation as quality criteria for clinical chemical test
Example: diagnosis of alcoholism (data from Stamm et al. (23, 24))
(transinformation for input probability $P(D) = 0.05$)

Test	Decision limit	Test capacity C in bit	Trans-information I in bit
GGT	≥ 28 U/l	0.153	0.029
	≥ 44 U/l (optimized)	0.203	0.046
AST	≥ 18 U/l	0.159	0.038
	≥ 19 U/l (optimized)	0.195	0.054
MCV	≥ 96 fl	0.081	0.015
	≥ 95 fl (optimized)	0.110	0.020
Urea	≤ 5.0 mmol/l	0.096	0.017
	≤ 3.9 mmol/l (optim.)	0.107	0.022

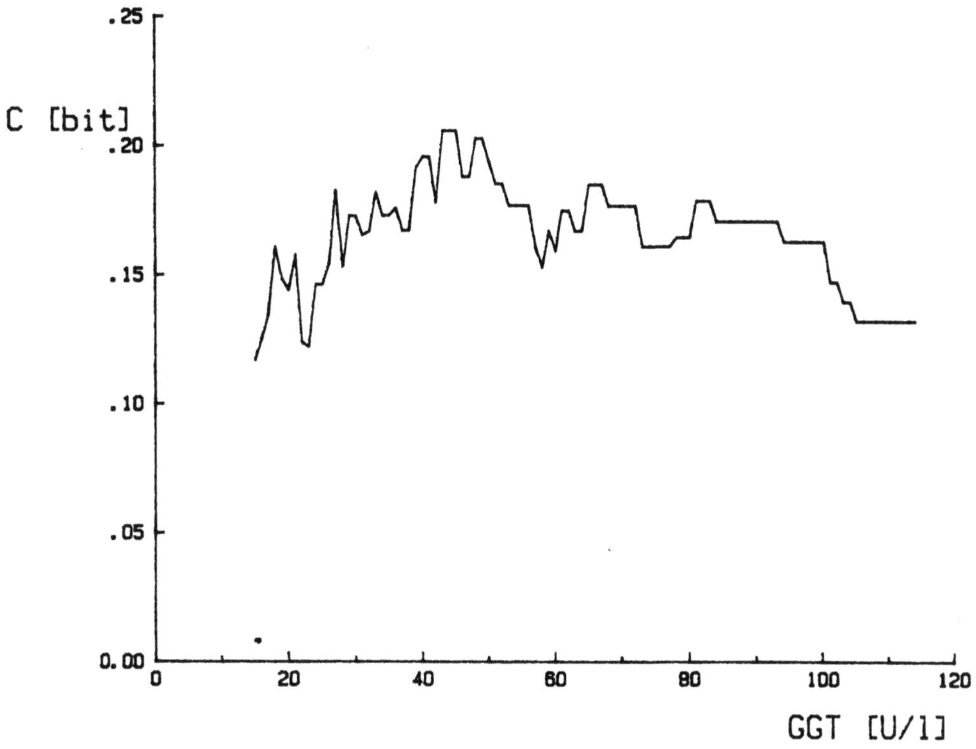

Fig. 9. Optimization of the decision limit of a quantitative test. Example: GGT in diagnosis of alcoholism (Data from Stamm et al. (23, 24)). Optimization by means of test capacity C.

suited to optimizing the decision limits for concrete questions. Once again, for the optimizing parameter the test capacity C can be used. In Figure 9 this is shown for the GGT test in the diagnosis of alcoholism. The decision limit was shifted stepwise and C calculated in each case. The maximum for C is not found at the conventional decision limit (28 U/l), but at 44 U/l. In this manner the optimized decision limits indicated in Table 8 were obtained. A comparison shows that after the optimization GGT provides more transinformation than the other tests, it is now better than AST.

9.3 Adaptation of a test to a given input probability

Clinical chemical tests are used in diagnostics under a great variety of conditions, for example for case finding in the doctor's office, for screening large groups of the population or for confirming a provisional diagnosis. In these cases the input probability is very different. To obtain the maximum information from a test it must be adapted to the given input probability (Section 4). The simplest way of making this adaptation is to shift the decision limit. This optimization is also possible with our test model if the transinformation I is used as optimizing quantity.

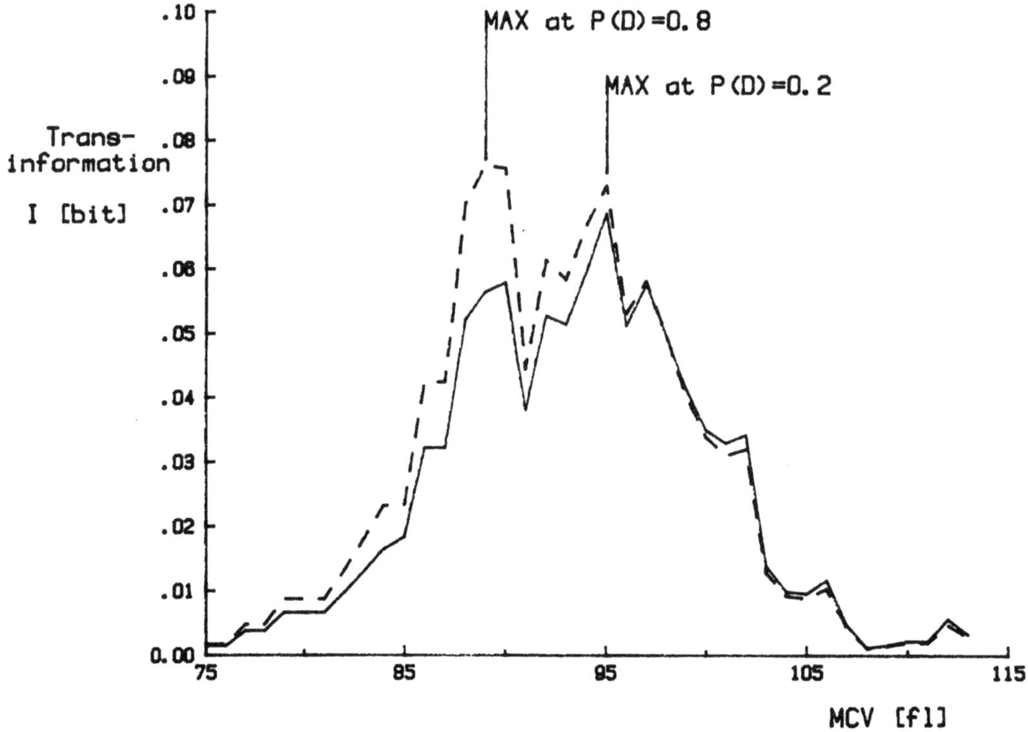

Fig. 10. Optimization of the decision limit of a test system with different input probabilities. Example: MCV in diagnosis of alcoholism (Data from Stamm et al. (23, 24)). $P(D) = 0.2$ and 0.8. Optimization by means of transinformation I.

Figure 10 shows as example the determination of the decision limit for MCV in the diagnosis of alcoholism for two different input probabilities. For $P(D) = 0.2$ the limit has an optimum at 89 fl and for $P(D) = 0.8$ at 95 fl.

9.4 Optimization of decision limits in combined tests

To increase the diagnostic efficacy, for a long time Clinical Chemistry has employed the combination of several tests. The development of multi-channel analysis has led to the tests to be combined mainly being carried out simultaneously, i.e. in parallel. In clinical practice strategies for combining tests are usually devised on the basis of empirical knowledge. Rational procedures are using statistical methods, e.g. discriminant analysis (see for a review Solberg (25)). The particular problem resides in that classical statistical methods make certain assumptions on the underlying distributions. The empirical data obtained in the evaluation of tests seldom shows defined distribution types. A test strategy must also observe that because of the mutual dependence of many test quantities on each other depending on the test combination chosen different decision limits must be used for the interpretation if a maximum diagnostic efficiency is desired. Recently Keller and Keller (26) have published a distribution-free multivariate discriminating method and Hansert et al. (22) have developed a distribution-free method which makes it possible to optimize such "multivariate decision limits". Among others the efficiency, that is the sum of sensitivity and specificity (Table 1) is used as optimizing quantity. The information theoretical test model, which – as outlined in Section 7 – can be extended in

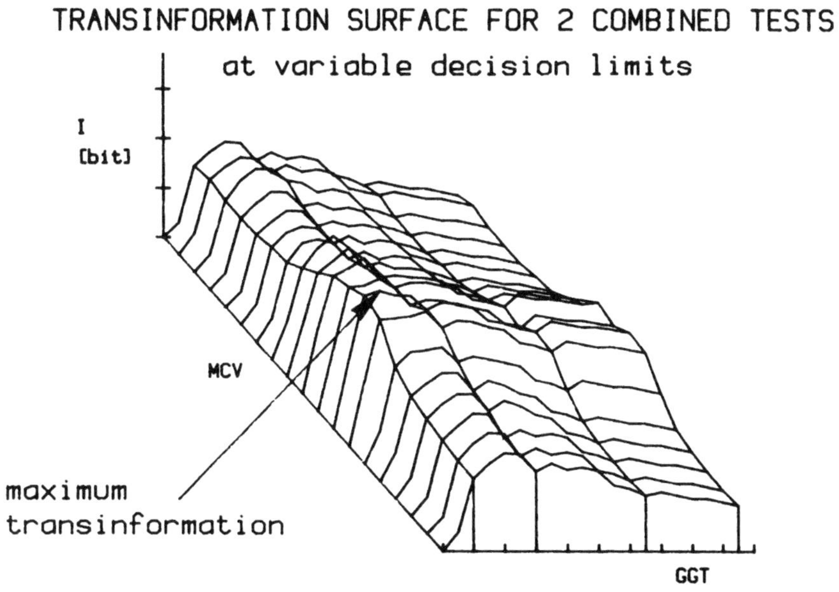

Fig. 11. Optimization of the decision limits of a test combination. Example: combination of GGT and MCV in diagnosis of alcoholism. The figure shows the "transinformation surface".

simple manner to test combinations as well, can be used for the same purpose because like the method of Hansert et al. (22) it makes no assumptions on the underlying distributions. In Figure 11, once again for the diagnosis of alcoholism, the mutual dependence of the decision limits for the tests GGT and MCV in parallel combination is represented as "transinformation surface". The optimum, i.e. the optimum "multivariate decision limit" for GGT and MCV, is marked by the arrow. An exact comparison with the results of Stamm et al. (23, 24) shows that substantially the same decision limits are obtained. We cannot go into detail here on fine differences which result from the relationships discussed in Section 8 between the quantities used by Hansert et al. (22) and the information theoretical quantities.

9.5 Selection of a decision scheme in the combination of several tests

In the clinical use of simultaneously executed diagnostic tests the choice of a decision scheme is a great problem. It has been pointed out by various authors (3, 27) that the current practice of subjecting the individual test results "unlinked" to the transversal evaluation carries considerable risks when the tests exhibit mutual

Table 9. Choice of a decoder for a combination of three quantitative tests using test capacity

Example: combination of GGT (= T), AST (= U), and MCV (= V) in the diagnosis of alcoholism (data from Stamm et al. (23, 24))

Population Groups *D:* alcoholism confirmed by outer criteria
−*D:* alcoholism excluded by outer criteria

input probability $P(D) = 0.5$, decision limits optimized
$P(e)$ = error probability (see eq. (36))

No.	Decoder	C	$P(e)$
1	pos: TUV neg: −TUV, T−UV, TU−V, −T−UV, −TU−V, T−U−V, −T−U−V	0.193	0.310
2	pos: TUV, −TUV, T−UV, TU−V neg: −T−UV, −TU−V, T−U−V, −T−U−V	0.326	0.195
3	pos: TUV, −TUV, T−UV, TU−V, −T−UV, −TU−V, T−U−V neg: −T−U−V	0.039	0.474
4	pos: TUV, T−UV neg: −TUV, TU−V, −T−UV, −TU−V, T−U−V, −T−U−V	0.286	0.207
5	pos: TUV, T−UV, TU−V, −T−UV neg: −TUV, −TU−V, T−U−V, −T−U−V	0.149	0.349
6	pos: TUV doubtful: −TUV, T−UV, TU−V neg: −T−UV, −TU−V, T−U−V, −T−U−V	0.396	

dependence. The method leads to a high rate of undesirable false positive interpretations. In Section 8 the problem of choosing a decision scheme ("decoder") has been dealt with in detail. In Table 9 a numerical example is given from the data of Stamm et al. (23, 24). For the parallel combination of three tests (GGT, AST, MCV) six different decoders are calculated. It is found that the decoder no. 2 ("at least two of the three results positive") has a very good test capacity *C*. However, decoder no. 6 is even better and operates as follows: The decision "positive" is only taken if the results of all three tests are above the decision limits. A decision "negative" is taken if two or three of the individual results are negative. All other results are considered "doubtful" and subjected to a check.

9.6 Calculation of the "optimum sets" in multiple tests

An important application of the information theoretical model is the determination of "optimum subsets" for a concrete medical question from a larger series of tests (28, 29, 30). This is taken to mean the minimum combination of tests supplying the desired information. Such considerations play a particular part when with regard to economy in health treatment superfluous, i. e. redundant, tests are to be avoided. The problem can only be solved with the multivariate approach because of the mutual dependence of many tests. For the reasons outlined above for this application as well the information theoretical model is particularly well suited. For the entire series of the tests an information quantity is calculated and the stepwise, continually checking the information quantity, a test is omitted until finally a given limit for the information is achieved. Norwich (30) has devised such a method to which we refer here.

9.7 Cost-benefit analysis of a test

In view of the limited resources of health systems, like other medical action clinical chemical tests must be subjected to a cost-effectiveness or cost-benefit analysis. Hitherto, such an analysis had to overcome the difficulty that the "effectiveness" or the "benefit" of a test cannot easily be quantified. The information theoretical quantities of our model are very suitable for use as input indicators for this purpose (31). A simple example is given in Figure 12. The effectiveness can be measured by the quotient transinformation I (bit) / unit costs per test (DM).

9.8 Performance of a clinical chemical test in the framework of the entire diagnostic process

In the last few decades the theoretical explanation of the diagnostic process has made considerable progress. A question of interest in this connection is the value which a clinical chemical test has within the framework of the entire diagnostic process. From the information theoretical point of view a very good answer can

EFFECTIVENESS OF CK–TEST vs. AST–TEST IN MYOCARDIAL INFARCTION

OUTPUT INDICATOR: TRANSINFORMATION [bit]

INPUT INDICATOR : UNIT COST PER TEST [DM]

	PREVALENCE 0.05	PREVALENCE 0.50
CK–TEST	0.029	0.171
AST–TEST	0.033	0.163

Fig. 12. Effectiveness of CK-Test versus AST-Test in myocardial infarction.

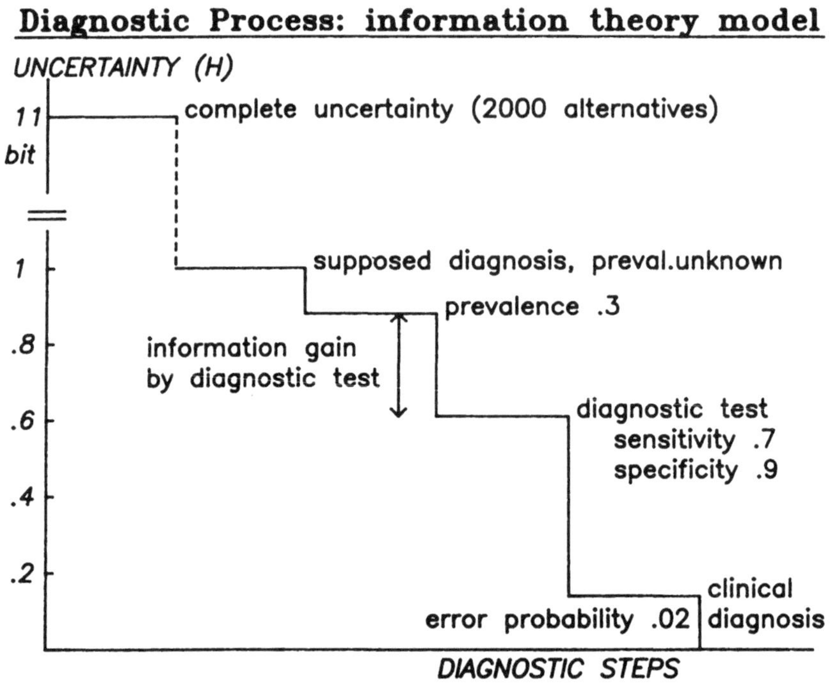

Fig. 13. Information theory model of the diagnostic process.

be given to this question. Proceeding from our definition in Section 2, according to which information is the elimination of uncertainty on the occurrence of possible events, and measuring the uncertainty by the information entropy, the course of a diagnostic process could be characterized by the scheme reproduced in Figure 13. The initially present complete uncertainty on the disease which a patient has is reduced to a few possibilities by drawing up a provisional diagnosis, with corresponding reduction of the entropy H (the assumptions in the example in Figure 13 have been made completely arbitrarily). If the prevalence of the assumed disease is known the uncertainty is reduced still further. A diagnostic test (with the test properties indicated) is now used. The entropy decrease which results is marked by the arrow. Finally, by a series of steps a clinical diagnosis is achieved which still has a slight error probability, for which reason the entropy does not drop completely to 0. The imaginary example outlined might stimulate more exact analysis of concrete diagnostic processes.

10 Prospectus

The information theoretical test model derived makes it possible to describe a test in a very general form and to measure its diagnostic efficacy or validity. The Shannon information measure, the information entropy, is particularly well suited to this purpose because it does not presume any assumptions on the distribution of the data. Thus, the model is excellently suited to a comparison of different tests. It is of no consequence whether qualitative or quantitative tests are to be compared. Even tests representing their results on ordinal scales (they have not been dealt with in this paper) can be evaluated and compared. It is also possible to deal without restriction with multivariate problems of the type which occur in the combination of tests. It must be pointed in particular that the complexity of the calculation remains within acceptable limits because it increases linearly with the number of variables. The formulae employed for the calculation are particularly clear because they all can be derived from the Shannon definition equation of information entropy (eq. (1)). The general assumption-free applicability of the model makes it easy to deal with not only with clinical chemical tests but also with diagnostic investigations of all types, even in combination with clinical chemical tests. An extension to medical decisions following a test is also conceivable. For this purpose the model must be linked to decision theoretical methods.

The Shannon information theory takes account only of the "technical" or "syntactic" relationship between signs or messages to be transmitted (4). The "semantic" aspect, i.e. the meaning or the value of the messages transmitted, is disregarded. The test model introduced keeps completely within this scope. Proceeding from Shannon's statement in particular Bar-Hillel and Carnap (32) have attempted to define the "semantic information" and draw up a corresponding theory. It is conceivable that our own test model on incorporation of these results could be extended to the semantic area as well.

Appendix

Calculation of the test capacity (channel capacity) C

For the calculation of the test capacity (the "channel capacity" in information theory) numerous methods have been proposed in the literature and three of them appear suitable for our purposes.

1. Silverman (33) derived for the general binary channel (number of inputs $m = 2$, number of outputs $n = 2$) a formula enabling a direct calculation from the transition probabilities. Using the abbreviated symbols $\alpha = P(T_1|D_1)$ and $\beta = P(T_1|D_2)$ we obtain

$$C = \frac{-\beta \cdot H(\alpha) + \alpha \cdot H(\beta)}{\beta - \alpha} + \mathrm{ld}\,[1 + 2^{\frac{H(\alpha) - H(\beta)}{\beta - \alpha}}] \tag{A1}$$

The optimum input probability $P(D)_{\text{opt}}$ can also be calculated in simple manner:

$$P(D)_{opt} = \beta(\beta - \alpha)^{-1} - (\beta - \alpha)^{-1} \cdot [1 + 2^{\frac{H(\beta) - H(\alpha)}{\beta - \alpha}}]^{-1} \tag{A2}$$

(in my former publication (7) the term for $P(D)_{\text{opt}}$ (eq. (6) on page 481) is wrong because of a printing error).

2. A more generally applicable method which can also be used if the number of inputs and outputs is greater than 2 has been described by Muroga (34). The method maximizes the transinformation utilizing Lagrange multipliers.

3. Using a theorem derived by J. Peters (10) a numerical method can be derived which is the one which has mainly been employed in the present work. This theorem states that in a test the transinformation for the entire test reaches its maximum (= test capacity) when the transinformation for the individual signs D_i of the message source, i. e. for the rows of the frequency table, is equal. A transcendental equation system results which with a zeroing method numerically furnishes the desired test capacity and the optimum input probability. The method will be briefly described below. A table is given with the transition probabilities

	T_1	T_2	....	T_n
D_1	$P(T_1 \mid D_1)$	$P(T_2 \mid D_1)$		$P(T_n \mid D_1)$
D_2	$P(T_1 \mid D_2)$	$P(T_2 \mid D_2)$		$P(T_n \mid D_2)$

The "row transinformation" I_1 and I_2 is calculated from

$$I_1 = \sum_{j=1}^{n} P(T_j \mid D_1) \cdot \mathrm{ld}\left(\frac{P(T_j \mid D_1)}{P(T_j)}\right) \tag{A3}$$

$$I_2 = \sum_{j=1}^{n} P(T_j \mid D_2) \cdot \mathrm{ld}\left(\frac{P(T_j \mid D_2)}{P(T_j)}\right) \tag{A4}$$

The output probabilities $P(T)$ can be substituted by

$$P(T_j) = P(D_1) \cdot P(T_j \mid D_1) + P(D_2) \cdot P(T_j \mid D_2) \tag{A5}$$

and furthermore $P(D_2)$ can be replaced by $(1 - P(D_1))$. This then gives the row transinformation I_1 and I_2

$$I_1 = \sum_{j=1}^{n} P(T_j \mid D_1) \cdot \mathrm{ld} \left(\frac{P(T_j \mid D_1)}{P(D_1) \cdot \{P(T_j \mid D_1) - P(T_j \mid D_2)\} + P(T_j \mid D_2)} \right) \tag{A6}$$

$$I_2 = \sum_{j=1}^{n} P(T_j \mid D_2) \cdot \mathrm{ld} \left(\frac{P(T_j \mid D_2)}{P(D_1) \cdot \{P(T_j \mid D_1) - P(T_j \mid D_2)\} + P(T_j \mid D_2)} \right) \tag{A7}$$

Since according to Peters' theorem (10) for the case of maximum transinformation

$$I_1 - I_2 = 0 \tag{A8}$$

by means of a zeroing method it is determined numerically for which value $P(D_1)$ eq. (A8) becomes zero. This value corresponds to the optimum input probability $P(D)_{\mathrm{opt}}$. The "row transinformation" I_1 then provides the desired test capacity C.

Acknowledgements

The author wishes to express thanks to Dr. E. Hansert, München for valuable criticism.

References

1. Büttner, J. (1978) Optimization of laboratory testing. In: Logic and Economics of Clinical Laboratory Use (Benson, E. S., and Rubin, M., eds.) Elsevier, New York, Oxford.
2. Büttner, H. (1980) Grundlagen der Bewertung klinisch chemischer Befunde. In: Validität klinisch-chemischer Befunde. Merck Symposium 1979 (Lang, H., Rick, W. and Büttner, H., eds.) Springer, Berlin, Heidelberg, New York.
3. Büttner, J. (1977) Die Beurteilung des diagnostischen Wertes klinisch-chemischer Untersuchungen. J. Clin. Chem. Clin. Biochem. *15*, 1–12.
4. Shannon, C. E. (1948) A mathematical theory of communication. Bell Syst. tech. J. *27*, 379–423, 623–656. Reprinted in: Shannon, C. E., Weaver, W. (1949) The mathematical theory of communication. Univ. of Illinois Press, Chicago.
5. Büttner, J. (1982) Information theory applied to the utilization of laboratory data. In: XI. International Congress of Clinical Chemistry (Kaiser, E., Gabl, F., Müller, M. M., Bayer, M., eds.) de Gruyter, Berlin, New York.
6. Büttner, J. (1984) Some theoretical aspects of test selection strategy. Scand. J. Clin. Lab. Invest. *44*, Suppl. 171, 83–95.
7. Büttner, J. (1982) Grundlagen der Anwendung der Informationstheorie auf qualitative klinisch-chemische Untersuchungen. Anwendung der Informationstheorie auf klinisch-chemische Untersuchungen, I. J. Clin. Chem. Clin. Biochem. *20*, 477–490.
8. Ash, R. (1965) Information Theory. Wiley, New York, London, Sydney.
9. Meyer-Eppler, W. (1969) Grundlagen und Anwendungen der Informationstheorie, 2nd. ed. Springer, Berlin, Heidelberg, New York.
10. Peters, J. (1967) Einführung in die allgemeine Informationstheorie. Springer, Berlin.
11. Ingels, F. M. (1971) Information and Coding Theory. Intext Educational Publishers, Scranton, San Francisco, Toronto, London.

12. McGill, W. J. (1954) Multivariate information transmission. Psychometrika *19*, 96–116.
13. Garner, W. R. & McGill, W. J. (1956) The relation between information and variance analyses. Psychometrika *21*, 219–228.
14. Woodward, P. M. (1953) Probability and Information with Applications to Radar. Pergamon Press, London.
15. Kullback, S. (1959) Information Theory and Statistics. Wiley, New York. Reprint (1978) P. Smith, Gloucester, Mass.
16. Miller, G. A., Madow, W. G. (1954) On the maximum likelihood estimate of the Shannon–Wiener measure of information. Report AFCRC-TR-54-75. Operational Applications Laboratory, ARDC, August 1954.
17. Kullback, S., Kupperman, M. & Ku, H. H. (1962) An application of information theory to the analysis of contingency tables, with a table of 2n ln n, n = 1 (1) 10000. J. Res. Nat. Bur. Stand. *66*B, 217–243. See also: Kullback, S., Kupperman, M. & Ku, H. H. (1962) Tests for contingency tables and Markov chains. Technometrics *4*, 573–608.
18. Woolf, B. (1957) The log likelihood ratio test (the G-test). Methods and tables for tests of heterogeneity in contingency tables. Ann. Human Genetics *21*, 397–409.
19. Yerushalmy, J. (1947) Statistical problems in assessing methods of medical diagnosis. With special reference to x-ray techniques. Publ. Health Rep. *62*, 1432–1449.
20. Galen, R. S. & Gambino, S. R. (1975) Beyond normality: The predictive value and efficiency of medical diagnoses. Wiley, New York, London, Sydney, Toronto.
21. Keller, H. & Gessner, U. (1982) Die logischen Grundlagen der Laboratoriumsdiagnostik. Schweizerische Med. Wochenschr. *112*, 914–926.
22. Hansert, E., Federkiel, H. & Stamm, D. (1984) A new procedure for discriminating between two patient populations using multivariate decision limits: Application in the detection and exclusion of alcoholism based on clinical laboratory findings. J. Clin. Chem. Clin. Biochem. *22*, 791–810.
23. Stamm, D., Hansert, E. & Feuerlein, W. (1984) Detection and exclusion of alcoholism in men on the basis of clinical laboratory findings. J. Clin. Chem. Clin. Biochem. *22*, 79–96.
24. Stamm, D., Hansert, E. & Feuerlein, W. (1984) Excessive consumption of alcohol in men as a biological influence factor in clinical laboratory investigation. J. Clin. Chem. Clin. Biochem. *22*, 65–77.
25. Solberg, H. E. (1975) Discriminant analysis in clinical chemistry. Scand. J. Lab. Clin. Invest. *35*, 705–712.
26. Keller, B., Keller, H. (1986) A distribution-free, multivariate discriminating method. J. Clin. Chem. Clin. Biochem. *24*, 601–609.
27. Rehpenning, W., Harm, K., Domesle, A. & Voigt, K. D. (1979) Falsch positive Werte bei der Vielfachanalyse: Die Abschätzung ihrer Häufigkeit mit der Sylvesterschen Formel und ihre Reduktion durch eine multivariate Testgröße. J. Clin. Chem. Clin. Biochem. *17*, 565–572.
28. Lee, R., C.-T. (1971) Application of information theory to select relevant variables. Mathematical Biosciences *11*, 153–160.
29. Kryspin, J. & Norwich, A. M. (1973) Application of information calculus to medical data analysis and reduction. Mathematical Biosciences *17*, 165–172.
30. Norwich, A. M. (1970) Application of information theory to medical data analysis and reduction. Thesis, M. Appl. Sci., Univ. of Toronto, typoscript.
31. Büttner, J. (1985) Messung der Effizienz und Effektivität bei klinisch-chemischen Untersuchungen. In: Effizienz und Effektivität medizinischer Diagnostik (Vogel, H. R., ed.) G. Fischer, Stuttgart.
32. Bar-Hillel, Y. & Carnap, R. (1953) Semantic Information. In: Communication Theory (Jackson, W., ed.) Butterworth, London.
33. Silverman, R. A. (1955) On binary channels and their cascades. IRE Transactions Information Theory *1*, 19–27.
34. Muroga, S. (1953) On the capacity of a discrete channel. J. Physical Soc. Japan *8*, 484–494.

Chapter 3.4
The Multivariate Approach

P. Winkel

1 Introduction

The laboratory produces a steadily increasing number of values on an ever-expanding array of assays. These values are considered in conjunction with other clinical data in order to aid in clinical decision making. Among the various quantitative approaches which can be used to optimize the utilization of laboratory data already present and assist in ordering additional laboratory measurements multivariate techniques play a predominant role.

Multivariate analysis may be defined as the collection of techniques appropriate for situations in which the random variation in several variates has to be studied simultaneously.

The emphasis in this chapter will be on the methodological aspects and examples of specific clinical applications of the various multivariate techniques will only be presented in so far as they illustrate methodological aspects. For reviews of clinical applications in general or confined to specific problem areas the reader is referred to Chapters 3.9, 3.10, 3.11 in this book and to (1, 2, 3, 4, 5, 6).

Multivariate analysis may be applied in the situation where each of several quantities is measured once in each of the patients studied (one observation per patient) or the same quantity (or set of quantities) is measured repeatedly in the same patient (multiple observations per patient). In the former situation we distinguish two major problem areas namely probabilistic diagnosis and the construction of new clinical classes which require quite different methodological approaches.

2 One Observation per Patient

2.1 Predefined clinical classes presumed

2.1.1 **Probabilistic diagnosis**

2.1.1.1 ***Bayes theorem for multiple variates***

The Bayesian theorem which is reviewed elsewhere in this book can be applied to a diagnostic problem where one evaluates a very large number of variates and many possible clinical classes such that the patient with certainty belongs to one and only one of these. For illustrative purposes, two binary variates: S_1 and S_2 and three clinical classes D_1, D_2, and D_3 will be considered. Figure 1 presents the values of S_1 and S_2 for a randomly selected subset of 25 patients from each of the three clinical classes. It is assumed that the distribution of values of S_1 and S_2 among these patients is identical with that of the total population. As a consequence, various conditional probabilities can be computed from the information presented in Figure 1, i. e., the probability that a patient will present certain values of S_1 and S_2, given that he is a member of a particular clinical class. The prior probabilities (prevalences) of obtaining a member of each clinical class when a patient is picked at random from the universe of patients, are referred to as $P(D_1)$, $P(D_2)$ and $P(D_3)$.

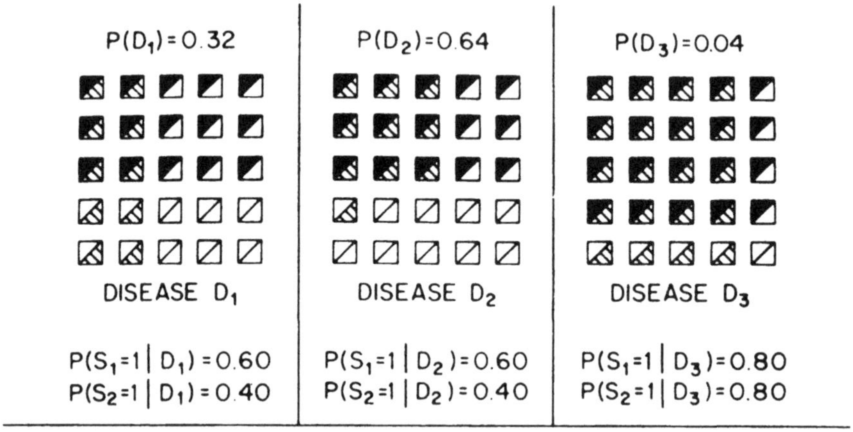

	CONDITIONAL PROBABILITY $P(S_1=a_1, S_2=a_2 \mid D_j)$			CONDITIONAL PROBABILITY $P(D_j \mid S_1=a_1, S_2=a_2)$ *		
$(S_1=a_1, S_2=a_2)$	D_1	D_2	D_3	D_1	D_2	D_3
$(S_1=0, S_2=0)$	0.24	0.36	0.04	0.249	0.746	0.005
$(S_1=1, S_2=0)$	0.36	0.24	0.16	0.419	0.558	0.023
$(S_1=0, S_2=1)$	0.16	0.04	0.16	0.615	0.308	0.077
$(S_1=1, S_2=1)$	0.24	0.36	0.64	0.231	0.692	0.077

$$*P(D_j \mid S_1=a_1, S_2=a_2) = \frac{P(S_1=a_1, S_2=a_2 \mid D_j) \times P(D_j)}{\sum_{i=1}^{3} P(D_i) \times P(S_1=a_1, S_2=a_2 \mid D_i)}$$

Fig. 1. The values of two discrete variates (S_1 and S_2) in each of 25 randomly selected patients from each of three clinical classes, (D_1, D_2 and D_3). The subjects are depicted as squares with a diagonal; $S_1 = 1$ is depicted by a solid shading in the upper left portion of the squares, and $S_2 = 1$ is depicted by lines drawn in the lower right portion of the square; when either of the variate values is zero, the appropriate portion of the square is left open. The prevalences, the conditional probabilities, and Bayes formula for computing the posterior probability are presented in this figure. (See text for explanation of the various symbols used in this figure.) (Reproduced from (6) with permission.)

This information is also presented in Figure 1. When the prevalence of the clinical classes and the prior conditional probabilities are known the posterior probabilities can be computed. The posterior probability can be symbolized as $P(D_j | S_1 = a_1, S_2 = a_2)$, that is, the probability of a patient being a member of class D_j, given that he presents the values a_1 and a_2 for the variates S_1 and S_2.

Two variates are considered to be independent when knowing the value of one variate does not affect the probability of the value that the other variate may assume. The variates S_1 and S_2 are considered to be independent within a given clinical class D_j if the following condition can be met for all possible values of a_1 and a_2:

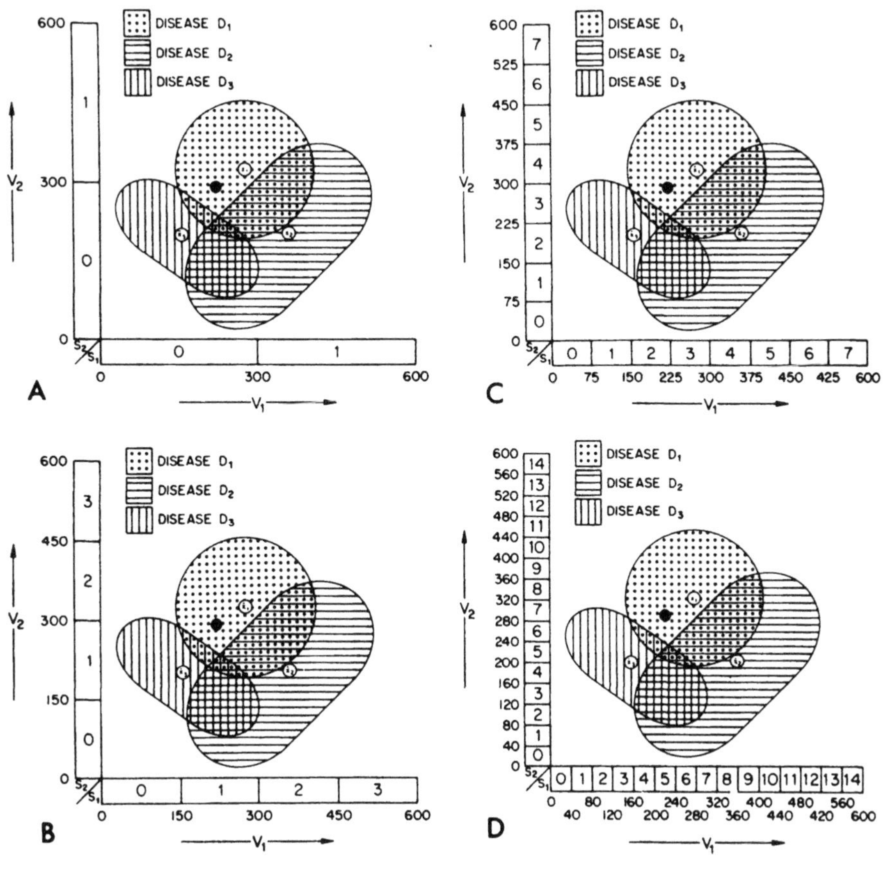

Fig. 2. The 99 per cent tolerance regions of the two continuous variates V_1 and V_2 are depicted for three clinical classes: D_1, D_2 and D_3. The circled values of X_1, X_2, and X_3 are sets of mean values of the classes D_1, D_2, and D_3, respectively. The closed circle corresponds to the case where $V_1 = 220$ and $V_2 = 290$. In each of the four figures in this series the continuous variates V_1 and V_2 are tranformed to the discrete variates S_1 and S_2. In A, the total number of possible values that each discrete variate can obtain is two; in B, the number is four; in C, the number is eight; and in D, the number is fifteen. (Reproduced from (6) with permission.)

$P(S_1 = a_1, S_2 = a_2 | D_j) = P(S_1 = a_1 | D_j)\ P(S_2 = a_2 | D_j)$ where (in this case) a_1 and a_2 can take on the values "0" or "1". The constructed examples in Figure 1 illustrate independence for variates S_1 and S_2 as observed in members of class D_1 and class D_3. Using $S_1 = 1$ and $S_2 = 1$ as one pattern, we note for example for class D_1:

$$P(S_1 = 1, S_2 = 1 | D_1) = P(S_1 = 1 | D_1)\ P(S_2 = 1 | D_1)$$

is a correct statement in that, 0.24 = (0.60) (0.40), and also for class D_3: $P(S_1 = 1, S_2 = 1 | D_3) = P(S_1 = 1 | D_3)\ P(S_2 = 1 | D_3)$ is a correct statement, in that 0.64 = (0.80) (0.80). However, as observed among the members of class D_2, there is a dependency of the variates, i.e., $P(S_1 = a_1, S_2 = a_2 | D_2) = P(S_1 = a_1 | D_2)$ $(S_2 = a_2 | D_2)$ for, as can be noted in Figure 1, the probability of finding a patient for whom $S_1 = 1$ and $S_2 = 1$ is greater than the product of the probability of finding a patient for whom $S_1 = 1$ and the probability of finding a patient for whom $S_2 = 1$. Specifically, based on the constructed example, $P(S_1 = 1, S_2 = 1 | D_2) > P(S_1 = 1 | D_2)\ P(S_2 = 1 | D_2)$ i.e., 0.36 > (0.60) (0.40).

If continuous variates (e.g. S-Sodium) are involved these may be transformed to discrete variates (Fig.2) and we can apply Bayes theorem to discrete variates having more than two values.

An alternative approach for computing probabilities for continuous variates is based on the "probability density function". Mathematically this is done by defining a function, (i.e., a function in two variates S_1 and S_2) which has the property that when integrated over all possible sets of values, the result is equal to 1.0; and when integrated over any particular region the result is equal to the probability that a subject picked at random from the population in question will have a set of values falling within the region. The Bayesian formula for continuous data then is derived from that of discrete data by replacing probabilities with density function values.

It is necessary to define a general algebraic formula according to which the value of the density function can be computed for all combinations of variate values. The formula includes a number of constants (parameters) which characterize the distribution of the data and which, therefore, have to be estimated from the data.

2.1.1.2 *Choice of probabilistic mode*

A model postulating independence among variates within clinical classes was originally applied for automating the diagnosis of congenital heart diseases (7). The advantage of this model is that in addition to the prior probabilities, we need estimate only as many probabilities for each disease as there are (binary) variates. From these probabilities all other probabilities may be computed. If we do not postulate independence, we must estimate the probability for each combination of variate values; i.e., we would estimate the probability as the number of patients with the disease who present the pattern of variate values divided by the total number of patients with the disease. In general, for a given disease, if we have n binary variates, there are 2^n possible patterns and therefore 2^n-1 probabilities to estimate when we do not assume independence among variates.

The performance of Bayes model when independence is incorrectly assumed deteriorates as compared to its performance when used on artificial materials which correspond exactly to the real data except that the condition of independence is fulfilled. This phenomenon is more pronounced the more variates are included in the model (8). The performance may be improved when a version of Bayes model is used which retains part of the correlation structure of the data (8, 9, 10). It is perhaps not so surprising that the performance of a model improves when applied to constructed data adhering more closely to the assumptions of the model than the real data. The problem however, is: given the real data which model should be used? A simplified one based on independence and therefore including relatively few parameters to estimate or a more complete model? It appears that in a number of situations where the assumption of independence is violated the distribution of the posterior probabilities becomes U-shaped and more so as the number of variates increases thus biasing the estimates of the posterior probabilities toward zero and one (11). This may not necessarily impact on the rate of misclassifications. However, when assessing the risks and benefits of various actions for an individual patient the posterior probability is critical.

On the other hand if a more complete model is used (be it a discrete one or a continuous one) more parameters have to be estimated to describe all the possible interdependencies and this clearly decreases the overall precision of the estimates. This may be the explanation why a number of simulation experiments as well as examples from medical research have shown that the (false) independence assumption results in classification error rates comparable or better than rates obtained by using the correct model (11, 12). Hilden (13) calls attention to another possible reason why the Bayes model derived by treating discrete symptom variates as conditionally independent (BFCI, Bayes formula based on conditional independence) often works so well in practice. He showed that BFCI does not require the independence assumption. It is a sufficient but not a necessary assumption for the validity of the derived formula. Thus the popular claim that BFCI must necessarily produce wrong results if discrete symptoms are mutually dependent is wrong.

Various models are available when one wants to apply Bayes theorem for continuous variates. The one most commonly used is that implied in the technique of linear discriminant analysis.

Linear discriminant analysis was originally proposed by Fischer (14) as a means of deriving a linear function which could best discriminate between two groups of items on which multiple observations were made. He derived a linear combination of the measurements called the linear discriminant function Z, which has the property that the univariate distributions of Z in the two groups are maximally separated. In the particular model of linear discriminant analysis to be presented here it is assumed that the values from each of the classes follow a multivariate Gaussian distribution. Furthermore, it is assumed that the distributions differ only with regard to location, i. e., differ with regard to mean values.

In principle, linear discriminant analysis amounts to the computation of density function values. If prior probabilities are known or estimated, Bayes formula may be applied subsequently. Owing to the assumptions made, the computations become very simple. Assume we have two classes: D_{10} and D_{20}, and p' variates: S_1, S_2, ..., S_p observed in each subject belonging to the two classes. The p variates involved in the linear discriminant analysis are weighted (multiplied) by (b_1, b_2, ..., b_p) and expressed mathematically as $Z = b_1 S_1 + b_2 S_2 + \ldots + b_p S_p$. This linear function Z,

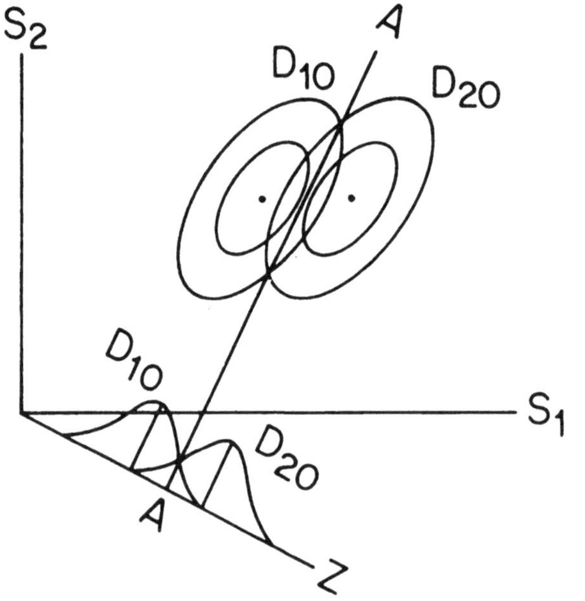

Fig. 3. Tolerance regions (90 and 95 per cent) of two continuous variates S_1 and S_2 for two clinical classes D_{10} and D_{20}. The classes have equal prevalences, and the corresponding bivariate distributions have equal variances and correlation coefficients. For further explanation of symbols, see text. (Reproduced from (6) with permission.)

a weighted sum, is termed the discriminant function. When the discriminant function is evaluated for a patient, the resulting number z is the discriminant score for that patient. Figure 3 presents a geometric interpretation of the discriminant analysis where we have two classes, D_{10} and D_{20}, and two variates, S_1 and S_2. The two variates S_1 and S_2 are positively correlated as illustrated. Each ellipse in the diagram is the focus of points of equal probability density for a particular class. For example, the outer ellipse for class D_{10} might define the region within which 95 per cent of class D_{10} lies. These elipses are called centours (centile contours). The two points at which corresponding centours intersect define the discriminator A, the line of equal probability density values for the two classes. If a second line Z is constructed perpendicular to line A and if the points in the two-dimensional space are projected onto Z, the overlap between the two groups will be smaller than for any other possible line. The linear function used for this projection is the discriminant function. It is noted that the discriminant function follows a univariate Gaussian distribution for each of the two populations. The two distributions have identical variances and their intersection falls on the line of equal density function values in the original two-dimensional distributions. The technique of linear discriminant analysis may be generalized to more than two groups (15, 16). For illustrative purposes a constructed example of the use of discriminant function analysis is presented in Figure 4.

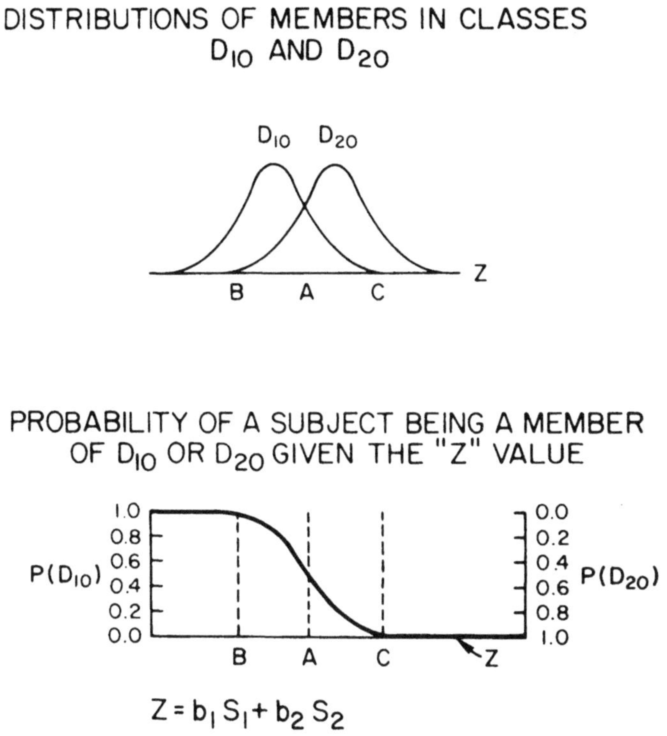

Fig. 4. Constructed example showing the frequency distribution of the Z (discriminant function) values for each clinical class D_{10} and D_{20}. It is assumed that the prior probabilities are equal; i. e., $P(D_{10}) = P(D_{20})$. As an example of two clinical groups, D_{10} could be the patients not alive two years after the first myocardial infarction and D_{20} the patients alive two years after the first myocardial infarction. The Z values can be transformed to probability values, e. g., the probability that a patient will be alive two years after the insult of myocardial infarction. It is noted that when $Z = B$, the probability of a patient being in group D_{20} is approximately 0.0; when $Z = A$ (discriminator value), the probability of a patient being in group D_{20} is 0.50; and when $Z = C$, the probability of being in D_{20} is approximately 1.0. When the prior probabilities (prevalences) are not identical, the discriminator value is changed to adjust for this fact. Thus, if $P(D_{10}) > P(D_{20})$ the discriminator value is shifted to the right, and if $P(D_{10}) < P(D_{20})$ it is shifted to the left. (Reproduced from (6) with permission.)

Discriminant methods other than linear discriminant function analysis have been introduced for those cases where the latter analysis is inappropriate. For instance, when the classes have unequal variances and different correlation values, a quadratic discriminant function is used (16). The geometric representation in this case is a curved surface. A major disadvantage of this model is that the number of parameters to estimate increases dramatically with the number of variates included in the model (4).

Discrimination by logistic regression presents an alternative to the linear and quadratic discriminant function analysis. In this model the probability of the occurrence of each of two (or more than two possible outcomes) is calculated. To be more precise a binary outcome variable y is predicted from a vector of covariates $X = (X_1, \ldots, X_p)$ using a linear function of the variables $I = a_0 + a_1X_1 + \ldots + a_pX_p$. One assumes that the prediction model has the logistic form i.e. $pr\ (y = 1\,|X) = \exp(I)/(1 + \exp I))$. In the clinical context to predict an outcome for a given patient corresponds to allocating him to a clinical class. Some of the advantages and disadvantages of logistic regression versus linear discriminant analysis are discussed in (17).

Various non-parametric models have been derived (18, 19, 20, 21). Some of them are supposed to be particularly well suited for binary data. Hand (22) compared the performance of one of these techniques (a so-called Kernel method) to that of the popular linear discriminant function analysis. He used six different binary data sets. The Kernel method was superior when the performance was assessed using the resubstitution method i.e. the data set which is used to derive the classifiction rule is also used to test its performance. However, when the performance was assessed using independent data the two methods showed the same error rates. The reason for this is probably that the Kernel method models the particular data used to derive the classification rule more closely than does the LDF.

As a multivariate analogous to the univariate reference interval which includes a certain fraction say 95% of the test results from healthy subjects it has been suggested to calculate a multidimensional reference region including a specified fraction of laboratory test profiles of a given type measured in healthy subjects (23, 24). This approach has been adopted for the classification of clinical laboratory profiles into diagnostic classes in that a multivariate reference region is calculated for each diagnostic class (25). A patient's profile is then classified as belonging to one diagnostic class, to more than one diagnostic class (i.e. the profile is located in an overlapping region of two or more classes), or to none at all. The advantage of this approach is that a patient who is an outlier with a strange laboratory profile is not automatically allocated to one of the diagnostic classes.

The SIMCA method (26, 27) combines this approach with the principal component analysis (see below). First the original variates are transformed into fewer by a principal component analyis which is done for each diagnostic class separately. Using the new variates the multivariate reference regions are then calculated etc. Outlying observations may also be detected by conventional discriminant function analysis provided special techniques are used which are designed to detect and compensate for the presence of outlying observations in multidimensional space (28).

Clinical problems usually involve a mixture of continuous and discrete variates. In this situation, one may transform the continuous variates to discrete variates and use a discrete model; or, alternatively, treat the discrete data as continuous data and use a continuous model, or, finally, formulate a model which may accommodate a mixture of discrete and continuous data (29, 30).

A detailed and technical review of the many different implementations of Bayes theorem and the advantages and disadvantages of each specific technique is beyond the scope of this chapter. The following, however, summarizes the author's conclusions:

1. Since it is probably impossible to derive some general guidelines which are sufficiently explicit and precise then for a given application one should keep an open mind and try out a number of different techniques and choose the one which turns out to be the best suited for the particular problem at hand.

2. In assessing the performance of a given methodology data which are independent of those used to derive the classification rule should be used.

3. The features to be assessed should be those enumerated in Section 2.1.1.4 of this chapter.

2.1.1.3 *Application of Bayes theorem*

The Bayesian model is used for computer aided diagnosis. Three different approaches may be distinguished. By the first one the values of all variates incorporated in the Bayesian model are obtained and then the patient is allocated into a disease class. By the second one (31) an initial set of variate values are first obtained. Then the values of additional variates are obtained one at a time. The current disease probabilities are calculated at each step and the process is terminated either when all observations incorporated in the Bayesian model have been obtained or when some arbitrary stopping point has been reached. The various implementations of this sequential model differ with regard to the approach taken to direct the selection of tests at each step, the choice of the set of tests selected initially, and the choice of stopping rule (3). The third approach is to extend the sequential model described above in that the costs and expected benefits of performing diagnostic tests and the relative costs of misdiagnosis are incorporated (31). For a thorough discussion of the advantages and disadvantages of the various models the reader is referred to (3).

2.1.1.4 *Assessing the performance of probabilistic diagnosis*

The development of a diagnostic system which produces probabilistic diagnoses requires a construction sample of patients. Based on the data observed in these patients the diagnostic system is derived. To test the performance of the system an evaluation phase is necessary. During this phase an evaluation sample of patients is obtained and the system is applied on the data of these patients. If the construction and evaluation sample of patients coincide an undesirable dependence (bias) between construction and evaluation is introduced resulting in a too optimistic picture of the performance of the diagnosis rule. In this situation it has been recommended to use various statistical cross-validation techniques (32, 33, 34, 35). In the "one-left-out" method, one patient is removed and the rule is rederived and used to classify the excluded patient. The patient's predicted state is compared with the true state. This process is repeated systematically for all patients. The misclassification rate is the fraction of the test patients who were incorrectly classified. This method and another method the bootstrap method which resembles the "one left out" method use the variability in the original data set to simulate the performance of the clinical prediction rule in a new population. By the split-sample technique the study population is divided into a training set and a test set. The training set may contain any proportion of the original sample; the test set contains the remaining patients. The rule, which was derived from the training set, is applied to each member of the test set. None of the mentioned techniques eliminates the effects of biases in selecting patients or collecting data since all the testing is carried out on patients from the training-set population. It is therefore preferable to use new patients in the evaluation phase. These new patients can be from the clinical setting in which the diagnostic system was derived or from a different clinical environment. The latter approach is preferable because it requires reproducible clinical methods and definitions. Since the study patients may not be representative of the clinical practice in which the prediction is to be used the best approach is to conduct a number of prospective validation studies in a wide variety of settings. Before a diagnostic system is generally accepted this latter kind of testing is essential. Since the nature of a given disease may change over time in the same setting and new clinical and laboratory tests continue to be developed a diagnostic system requires regular updating. This type of updating calls for skillful organisation and international collaboration. The need to emphasize the evaluation phase is supported by various studies. For instance the study by Harris (36) showed that the sensitivities and specificities of five out of seven commonly used tests varied significantly. For xeromammography the sensitivity ranged from 0.64 to 0.98 and the specificity ranged from 0.37 to 0.92. Wasson et al. (37) evaluated the methodological standard of 33 reports of multivariate prediction rules published in leading journals. Forty-two percent contained adequate description of the prediction rule, the patients, and the clinical setting. The misclassification rate of the rule was measured in only 34 percent and the effect of the rule on patient care only in 6 percent of the reports.

In assessing the value of a diagnostic system it is necessary to distinguish the following concepts:

1. The reliability of the diagnostic statements generated by the system.
2. The discriminatory ability of the system and
3. The patient benefit which may be achieved by the system.

In a series of very thorough and complete papers Habema, Hilden, and Bjerregaard analyzed the concepts of reliability, discriminatory ability and utility (38, 39, 40, 41). They also presented a number of statistical test quantities and graphical techniques useful to assess the various aspects of the reliability and discriminatory ability of a diagnostic system. The interested reader should consult these papers.

2.1.2 Studying the relationships between variates and between variates and clinical classes

The purpose of using linear discriminant analysis and similar techniques in clinical medicine is to utilize the discriminatory ability of the variates which is due to their interrelationship. However, often times a multitude of variates is available, and one must therefore choose from among the variates those which discriminate the best. In the clinical setting, we are faced with the problem of selecting the smallest subset of variates which contains all the discriminatory information from the total number of variates available. One approach might be to perform a univariate significance test for each of the variates and then elminiate all variates for which the test result is insignificant. Figure 5 illustrates the fallacy of this approach. It shows the joint distribution of two laboratory tests, S_1 and S_2, for each of two clinical classes, D_1 and D_2, for two cases. Evaluating the variates S_1 and S_2 univariately, we note that

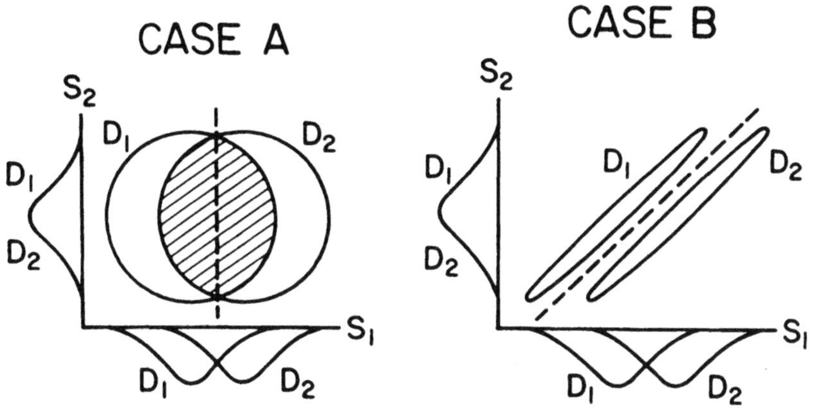

Fig. 5. Bivariate tolerance regions of two continuous variates S_1 and S_2 for two clinical classes D_1 and D_2. In both A and B, the univariate distributions of S_1 and of S_2 are the same for both classes. There is a definite separation of values for S_1, but the univariate distributions for S_2 are identical. In case A, there is no correlation between S_1 and S_2 and thus there is much overlap of the two tolerance regions. In case B, there is a very strong (positive) correlation between S_1 and S_2 such that there is much less overlap of the two tolerance regions.

the univariate distributions for classes D_1 and D_2 are comparable in both cases and S_2 does not discriminate very well in either case. However, if we decide to ignore S_2 as one of the variates on the basis of a non-significant t-test, the discrimination in case A will remain unchanged while it will decrease dramatically in case B.

Because of the correlations among variates, ideally one should examine all possible combinations of variates even if by univariate tests the variates are poor discriminators (42). In theory one could do this; but unless the number of variates is relatively small, this approach requires a very large amount of computer time. Two approaches which have been commonly used are the stepwise forward selection procedure and the stepwise backward selection procedure. By the former procedure the single variate which gives the best discrimination (smallest number of misclassifications) is first selected (43). At each of the following steps a new variate is selected, namely the variate which, combined with the previously selected variates, gives the best discrimination. After each step, the previously selected variates are reevaluated and if a variate no longer contributes significantly to the discrimination, it is eliminated. The discriminatory ability of each variate combination may be compared with that of all the variates when used in combination (44). When this difference is no longer significant, the variates not yet entered in the analysis are declared redundant. If the problem is to decide whether or not a particular test is redundant, the discriminatory ability of all tests is computed and compared with that obtained when the test in question is excluded.

By the backward procedure all variates are included initially and then systematically eliminated one by one. At each of these steps it is assessed if the reduced set of variates performs as well as the entire array. When this is no longer the case the procedure stops. Whether the forward or the backward procedure is applied a criterion must be used to assess the performance of the subsets of variates derived at each step i.e. each subset must be compared to the entire array. The options available for such comparisons, however, depend on the model used. For the quadratic discriminant analysis the error rate cannot be determined analytically nor are any statistics derived that are related to it so far (45). For the non-parametric Kernel method the estimated error rate can be calculated relatively easily and in case of the linear discriminant analysis various statistics related to the error rate are available (45). Goldsmith et al. (45) call the attention to the so-called peaking phenomenon (46). If poorly discriminating quantities are added to a given set of discriminating quantities the score of correctly classified patients may actually drop indicating that the additional estimation error introduced exceeds the gain in discriminating power, so to speak.

Beck et al. (47) used the classification of symptoms by numerical taxonomy as a means to eliminate redundant tests.

Instead of reducing the number of quantities measured a reduced number of quantities may be derived from those originally observed. The principal components analysis (48) is an example of a technique which may be used to achieve this goal. By this technique a number of linear combinations (principal components) of the variates are formed. If N variates are observed originally N principal components are formed. The first component is the linear function of the original variates which best fit the observations and explains more of the variation of the data than any other linear function. The second and subsequent components are derived similarly and sequentially, each explaining as much as possible of the variation unexplained by the preceding components. Each component is independent of the preceding ones and explains progressively less of the total variation. The data reduction is achieved by selecting a few of the principal components which explains the major part of the total variation and then only use these derived quantities in the subsequent analysis. Since the components depend on the scales in which the original quantities are measures it is necessary first to standardize each variate to unit variance. Another weakness of the technique is perhaps that the biological meaning of the components may be lost the risk being that ultimately clinical reasoning is reduced to a numbers game. However, it does appear that relatively meaningful quantities have been derived in some studies applying this technique (49, 50).

When the techniques described above are used for predictive purposes the interest lies in the predictive value of a subset of the variates and we are not interested in the exact relationships among the variates. The mentioned techniques, however, may also be used for interpretive purposes in that the absolute value and sign of various correlations or regression coefficients may serve to support or refute a pathophysiological hypothesis. For instance one may want to determine if a quantity discriminates between subjects who do develop ischemic heart disease within five years and subjects who don't when the discriminatory ability of all known major risk factors have been taken into account. The hypothesis being that the quantity is a risk factor itself. When the techniques are used in this way some caution has to be exercised. If highly intercorrelated variates are included in the analysis the various techniques may produce drastically different results at least when the multivariate logistic model is used (51) and in all likelihood also when other multivariate models are used. Furthermore if the analytical variation is not negligeable as compared to the biological variation it may impact on the results in an unpredictable way (see below).

Even if one disregards the impact of analytical variation a correlation coefficient is still an ambiguous concept. Thus Harris (52) pointed out that the biological correlation coefficient reflects the average intraindividual correlation as well as the correlation of mean values across individuals.

2.1.3 **The effect of random analytical variation**

A laboratory result may be considered as a sum of two components, a biological and a random analytical component. The values of each of these components vary from one result to the next one. Assume that we are observing two laboratory quantities measured simultaneously in each of a number of subjects and we want to study the correlation of these two quantities.

We want to measure the correlation between the two biological components (the biological correlation). What we are measuring, however, is the correlation (the observed correlation) between the observations each of which is a sum of a biological and an analytical component. If neither the biological and analytical nor the two analytical components are correlated the observed correlation will be equal to the biological one multiplied by a factor which is less than one but larger than zero (53). The larger the variability of each analytical component is relative to that of the corresponding biological component the smaller will this factor be. In the slightly more complicated case where we want to study a partial correlation coefficient the potential impact of the analytical variability is much more serious. If the analytical variation present when measuring the variate partialled out is substantial the result is unpredictable (53). For instance the observed correlation may turn out to be negative when in fact the true biological correlation is positive. Therefore it is always wise to assess the relative magnitudes of the biological and the analytical components of variation before a multivariate analysis is carried out in particular if one wants to interpret the results i. e. attempt to attach meaning to the parameters estimated for the multivariate model. If deemed prohibitively large the random analytical variation may be reduced by obtaining replicate measurements of each specimen and then averaging them.

If the analytical components are correlated it is generally impossible to determine the extent to which an obtained correlation coefficient reflects associations between the biological components and the extent to which it reflects associations between the analytical components. In this context it is worth noting that the use of multitest analyzers may very well introduce correlations among the analytical components of variations. For instance the correlation between the analytical errors of S-K and S-Ca demonstrated in (54) could be due to simultaneous assay of these cations in the same flamephotometer since variations in flow of gas influence cations in the same direction.

Techniques useful to break the correlations between analytical errors are given in (54, 55).

2.2 Constructing new clinical classes

2.2.1 Numerical taxonomy

The probabilistic approach used to assign a given patient to a particular clinical class is based in part upon well-accepted definitions of the various clinical classes. These definitions are made by clinicians or pathologists who have studied the consistent observations present in a number of patients. Clinical classifications are of two types, "monothetic" and "polythetic". A monothetic class includes only members that share the same attribute or attributes implied by the definition of the clinical class; for example, all patients with nephrotic syndrome must demonstrate peripheral edema, hypoalbuminemia, and proteinuria. By contrast, a polythetic class is a class for which it is not required that any one attribute (symptom, sign, or laboratory result) be shared by all patients making up the class. Consider four patients possessing the following attributes respectively: S_1 S_3 S_4, S_1 S_2 S_4, S_1 S_2 S_3, and S_2 S_3 S_4. Although none of the attributes is shared by all four patients, each patient is similar to the other three by having two attributes in common. According to the definition of polythetic classes, these four patients may very well fall into one clinical class. This section will present various techniques of automated polythetic classifications (techniques of numerical taxonomy or cluster analysis) and the criteria used to test their validity. The clinical meaningfulness of the classes formed by these techniques will be discussed. The interested reader is referred to (56, 57, 58) for extensive reviews of the theoretical and computational problems of numerical taxonomy and to (59) and chapter 3.5 in this book for a review of the applications in clinical medicine.

2.2.1.1 *The techniques of numerical taxonomy*

Generally speaking, cluster analysis covers a multitude of methods all of which aim at sorting items on the basis of multiple observations obtained from each item. Items (or in this case, patients) within the same group should resemble each other as much as possible while patients belonging to different groups should be as dissimilar as possible.

Two different approaches to the problem may be distinguished.

By the first approach (the hierarchic methods), all possible pairs of patients are first formed and the dissimilarity in terms of the observations obtained from the patients is determined for each pair. Based on these dissimilarities, the patients are then sorted into groups. By the second approach, the patients are sorted into groups directly to optimize some criterion, for instance, the variation among groups.

The hierarchic techniques are always based on a dissimilarity coefficient. A dissimilarity coefficient is a multivariate function used to compare the observations made in two patients. Usually a very simple dissimilarity coefficient is used in clinical applications namely the so called matching coefficient which is equal to the number of matching symptoms over the total number of symptoms compared between two patients. Figure 6 shows a constructed example where this coefficient is used to compute the dissimilarities of four patients each of whom have been examined for the presence or absence of 10 symptoms. If continuous variates are used say K variates the geometrical distance between the data points in K-dimen-

1 DATA

SYMPTOMS

PATIENTS

	1	2	3	4	5	6	7	8	9	10
A	0	0	1	1	1	1	1	1	1	1
B	1	1	1	1	1	1	1	1	1	0
C	0	0	0	0	0	0	1	1	1	1
D	1	0	0	0	0	0	0	0	0	0

1. DENOTES THE PRESENCE OF A SYMPTOM

0. DENOTES THE ABSENCE OF A SYMPTOM

2 DISSIMILARITIES

DISSIMILARITY COEFFICIENT (D)

$$D = \frac{(N - X)}{N}$$

N IS THE NUMBER OF SYMPTOMS OBSERVED IN BOTH OF TWO PATIENTS

X IS THE NUMBER OF MATCHES IN THE TWO PATIENTS

MATRIX OF DISSIMILARITIES

	A	B	C	D
A	0	03	04	09
B	02	0	07	08
C	04	07	0	05
D	05	08	05	0

Fig. 6. Constructed example showing the computation of a matrix of dissimilarity coefficients from data offered from four patients each of whom have been examined for the presence or absence of 10 symptoms.

sional space may be calculated and used as the dissimilarity coefficient. The further apart two data points are the longer the distance is and the more dissimilar are the corresponding two patients. Other dissimilarity coefficients have also been suggested (60). In performing a hierarchical classification a matrix of dissimilarity coefficient values is first computed. Then one proceeds to classify the items (patients) (Fig. 7) in the following steps: At the first step (at level zero) each patient is considered a

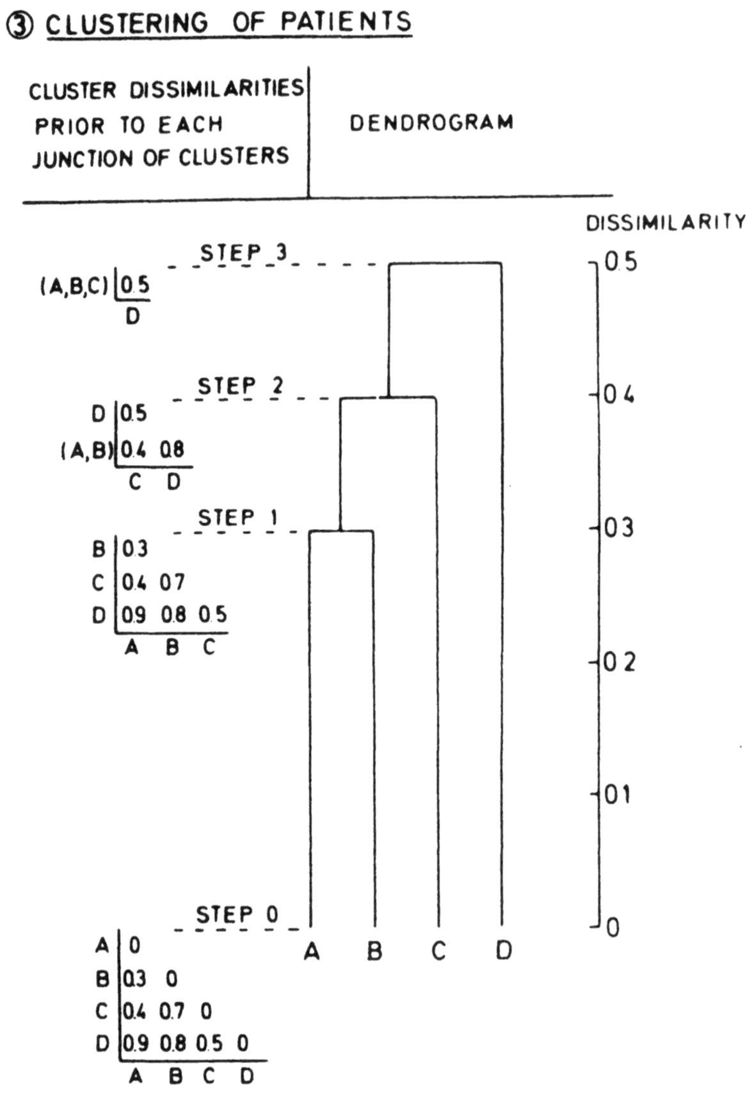

Fig. 7. Constructed example showing the hierarchic classification of four patients using the single linkage technique (see text). First, we have four clusters in that each patient forms a cluster of his own. The dissimilarity of the two most similar patients, i.e., the patients with the smallest dissimilarity is equal to 0.3. At this level, therefore, we fuse A and B. Next we measure the dissimilarity between this group and C as well as with D and compare these dissimilarities as well as that between C and D. To find the dissimilarity between the group comprising the patients A and B and the group comprising patient C we compare all the dissimilarities of patients belonging to different groups. That is, we compare the dissimilarity between A and C and the dissimilarity between B and C. The smallest of these dissimilarities is that between A and C which is 0.4. Therefore, the dissimilarity between the two groups is 0.4. In the same way, we can compute the dissimilarity between the group comprising the patients A and B and the group comprising the patient D. It is noted that it is 0.8. Finally, we note that the dissimilarity between C and D is 0.5. The two groups that have the lowest dissimilarity then are the group comprising the patients A and B and the group comprising patient C. These groups are, therefore, fused at the dissimilarity level 0.4 to form the group comprising the patients A, B, and C. This group is then finally fused with D at level 0.5. Thus, we have a hierarchy of partitions of the patients spanning from the partition where each patient forms his own group to the other extreme where all patients are considered as one big group.

single member cluster. The clusters are then joined in successive fusion steps by combining at each step those two clusters which have the lowest dissimilarity. The various methods differ in respect to the definition of the dissimilarity between two clusters. In the single linkage method (61) the dissimilarity is defined as the lowest value of the dissimilarity coefficient between two patients belonging to different clusters.

The classification process may be represented by a dendrogram in which the fusions are indicated by horizontal lines. The dissimilarity levels of the fusion steps may be read from the scale of the dendrogram. This is illustrated by the example explained in Figure 7.

An example of the second classification approach is the method developed by Friedman and Rubin in the sixties (62). By this method, it is assumed that the data comprise a mixture of k multivariate Gaussian distributions which differ with regard to their locations but otherwise are identical, k being the number of groups. The user of the method must decide in advance on the value of k, that is, the number of populations he believes are represented in the data.

In principle, all possible divisions of the data material corresponding to k groups of patients may be performed and the best one chosen. In order to decide what is an optimal division of the data material, the user has to decide on some criterion to use in assessing each division of the material. Friedman and Rubin used the within group variance or rather the determinant of the common within group covariance matrix. The optimum division of the material would be the one for which this criterion attains its minimal value.

Another example is the technique used by Stanfel (63). He defines a clustering problem as one which requires the partitioning of a finite set of objects into subsets so as to optimize a given objective function. The function used by him is the average of the intersubject dissimilarities within clusters (W) minus the average of the distances between objects belonging to different clusters (B) multiplied by some factor. The factor with which B is multiplied must be defined by the user of the method since it is not clear how W and B should be weighted relative to each other.

2.2.1.2 *Unresolved problems in numerical taxonomy*

There are a number of unresolved problems in relation to the definition of a dissimilarity coefficient as well as the actual sorting of the patients into groups. An important but difficult problem in the computation of dissimilarities is the decision on the weights to be assigned to the variates (e. g. symptoms and laboratory quantities). According to this author if a classification is based on truly qualitative phenomena it would be reasonable in order not to prejudice the issue to let them enter the analysis with equal weights, as it is advocated by the numerical taxonomists (60). However, in case the variates are quantitative it appears that it is not very useful to aim at equal weighting. This would mean that the quantity values would enter the analysis directly and consequently it would become crucial which units were used originally. Therefore, some procedure for an *a priori* assignment of weights to the raw data must be established.

Usually each variate is divided by its standard deviation as computed from the data to be classified. Such a procedure is invariant to linear transformations of the original quantity values but it is not ideal. Assume that we have a mixture of two or more distinct groups. Then for a given variate its variance would depend on its effectiveness in discriminating among the groups, the better it discriminates the greater the variance and the less weight it would be given. Furthermore, the variance in such a mixture depends on the relative representation of the groups i. e. the weighting depends on the *a priori* probabilities. A better choice might be to define the weighting system independently of the material which is to be classified i. e. to use some kind of reference population relevant for the classification problem (64).

The hierarchic methods produce a tree structure. For some applications such a tree structure may be desirable. In clinical medicine, however, a partition of the patient material into clinically relevant groups is needed. In this case it is a problem which of the many possible classifications one should choose. In the example in Figure 7 for instance should one choose the classification obtained at step 1 or that obtained at step 2? There is no generally acceptable objective solution to this problem. The same is true with regard to choice of clustering technique. Over the years several hierarchic cluster methods have been suggested. Then during the late sixties a systematic derivation and evaluation of such methods within the framework of a mathematical model was attempted. The model is based on the existence of a one to one relationship between a hierarchic dendrogram and a so called ultrametric dissimilarity coefficient (65, 66). Thus any hierarchic cluster method may be viewed as a transformation from the dissimilarity coefficient of the data to the ultrametric dissimilarity coefficient of the dendrogram i. e. as a real-valued function (56). It may be shown that within the model the single linkage method is the only known hierarchic method that satisfies a number of criteria for instance an elementary condition such as that of continuity (66). The latter quality ensures that small changes in the data due to rounding errors and random variation only result in similar small changes in the dendrogram.

Although the single linkage method has a large number of satisfactory mathematical properties (66), it suffers from one defect called chaining. The meaning of this notion appears from Figure 8 depicting two clusters which are formed at the dissimilarity level 3. The left cluster is a chain while the right cluster shows a high degree of internal cohesion evidenced by the many links among its members. Various modifications of the single linkage method have therefore been suggested by which only such clusters are selected that satisfy specified conditions of internal cohesion and/or isolation (67, 68). Empirical methods to compare the various clustering techniques have also been used (69, 70, 71). If a variety of data sets are generated with each one having a particular structure the solutions obtained by applying the various hierarchical methods on the data sets may be compared with the generated structures. Such studies indicate that there are more useful methods than the single linkage method for instance the group average method due to Ward (72).

By the techniques where the solution does not portray hierarchical relationships among the entities the material is partitioned into a number of groups by attempting to optimize some predefined numerical measure which reflects a desirable clustering solution. Unfortunately these techniques also present a number of problems. Some of the techniques for instance that of Friedman and Rubin require that the number of groups be defined in advance. However, there is no generally acceptable objective way of assessing which is the best number of groups to choose. In principle the validity of all of the methods may be questioned simply because it is not possible

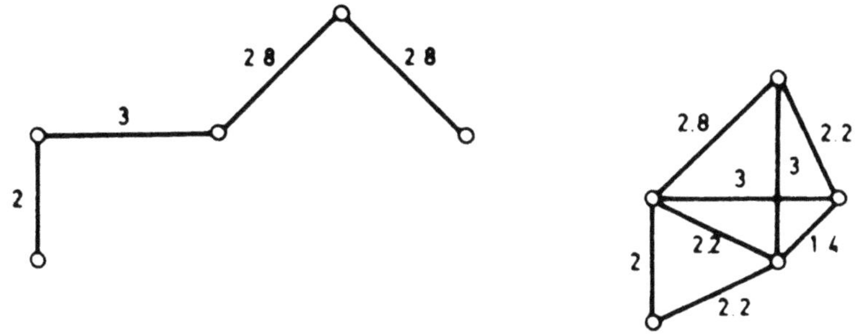

Fig. 8. Two clusters formed at the same dissimilarity level by the single linkage method. The points symbolize patients and a line connecting two points indicates that the dissimilarity between the corresponding patient patterns is equal to or less than 3.

to define a single best criterion to optimize. Last but not least once a criterion has been selected one has no assurance that the solution found is really a global optimum because it is not possible to examine all possible partitions of a reasonably sized material even when a computer is used.

It is usually not possible to give a simple explicit characterization of the groups isolated by numerical taxonomy so that others may identify subjects belonging to a given group. As a solution to this problem it has been suggested that the various symptoms rather than the patients, be classified. The technique then is to cluster positively correlated symptoms to delimit syndromes within a heterogeneous disease entity. Bouchaert et al. (73) studied patients with goiter using this approach. Instead of correlating the actual test values Goldsmith and Leijten (74) have proposed a transformation of the test values whereby each value is scored according to the attention it attracts among the requesting physicians. Then the tests are clustered on the basis of the correlation of attention function scores.

2.2.1.3 *Considerations when using numerical taxonomy*

It should be apparent from the above that unless one is very careful extremely misleading results may be obtained when applying a clustering technique. The results must be both meaningful and reproducible.

A numerical taxonomic analysis should generate some useful hypothesis which need not necessarily be a clinical hypothesis (75).

For a clinical classification to be meaningful, it must segregate the various classes on the basis of different etiologies, prognoses, and/or responses to therapy (Fig. 9) (76, 77). If the classification does not allow such a distinction, the classes formed by numerical taxonomy will be of no immediate value to the clinician.

It should be possible for others to reproduce a classification granted of course that the patient populations considered are compatible.

The solution of this problem requires first of all that the patient material to be classified is well defined in the sense that it is possible for others to obtain a similar material. Thus, the patient population from which the patients are recruited should be clearly delimited i. e. the clinical demographic definition of patients entering the study should be explicit and clear and consecutive patients from a well defined

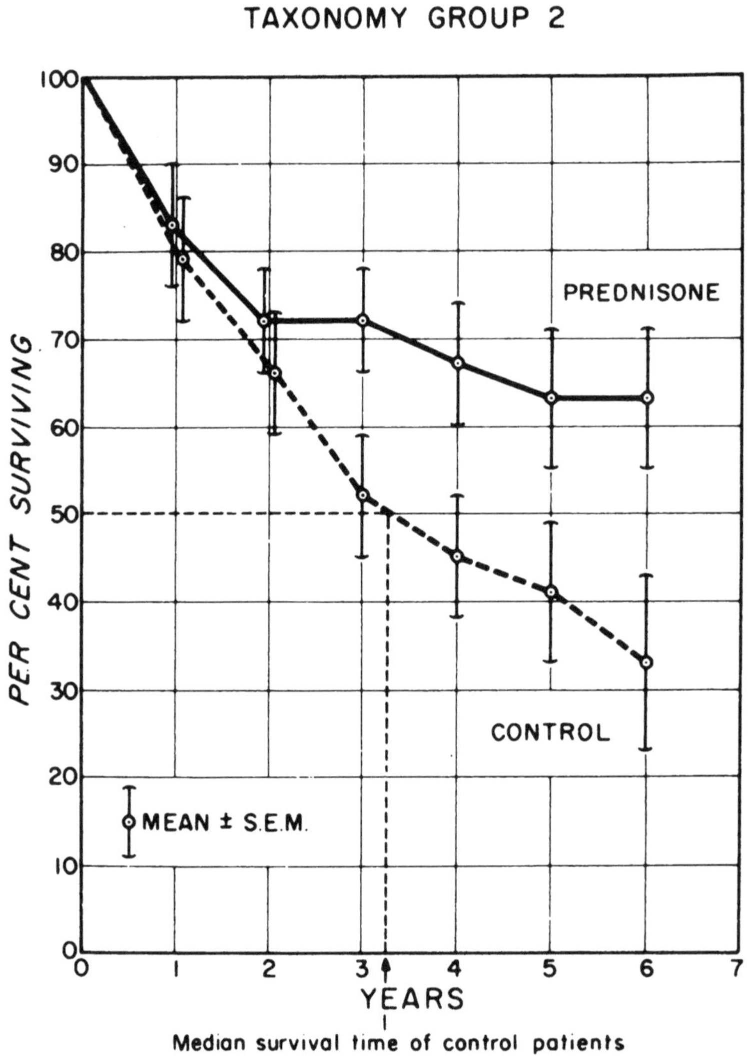

Fig. 9. The survival curves for 46 placebo-treated and 58 prednisone-treated patients with cirrhosis of the liver. The patients (taxonomy group 2) were isolated from a total body of 442 patients with cirrhosis by using a technique of numerical taxonomy. (Reproduced from (76) with permission.)

geographical area should be obtained. Once the patient material to be classified is well defined and can be reproduced it should be verified that the classification of one material can be reproduced in another independent material (78).

In the hierarchic methods there is the added problem that one has to make a choice among alternative classifications of one and the same material. The following procedure appears reasonable: In an initial study one partition of the material corresponding to a fixed level of dissimilarity is chosen because it appears to be

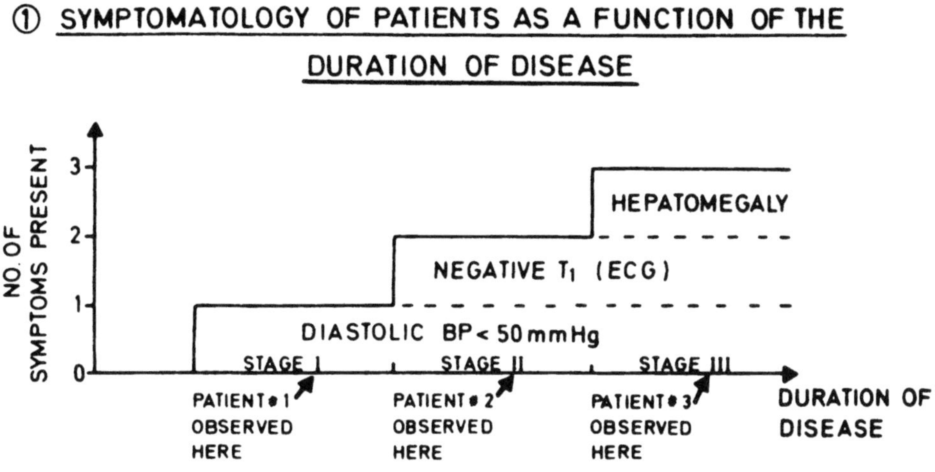

② FREQUENCY OF SYMPTOMS IN THE SAMPLE OF THREE PATIENTS

SYMPTOMS	Pt. # 1	Pt. # 2	Pt. # 3	FREQUENCY
DIASTOLIC BP< 50 mm Hg	+	+	+	1.00
NEGATIVE T_1 (ECG)		+	+	0.67
HEPATOMEGALY			+	0.33

Fig. 10. A simplified version of the typical course of the disease, aortic valvular incompetence. In stage I diastolic BP is $\leq$ 50 mmHg. In stage II diastolic BP is $\leq$ 50 mmHg and T_1 has become negative and in stage III in addition to the two symptoms the liver is enlarged. The distribution of symptoms in a constructed sample comprising three patients in different stages is shown above. (Reproduced from (59) with permission.)

meaningful for one reason or another. A second study using independent data is then performed and the classification (partition) formed at this same level of dissimilarity is compared to the first (79). Another problem often encountered when hierarchic methods are used is that the majority of patients are left unclassified in the sense that sufficiently large groups are not formed. This is a rather disturbing problem which is characteristic of many applications and it concludes the long list of potential problems which the user of numerical taxonomy may be confronted with.

2.2.2 Multivariate prognostic stratifications

When a classification has a specific purpose (e.g. the delimitation of groups with distinct five-year survival rates), more direct methods than those of numerical taxonomy should be used.

Much systematic work with specific goal-oriented classifications has been done by Feinstein (80, 81, 82). He distinguishes between "predictor variates" and a "target variate," where the latter may be e. g. the six-month survival after diagnosis of lung cancer and the former e. g. clinical and laboratory observations describing the initial state of patients at the time of diagnosis. By combining the predictor variates, various groupings of a patient sample may be obtained. If the resulting groups have different likelihoods for the target event, they comprise prognostic a stratification.

An alternative approach may be to define a model of the typical course of a given disease and then deduct the procedure necessary to construct a multivariate prognostic stratification. A simple model is shown in Figure 10. Based on this model, a method for the derivation of disease phases may be deduced (83, 84).

3 Multiple Observations per Patient (Time Series Analysis)

Statistical analysis may involve an analysis of values of a certain quantity (or set of quantities) obtained at successive points of time. In the following I will refer to such a series of values as a time series. One example is daily measurement of S-creatinine made in a renal dialysis patient.

The problems studied by analysis of time series are usually not just simple ones where one may perform a regression of the quantity on time. Often the general trend of a time series cannot be represented by the simple functions usually used in regression analysis. Furthermore the random fluctuations of individual observations relative to a trend may be correlated. This type of correlation is referred to as serial correlation. Time series analysis has not been used much in medical research perhaps because many baseline observations under stable conditions are seldom available.

The reason to analyse a time series may be to gain some insight into the mechanism underlying the variation or to predict future readings from a previous series.

A systematic review of all the various types of statistical time series models is beyond the scope of this chapter. For a brief introduction to the topic the reader is referred to (85). The standard textbook, however, is that of Box and Jenkins (86).

In the following I will review a few simple models and discuss some of the difficulties involved in using time series models in clinical practice.

The Markov chain (87) (a so called stochastic process model) is a simple model which is quite useful for a number of applications. A stochastic process represents the patient at any given point in time as being in one of a number of states. Time is a discrete variable in the model and is measured using a predefined unit. The study by Myers et al. (88) is a good illustration of the application of this model. They used it to represent changes in advanced prostate cancer patients' prognosis. Based on serum alkaline and acid phosphatase values which were characterized as being normal or elevated they defined six states which included death. The time

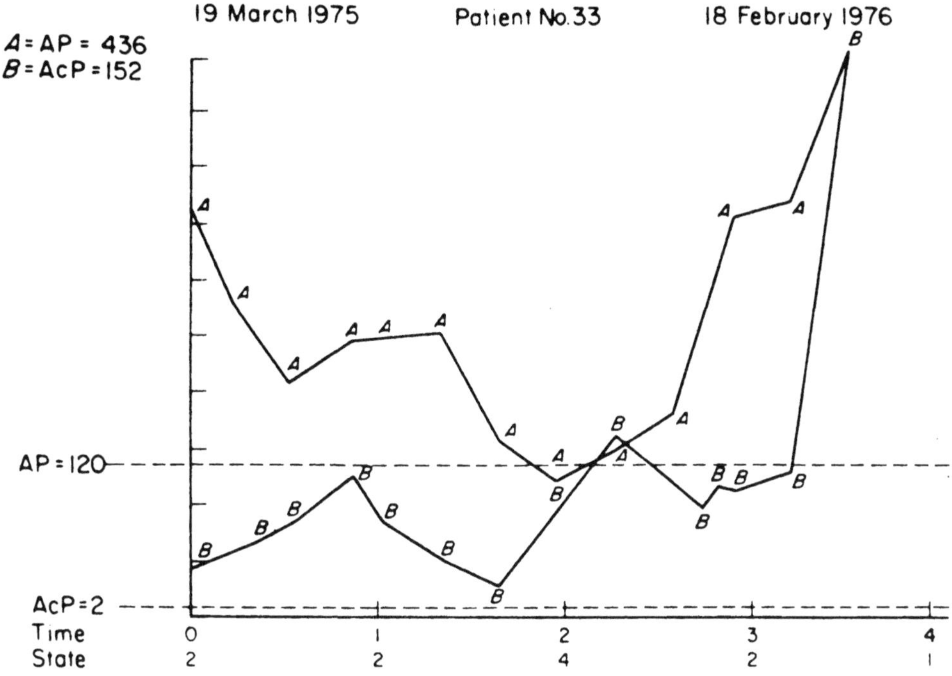

Fig. 11. The definition of the states for a patient during four time intervals (90-days intervals). The two curves depict the results of A (serum alkaline phosphatase) and the results of B (serum acid phosphatase). The states are defined as follows:

1. Dead
2. $A > 120$ and $B > 2$ (both unfavorable)
3. $A > 120$ and $B \leq 2$
4. $A \leq 120$ and $B > 2$
5. $A \leq 120$ and $B \leq 2$ (both favorable)

At time 0, both phosphatase measurements exceed their thesholds, so that the patient enters in state 2. He remains there throughout the first 90-day interval, but in the second interval, his alkaline phosphatase drops below 120, while his acid phosphatase remains above 2, so that he moves (makes a transition) to state 4 at time 2. He then regresses to state 2 at time 3 and is dead (state 1) at time 4. (Reproduced from (86) with permission.)

unit used was 90 days. The various states are defined in the legend to Figure 11 which follows the states of a given patient during four time units i.e. during a period of one year. Given any history (i.e. sequence of states) a patient can either remain in this present state or make a transition to one of the other states. The likelihood of undergoing a particular change of state, given the preceding sequence of states, is called a transition probability. In the Markov chain model the following simplifying assumptions are made:

1. At any point in a patient's history the transition probability is independent of the patient's past history except the most recently occupied state.

2. Transition probabilities depend only on the states of departure and arrival, and not on the time of departure i.e. the transition probabilities are stationary.

Based on these assumptions the transition probability of going from one state (e.g. the state where both enzyme activities are elevated (state 2 in the figure)) to another state (e. g. the state of death (state 1 in the figure)) is estimated as the number of times this particular transition is observed in the patient material over all observed transitions with the former state being the state of departure. In the figure the transition from state 2 to state 1 is observed once and a transition with state 2 being the state of departure is observed three times.

These simplifying assumptions reduce the number of parameters to be estimated considerably but they may not always be realistic from a clinical point of view. Fortunately they may be tested quite easily and even if they do not hold this information may be valuable in of itself.

Harris (89, 90) reviewed three major statistical time-series models which belong to the class of so called "auto-regressive models". Two extreme models, the "homeostatic model" and the "random-walk model", as well as an "intermediate (more general) model" were introduced. For many analytes, we may assume that the values obtained from healthy subjects follow the homeostatic model. In the homeostatic model it is assumed that the observed values fluctuate at random around a fixed set point. It is also postulated that these fluctuations are independent of each other. Random biologic fluctuations relative to a fixed set point take place. However, the organism responds to these fluctuations by changing the concentration back to the set point. The salient point is that this counter-regulation is so fast relative to the time interval between consecutive measurements that the fluctuation reflected in any given measurement is without influence on the value of the subsequent measurement. However, if we shorten the length of the time interval sufficiently (i. e., increase the frequency of timed measurements), we will eventually reach a point at which measurements are no longer independent.

Once a model has been chosen, the next step is to predict future measured values on the basis of values already observed in a given subject. The clinician may have introduced a factor, e. g., a new therapeutic maneuver, that may cause a change of the measured value in addition to that induced by random biologic and analytical factors, or alternatively he may suspect that a significant shift has occurred in a patient secondary to a change in the course of the disease. In either of the two situations, the observed value would be compared with that predicted by the model chosen and a decision made whether or not one should act on the assumption that the observed change is due to the influence of random biologic and analytical factors alone. Before the results of such computations are accepted at least two conditions should be fulfilled: 1. a well-defined hypothesis alternative to that implied by the model should be formulated by the clinician, and 2. the preparation of the patient and the specimen, as well as the analytical procedure, should be under good control. When assuming either the random-walk model or the intermediate model, it is necessary for the time intervals between consecutive measurements to be constant. Furthermore, in the case of the intermediate model a considerable number of observations must be obtained because one must estimate a parameter specifying the dependency among observations in addition to estimating the usual parameters.

In the clinical situation, usually few observations are available and the time intervals between observations may often vary considerably. From a practical point of view therefore, the homeostatic model will probably have to be the model of choice in most situations. This model has been modified (91) for a specified clinical application, namely, the postoperative monitoring of plasma carcinoembryonic

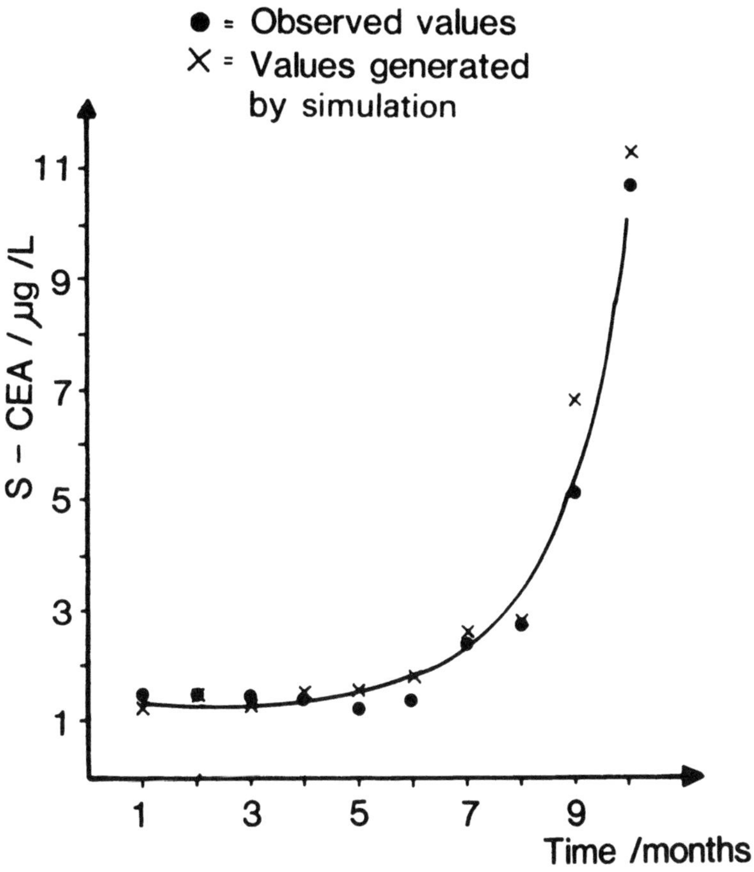

Fig. 12. Postoperative serum CEA values (●) are presented from a patient who underwent surgery for primary breast cancer. Clinically overt recurrence was noted shortly after the time when the last serum specimen was obtained from the patient. The curve is fitted to the observations according to a mathematical model which states that the serum CEA level is a sum of two components, one which is not related to the tumor and one which reflects the size of the tumor. If one makes the additional assumption that the growth of the tumor is exponential, the following equation is reached:

$$Y(t) = \alpha + kT(t_0)\,e^{\beta(t-t_0)}$$

where $Y(t)$ is the CEA concentration at time t, α the constant component, and $kT(t_0)\,e^{\beta(t-t_0)}$ the variable component (that is, the trend). In the latter expression, k is the (unknown) proportionality constant relating tumor size to plasma CEA level, $T(t_0)$ is the tumor size at some arbitrary time (t_0), and β is the growth rate of the tumor. Values generated by computer simulation (×) and based on this model are also shown. (Reproduced from (89) with permission.)

antigen (CEA) measured in patients operated on for primary breast cancer. The significance test procedure was modified to make the model more sensitive to persistent deviations believed to be unique for the development of recurrence and less so to unspecific aberrations. A mathematical model describing the alternative hypothesis, i.e., the growth of a CEA-secreting metastasis was also specified

(Fig. 12). Examples of medical applications of time series analysis in addition to those already presented include the monitoring of pregnancy to predict abortion (92), prediction of platelet utilization (93), longitudinal studies of blood pressure (94), monitoring of patients with myocardial infarction (95, 96) and animal studies of the relationships among various sex hormones (97, 98, 99). This list clearly is not complete but it may be useful for the reader who wants to study the theory of time series and needs to relate theory to relevant medical applications.

References

1. Solberg, H. E. (1978) Discriminant analysis. CRC Crit. Rev. Clin. Lab. Sci., 209–42.
2. Goldberg, D. M. & Ellis, G. (1978) Mathematical and computerassisted procedures in the diagnosis of liver and biliary tract disorders. Adv. Clin. Chem. *20*, 49–128.
3. Wardle, A. & Wardle, L. (1978) Computer aided diagnosis – a review of research. Meth. Inform. Med. *17*, 15–28.
4. Brown, G. W. (1984) Discriminant analysis. Am. J. Dis. Child *138*, 395–400.
5. Beck, J. R., Meier, F. A. & Rawnsley, H. M. (1981) Mathematical approaches to the analysis of laboratory data. In: Progress in Clin. Path. VIII, (Stefani, M. & Benson, E. S., eds.) pp. 67–100. Grune and Stratton, New York.
6. Statland, B. E., Winkel, P., Burke, M. D. & Galen, R. S. (1979) Quantitative approaches used in evaluating laboratory measurements and other clinical data. In: Clinical Diagnosis and Management by Laboratory Methods (Henry, J. B., ed.) pp. 526–555.
7. Warner, H. R., Toronto, A. F., Veasey, L. G. & Stephenson, R. (1961) A mathematical approach to medical diagnosis. J. Amer. Med. Ass. *177*, 177–183.
8. Fryback. D. G. (1978) Bayes' theorem and conditional nonindependence of data in medical diagnosis. Comput. Biomed. Res. *11*, 423–434.
9. Lincoln, T. L. & Parker, R. D. (1967) Medical diagnosis using Bayes theorem. Health Serv. Res. *2*, 34.
10. Norusis, M. J. & Jacquez, J. A. (1975) Diagnosis I. Symptom nonindependence in mathematical models for diagnosis. Comput. Biomed. Res. *8*, 156–172.
11. Russek, E., Kronmal, R. A. & Fisher, L. D. (1983) The effect of assuming independence in applying Bayes' theorem to risk estimation and classification in diagnosis. Comput. Biomed. Res. *16*, 537–552.
12. Ott, J. & Kronmal, R. A. (1976) Some classification procedures for multivariate binary data using orthogonal functions. J. Amer. Stat. Ass. *71*, 391.
13. Hilden, J. (1984) Statistical diagnosis based on conditional independence does not require it. Comput. Biomed. Res. *14*, 429–435.
14. Fisher, R. A. (1936) The use of multiple measurements in taxonomic problems. Ann. Eugenics *7*, 179–188.
15. Anderson, T. W. (1958) An Introduction to Multivariate Statistical Analysis. Wiley, New York.
16. Lachenbruch, P. A. (1975) Discriminant Analysis. Hafner Press, New York.
17. Press, S. J. & Wilson, S. (1978) Choosing between logistic regression and discriminant analysis. J. Amer. Stat. Assoc. *73*, 699–705.
18. Aitchison, J. & Aitken, C. G. G. (1976) Multivariate binary discrimination by the Kernel method. Biometrika *63*, 413–420.
19. Hall, P. (1981) On nonparametric multivariate binary discrimination. Biometrika *68*, 287–294.
20. Hand, D. J. (1982) Kernel Discrimination Analysis. Letchworth. Research Studies Press, England.
21. Titterington, D. M. (1980) A comparative study of Kernel-based density estimates for categorical data. Technometrics *22*, 259–268.
22. Hand, D. J. (1983) A comparison of two methods of discriminant analysis applied to binary data. Biometrics *39*, 683–694.
23. Grams, R. R., Johnson, E. A. & Benson, F. S. (1972) Laboratory data analysis system: III. Multivariate normality. Am. J. Clin. Pathol. *58*, 188–200.

82. Feinstein, A. R. (1972) Clinical biostatistics. XVI. The process of prognostic stratification (Part 2). Clin. Pharmacol. Ther. *13*, 609.

83. Winkel, P., Lyngborg, K., Olesen, K. H., Meibom, J. & Fritz Hansen, P. (1972) A method for systematic assessment of the relative prognostic significance of symptoms and signs in patients with a chronic disease. I. Description of model and selection of prognostically significant symptoms and signs. Comput. Biomed. Res. *5*, 576.

84. Winkel, P., Lyngborg, K., Olesen, K. H., Meibom, J. & Fritz Hansen, P. (1973) A method for systematic assessment of the relative prognostic significance of symptoms and signs in patients with a chronic disease. II. Analysis of validity of model and derivation of major prognostic stages. Comput. Biomed. Res. *6*, 457.

85. Metzger, B. L. & Schultz, S. (1982) Time series analysis: An alternative for nursing. Nurs. Res. *31*, 375–379.

86. Box, G. E. P. & Jenkins, C. M. (1970) Time Series Analysis. Forecasting and Control. Holden-Day, San Francisco.

87. Kemeny, J. & Snell, J. L. (1976) Finite Markov Chains. Springer, New York.

88. Myers, L. E., Paulson, D. F., Berry, W. R., Cox, E. B., Laslo, J. & Stanley, W. (1980) A time-dependent statistical model which relates current clinical status to prognosis: Application to advanced prostatic cancer. J. Chron. Dis. *33*, 491–499.

89. Harris, E. K. (1975) Some theory of reference values. I. Stratified normal ranges and a method for following an individual's clinical laboratory values. Clin. Chem. *21*, 1457–1464.

90. Harris, E. K. (1976) Some theory of reference values. II. Comparison of some statistical models of intra individual variation in blood constituents. Clin. Chem. *22*, 1343–1350.

91. Winkel, P., Bentzon, M. W., Statland, B. E. et al. (1982) Predicting recurrence in patients with breast cancer based on cumulative laboratory results. A new technique for the application of time series analysis. Clin. Chem. *28*, 2057–2067.

92. Winkel, P., Gaede, P. & Lyngbye, J. (1976) Method for monitoring plasma progesterone concentrations in pregnancy. Clin. Chem. *22*, 422–428.

93. Critchfield, G. C., Connelly, D. P., Ziehwein, M. S., Olesen, L. S., Nelson, C. E. & Scott, E. P. (1985) Automatic prediction of platelet utilization by time series analysis in a large tertiary care hospital. Am. J. Clin. Pathol. *84*, 627–631.

94. Darby, S. C. & Fearn, T. (1979) The Chatham Blood Pressure Study. An application of Bayesian growth curve models to a longitudinal study of blood pressure in children. Int. J. Epidemiol. *8*, 15–21.

95. Albert, A., Chapelle, J. P. & Smeets, J. P. (1981) Stepwise probit discrimination with specific application to short-term prognosis in acute myocardial infarction. Comput. Biomed. Res. *14*, 391.

96. Albert, A., Harris, E. K., Chapelle, J. P., Heusghem, C. & Kulbertus, H. E. (1984) On the interpretation of serial laboratory measurements in acute myocardial infarction. Clin. Chem. *30*, 69.

97. Marks, R. G. (1982) Measuring the correlation between time series of hormonal data. Statistics in Medicine *1*, 49–57.

98. McLeod, B. J. & Craigon, J. (1985) Time series analysis of plasma LH and FSH concentrations as a method of assessing episodic secretion. J. Reprod. Fert. *74*, 575–587.

99. Murdoch, A. P., Diggle, P. J., Dunlop, W. & Kendall-Taylor, P. (1985) Determination of the frequency of pulsatile luteinizing hormone secretion by time series analysis. Clin. Endocrinol. *22*, 341–346.

Chapter 3.5
Cluster Analysis as a Help in Interpretation of Complex Data

W. Vogt, D. Nagel and H. Sator

1 Introduction: On the Theory of Diagnosis and Diagnosing

In the beginning of this contribution on a mathematical means as a help in diagnosing some thoughts on the theoretical bases of diagnosing and diagnoses are useful in our opinion. During the daily work of routine or science elementary questions on the fundamentals are put away resulting in an imperturbable faith in the certainty of the underlying grounds. Unfortunately these are fragile and are conventions at best.

For a more profound information on this subject the reader is referred to the books of Feinstein (1967), Wieland (1975) and Gross (1969).

With the possibilities of electronic data processing arising about 30 years ago the occupation with the theoretical basis of diagnosing and diagnosis gained particular importance again. As a consequence of the efforts to algorithmize the diagnostic process and so strictly to formalize it, the analysis of medical decision making and medical action has got a new and more common weight.

At the beginning there is the banal charge of a patient to a person skilled in medicine: *The patient suffers and wants cure.* At the same time, but in second order he is interested in the reasons for his complaints. To cure is to act. For about 300 years the rational basis of this action is a diagnosis, the means hereto is diagnosing.

It should be mentioned, however, that just in emergency and intensive care medicine, action is oriented rather on vital functions and their disturbances, than on the basis of diagnoses in a strong sense. However, this is the exception, the central position of diagnosis is not unattacked but it is still valid.

1.1 Taxonomy in medicine

Diagnosis is inseparably connected with the repertoire of disease terms. Since diagnosis is the basis of medical action, these concepts and the principle classifying them have to be conformed to action.

Again the usually accepted basis rests on categories of thinking of the last two centuries, namely on the belief, that a taxonomy of diseases would be possible just as in botany. But nosology is not a system, since the minimal condition, a uniform criterion of classification, is not fulfilled (Wieland, 1975). Also a classification of diseases according to etiology is not successful, since a last reason in a chain of reasons will not exist. Additionally this chain is not unicausal. It branches out the more we go away from the situation to be cleared. The further the reasons are away from the effect observed, the more unspecific they become (Wieland, 1975). Thus the rating, if something is clarified etiologically, is only a question of demand. Actually disease names are rather practical sorting terms (Koch, 1920) inducing a physician to his acting and behaving.

1.2 Diagnosis as basis of action

Consequently the question arises whether the comprehensive knowledge of the conditions is needed to allow correct acting. We think this question can be answered with a "no" unequivocally. The identification of signs and the subsumption under one term, which allows to deduct the therapeutic rules valid for the special case at that special moment, are essentials of applied medicine. The signs used diagnostically do not have to reflect, however, the pathobiochemical or pathomorphological cause. They can be helpful nevertheless. Transaminases are surely not in the centre of the pathological process of an acute hepatitis, but they show the way in the diagnostic process. Neither urea nor creatinine are causative for uraemic coma, but they are leading to diagnosis. It is not the aim of diagnosing to retrace the way of discovery of a disease, to detect the pathological, pathophysiological or pathobiochemical reasons anew.

"Whatever helps is allowed." Thus the practical medicine differs from sciences substantially. In practical medicine the purpose of all acting has to be related to the individual in contrast to the scientific side of medicine.

1.3 Medical action and diagnostic procedure

The basis of our therapeutical acting is a diagnosis. Diagnosis thus is not an end in itself. In applied medicine a diagnosis is a statement concerning one individual person. This statement is based on decisions, and these decisions are founded on previous medical actions.

Generally speaking a decision is followed by an action. The result of it is interpreted directly and then in context with all other already gained results of actions. This strong sequence "decision – action – interpretation" is a unit, an action unit. Two or more such units build up a chain which also can be branched. Action and interpretation are inseparably connected: Whoever acts, has to interpret the results of this action! Only he himself knows the circumstances and conditions of it. By the way, there exists no basic difference between diagnostic and therapeutical action units.

The decision step contains the question whether to continue or to stop the action unit chain. In case of continuing the following diagnostic or therapeutical actions are chosen then on the basis of the expected benefit.

The action step is the act itself. It can be diagnostic or therapeutic, direct as asking the history from a patient, auscultation, percussion, palpation or indirect as performance of an endoscopy, of an electrocardiogram or a laboratory examination (Gross, 1969).

The interpretation step, now restricted to the diagnostic part is a comparison of an action's result with previous results of interpretation, with hypotheses, with groups of symptoms with uniform origin (pathogenesis) or uniform cause (etiology) or with a group of similar symptoms and signs without known cause (Gross, 1969).

Among the usual forms of diagnosing a procedure along branched structures as well as the hypothetico-deductive method (Sackett, 1979) lead to a sequential course of investigation. Due to a clear underlying principle and not least even to economic reasons, these two methods should be preferred to others unequivocally. Both can be supported by computer based diagnostic strategies.

1.4 How clinical chemistry can help

The maximal information content is achieved if every possible clinical chemical parameter is determined (and the results are sent to a physician at the patients side). This corresponds to the exhaustive procedure (Sackett, 1979) with high redundancy and high costs. Furthermore it is impossible simultaneously to detect interrelationships between more than three to five data. The information of a single value of a parameter can be clearly realized, but the information due to the combination of different quantitative values cannot.

Therefore it is important to find procedures helping a physician to choose suited clinical chemical parameters for the solution of a certain diagnostic problem first, and secondly to transform the information contained in the single values into a general information (Vogt et al., 1984).

2 Data Reduction as an Aid in Interpretation

Some methods are known which solve the described problems at least partly. However, even now, we mention a drawback of commonly used methods: Most of them are based on the assumption of clearly defined and existing diseases and on the possibility of a doubtless assignment to these entities.

2.1 Commonly used methods

Parameters can be considered univariately or multivariately. Even today and in spite of convincing arguments to the contrary, the interpretation of levels of the constituents of biological fluids is largely based on *reference ranges of healthy people*. "Normal" is used so often as a synonym for "healthy" that it has become customary to interpret values lying beyond the so-called "normal" range, as having been caused by a disease. This has been contradicted and discussed excellently by Galen and Gambino (1975).

Applying the *model of predictive value* (Vecchio, 1966, Werner et al., 1973, Galen and Gambino, 1975) instead of reference ranges, is clearly a step forward. It is essential above all to mention, that here comparisons are usually not made with values based on a population of healthy people. Nevertheless, there are real objections to this model (Vogt et al., 1981). Data are reduced at an extremely early stage. The full information contained in the measured value of a particular parameter is narrowed down to decision limits and is reduced to a binary information, a Yes- or No-answer. This means, that the results of an analytical method, which have been optimized with considerable trouble, are truncated as far as their diagnostic significance is concerned. Quantitative values become in effect qualitative values! This loss of information is especially serious when several parameters are combined. The decision limits are rigid once they have been established. In addition, a correct and complete final diagnosis is an absolute prerequisite for the sampled group. This final diagnosis, which is necessarily the basis for the sampled group, can only be made without knowledge of values measured in the laboratory. If a laboratory measurement is used with a decision limit, no matter what type, the previously inserted information is returned merely as an echo of the original results of the analysis. On the other hand, if it is possible to make a diagnosis without knowledge of the laboratory measurements and if other more elaborate clinical or technical procedures are not replaced by clinical chemical data, then what is the use of laboratory assays in such cases?

Other means as the use of the likelihood ratio (Keller and Gessner, 1981) or the adoption of an information theoretical model (Büttner, 1978) lead to similar or equal results and show the same drawbacks.

Discriminant analysis is suited to answer two questions:

1. How well can two or more groups be separated with the aid of clinical-chemical measurements, and
2. with what probability can a patient with a particular constellation of values be assigned to a specific group of patients with the same disease?

The first point is suited to answer the question whether a clinical chemical parameter has discriminating power, the second one deals with the individual diagnostic problem. Discriminant analysis has a considerable advantage over the univariate procedures previously mentioned. Reduction of data is carried out at the last possible moment, so that the information embodied in the data is preserved as long as possible. A grave disadvantage is, however, the previously mentioned problem of an echo, namely of choosing a diagnosis as a grouping characteristic.

2.2 Pattern cognition by cluster analysis

The objections which can be raised against all the procedures mentioned induced us to seek for an alternative. In our opinion a suited model for a diagnostic support should meet two requirements:

1. As pointed out already the information of the single data should be condensed to a total information. This should facilitate and ameliorate the diagnostic decision process of the physician in charge of the patient.

2. From our critical remarks concerning commonly used methods it follows that the procedure had to be independent of symptoms and signs and in particular of the diagnosis. This means a restriction exclusively to clinical chemical data for definition of entities.

Therefore the only viable method seemed us to be assessing entities based only on biochemical and clinical chemical data. This can be done by pattern cognition using cluster analysis. With this conception methods are summarized which are suited, more or less, to detect structures of a given, multidimensional data set, empirically by classifying objects into groups. Because cluster analysis itself defines the different entities, this procedure is independent of diagnoses.

One has to be aware that by this way information never can be increased from an information theoretical point of view. On the contrary, every condensation of information to a general one causes a loss of information. But this reduced information content is more intelligible to the physician, and therefore can be more helpful.

Since cluster analysis is not a clearly defined theoretically founded mathematical procedure, there are many problems associated with the application of clustering techniques, for example the scaling of variables, the choice of clustering method, the testing of stability and validity of clusters found and so on.

The potential user should be aware of the fact that most of these problems have no generally accepted solution. The reader interested in more details is referred to our monography (Vogt et al., 1987), or if he is interested in a more comprehensive treatment of cluster analysis to the books of Bock (1974), Hartigan (1975), Everitt (1974), and Späth (1977).

3 A Cluster Based Model as a Help in Clinical Judgement

Diseases of the thyroid gland appeared to be a particularly well suited model for such an objective computer assisted diagnostic procedure. This is because the significance of clinical chemical parameters is undeniable in the diagnosis of this organs functional state. Furthermore the interpretation is relatively simple because of the clearly understood pathobiochemical mechanisms.

3.1 Collection of data

In a prospective study all 592 patients, who were referred to the outpatient section of the hospital Augustinum in Munich during the period of April to November 1978 for clarification of a thyroid disease were included. No selection was made.

Using a clearly defined procedure they were always examined by the same doctor[1]. The patients themselves filled in a questionaire used in this clinic. The history was asked again by the doctor and the clinical status was specified. The doctor made a provisional diagnosis. This was followed by a TRH test, collection of blood samples and a ^{99}Tc-scintigram. All the data for each patient were assembled and a final diagnosis was made. Because the doctor had many years of experience, no trend of preference for certain diagnoses was observed during the course of the study.

As some clinical chemical methods changed considerably during the following years the corresponding parameters were determined a second time at the end of 1981 to the beginning of 1982. In the intervening period the sera were preserved at −75°C. In this connection the number of patients was reduced to 552 because the sera of 40 patients didn't suffice to measure all needed parameters.

3.2 Selection and number of the clinical chemical parameters

The first step in cluster analysis – and a very important one – is always the choice of the particular set of variables used to describe the objects with respect to the classification. This choice of variables reflects the investigators judgement of their relevance for the purpose of classification. Here no mathematical or statistical guidelines can be given. Hence the choice of appropriate clinical chemical parameters is determined mainly by medical aspects by the statement of the medical problem. In our example the problem is to diagnose thyroid diseases with special respect to the functional state. Once defined the parameters produce a vector of m attributes for every patient. In this way every patient is completely described with respect to the diagnostic model.

The problem of choosing the "right" variables should not be solved by choosing as much variables as possible.

First, the computing time usually increases with increasing number of variables usually not only linearly.

Second, too many variables may impair the stability of clusters (Section 3.5). The more variables are used the higher the dimension of the space and the more objects are apparently needed to obtain a stable classification. Of course there are limitations in practice. We regard the number of 552 patients in a space of 6 dimensions as used in the example described here as a minimum.

Taking these remarks into account the following six clinical chemical parameters were chosen.

(TBI)	T3-uptake test
(TBG)	Thyroxine binding globulin
(T3)	Triiodothyronine
(T4)	Thyroxine
(TSH_0)	Thyroid stimulating hormone before 200 μg TRH i.v.
(ΔTSH)	Difference $TSH_{30} - TSH_0$ (TSH_{30}: TSH thirty minutes after 200 μg TRH i.v.)

---

[1] We are thankfully indebted to Dr. P. Schwarzfischer for his collaboration.

3.3 Transformation of data

3.3.1 **Linear transformation**

In a typical data set the variables are measured in a variety of units. Prerequisite of nearly all cluster methods is a measure of similarity or distance between two different objects. Because most of these measures are not scale-invariant it is important to normalize the data. The most usual way is to standardize all variables to zero mean and unit variance.

Though standardization as well as other methods of normalization has a drawback in that it may lead to decrease of differences between groups, it is strongly advised in most cases.

Once the scales of variables are equalized the investigator can apply different weights to the variables according to his estimation of their relevance. The answer to the question whether the weighting of variables is acceptable or not has always been controversal. In our opinion it should be avoided since the applied weights can only be based on intuitive judgement which usually reflects existing classifications of the data. In this context it is also warned against using variables which are calculated from others applied in the model, since this implies a stronger weight of these variables. Certainly the decision whether to weight variables or not depends on the objectives of classification.

3.3.2 **Nonlinear transformation**

Sometimes the investigator may wish to apply a nonlinear transformation to the data. Especially clinical chemical data often show a bias and a noticeable tail formation. In such cases the great difference between two very high and pathological values may not have the same significance as the difference between one high and an average value. Therefore the extreme values influence the classification to a higher degree than it might be significant with respect to medical aspects. Especially if a standardization is executed the occurance of extreme values causes the compression of most data into a very small interval.

In our example the distributions of T4, T3, TBI, and TBG are slightly skewed, TSH_0 and ΔTSH, however, show an extremely skew distribution. The existence of some values with an exceptionally high variation from the mean-value is common to all parameters. The density of TBI shows such a behaviour for both low and high values. The densities of the other parameters show this variation for high values only. Especially strikingly this can be observed at TSH_0 and ΔTSH for high values.

In the first instance looking at the histograms (for example Fig. 1) it seems reasonable to employ a logarithmic transformation for eliminating the skewness of the distributions. This, however, is not necessary. The purpose of a transformation is not the construction of symmetrical or even normal distributions. There is no reason to believe that the classification results of normal distributed values are better than those of data from skewed distributions.

On the contrary, since every transformation is a more or less arbitrary manipulation of the original data, the data should be varied as little as possible. Hence,

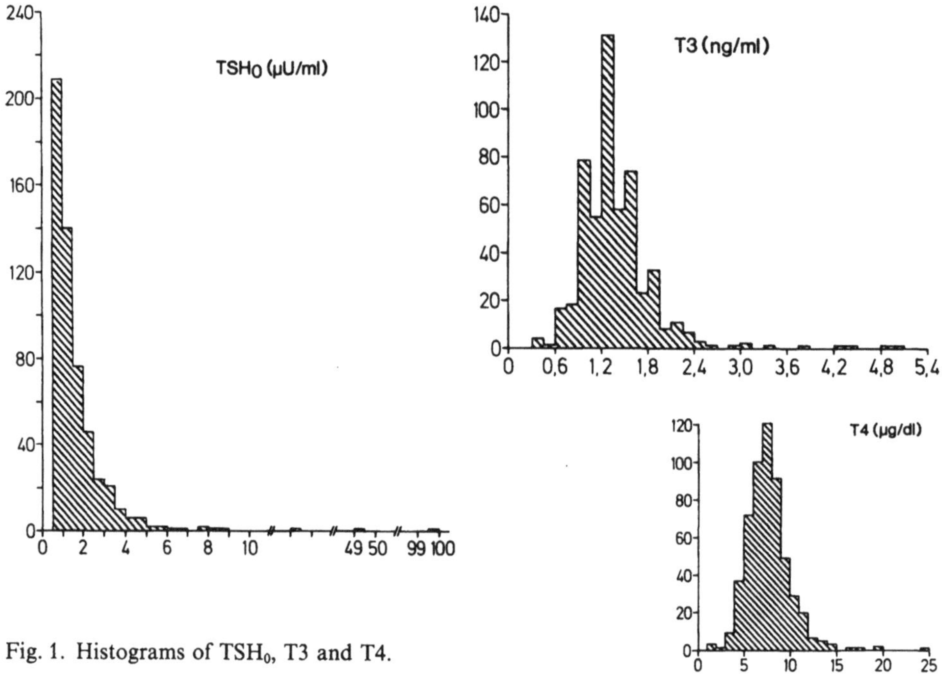

Fig. 1. Histograms of TSH_0, T3 and T4.

a transformation should be performed only on the extreme values, if it is necessary at all.

To choose the transformation as objective as possible two other points should be regarded: First, if possible all parameters should be transformed with the same method. Second, though the question whether a transformation is necessary can only be answered after the classification has been performed, the choice of the suited transformation has to be made before classification. It is not allowed to test various transformations until the result of the classification pleases the expectations.

Generally we suggest the following form of transformation:

$$f(x) = \begin{cases} -\ln(a + 1 - x) + a & \text{for } x \leq a \\ x & \text{for } a < x < b \\ \ln(x - b + 1) + b & \text{for } b \leq x \end{cases}$$

where a and b are chosen as some specified fractiles of the distribution. The function leaves the range between the specified fractiles unchanged and pushes together the extreme values from below or from above or both according to the values of a and b.

Figure 2 shows an example of this function in the case where the extreme values on both sides are pushed together symmetrically.

The user should keep in mind that the clusters he will obtain are in a transformed space and not in the original space. So, the only general rule that can be given is to manipulate the data as slightly as possible.

In the example discussed here the 10%, respectively 90% fractile was chosen. As for the densities of TSH_0 and ΔTSH only the high values are changed, because

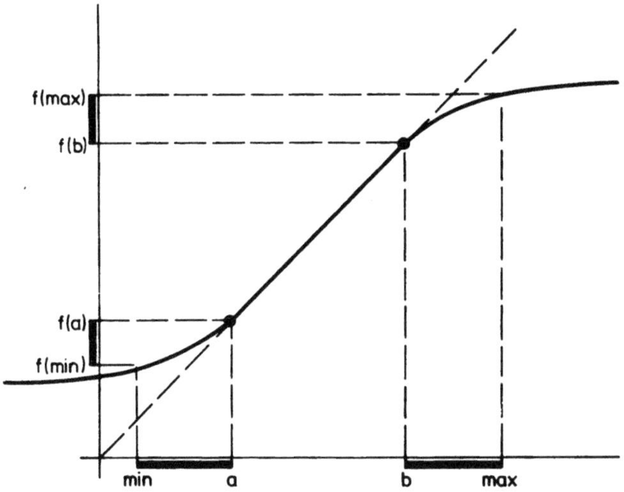

Fig. 2. Transformation function (for further explanation see text).

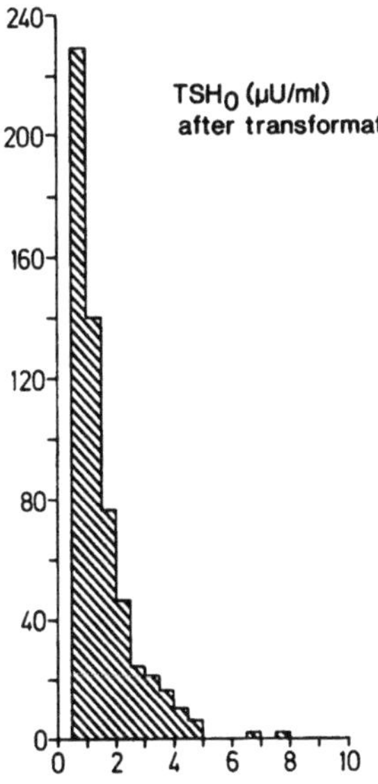

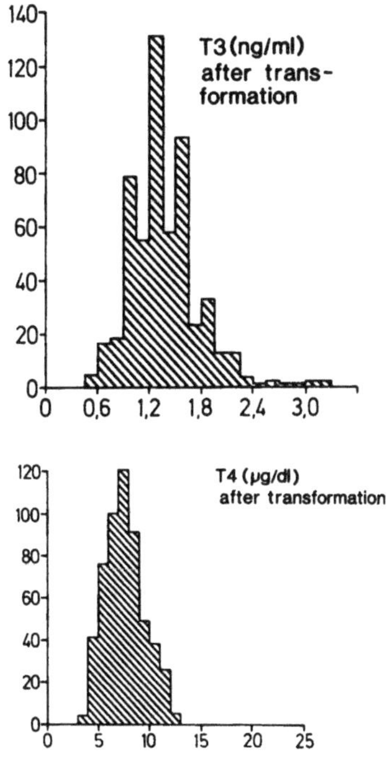

Fig. 3. Histograms of transformed TSH_0, T3 and T4 according to the function defined in text.

more than 10% of the low values lie beyond the detection limit and therefore assume the same previously defined value (Fig. 3).

As mentioned the decision whether a transformation is medically meaningful can only be taken after the classification has been performed. Therefore both data records, the original and the transformed data were classified.

Contrary to our expectations, however, from a medical point of view the results of the classification with the original data were as satisfying as the results obtained with the transformed data. Probably this is due to the fact that the structure is debilitated by the transformation which eliminates extreme values. To avoid manipulation of the data, therefore the final decision was taken in favour of the original data. Previous to this final decision, however, the calculations for deciding the correct number of groups and estimating the stability had to be performed for both data sets.

3.4 Choice of classification method

Hundreds of different methods of clustering have been proposed (Bock, 1974, Späth, 1977). The major work in the field of classification consists of constructing new methods and algorithms. The intuitive and subjective nature of the clustering problem has prevented so far a mathematical comparison of all clustering techniques. However, a potential user needs some rational basis for choosing among different clustering procedures.

Essentially there are two ways of analyzing the performance of a clustering method.

Some authors propose mathematical criteria and recommend those methods which meet these criteria (Jardine and Sibson, 1967, Fisher and Van Ness, 1971). Unfortunately there is no generally accepted rule what qualities a clustering method should have. Once an investigator has decided on some mathematical criteria there may not even be a method which satisfies them all. Thus, this way of testing the suitability of a clustering method is troublesome and may not lead to a unique solution.

Other investigators use Monte Carlo methods as specified below. They perform comparisons among different clustering methods by using computer generated mixtures of probability distributions with known parameters as input data in their simulation experiments. The task of cluster analysis is defined as the resolution of these mixtures into its components. The degree of correspondence between the obtained clusters and the underlying populations in the mixture is a measure for the accuracy of the clustering solution.

The results obtained from these empirically oriented Monte Carlo studies are not consistent with those derived from theoretical considerations. The reason for this disagreement probably is that some methods perform very well if the groups are clearly distinctive but have undesirable properties if the groups are not so clearly separated.

Unfortunately, in many practical applications groups are not clearly distinctive. This is also true for the special case we are dealing with – patients have to be classified by means of clinical chemical parameters. Here a classification result is more likely, where only some clusters are very clearly distinctive of the others, while a certain range exists, where the structure is not as clearly separated. For this reason and because of the difficulties of theoretical proceeding stated above we think that the second method of establishing the performance of a clustering method is the more realistic one at least for practical applications.

The best known and mostly used clustering methods may be subdivided into hierarchical methods and partitioning methods. While hierarchical methods build up a hierarchy of partitions of a data set, partitioning methods produce clusters by optimization of a criterion function.

Hierarchical methods have been studied extensively by means of Monte Carlo methods (for example Blashfield, 1976, Kuiper and Fisher, 1975, Mojena, 1977, Milligan and Isaac, 1980, Edelbrock, 1979, and Edelbrock and McLaughlin, 1980). However, there is a drawback common to all hierarchical methods: An object which has been misclassified at an early stage never can change the branch of the tree on which it has been placed once. Especially with a large sample, this can leed to considerable misclassifications. So, we would recommend hierarchical methods only if an investigator is interested in a hierarchical structure of his data or as a starting partition for one of the optimizing techniques proposed below.

Unfortunately, there are only few studies comparing partitioning techniques (Bayne et al., 1980, Friedman and Rubin, 1967 and Ganesalingam and McLachlan, 1980). Furthermore the conclusions that can be drawn from these studies are by no means unique. Hence any application of a clustering algorithm should be accompanied by other validating procedures (Section 3.5).

We decided to use the criterion of variance for classification. That partition of a data set is chosen which minimizes the variances within each group. On an average, this criterion performed quite well in the cited studies. Further essential advantages are its computational simplicity and low storage requirements.

A partition which optimizes a predefined criterion cannot be obtained by enumerating all possible partitions. Therefore algorithms are needed which start with any partition of the data set and try to find at least a local optimum by sifting every object with regard to its cluster membership and shifting it, if necessary, into another cluster. Having chosen the variance-criterion as a clustering method there are numerous algorithms for optimization available. The changeable components are the starting partition, the updating rule and the options for deciding the right number of clusters. We propose to use a pre-optimized starting partition and decide for the correct number of clusters after running the algorithm several times for different numbers as described later (Section 3.5). We used the K-MEANS algorithm (see for example Späth, 1977), because it produces a minimal-distance-partition (Bock, 1974). Additionally there is some consensus that this algorithm is very efficient (Hartigan, 1975, Späth, 1977). To get a pre-optimized starting partition we used the hierarchical method of Ward (1963), which seeks to optimize the same criterion and performed sufficiently well in the studies cited above.

3.5 Classification: Description of reality or product of chance?

Clustering algorithms have the property of finding clusters whether or not the data exhibit a grouping structure in reality. Cluster validity is concerned with the investigation of the result of a clustering algorithm and tries to separate true structures from artifacts of the algorithm. The problem with evaluating a cluster solution is that in applied research clustering methods are used on data sets for which the "true" classification is not known. Though there is no direct measure how well a cluster solution coincides with the true structure of the data set, some necessary conditions for a valid solution can be stated.

1. The data set is not homogenous but some clustering tendency exists.

2. The partition obtained is stable in the sense that it will be found for different samples from the same general population. We call this concept internal stability or replicability.

3. The partition obtained is stable in the sense that small changes in the input data lead to only small changes in the classification achieved. We call this concept external stability. This is on the one hand a measure for the quality of the algorithm that it is not too sensitive to perturbations in the input data, on the other side also it is a measure for the structural quality of the data set. Therefore the concept of external stability serves to achieve the validity of single clusters as well.

These conditions are only necessary not sufficient. Nevertheless if they are fulfilled and procedures underlying conditions 2. and 3. lead to the same partition there is good reason to believe that the data are really structured.

The general field of significance testing in cluster analysis includes tests of significance for a partition obtained by some clustering method, the fit of a hierarchical structure to a given data set, and tests about the "clustering tendency" of a data set. A survey of existing approaches is provided for example by Dubes and Jain (1979) and Perruchet (1983). Tests for the hypothesis of a "homogenous" population against the alternative of a "heterogeneity" in the data can be found in Bock (1985). One major problem with all these tests is the formulation of a null hypothesis. At present it seems to be impossible to define only one null hypothesis which expresses non classifiability and is equally useful for all practical applications.

All proposals known to us suffer from difficulties with respect to practical applications. In addition one single test for all applications cannot be recommended.

Often the investigator has some prior knowledge whether his data exhibit a grouping structure or not for example from the examination of the distributions of the single attributes. Hence, for the moment and from a pragmatic point of view it seems to be the more economic way to apply a clustering method to a data set at first and test the validity of the obtained solution afterwards.

3.5.1 Internal stability of the structure found

The question whether there is a cluster structure at all can be answered to a certain degree by measuring the stability of a partition. A second problem common to all clustering techniques, the difficulty of choosing an appropriate value for the number of groups should be considered in context therewith. The idea is that the stability of a solution is probably different for different numbers of clusters. While a good stability indicates an inherent structure within the examined sample a bad one suggests that the clusters are only random groupings formed by cluster analysis. So the number of clusters with the highest stability is supposed to reveal the structure of the data in the best way.

Therefore the number of clusters should not be decided in advance but the optimization algorithm should be run for several different values. First the user chooses a value k_{min} for the lowest value and a value k_{max} for the highest number of groups he is interested in. Then the optimization algorithm is run for all values of k for $k_{min} \leq k \leq k_{max}$.

To measure the stability the following procedure is proposed: Take randomly half of the examined sample and cluster it in the same way as the original one. Compare this new clustering with the original one of the selected objects. Repeat this procedure several times. For comparing these partitions one of the measures cited in the appendix can be used. Calculate the median values of these repeated measurements. It is suggested to calculate the median values instead of the mean values because the distributions of the similarity values proved to be asymmetrically.

Among others the number of repeats depends on the computational facilities available. In our experience a number of 100 repeats gives a very similar result to only 30 repeats. So it can be recommended to use about 30 repeats. Possibly, however, the variation of these repeated measurements depends on the number of objects as well as on the distinctiveness of the structure.

It is proposed to measure the stability in this way for the previously chosen range of possible numbers from k_{min} to k_{max} clusters. The interpretation of the resulting curve depends on the characteristics of the measure chosen. The question, is a measure of similarity of 0.8 for two partitions good or is it not, cannot be interpreted from such a single value. It can only be answered if we know which value an unstructured data set yields. Thus a reference curve is needed for the interpretation of the values.

A multivariate uniform normal distribution of all parameters is a good example for a data set having no cluster structure. In this case the structure of the data indicates just a single group. We simulated such a data set for 552 objects and 6 parameters. This data set also was clustered by the method of Ward followed by the K-MEANS algorithm (Section 3.4). The stability for the cluster numbers 2–21 was measured as described above. The results for the measure of Rand (1971) and for the measure of Rand corrected for chance (see appendix for definitions) are shown in Figures 4 and 5.

If the stability curve increases monotonically there is no cluster structure in the data. If there are peaks or shoulders the corresponding cluster numbers indicate for the correct ones. The question which measure to use for constructing the stability curve cannot be answered in an unequivocal way. The measure of Rand has the advantage that the reference curve is very smooth so that peaks or shoulders can be easily detected. But it has the disadvantage that structures with more than 10 clusters will probably not be discovered due to its increase with increasing number of groups.

For our data record twenty partitions with the numbers of groups from 2 to 21 obtained by the method of Ward were used as starting configurations for the algorithm K-MEANS. The procedure described was performed 35 times for these 20 partitions. For comparing the clustering of the selected patients with the original one the measure of Rand corrected for chance was chosen.

The median values of these 35 similarities were computed for all twenty numbers of groups and plotted as well for the non-transformed data and the transformed data (Figs. 6 and 7). In Figure 6 also the reference curve from Figure 5 is included. Taking into consideration the discourses about the reference curves a first look at

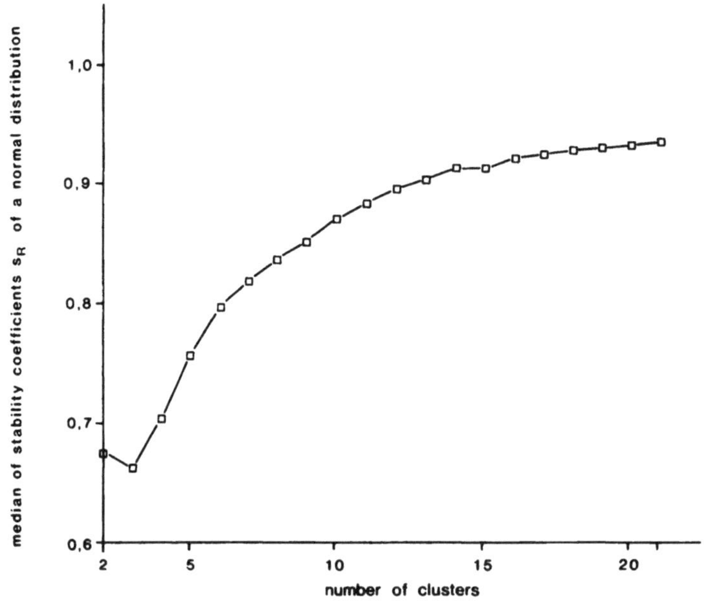

Fig. 4. Stability curve using Rand's similarity measure for a unimodal normal distribution with 552 objects and 6 parameters.

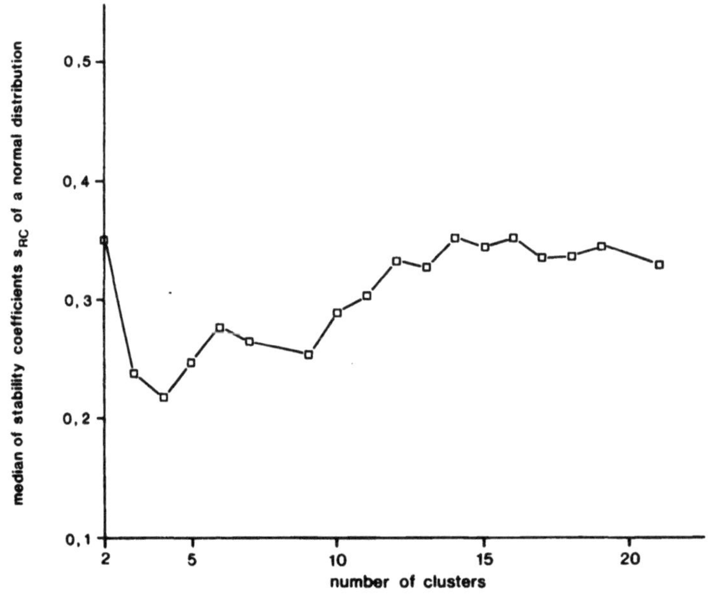

Fig. 5. Stability curve using Rand's similarity measure corrected for chance for a unimodal normal distribution with 552 objects and 6 parameters.

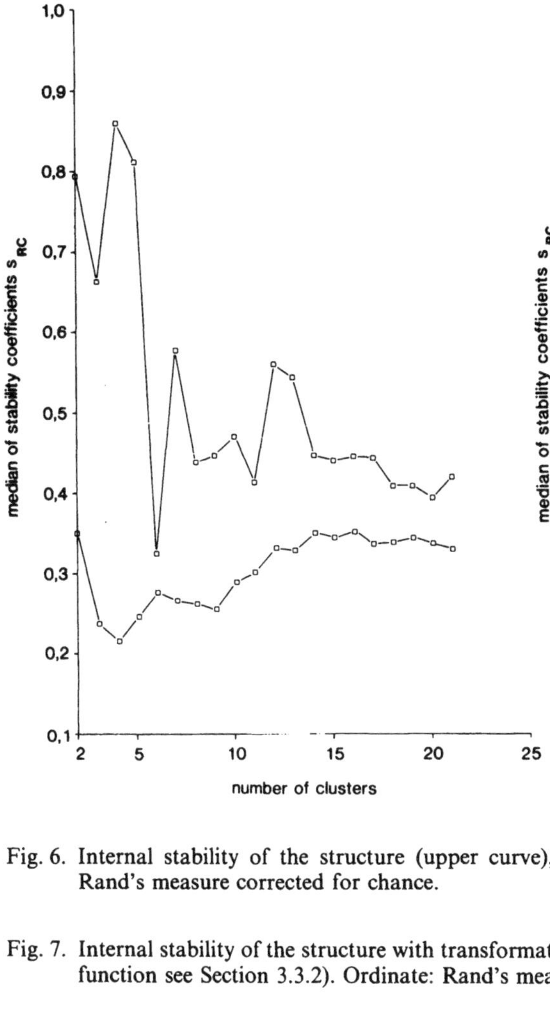

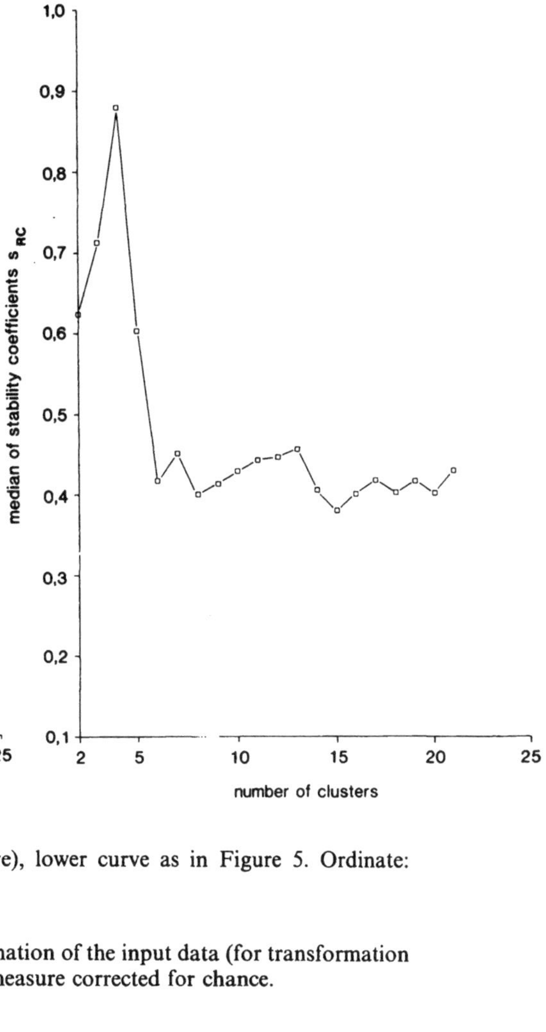

Fig. 6. Internal stability of the structure (upper curve), lower curve as in Figure 5. Ordinate: Rand's measure corrected for chance.

Fig. 7. Internal stability of the structure with transformation of the input data (for transformation function see Section 3.3.2). Ordinate: Rand's measure corrected for chance.

the median curve shows that the stability decreases with increasing numbers of clusters.

Possibly, a strong structure which can be represented by more than twelve clusters really does not exist. But it is more plausible that the number of 552 patients is not big enough to detect more detailed groups in the example discussed here.

A very high stability is found at two, four and five groups with the original data record. A relatively good stability with respect to the environment is found at seven and twelve groups.

A unique result is obtained with the transformed data record. The clustering with four clusters turns out to be the most robust one.

The examination of the stability of the structure shows that the choice of an adequate number of clusters is an important prerequisite for obtaining a sufficiently stable solution. Of course there may be indicated more than one "correct number" of clusters.

3.5.2 **Measuring the external stability**

Measurement of the amount of change between different classifications as a consequence of small variations in the input data is fundamental to the assessment of stability. Perturbations to the input data occur as a result of errors made. This kind of perturbations can be lowered to a certain degree but not totally avoided.

Two different kinds of errors may occur. First, systematical errors affect the accuracy of the measured values. Usually these are small and can be understood as a linear transformation mostly as a shift of the correct values. Hence they are not taken into consideration with regard to the classification results. Second, random errors affect the precision of the measured values. These are more serious with respect to classification, because large dispersions round the correct values may considerably influence the classification result.

When dealing with clinical chemical data there is the advantage that the quantity of analytical errors is known at least approximately. It can be assumed that these are normally distributed with zero mean and an empirically known variance s. This value is taken from the daily routine precision measurements. If n is the number of objects in the studied sample, n values coming from a normal distribution with zero mean and standard deviation s are generated randomly. This simulated distribution represents the analytical errors. These "errors" are added randomly to the original data. It will be often necessary to divide the range of measurement

Table 1. Imprecision of clinical chemical parameters used

Dividing the measurement-range into three intervals

	Interval 1	Interval 2	Interval 3
TBI	– 0.95	0.95–1.1	1.1–
TBG (g/ml)	–16.0	16.0–35.0	35.0–
T3 (ng/ml)	– 0.9	0.9– 1.8	1.8–
T4 (g/dl)	– 4.0	4.0–10.5	10.5–
TSH (U/ml)	– 5.0	5.0–18.0	18.0–

Standard deviation of the parameters in the three intervals

	Interval 1	Interval 2	Interval 3
TBI	0.020	0.029	0.040
TBG (g/ml)	0.912	1.713	1.953
T3 (ng/ml)	0.097	0.136	0.107
T4 (g/dl)	0.187	0.211	0.215
TSH (U/ml)	0.233	0.415	2.837

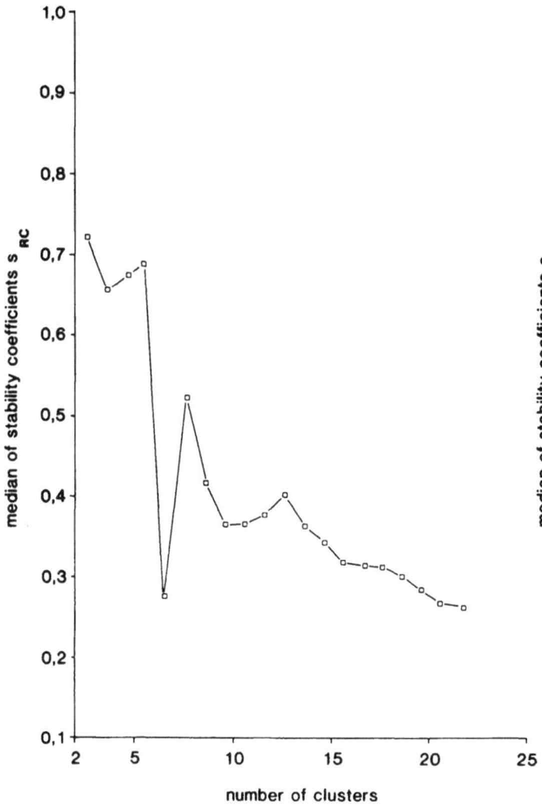

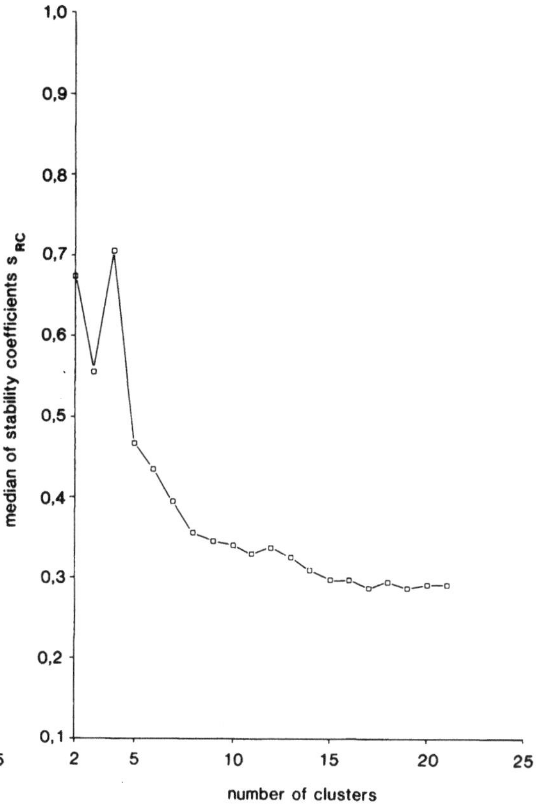

Fig. 8. External stability of the structure. Ordinate: Rand's measure corrected for chance.

Fig. 9. External stability of the structure with transformation of the input data (for transformation function see Section 3.3.2). Ordinate: Rand's measure corrected for chance.

into several regions because the errors depend on the relative position of the measured values. In this case the procedure is similar.

Two different procedures are recommended. The second one will be considered in Section 3.5.3. The first one is similar to that one proposed to measure the internal stability. The data with added "errors" are clustered in the same way as the original ones and the results are compared with the original clustering. The whole procedure – simulating errors, adding these errors to the original data, clustering these new data, and measuring the similarity with the original clustering – shall be repeated several times for every number of clusters of a certain range as described. Concerning the number of repeats again we recommend a value of about 30. Calculating the medians of these similarity values a curve similar to the inner stability curve will result. Also the interpretation of the curve is to be performed in a similar way.

Because the variances in our example are not the same for the whole range of measure, it was divided into three intervals for every parameter. The standard deviations are shown in Table 1.

To compare the original groupings with the new ones we also chose the measure of Rand corrected for chance used in Section 3.5.1. The medians of 35 comparisons were plotted into a diagram for the groupings with the number of clusters from 2 to 21. The resulting curves for the original and the transformed data record are shown in Figures 8 and 9.

For the original data record the groupings with the numbers of clusters 2, 5, 7 and 12, for the transformed data record the groupings with 4 and 12 clusters point at a relative high stability.

From Figures 8 and 9 again it can be seen clearly that the stability decreases with increasing number of groups. The decision for the number of clusters which shows a sufficient stability with respect to the errors of measure, however, cannot be answered generally. According to the circumstances one has to decide which grouping is stable enough to obtain satisfactory results with respect to the medical problem to be solved.

3.5.3 **Reclassification: Reliability despite of analytical errors?**

Concerning the second method for measuring the external stability a new question which is closely related to the problem of classification has to be discussed at first: Given the existence of groups, how do we set up a rule to allocate further individuals to these groups. Such a rule or algorithm will be called reclassification algorithm. The assignment itself is called reclassification.

Since we are only interested in cases where the groups under consideration have been obtained by cluster analysis, we search for a reclassification or discrimination technique, which is "compatible" with the cluster method.

It is a reasonable question to ask why a problem exists at all. One could argue that it is possible to proceed in the same way as the existing groups were obtained. Essentially there are two reasons for the requirement of a special reclassification rule. The first reason is missing information. It should be possible to assign an individual to one of the groups even if not all measurements have been performed. Second, for economical reasons it is not possible to perform a new cluster analysis whenever a new individual shall be assigned to the groups.

As has been said such a reclassification algorithm shall be compatible with the classification method in the sense that objects of the sample used for cluster analysis are assigned to the same groups by the classification and reclassification algorithms. Compatibility, however, is only a necessary but not a sufficient condition because the reclassification rule should also take into account the specific properties of the particular clustering algorithm used.

Therefore no general reclassification algorithm compatible with all classification methods can be recommended. Fortunately, at least for algorithms which produce a minimal-distance-partition a very simple rule can be given. Since these partitions have the property that the distance of all objects to the own centroid is the smallest one it is natural to choose the following compatible reclassification algorithm:

An object is assigned to that group for which its distance to the centroid is a minimum.

If the second method mentioned is used measuring the external stability makes use of this reclassification rule. We assume that the partition $G = \{g_1, \ldots, g_k\}$ found for the original data is fixed. If an object x belonging for example to g_j is reclassified to the same cluster when analytical errors have been added to x, we can say x has been correctly reclassified. The percentage of correctly reclassified objects is a very concrete measure for external stability.

If we calculate the percentage of correctly reclassified objects for each cluster separately this has an additional advantage compared with the procedure described in Section 3.5.2, namely that the validity of the single clusters can be judged. It is also interesting to calculate for each pair g_i, g_j of clusters the percentage of objects belonging to g_i and reclassified to g_j and vice versa thus obtaining a measure of similarity between the clusters. This shows the interrelationship between the clusters in space very clearly.

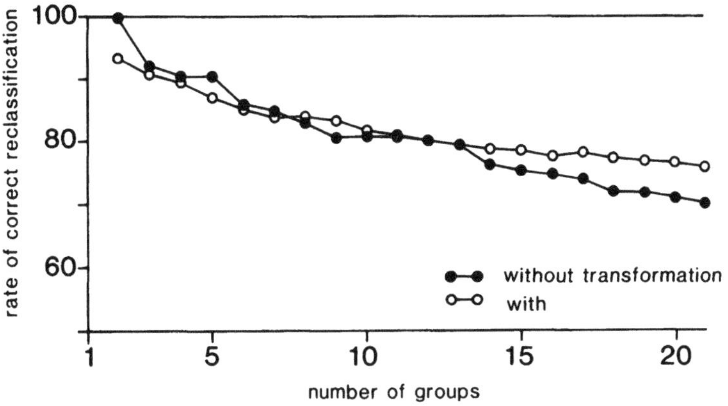

Fig. 10. Mean rate of correct reclassification for data with random errors added (for transformation function see Section 3.3.2).

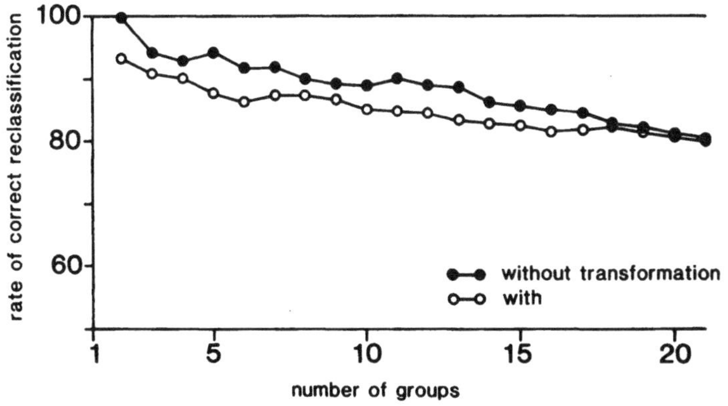

Fig. 11. Mean rate of correct reclassification rate per cluster for data with random errors added (for transformation function see Section 3.3.2).

The procedure of adding errors and reclassifying the patients was done 35-times. The mean of the correctly reclassified patients is shown in Figure 10 for the numbers of groups from 2 to 21.

In order not to weight the big groups excessively another method seems to be even more appropriate. The mean of correct reclassification is computed for every group at first and afterwards the mean is taken from these results. Doing this we obtain the curve shown in Figure 11.

The better results of the original data in Figure 11 are due to the fact that extreme values produce more groups with few patients, which can be clearly separated from the clusters with the euthyroid patients. This structure is softened by the transformation which tends to classifications with groups of same size. The high negative correlation of the size of groups with the percentage of correct reclassification can impressively be seen in Figure 12.

Furthermore the method described above is suited to examine the influence of analytical errors on the percentage of correct reclassification. The random errors are reduced gradually multiplying them with a factor minor 1. Afterwards the reclassification is repeated with the resulting data record. Plotting the mean of the correct reclassification per group against the factor, a monotone decreasing curve is obtained for every number of groups. For numbers 2, 4, 6, 12 and 20 the curves are shown exemplaryly in Figure 13.

From these curves one can derive whether the analytical errors are small enough to obtain a given percentage of correct reclassification. Of course the same method can be used to examine single parameters and parameters with different factors. Looking upon the ratio of correct reclassification of the chosen grouping with 12 groups (see next section) we notice that the stability with respect to analytical errors

Fig. 12

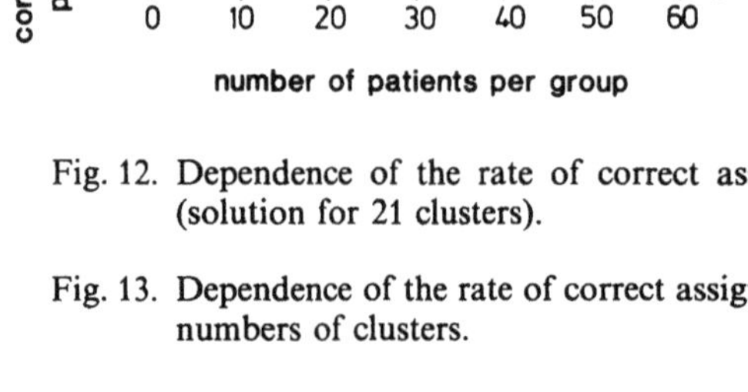

Fig. 13

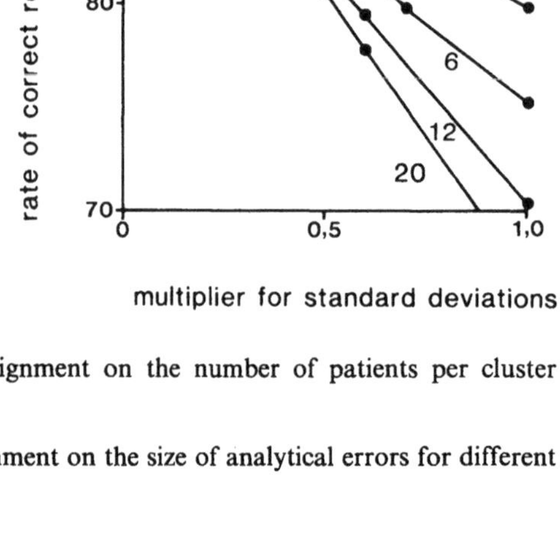

Fig. 12. Dependence of the rate of correct assignment on the number of patients per cluster (solution for 21 clusters).

Fig. 13. Dependence of the rate of correct assignment on the size of analytical errors for different numbers of clusters.

Table 2. Ratio of correct reclassification with data changed by addition of random errors for the various clusters. n = number of patients in the cluster

Cluster	Ratio (%)	n
1	83.23	84
2	89.95	31
3	89.94	23
4	73.05	129
5	80.33	95
6	100.00	4
7	83.32	37
8	90.59	34
9	74.81	108
10	100.00	2
11	100.00	2
12	100.00	3

is sufficient (Table 2). The results for the groups with many patients certainly are not as good as for the groups with few patients, but exactly at these groups the stability is less important with respect to a medical point of view.

3.5.4 Choosing the number of clusters

Generally the problem of choosing an appropriate value for the number of groups is solved by user evaluation under only subjective criteria. The investigator chooses that number of groups which suits his aims in the best way from a pragmatic point of view.

Though various methods for a more objective choice of the best number of groups have been proposed up to now the problem is not solved in a satisfactory way. Therefore in our opinion two different methods should be used for this purpose.

The method we proposed in Section 3.5.1 has the advantage that it also gives hints for the validity of the cluster solution. A second method which is restricted to the case where an optimization algorithm has been used for clustering, but has the advantage of being very simple, is shortly discussed now.

Several authors suggest that a plot of the criterion value against the number of groups will indicate the correct number (for example Späth, 1977, Friedman and Rubin, 1967). Usually this function is monotonously increasing if the criterion is to be minimized or decreasing if the criterion is to be maximized. The value of k for the number of groups should be chosen in that way that the difference of the function values at k and $k-1$ is significantly greater than the difference of the function values at k and $k+1$. The reason is that a great difference between the values for k and $k-1$ indicates that the partition at $k-1$ clusters contains still heterogenous groups while a small difference between k and $k+1$ indicates an only artificial refinement. In practice, however, the decision whether such a plot contains a sharp step is quite subjective. Moreover this procedure does not always lead to a solution.

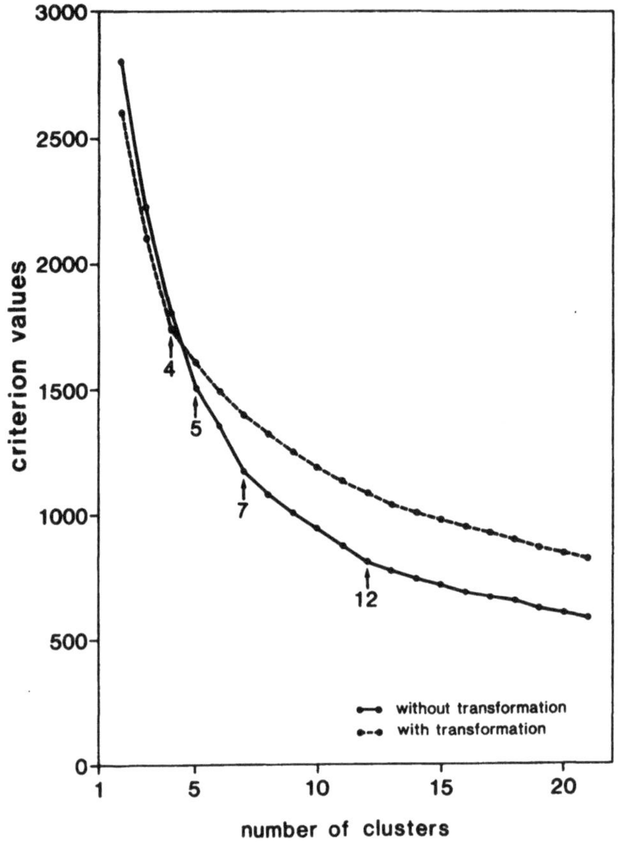

Fig. 14. Criterion values for the variance criterion. Arrows indicate local minima (for transformation function see Section 3.3.2).

Table 3. Suited numbers of clusters obtained by the different methods described

	Number of clusters						
	Data not transformed					Data transformed	
Results of optimization function			5	7	12	4	
Stability of the structure found	2	4	5	7	12	4	
Stability with respect to analytical errors	2		5	7	12	4	12

In Figure 14 the transformed data show a clear break at four clusters. The curve of the original data shows three indistinct breaks at 5, 7 and 12 clusters.

The results together with those from Sections 3.5.1 and 3.5.2 are summarized in Table 3. The conformity of all three methods with respect to the right number of clusters is remarkably good. Because of this agreement we accepted the solution of

4 clusters for the transformed data and the solutions of 5, 7 and 12 clusters for the nontransformed data.

Due to the decision against the transformation (Section 3.3.2) only the numbers of groups 5, 7 and 12 are taken into consideration. Though the partitions with number 5 and 7 show better stability the final decision was made in favour of number 12. There were two reasons to do so.

First, there are medical aspects. It is less dangerous to distinguish groups which possibly have no various meanings than to join groups with different medical significances.

Second, the more important groups of the twelve groups classification have a higher stability than the results of Sections 3.5.1 and 3.5.2 would have us believe. The stability is reduced by the large groups with euthyroid patients which cannot be separated clearly. But just at these groups the stability has no decisive importance.

3.6 Representation of the groups

The resulting groups of the diagnostic model can be graphically illustrated as a polar diagram. The axes correspond to the various parameters (Fig. 15). The means of the normalized parameters of each group define the length of these axes (Fig. 16). The means of the parameters of all patients are marked by a circle which is plotted into every diagram. It can be clearly seen that the groups with few patients are those, which have extremely abnormal shapes, namely groups 6, 10, 11 and 12 (Fig. 16).

The connections between the various groups cannot be seen in Figure 16. To show these connections we have plotted a geometric figure (Fig. 17).

Of course the reality cannot be exactly reproduced, because it is impossible to represent a six dimensional space by a "three dimensional pattern" on a two dimensional sheet of paper. Nevertheless this picture gives an approximative idea of the relations between the groups. The size of the globes corresponds to the

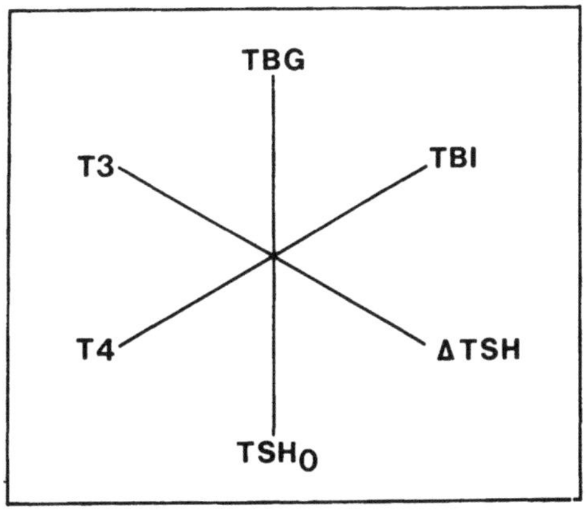

Fig. 15. Meaning of the six vectors of Figure 16.

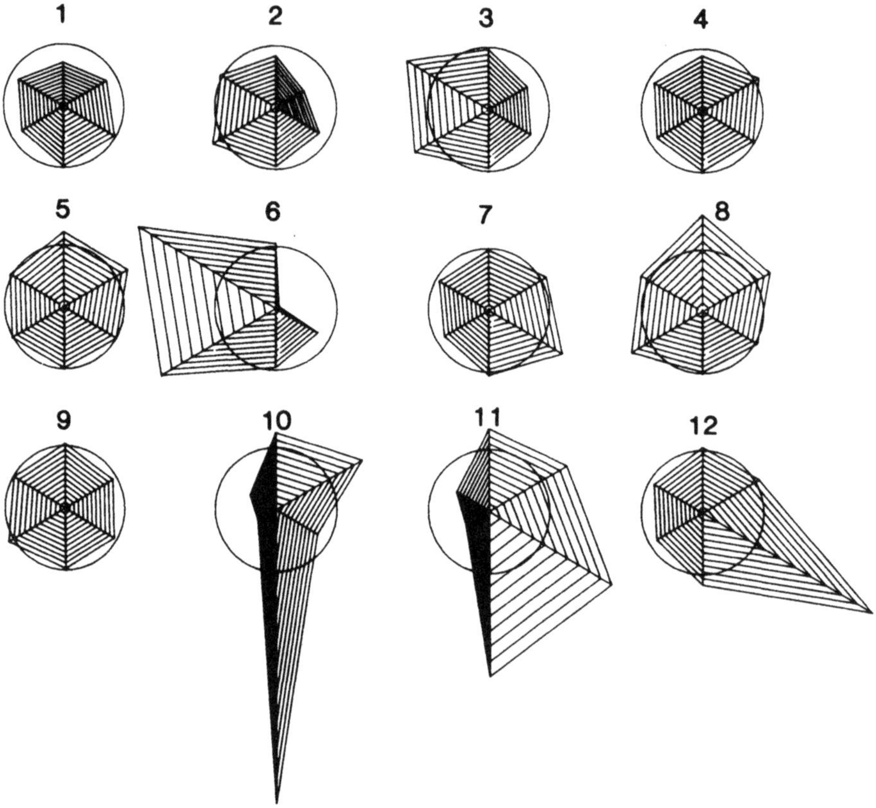

Fig. 16. Vector diagrams of the six clinical chemical parameters for the twelve cluster solution (for further explanation see text).

Table 4. Matrix of right or wrong classifications (values are in %). Right classification corresponds to the diagonal. n = number of patients in the cluster which they belong to

Cluster to which the patient belongs	Cluster to which the patient is assigned												
	1	2	3	4	5	6	7	8	9	10	11	12	n
1	83	2		11			1		4				84
2	3	90	2						5				31
3		5	90					1	4				23
4	12			73	6		1		8				129
5				6	80		1	4	8				95
6						100							4
7	3			6	5		83		3				37
8			1		7			91	1				34
9	4	3	1	6	11				75				108
10										100			2
11											100		2
12												100	3

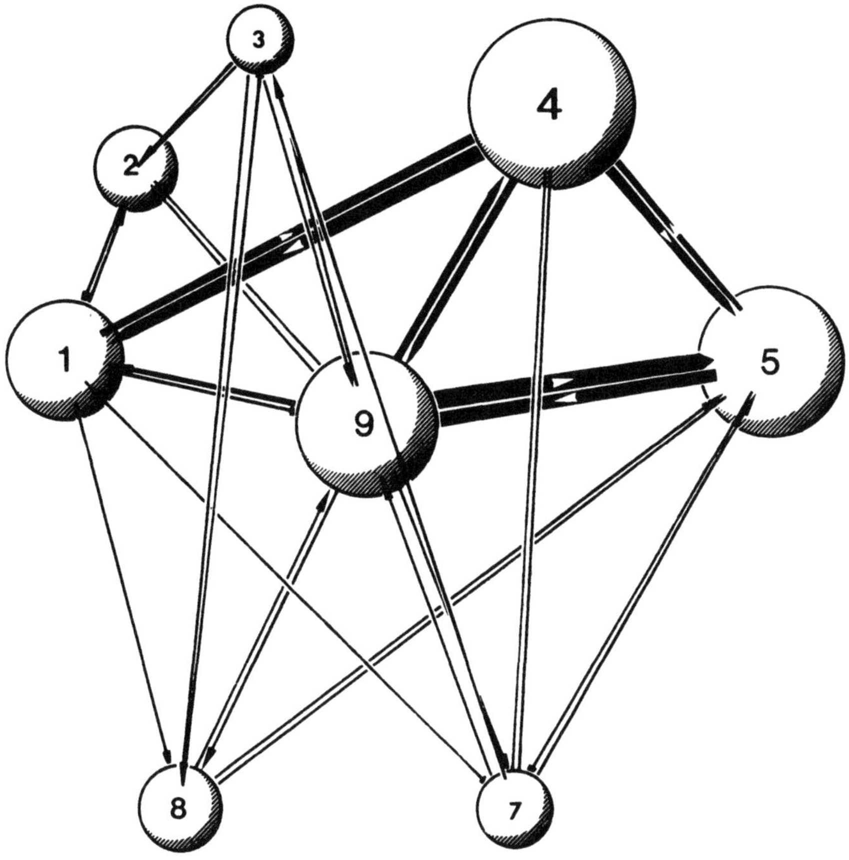

Fig. 17. "Three-dimensional" model of the interrelations and interchanges between clusters. Clusters 6, 10, 11 and 12 are totally isolated and therefore omitted.

number of patients in the group, the thickness of the vectors corresponds to the interchange between the groups.

Basis of this picture is the examination of stability with respect to reclassification, described in Section 3.5.3. The results for the grouping with 12 clusters are shown in detail in Table 4. Ninety percent of the patients of group 3, for example, were assigned to group 3 again, after random errors were added to the original values of the parameters. Five percent were assigned to group 2, one percent to group 8 and four percent to group 9. On the other side two percent of the patients of group 2 were assigned to group three and so on.

3.7 Clinical significance

Due to the fact that medicine at the patient's side has nothing to do with natural science but is a practical discipline, every medical procedure has to be proven with respect to its benefit at any time. Benefit means the procedure chosen leads to a more accurate prognosis or a better founded treatment. In Section 5 it will be tried to estimate the usefulness of the model proposed. Independent of that, however, it is also of interest to look if these clusters found show a connection with the established, present diagnostic conceptions. But it should be pointed out once more that it was not the aim to correlate exactly the groups with diagnoses and it should be emphasized that the "final" diagnosis and the "laboratory" diagnosis are working hypotheses of a physician and nothing more.

The information the model gives is, for example, that a patient belongs to cluster 4 of 12 possible groups. This information shall not anticipate any diagnosis by the clinician but shall support him to find it. It is possible that it has varying significance for the individual physician according to the experience he has made with the model. The following interpretation shall demonstrate that the classifications found are not medically meaningless.

3.7.1 Comparison of the results of classification with the laboratory diagnosis

Laboratory diagnosis means the estimation of the functional state of the thyroid gland based only on the knowledge of the six clinical chemical parameters determined thyroxine, triiodothyronine, TBG, TBI, TSH before 200 μg TRH intravenously and ΔTSH. This commitment corresponds partially to a pattern cognition done by the physician. It was made independently by two experienced doctors, who could choose between six possibilities: euthyroid, hyperthyroid in question, hyperthyroid, hyperthyroid with triiodothyronine elevated only (T3-thyrotoxicosis), hypothyroid in question (latent hypothyroidism), hypothyroid (primary or secondary). It is interesting but not surprising that in 7,6% of all cases a different meaning of the two doctors could be stated. The discrepancies naturally referred to the borderline cases.

Table 5. Clusters versus laboratory diagnoses (values are in % with respect to the size of the cluster). Code for laboratory diagnosis of the functional state of the thyroid gland: 10 = euthyroid; 20 = hyperthyroid; 25 = latent hyperthyroid; 26 = T3 hyperthyroid; 30 = hypothyroid; 35 = latent hypothyroid. *n* = number of patients in the cluster

	Laboratory diagnosis						
Cluster	10	25	20	26	35	30	*n*
1	100						84
2	42	45	13				31
3	17	30	43	9			23
4	99					1	129
5	99				1		95
6			100				4
7	70				30		37
8	94	3			3		34
9	98	2					108
10						100	2
11						100	2
12					100		3

For comparison with the automatic classification in such doubtful cases the less pathological rating of the doctors was chosen. Table 5 shows the composition of the single groups with respect to the laboratory diagnosis.

As can be seen the agreement is very good for groups 1, 4, 5, 6 and 8 to 12. The remaining groups 2, 3 and 7 are clusters with medium size. Besides 42% euthyroid patients, cluster 2 contains 45% hyperthyroid in question and 13% hyperthyroid cases. Similarly in cluster 3 30% are questionably hyperthyroid and 52% hyperthyroid (a fifth of them with T3 elevated only). About a third is latent hypothyroid in cluster 7.

3.7.2 **Comparison of the results of classification with the final diagnosis**

The final diagnosis was made with knowledge of all findings including ^{99}Tc-scintigrams. It was the basis of further therapeutic management by the physician in charge. This final diagnosis was partially summarized to a so called "reduced final" diagnosis as shown in Table 6.

Table 6. Clusters versus final diagnoses (values are in % with respect to the size of the cluster). Code for final diagnosis: 1 = hyperthyroidism; 2 = non-toxic goiter; 3 = euthyroid function, normally sized thyroid; 4 = euthyroid function, after thyroid surgery; 5 = primary hypothyroidism; 6 = subacute thyroiditis; 7 = patients with sufficient thyroid therapy. n = number of patients in the cluster

	Final diagnosis							
Cluster	1	2	3	4	5	6	7	n
1		62	35	2			1	84
2	16	71	13					31
3	74	26						23
4		68	28	3	1*			129
5		62	32	5		1		95
6	100							4
7		27	41	30	3			37
8	6	47	44	3				34
9	2	50	42	5			2	108
10					100			2
11					50		50	2
12			67		33			3

* secondary hyperthyroidism

As one could expect the correlation between reduced final diagnosis, which includes also morphologic and other clinical criteria besides laboratory findings, and clusters is poorer than with the laboratory diagnoses.

The hexagones in Figure 16 show the similarities or differences of the individual clusters impressively. Cluster 1, 4, 5 and 9 represent patients with non-toxic goiter or with normally sized euthyroid gland. Clusters 2 and 3 contain patients with hyperthyroidism or with non-toxic goiter the latter under sufficient thyroid hormone therapy. In cluster 7 patients are combined with normally functioning and sized thyroid gland or euthyroid patients after successful thyroid surgery. 80% of the patients took estrogens in cluster 8. Cluster 6 contains exclusively patients with proven thyrotoxicosis. T4 and T3 are markedly elevated, the TSH response to TRH is suppressed and TBI is low. In contrast to this cluster 10 includes only hypothyroid patients and shows also a very typical pattern, low T4 and T3, very high TSH and increased TBG.

3.7.3 **Comparison of the laboratory diagnosis with the final diagnosis**

A comparison of laboratory diagnosis with reduced final diagnosis shows also marked differences (Table 7). 3% of hyperthyroid patients have been classified as euthyroid, 3% of euthyroid patients as hyper- (2%) or latent hypothyroid (1%). 40% of patients with primary hypothyroidism have been sorted as euthyroid. In all these cases the clinical chemical parameters lay in the borderline region.

Table 7. Final diagnosis versus laboratory diagnosis (values are in % with respect to the size of the cluster). Codes see Tables 5 and 6. n = number of patients in the cluster

	Laboratory diagnosis						
Final diagnosis	10	25	20	26	35	30	n
1	3	17	69	11			35
2	92	6	1		1		324
3	98	2					187
4	94	3			3		36
5	40				20	40	5
6	100						2
7	100						3

4 A Diagnostic Strategy

The best way to have an unbiased look at a plenty of clinical chemical parameters multivariately is to define biochemical entities by performing a cluster analysis.

Of course this is only the first step. Once these entities or groups have been defined for a "teaching sample" the goal of the diagnostic process, as far as it is considered in this place, consists in assigning new patients to these groups. Here the problem is considered from an economical point of view. Every determination of a clinical chemical parameter causes workload and costs. Hence the following question is raised: Is it possible to assign patients correctly to the groups without all clinical chemical parameters having to be measured?

This question will be answered in the following by the development of a diagnostic strategy.

The model discussed here is not restricted to that case where the groups in consideration are clusters. However, it only makes sense if the groups are entirely or at least nearly entirely determined by the clinical chemical parameters in consideration.

4.1 Outline of the problem

For a specified diagnostic problem a number of clinical chemical parameters is chosen. These parameters are measured for a number of patients, the "teaching sample", and a cluster analysis is performed. The question has been raised, if it is possible to assign a new patient to the correct cluster with less parameters. "Correct" in this context can only mean that the patient is assigned to the same group to which he would have been assigned by means of all clinical chemical parameters.

An obvious method is to assign the patients of the teaching sample with all possible combinations of parameters to the groups. Then there are two ways to assign new patients. First, given the maximal number of parameters, that combination with the highest rate of correct assignment is chosen. Second, given the maximal tolerated rate of misclassification, that combination with the smallest number of parameters is chosen. Here the second method seems to be more meaningful. However, both procedures are not optimal. For different groups the optimal combination of parameters is different. Since it is not known in advance to which group a patient belongs, it is not possible to choose the group-optimal combination at that time.

The alternative is a stepwise procedure. The parameters should not be determined in parallel but sequentially. Hence the above question is posed in another way: Can diagnostic strategy models be formulated for the step-wise application of clinical chemical tests thus using them in an economical way? Summarized such a diagnostic strategy should meet the following requirements:

1. The optimal combination of clinical chemical parameters should be provided individually for every patient.

2. With the object of maximum flexibility it should be possible to assign a patient to one group out of the *total* number in every step. This means the diagnostic action should not follow the branches of a hierarchy.

3. It should be possible to choose the criterion of termination. For example the procedure should be stopped if a patient had been assigned to a group with a predefined probability.

4. The assignment rule should be compatible with the classification algorithm.

A diagnostic strategy taking these four points into account looks as follows:

1. Choose a reclassification rule which is compatible with the classification algorithm.

2. Choose a criterion of termination.

3. Choose one parameter as the starting value for all patients.

4. Assign a patient to one of the groups by means of the up to now measured parameters.

5. Decide whether the criterion of termination has been achieved.

6. If the criterion of termination has not yet been achieved choose the next parameter depending on the parameters measured up to now and the cluster to which the patient has been assigned.

7. Continue with step 4.

While steps 1 to 5 are rather straightforward the main difficulty occurs in step 6.

4.2 A simple example

Consider the following simple situation: Patients shall be assigned to four diagnostic groups 1, 2, 3, 4 by means of three clinical chemical parameters a, b, and c. These diagnostic groups should be the result of a cluster analysis using 20 patients.

Following the 7 steps of Section 4.1 first the assignment rule from Section 3.5.3 is chosen. As a criterion of termination it is natural to require that the probability of misclassification is smaller than for example 5%. This is equivalent to the request that the predictive value of the positive test result shall be greater than 0.95. These predictive values (*PVP*) are calculated for every possible combination of parameters in the following way: Assigning all 20 patients of the teaching sample to the four clusters using any combination of the parameters *a*, *b*, *c* gives a reclassification matrix as shown in Figure 18 for the parameter *a*.

The predictive value which is the probability that a patient who is assigned to cluster *i* really belongs to *i* can now be calculated for each cluster by:

$$PVP(pc, i) = \frac{n_p \text{ assigned correctly to } i \text{ by means of } pc}{n_p \text{ assigned at all to } i \text{ by means of } pc}$$

Fig. 18

Pat. belonging to ↓ / Pat. assigned to →	1	2	3	4	Σ
1	1	2	1		4
2		4	1	1	6
3	2		1	2	5
4			3	2	5
Σ	3	6	6	5	20

Fig. 19

combinations of parameters / cluster	1	2	3	4
a	0.3	0.7	0.2	0.4
b	0.4	0.6	0.8	0.8
c	1	0.8	0.6	0.6
ab	0.6	0.8	1	1
ac	1	0.8	0.6	0.6
bc	1	1	1	1
abc	1	1	1	1

Fig. 18. Reclassification matrix based on determination of parameter *a*.

Fig. 19. Matrix *A* containing the predictive values of the positive test results as elements.

Here *pc* denotes any combination of parameters *a*, *b*, *c* and *i* denotes any cluster of {1, 2, 3, 4}; n_p means number of patients. Figure 19 shows the matrix of these predictive values for the example.

To begin with "*a*" is chosen as a starting parameter at random. Let us now assume that a patient has been assigned to cluster 3 by means of parameter *a*. Matrix *A* shows that the probability for correct classification is only 0.2. Hence at least one further parameter has to be measured. To choose this next parameter a "utility" for each cluster and each combination of parameters shall be defined. This utility shall be greater than 0 or equal to 0 and will be denoted by $U_t(pc, i)$, where *pc* denotes a parameter combination, *i* a cluster-number and *t* the step-number, i. e. the number of parameters which have already been determined. Furthermore it is requested that the utility is a constant for all clusters, whenever *pc* denotes the combination of all possible clinical chemical parameters.

It is natural but not compelling to choose again the probability of correct classification for this purpose. For example also any function containing the predictive value and the sensitivity is possible. The sensitivity which is the probability that a patient who belongs to cluster *i* is assigned to cluster *i* can be calculated by:

$$S(pc, i) = \frac{n_p \text{ assigned correctly to } i \text{ by means of } pc}{n_p \text{ belonging to } i}$$

For the moment in the example we confine ourselves to the predictive value of the positive. Hence matrix *A* is also the "utility-matrix".

To determine the next parameter to be measured the question is which parameter will yield the greatest utility in the next step. Since it is not known to which cluster the patient will be assigned, the respective transition probabilities for each cluster and parameter combination *a*, *b* and *a*, *c* have to be calculated. The probability that a patient who has been assigned to cluster 3 by parameter *a* will be assigned to cluster *i* by parameter combination *pc* is given by:

$$P_{a;pc}(3, i) = \frac{n_p \text{ assigned to 3 by } a \text{ and to } i \text{ by } pc}{n_p \text{ assigned to 3 by } a}$$

It seems meaningful to choose that parameter next for which the expected value of the utility, $E(U_2)$, in the following step is largest. This means the sum of the utilities for all 4 clusters weighted with the conditional probability that the patient will be assigned to this cluster shall be maximized:

$$E(U_2) = \sum_{i=1}^{4} PVP(pc, i)\, P_{a;pc}(3, i) = \text{max!}$$

In the simple example considered the respective probabilities are given in Figure 20.

For the two combinations of parameters coming into question *a*, *b* and *a*, *c* calculating the respective expected values of U_2 gives: $0.4 \cdot 0.8 + 0.6 \cdot 0.6 = 0.68$ for combination *a*, *c* and $0.1 \cdot 0.6 + 0.3 \cdot 0.8 + 0.4 \cdot 1 + 0.2 \cdot 1 = 0.9$ for combination *a*, *b*. Hence parameter *b* is to be measured next.

If the patient is now assigned to cluster 3 or 4 matrix *A* (Fig. 19) yields that the procedure can be stopped. Otherwise the remaining parameter *c* also has to be measured.

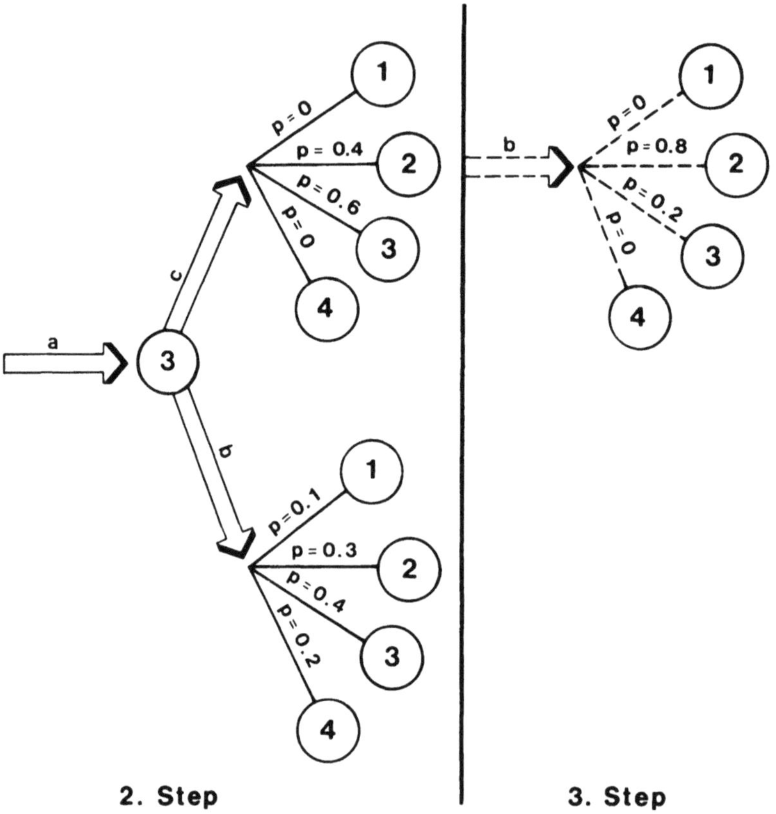

Fig. 20. Illustration of the calculation of matrix *B*. The respective transition probabilities are written on the lines. *a*, *b* and *c* represent clinical chemical parameters. The figures in the circles represent clusters. The 3. step presents one possible continuation.

combinations of parameters	cluster 1	2	3	4
a	c	b	b	c
b	a	a	a	c
c	b	a	b	b
ab	c	c	c	c
ac	b	b	b	b
bc	a	a	a	a
abc	–	–	–	–

Fig. 21. Matrix *B* containing the parameters as elements which have to be measured in the next step.

Performing the above calculation for all combinations of parameters and clusters yields a matrix B shown in Figure 21 which contains the parameters to be measured next as elements.

Up to now the starting parameter has been determined by random. Now it is shown that it may be determined in the same way as the other parameters: If no parameter has been measured the *apriori* probability of each cluster is given by:

$$P_p(i) = \frac{n_p \text{ assigned to } i \text{ by } p}{\text{total } n_p}$$

Hence that p can be chosen as a starting parameter for which

$$E(U_1) = \sum_{i=1}^{4} P_p(i)\, PVP(p, i) = \text{max!}$$

It will be discussed later whether this method for choosing the starting parameter generally can be recommended.

Once the matrix B has been calculated it is very simple to apply the strategy for new patients:

1. Assign the patient to a cluster with all parameters measured up to now.
2. If in matrix A the criterion of termination has been achieved the procedure is stopped. If not, matrix B gives the parameter which has to be measured next.

Alternate steps 1 and 2 until the criterion of termination has been achieved.

This would be the simplest form of a strategy. Now we modify it slightly. In the choice of the optimal path for an individual we need to take into account not only the point where we find ourselves but also how many steps there remain to be done.

If m clinical chemical parameters are to be determined, we proceed recursively: If $m - 1$ parameters have already been measured only one parameter is left. Hence the lowest lines of matrix B are always fixed. If $m - 2$ parameters have been measured we proceed as shown in the above example to determine the next parameter. The expected utility of this step will be called $E(U_{m-1})$. Now in the next step if $m - 3$ parameters have been determined instead of maximizing $E(U_{m-2})$ we maximize the expected value of the sum $E(U_{m-1} + U_{m-2})$ and so on. Hence at step number t for $2 \leq t \leq m-1$ pc is chosen as that parameter-combination for which the following sum is a maximum:

$$E\left(\sum_{l=t}^{m-1} U_l\right) = \text{max!}$$

or equivalently because $E(U_m)$ is a constant:

$$E\left(\sum_{l=t}^{m} U_l\right) = \text{max!}$$

This procedure better takes into account that there may be not only one step left but several ones. Therefore the latter method is to be preferred generally.

4.3 Problems of practical application

One of the problems occurring if the strategy is used in practice is arbitrariness of the choice of the utility function U and of the criterion of termination. Both depend on the objectives of the investigation.

As for the criterion of termination generally it appears advisable to require that a given percentage of misclassification shall not be exceeded. However, since these probabilities are based on the teaching sample strictly speaking they are only valid for this group. Their estimation will be the better the greater the teaching sample is. Also a very disproportionate distribution of the patients in the different clusters will have an adverse effect. A way out of this dilemma is to choose as a stop criterion a probability of misclassification which is lower than the actually accepted one.

To examine the quality of a utility function essentially two criteria are valid. First, saving in the number of parameters shall be as large as possible. Second, the number of actually misclassified patients shall be as small as possible. The latter does not only depend on the criterion of termination but also on the utility function. Generally these two criteria will be conflicting. Hence the user has to decide in his individual case. For the data base of our clinical example considered in Section 3 we made several trials with different utility functions namely $U = PVP^i \cdot S^j$ and $U = PVP^i + S^j$ (see Section 4.2 for definitions of PVP and S) for different positive values of i and j. We did not find great differences concerning as well the number of parameters needed and the number of misclassified objects.

The second problem, the determination of the starting parameter, actually is not fundamental. Essentially there are two ways: First, the starting parameter can be calculated in the same way as all other parameters by means of the strategy. Second, it can be determined by means of trial and error. At a glance the first method seems to be the more economic one. However, the saving of computing time using the first method is negligible. The reason is that the major amount of computing time is due to the calculation of matrix B (Section 4.2). Since this calcualtion is recursively the starting parameter can only be determined after the whole matrix has been calculated. As has been said once the matrix B is known the use of the strategy is very quick. Therefore it is not problematic to try the strategy for a number of starting parameters for the teaching sample. In our experience the results of these trials and the results of the above mentioned calculations slightly differed. Hence the method of trial and error seems to be preferable.

Only in the unlikely case when the starting parameter is known in advance there is a saving of computing time because all lines of matrix B which do not contain this starting parameter are superfluous. This yields a saving of computing time around a factor of two.

This leads to the major problem for developing the basis of strategy, the computing time. The amount of computing time necessary is mainly due to the calculation of matrix B. Once it is calculated the assignment of new patients is very simple and quick. For the calculation of matrix B the number of possible parameters is essential. If m is the number of parameters the number of combinations is $2^m - 1$. The number of calculations of a reclassification matrix is therefore $2^m - 1$. Of course for the calculation of the reclassification matrix also the number of patients is a factor in the computing time. However, here the increase is only linear. The number of calculations of a mean utility is

$$\sum_{t=1}^{m-2} \binom{m}{t} \cdot (m - t) \cdot k,$$

where k is the number of clusters. Hence also the number of clusters has an influence on the computing time but in this case as well the increase is only linear. Since one more parameter doubles the number of possible parameter combinations the number of parameters is the strongest factor in the computing time. Possibly a number of 10 parameters is a critical value (1023 combinations).

If more parameters shall be used a way out of this dilemma is to use combinations of some parameters at the same time instead of single parameters. This means at every step not a single parameter is to be measured but a combination of two or more. These combinations are determined in the same way as has been described for single parameters in Section 4.2. If for example two parameters a and b have been measured as starting values and in the second step also two parameters are to be determined, the total expected utility is calculated for every combination of four parameters containing a and b. That combination for which this value is highest is chosen next. Using combinations of two parameters for example has the same effect than using one parameter less. In addition this procedure has the advantage that it leads to a decision more quickly.

Generally speaking the number of reclassification matrices to be calculated decreases the more parameters are calculated in parallel. This positive effect is partly undone by the fact that the number of mean utilities to be calculated increases. A good way is perhaps to combine these methods. For example if altogether a number of 12 parameters is to be measured, it may be started with a combination of three parameters, for the following three steps a combination of two parameters is determined in each case, and the three last parameters are determined individually. Here the user can proceed as he likes and his computational facilities allow. In any case the basic principle of the strategy is not affected by this procedure.

4.4 Reduction of clinical chemical parameters needed

4.4.1 **Stepwise diagnostic strategy**

Classifying new patients is as confident as the model allows it. In practice this is not economic. Can the patients be assigned to the groups with sufficient accuracy without all parameters having to be measured? The diagnostic strategy proposed allows to find the next parameter to be determined for every patient depending on the parameters known up to now. Even if the optimal combination of parameters cannot always be found this method reduces the probability that unnecessary parameters are determined in this patient. It shall be mentioned that the parameters TSH_0 and ΔTSH were used in the same way as the others. In fact the determination of ΔTSH implies that TSH_0 has been measured before. However, to demonstrate how the strategy works in a concrete example it seemed to be not necessary to take this fact into account.

The strategy was used in the following way: The assignment rule from Section 3.5.3 was chosen. For the criterion of termination it was required that the probability of misclassification should be less than 5%. The utility-function to be optimized at each step should depend as well on the predictive values and on the sensitivity whereby the former should be of greater weight. Hence the function $U = PVP^2\ S$ was chosen (for definitions of PVP and S see Section 4.2). T4 proved to be the most appropriate starting parameter.

In Figure 22 the procedure is shown for two patients A and B. With T4 as starting parameter patient A is assigned to cluster 2. The determination of TBI in the next step leads to grouping into cluster 5. During the third step TBG has to be measured. The patient is now assigned to cluster 9. The value of T3 causes grouping into cluster 5. Here the given probability of 95% is nearly reached. Measuring the next parameter ΔTSH the patient again is assigned to cluster 9 and a probability of 100% has been achieved. This was one of the cases with a long sequence of steps. The diagnosis of this patient was a diffuse goiter with euthyroid function of thyroid gland. Patient B with the diagnosis toxic adenoma was assigned very rapidly by measuring only T4 and T3. From Table 8 showing the combinations of parameters chosen it can be seen that these two parameters were sufficient for all patients of cluster 6. The most frequently chosen combinations with respect to the whole sample were T4, T3, TBI, TBG and T4, T3, TBI, TBG, ΔTSH 213- and 168-times respectively. The hypothyroid patients of cluster 10 were classified by T4 and TSH_0 only.

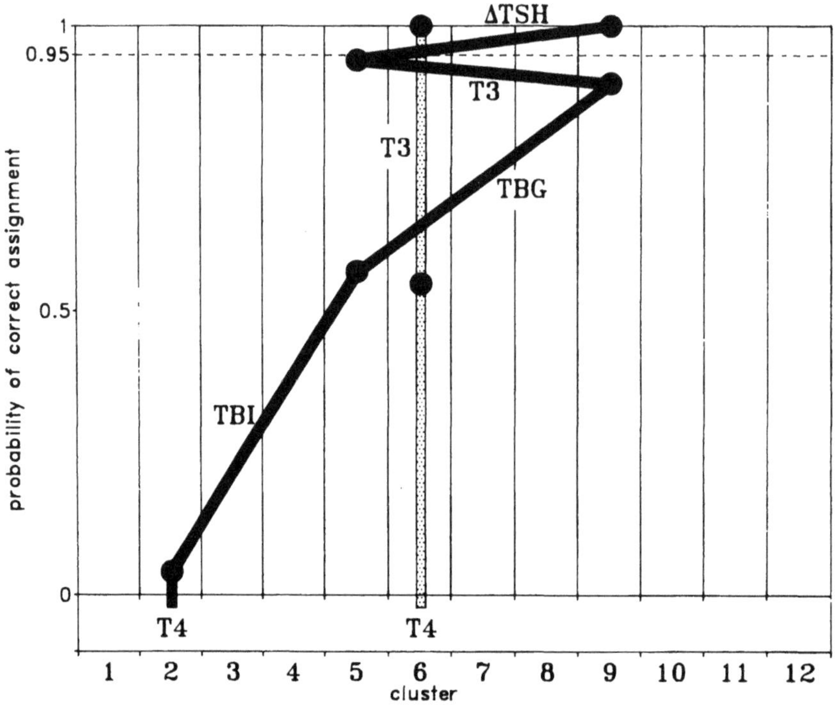

Fig. 22. Example for the stepwise individual strategy in case of two patients (●——● and ●···●).

Table 8. Combination of clinical chemical parameters which led to an assignment to the respective clusters. Column m states how often this combination has been chosen

	Parameters needed						Patients		Clusters											
m	1	2	3	4	5	6	*n*	Σ*n* (%)	1	2	3	4	5	6	7	8	9	10	11	12
2	T4	T3					4	0,7						4						
	T4		TSH_0				3	1,3										2	1	
3	T4	T3			TBI		15	4,0			14					1				
	T4	T3				TBG	10	5,8								10				
	T4	T3		ΔTSH		TBG	2	6,2												2
	T4				TBI	TBG	98	23,9				1	75		2	19				1
4	T4	T3		ΔTSH		TBG	18	27,2							17				1	
	T4	T3			TBI	TBG	213	65,8	72	31	8				5	4	93			
	T4			ΔTSH	TBI	TBG	20	69,4				2	18							
5	T4	T3		ΔTSH	TBI	TBG	168	99,8	12		1	126	1		13		15			
	T4		TSH_0	ΔTSH	TBI	TBG	1	100,0			1		1							
Σ	552	428	4	209	515	530	552	100,0	84	31	23	129	95	4	37	34	108	2	2	3

Table 9 shows the number of parameters needed for each cluster in more detail. It can be seen that indeed there is an economy in measuring the parameters. In mean it is 33%, i.e. instead of 6 parameters only 4 have to be measured. As expected the saving is more distinct at small groups containing those patients with the more pathological diagnoses. The result is in accordance with our experience from the traditional procedure.

Though the given threshold for correct classification was 95% the ratio of correct reclassified patients was in fact 97,5%. Table 10 shows the number of correct reclassified patients for each cluster.

Table 9. Number of parameters needed for correct reclassification with $p \geqslant 0.95$. n = number of patients in the cluster

Cluster	Number of parameters needed			n
	minimum	mean	maximum	
1	4	4.14	5	84
2	4	4.00	4	31
3	3	3.43	5	23
4	3	4.97	5	129
5	3	3.23	5	95
6	2	2.00	2	4
7	3	4.30	5	37
8	3	3.12	4	34
9	4	4.14	5	108
10	2	2.00	2	2
11	2	3.00	4	2
12	3	3.00	3	3

Table 10. Matrix of reclassification with $p \geqslant 0.95$. n = number of patients in the cluster

Cluster	n	Number of wrongly classified patients	Number of correctly classified patients (%)
1	84	0	100.0
2	31	1	96.8
3	23	0	100.0
4	129	3	97.7
5	95	1	98.9
6	4	0	100.0
7	37	7	81.1
8	34	1	97.1
9	108	0	100.0
10	2	0	100.0
11	2	0	100.0
12	3	1	66.7
	552	total 14	mean 97.5

4.4.2 **Diagnostic strategy using combinations of parameters**

In Section 4.3 the possibility of using combinations of parameters instead of single parameters in the strategy was discussed. There the main reason was saving in computing time. This is of minor importance if only six parameters are used. Another aspect, however, has to be considered, the time needed to come to a medically effective decision. If the determination of the parameters is very time consuming it is reasonable to measure two or more parameters in parallel.

Even if altogether only six parameters are under consideration there are many possibilities to do so. Table 11 summarizes the results.

Table 11. Dependence of number of parameters needed and wrongly classified patients on various parameter combinations

Start with	Number of parameters measured					Number of parameters needed			Number of wrongly classified patients
	step 2	step 3	step 4	step 5	step 6	minimum	mean	maximum	
T4	1	1	1	1	1	2	4.05	5	14
T4, TBI	1	1	1	1		3	4.09	6	13
TBI, TBG	2	1	1			4	4.49	5	8
TBI, TBG	2	2				4	4.97	6	8
T4, TBI, TBG	1	1	1			3	4.10	5	14
T4, T3, TBG	2	1				3	4.85	6	1
T4, TBI, TBG	3					3	5.34	6	5
T4, T3, TBI, TBG	1	1				4	4.49	6	8
T4, T3, TBI, TBG	2					4	4.97	6	8

It is not surprising that the more parameters are combined the more parameters are needed in mean. On the other hand the more parameters are used the more reliable is the reclassification.

4.4.3 **Considering analytical errors in the strategy**

It has been referred already to the fact that clinical chemical data are subject to random errors. Hence the question is posed how to take into consideration these errors when using the strategy. Essentially there are two means to do this.

Randomly produced errors are added repeatedly to the values of the patient, for example 100 times, according to the known variance distributions of the various parameters. The simplest way now seems to use the strategy in each instance, that means 100 times for every patient. However, since in many cases the patient will not be assigned to one group but to several groups in each step this possibly leads to different parameter sequences to be determined. Hence as a final result there might be no saving of parameters at all.

The second method also starts with assigning randomly produced errors to the patient's values. At each step of the strategy, whenever a cluster membership has to be calculated this is done using the new values obtained (for example 100). In many cases the patient will not be assigned to one group but to several groups with various frequencies. The cluster with the highest probability, that means that one to which the patient is assigned most times is considered to be the correct one and is therefore taken as the basis for the next step in the strategy. Trying this alternative strategy for our example we did not find better results. Neither the saving in parameters to be measured was greater nor was the number of wrongly classified patients smaller. Since this kind of strategy requires a higher amount of computing time it was not further used.

5 Experience with the Model in Clinical Practice

Besides the clinical usefulness of such a model the acceptance by the physician at the patient's side is a very important aspect, if one proposes a certain change of the well known and paved ways of medical action. Just a model delivering the clinician with a very abstract number for a patient's cluster membership is suited not to be accepted, all the more as the meaning of this group membership is unknown yet in the beginning.

Therefore we were pleasantly surprised, that a first study concerning the effect of the knowledge of a patient's cluster membership on clinical diagnosis showed encouraging results.

5.1 Design of the study

The investigation was realized in collaboration with Medizinische Klinik II of our university hospital. All consecutive patients with a tentative diagnosis of thyroid diseases delivered to the endocrinological out-patient department underwent the procedure described in the following. Only those 146 patients who visited the physician for the first time during the course of our study are included in the following investigations. The composition of the studied patient group can be seen from Table 12.

The doctor[2] who was experienced in endocrinology for a long time remained the same during the investigation. For a first help she was informed how the various final diagnoses were distributed over the twelve clusters derived from our first study (Section 3.7.2). The development of her diagnostic judgement was documented stepwise depending on her respective informational state (Table 13).

The physician in charge asked the history from the patient and examined him. Then she had to fix her opinion in a clinical diagnosis (CD), separated in a judgement of the functional state of the thyroid gland and a tentative diagnosis.

---

[2] We are thankfully indebted to Dr. E. Schineis for her collaboration.

Table 12. Diagnoses of the patients examined ($n = 146$)

Hyperthyroidism		6.1%
diffuse toxic goiter	2.7%	
nodular toxic goiter	2.7%	
toxic adenoma	0.7%	
Non-toxic goiters		54.1%
diffuse	26.0%	
nodular	28.1%	
Euthyroid function, normally sized thyroid		13.7%
Euthyroid function, after thyroid surgery		8.9%
Patients under therapy		17.2%
sufficiently managed	11.0%	
thyrostatic therapy, but patient hyperthyroid	5.5%	
hormonal substitution therapy, but patient hypothyroid	0.7%	

Table 13. Sequence of diagnostic commitment

1. Step	History + symptoms + signs clinical diagnosis (CD)
2. Step	CD + Cluster 2.D.
3. Step	2.D. + values leading to cluster assignment 3.D.
4. Step	3.D. + all six clinical chemical parameters 4.D.
5. Step	4.D. + all other informations available Final diagnosis (FD) = 5.D.

In a second step the colleague got the numbers of the clusters to which the patient had been assigned. Since the values of the measured parameters of new patients are also subject to analytical errors, we considered it necessary to take into account this fact. It is of great interest whether a certain patient is assigned randomly to a group because he is located between two groups or whether he belongs unambiguously to one group. Hence we proceeded in the following way: Randomly produced errors were added 100-times to the values of the patient according to the known distributions of the various parameters as described in Section 3.5.2. In many cases the patient was not assigned to only one group but to several groups with various frequencies. These frequencies can be given as percentages of assignment to a group. A medical report looked for example as shown in Table 14. It

Table 14. Example for a medical report

The clinical chemical analyses at your patient		
patient name:		John Smithson
patient number:		03 25 11 X
date of birth:		11/ 06/ 36
led to an assignment to the following clusters:		
	cluster	(probability)
1.	3	(50%)
2.	9	(31%)
3.	8	(17%)

contains more information than a single group number obtained by the assignment of the measured data only, since the physician can estimate the certainty of membership to a certain group. Based on this and the clinical picture she had to decide for a change of her judgement (2.D).

It has been pointed out already in Section 4.4.3 that it is complicated to take into account random errors when using the strategy. Hence the cluster numbers of step two are those to which the patient was assigned without using the strategy, i.e. by means of all six clinical chemical parameters. To estimate, however, the effect of the loss of information due to the strategy in a third step she got the quantitative values of the parameters, which had led to the assignment to this cluster by the strategy. Again she had to give a diagnostic rating (3.D).

At the fourth step she got all six values of the parameters measured (T4, T3, TBI, TBG, TSH_0 and TSH_{30}) (4.D). At the fifth and last step she had to fix the final diagnosis including all information available, also from cytology, scintigram, sonogram and so on. At this point the collegue had to give her subjective rating, whether the declaration and knowledge of cluster membership was helpful to her or not.

From these statements one can form a more objective and a subjective opinion on the usefulness of our procedure and it's acceptance. Even if we have as far as possible excluded the clinical diagnosis which is a mirror of the medical knowledge of the day and even if the only validation of our or every similar new procedure would be a therapeutical or prognostical progress, this last mentioned demand can hardly be realized. At present unfortunately a diagnosis is the measure after all.

5.2 Subjective estimation of the usefulness of the knowledge of cluster membership

At the beginning of the study the physician in charge of the patient specified that the cluster information was helpful only in 40%. With increasing familiarity with these numbers of clusters and with developing a feeling for their meaning the positive vote came up to 80% (Fig. 23).

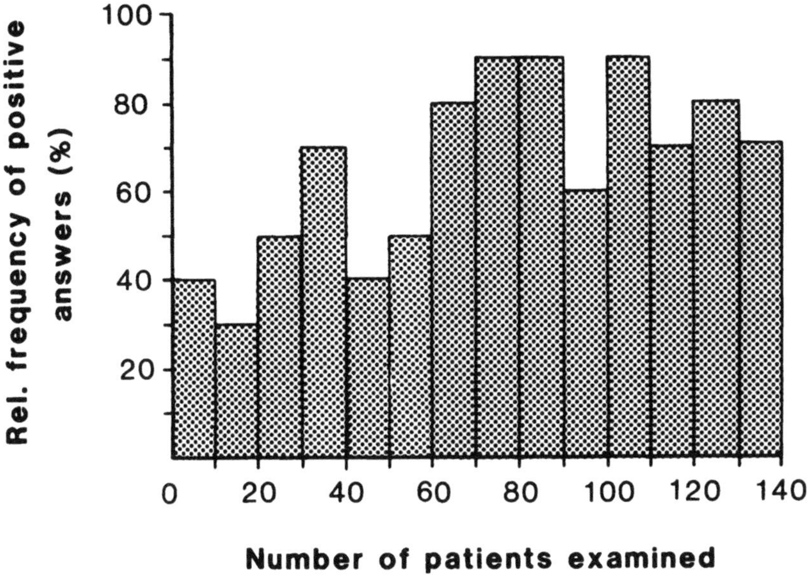

Fig. 23. Change of subjective estimation of usefulness depending on increasing experience with the model (i. e. the number of patients examined).

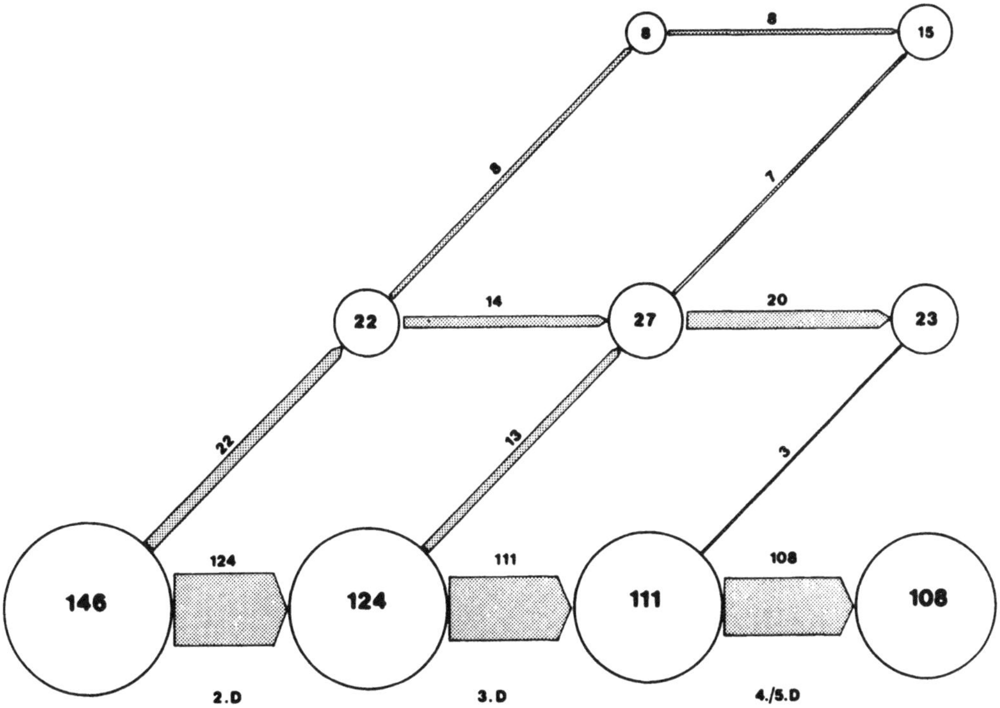

Fig. 24. Changes at the various steps of diagnostic judgement. Horizontal arrows represent no change in diagnosis. Oblique arrows represent changes in diagnosis.

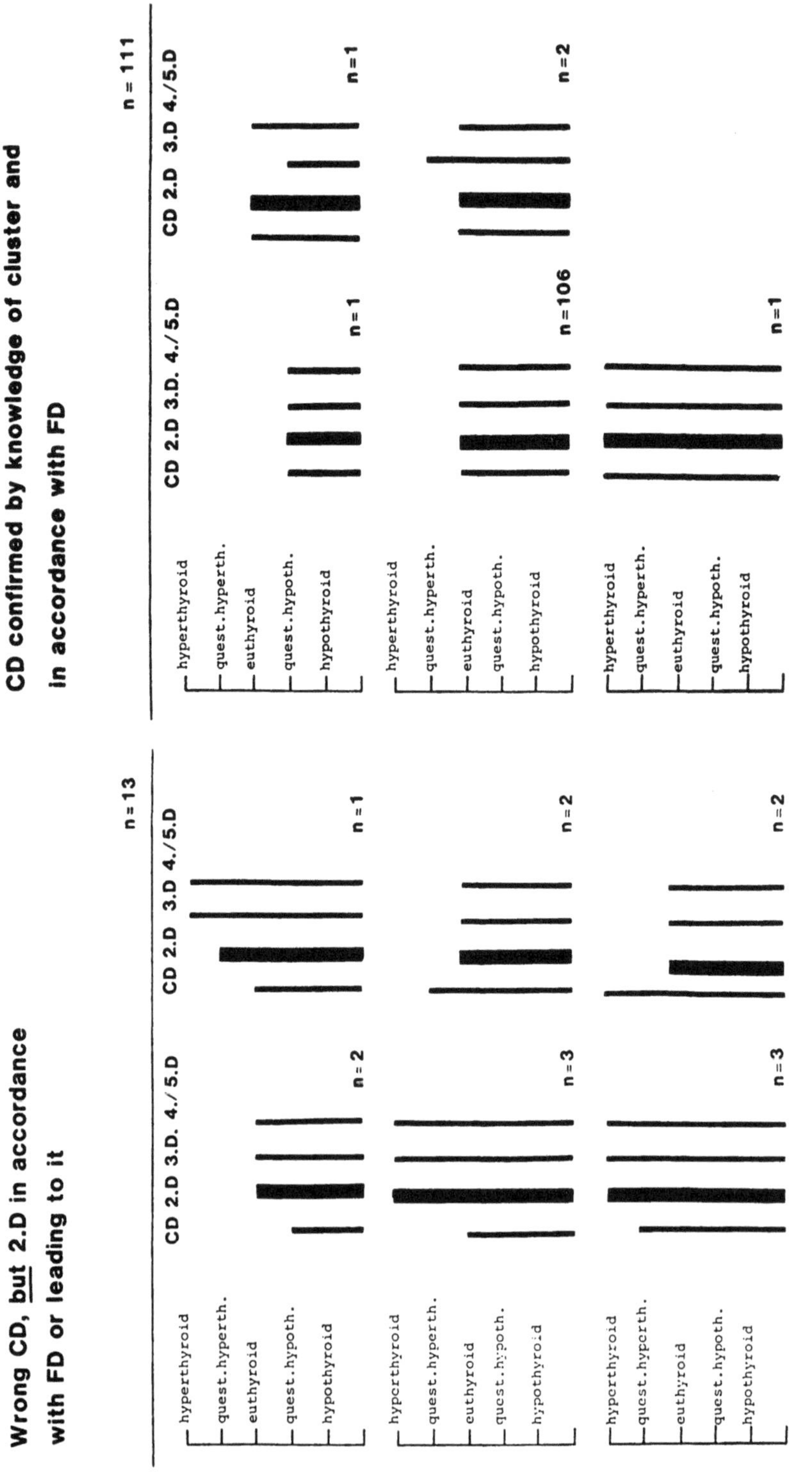
Wrong CD, but 2.D in accordance with FD or leading to it
CD confirmed by knowledge of cluster and in accordance with FD
CD 2.D 3.D. 4./5.D
CD 2.D 3.D 4./5.D
hyperthyroid
quest.hyperth.
euthyroid
quest.hypoth.
hypothyroid
n = 13
n = 111
n = 1
n = 2
n = 3
n = 106

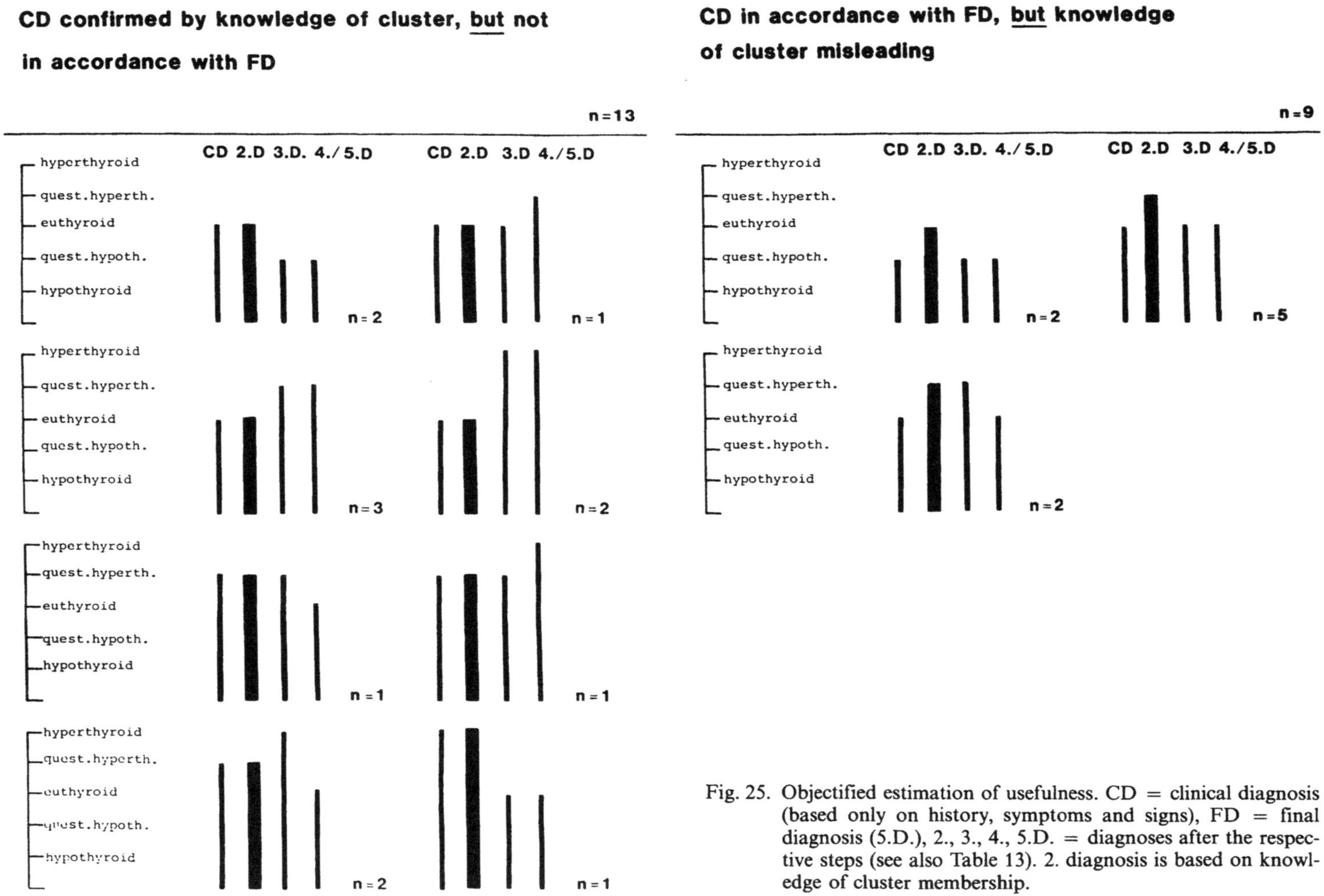

Fig. 25. Objectified estimation of usefulness. CD = clinical diagnosis (based only on history, symptoms and signs), FD = final diagnosis (5.D.), 2., 3., 4., 5.D. = diagnoses after the respective steps (see also Table 13). 2. diagnosis is based on knowledge of cluster membership.

5.3 Objective evaluation of the usefulness of the knowledge of cluster membership

The change of previously done diagnostic judgement from step 1 to step 2 and the accordance with the final diagnosis can give some references to the utility of the knowledge of cluster membership.

Figure 24 gives an overview on changes at the various steps of diagnostic judgement. 108 (74%) of 146 cases remained unchanged from the first clinical impression to the "final diagnosis". With 23 patients (16%) one change, with 15 (10%) two changes could be observed during the course of medical decision process. At the second step 22 (15%), at the third step 13 + 8 = 21 (14%) and at the last step 3 + 7 = 10 patients (7%) were classified differently from the previous decision.

The estimation of the functional state of the thyroid gland is shown in Figure 25 for diagnostic steps 1 to 4. Thirteen times the clinical diagnosis was changed by the knowledge of cluster membership, and this 2. diagnosis agreed with the final diagnosis. At 111 patients the diagnostic assignment remained unchanged until the final diagnosis. In 13 cases the knowledge of cluster membership confirmed the clinical picture but was not in accordance with the final diagnosis. In 9 cases clinical and final diagnosis were the same, but the knowledge of cluster membership led to a misclassification of the patient. Both last mentioned kinds of changing diagnoses showed a timedependent behaviour. They occured eighteen times during the first half of the study, in the second part, however, only four times.

Therefore including the subjective validation one can conservatively conclude that after a phase of adaption and accustoming the knowledge of cluster membership was helpful to the physician in charge of the patient in fact.

6 Outlook

6.1 Dependence of the model on patient sample used

The results of classification are based on data of the once chosen patient sample. Kind and frequency of diseases in this random sample depend on many factors like geographical location, type of hospital, moment at which data have been collected and so forth. Therefore due to such dependences a model cannot be transferred safely to other hospitals. Diminishing the influence of random sampling can be achieved by repeating classification with new data from time to time and then by correction of the model if necessary. But this means a considerable expenditure. In addition a further and greater difficulty is caused by the newly defined groups with possibly other meaning forcing the clinician to change his judgement rules. Understandably this would lower the acceptance of the model drastically.

6.2 Dependence of the model on analytical methods

A further dependence exists on the actually used analytical methods. This is not surprising. So we found statistically significant systematical differences between a first and a second determination of thyroid parameters three years later out of the same sera. During this time the analytical methods have changed partially. As a consequence of the technical and chemical progress just in clinical chemistry changes of methods and equipment are frequently to be expected. Thus classification has to be rerun with all the disadvantages described in the last section. A solution would be only further efforts in standardization of analytical methods to achieve a statistically acceptable long-term stability. But in endocrinology a long way is still to go to reach this goal.

6.3 Aspects for the future

Till now informations are missing on the dynamic aspects of group membership depending on time. It would be interesting to investigate the change from group to group during therapy, development of disease or course of time. Possibly new knowledge could be gained by this means.

Basically, however, can such models like this succeed? Usually the mind of a clinician is case-related, concrete, some times anecdotal, in any case mostly not abstract, mathematical. Therefore it is difficult for any mathematical-statistical, strictly speaken not imaginable model to be accepted for daily routine. That is an unpleasant insight, but it corresponds yet to reality.

Appendix: Comparison of Classifications

The problem of measuring the similarity between two different clusterings of the same data set is very important. For example, when dealing with stability of a cluster solution a measure of correspondence between classifications is needed.

There are two measures which have been mostly used for comparing two partitions: Cohen's kappa (1960) and Rand's measure (1971).

Given a set S of n objects $S := \{O_1, \ldots, O_n\}$ let $G := \{g_1, \ldots, g_k\}$ and $H := \{h_1, \ldots, h_l\}$ represent two different partitions of S. Let n_{ij} denote the number of objects classified in g_i in partition G and in h_j in partition H. If the numbers of clusters are equal in both partitions, i.e. $k = l$, then the mean observed agreement between the two partitions, P_0, and the mean agreement expected by chance – given the number of objects in each group –, P_c, are given by:

$$P_o = \frac{1}{n} \sum_{i=1}^{k} n_{ii}, \; P_c = \sum_{i=1}^{k} \frac{n_{\cdot i}}{n} \frac{n_{i\cdot}}{n}.$$

The kappa index of Cohen is now defined to be:

$$S_c(G, H) = \frac{P_o - P_c}{1 - P_c}$$

The statistic has upper bound 1 when there is total coincidence. It has no fixed lower bound since the observed coincidence may be even smaller than the coincidence expected by chance. However, at a chance level of agreement it takes the value 0. Some problems are associated with this measure. First, it can be used only for partitions with equal numbers of clusters. Even more serious is the problem that it has no unique values because each cluster of partition G has to be assigned to a single cluster of partition H and there are $k!$ ways to do this.

In contrast Rand's measure has the advantage that it is based on pairs of objects alone. Consider two objects O_i and O_j from the set S. Four cases are possible (Hubert and Arabie, 1985):

1. Objects O_i and O_j are placed in the same group in G and in the same group in H.
2. Objects O_i and O_j are placed in different groups in G and in different groups in H.
3. Objects O_i and O_j are placed in the same group in G and in different groups in H.
4. Objects O_i and O_j are placed in different groups in G and in the same group in H.

While object pairs of type 1 and type 2 represent agreements in both classifications, object pairs of type 3 and type 4 represent disagreements in the classifications. If A denotes the sum of object pairs of type 1 and type 2, then A is the total number of agreements. Analogously D – the sum of object pairs of type 3 and type 4 – represents total disagreement. Now Rand's measure is defined by:

$$s_R(G, H) = \frac{A}{\binom{n}{2}} = \frac{\binom{n}{2} - D}{\binom{n}{2}}$$

The measure has the drawback that at a chance level of agreement it yields a positive nonzero value determined by the relative size of clusters and the number of clusters.

Figure 26 shows the dependency of Rand's measure on the number of clusters. 2–21 groups were build randomly by arbitrary assignment of 552 objects. This was done 36 times, and the first "clustering" was compared with the 35 remaining clusterings where only partitions with the same numbers of clusters were compared. The figure shows the median values of the 35 similarities for partitions with 2–21 clusters. We could not find a dependency on the size of the population. Attempts with a number of 100 objects as well as attempts with a number of 1000 objects resulted in almost the same curve.

It has to be stressed that the curve is not an absolute lower bound for the Rand measure. Because a value of 0 is not possible when two partitions with the same number of clusters are compared the absolut lower bound is greater than 0 but it cannot be exactly computed in reality.

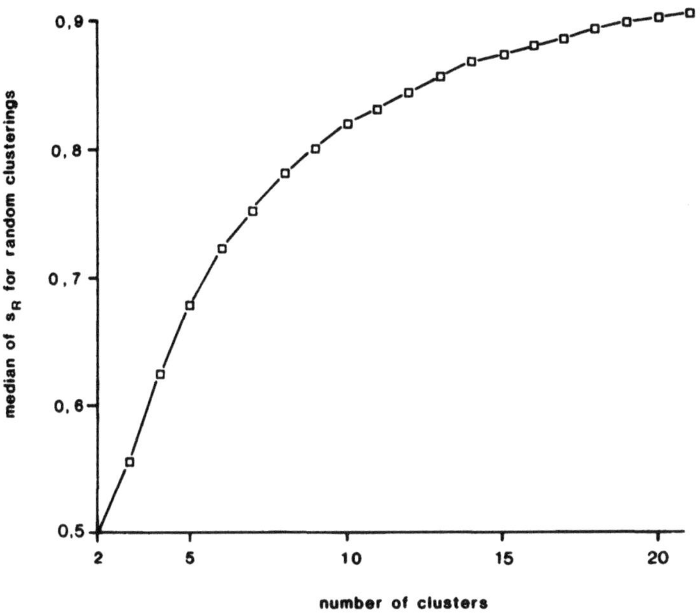

Fig. 26. Median values of Rand's measure for 35 comparisons of randomly build clusters.

Since a random assignment yields clusters of approximately equal size there might result a slightly lower curve if one assigns objects randomly subject to certain constraints on the numbers of objects in each cluster. However, the strong dependency of Rand's measure on the number of groups shown by the curve would not be affected by such a procedure.

An attempt to overcome this property has been made by Hubert and Arabie (1985). They suggest a correction for chance for Rand's measure, which analogously to Cohen's kappa is defined by:

$$s_{RC}(G, H) = \frac{s_R(G, H) - E(s_R(G, H))}{1 - E(s_R(G, H))}$$

where $E(s_R(G, H))$ denotes the expected value of Rand's measure, or:

$$s_{RC}(G, H) = \frac{\sum_{i=1}^{k} \sum_{j=1}^{l} \binom{n_{ij}}{2} - \left(\frac{1}{\binom{n}{2}}\right) \sum_{i=1}^{k} \binom{n_{i\cdot}}{2} \sum_{j=1}^{l} \binom{n_{\cdot j}}{2}}{\frac{1}{2}\left(\sum_{i=1}^{k} \binom{n_{i\cdot}}{2} + \sum_{j=1}^{l} \binom{n_{\cdot j}}{2}\right) - \left(\frac{1}{\binom{n}{2}}\right) \sum_{i=1}^{k} \binom{n_{i\cdot}}{2} \sum_{j=1}^{l} \binom{n_{\cdot j}}{2}}$$

Here $n_{i.}$ and $n_{.j}$ denote the number of objects in cluster g_i and cluster h_j respectively. The measure as well as kappa has upper bound 1 when the two partitions are identical and no fixed lower bound. At a chance level of agreement it has value 0.

Some other variations of Rand's measure have been suggested. A useful survey is given by Hubert and Arabie (1985). Which measure to use depends on the objectives of the comparison. Though the measures itselves are very different, fortunately, results obtained by them do not differ very much. For example although kappa and Rand's measure represent totally different approaches to measure similarity Edelbrock and McLaughlin (1980) as well as Milligan and Isaac (1980) found a high degree of agreement when using these two measures for estimating the accuracy of a clustering technique. This is in agreement with our experience. Using the measures s_{R}, s_{RC} as well as two other ones (Wallace, 1983 and Fowlkes and Mallows, 1983) to describe the stability of cluster solutions (Section 3.5) the results did not differ regardless which measure was used.

Hence it is probably not so important which measure is used but it is essential that the characteristic features of the measure one is willing to apply are known and taken into consideration.

References

1. Bayne, C. K., Beauchamp, J. J., Begovich, C. L. & Kane, V. E. (1980) Monte Carlo comparisons of selected clustering procedures. Pattern Recognition *12*, 51–62.
2. Blashfield, R. K. (1976) Mixture model tests of cluster analysis: Accuracy of four agglomarative hierarchical methods. Psychological Bulletin *83*, 377–388.
3. Bock, H. H. (1974) Automatische Klassifikation, Vandenhoeck & Ruprecht, Göttingen.
4. Bock, H. H. (1985) On some significance tests in cluster analysis. Journal of Classification, *2*, 77–108.
5. Büttner, H. (1978) in: Logic and Economics of Clinical Laboratory Use (Benson, E. S. & Rubin, M., eds.) pp. 91–102, Elsevier, New York.
6. Cohen, J. (1960) A coefficient of agreement for nominal scales. Educational and Psychological Measurement *20*, 37–46.
7. Dubes, R. & Jain, A. K. (1979) Validity studies in clustering methodologies. Pattern Recognition *11*, 235–254.
8. Edelbrock, C. (1979) Mixture model tests of hierarchical clustering algorithms: The problem of classifying everybody. Multivariate Behavioral Research *14*, 367–384.
9. Edelbrock, C. & McLaughlin, B. (1980) Hierarchical cluster analysis using intraclass correlations: A mixture model study. Multivariate Behavioral Research *15*, 299–318.
10. Everitt, B. (1974) Cluster Analysis, Heinemann Educational Books, London.
11. Feinstein, A. R. (1967) Clinical Judgement, Robert, E. Krieger Publ. Co., Huntington.
12. Fisher, L. & Van Ness, W. (1971) Admissible clustering procedures. Biometrika *58*, 91–104.
13. Fowlkes, E. B. & Mallows, C. L. (1983) A method for comparing two hierarchical clusterings. J. Amer. Stat. Ass. *78*, 553–569.
14. Friedman, H. P. & Rubin, J. (1967) On some invariant criteria for grouping data. J. Amer. Stat. Ass. *62*, 1159–1178.
15. Galen, R. S. & Gambino, S. R. (1975) Beyond Normality: The Predictive Value and Efficiency of Medical Diagnoses, Wiley, New York.
16. Ganesalingam, S. & McLachlan, G. J. (1980) A comparison of the mixture and classification approaches to cluster analysis. Commun. Stat. Theory Methods *A9*, 923–933.
17. Gross, R. (1969) Medizinische Diagnostik – Grundlagen und Praxis, Springer, Berlin, Heidelberg.
18. Hartigan, J. A. (1975) Clustering Algorithms, Wiley, New York.
19. Hubert, L. J. & Arabie, P. (1985) Comparing partitions. Journal of Classification *2*, 193–218.
20. Jardine, N. & Sibson, R. (1968) The Construction of hierarchic and non-hierarchic classifications. Computer Journal *11*, 177–184.
21. Keller, H. & Gessner, U. (1981) Zur Frage der Validität klinisch-chemischer Befunde. Med. Lab. *34*, 3–7.

22. Koch, R. (1920) Die ärztliche Diagnose, Beitrag zur Kenntnis des ärztlichen Denkens, 2. ed., Wiesbaden, cited in: Wieland (1975), p. 119.
23. Kuiper, F. K. & Fisher, L. A. (1975) A Monte Carlo comparison of six clustering procedures. Biometrics *31*, 777–783.
24. Milligan, G. W. & Isaac, P. D. (1980) The validation of four ultrametric clustering algorithms. Pattern Recognition *12*, 41–50.
25. Mojena, R. (1977) Hierarchical grouping methods and stopping rules: An evaluation. Computer Journal *20*, 359–363.
26. Perruchet, C. (1983) in: Numerical Taxonomy (Felsenstein, J., ed.) pp. 199–208, Springer, Berlin, Heidelberg, New York, Tokyo.
27. Rand, W. M. (1971) Objective criteria for the evaluation of clustering methods. J. Amer. Stat. Ass. *66*, 846–850.
28. Sackett, D. L. (1979) in: Clinician and Chemist (Young, D. S., Nipper, H., Uddin, D., Hicks, J. & King, J. S., eds.), pp. 23–30, the American Association for Clinical Chemistry, Washington, Winston Salem.
29. Späth, H. (1977) Cluster-Analyse-Algorithmen zur Objektklassifizierung und Datenreduktion, R. Oldenbourg, München, Wien.
30. Vecchio, T. J. (1966) Predictive value of a single diagnostic test in unselected populations. New Engl. J. Med. *274*, 1171–1173.
31. Vogt, W., Sandel, P., Schwarzfischer, P., Braun, S. L., Langfelder, C. & Knedel, M. (1981) Cluster-oriented discriminant analysis; taxonomic classification of the thyroid function. Clin. Chim. Acta *112*, 213–223.
32. Vogt, W., Sator, H. & Nagel, D. (1984) Computer assisted clinical chemical diagnosis, pattern cognition and recognition by cluster analysis. Trends in Analytical Chemistry *3*, 166–171.
33. Vogt, W., Nagel, D. & Sator, H. (1987) Cluster Analysis in Clinical Chemistry: A Model, Wiley, New York.
34. Wallace, D. L. (1983) Comment. J. Amer. Stat. Ass. *78*, 569–579.
35. Ward, J. H. (1963) Hierarchical grouping to optimize an objective function. J. Amer. Stat. Ass. *58*, 236–244.
36. Werner, M., Brooks, S. H. & Wette, R. (1973) Strategy for cost-effective laboratory testing. Hum. Pathol. *4*, 17–30.
37. Wieland, W. (1975) Diagnose: Überlegungen zur Medizintheorie, de Gruyter, Berlin, New York.

Chapter 3.6
Quasi Distribution-Free Assessment of Multivariate Data of Clinical Chemical Tests with the Aid of the X-Transformation

W. Rehpenning

1 Introduction

The clinical chemist is confronted with the problem of having to assess multivariate clinical chemical data vectors when several chemical quantities (blood constituents, etc.) are to be considered together. "To assess" in this context usually means to compare the set of the measured values of a patient with those of an appropriate collective of reference persons. The goal then is to determine whether differences exist between the measured values of the person in question and the reference data and the significance of possible differences. In the special case where the reference collective consists of healthy persons the task is to find out whether the given data indicate a pathological condition.

The easiest procedure is to examine each component of the data vector one at a time with respect to its deviation from the reference values, without considering the other values. There are well worked-out methods for this univariate approach (1). This procedure however, suffers from certain disadvantages. Firstly, with the increase in number of components in a data set, the probability of falsely positive decisions increases (1, 2, 4, 5). Secondly, correlations between different chemical quantities in a data set cause redundancy in the data, as is well known from information theory. Not taking into account this redundancy affects the sensitivity of a data-based decision in an unnecessary way (6). As a result a *multivariate approach* is indicated which would take account of the various correlations between the different quantities and their relationship to one another. This is usually done by the computation of multivariate statistical test quantities which are evaluated from all the measurements (5–7). If this statistical test quantity exceeds a given limit, which depends on the number of the variables and on the given error probability, then the data are considered to be pathological.

The setting up of such test quantities and the computation of their theoretical distributions and confidence limits requires intensive mathematical preparation and modelling which shall not be described here in detail. Basically, a distinction can be made between *parametric* and *nonparametric* (distribution-free) methods. The parametric approach makes use of the general assumption that the data follow a multivariate normal distribution. This means that the common probability density function of (say) k clinical quantities can be written as:

$$f(x_1, \ldots, x_k) = C \exp\left(-\frac{1}{2}(L(x_1, \ldots, x_k) + Q(x_1, \ldots, x_k))\right)$$

Here C is an appropriate constant, L is a linear expression and Q is a positive definite quadratic form. From the above assumptions it can be inferred that all the single quantities follow univariate normal distributions and that their probability density functions can be described by the well-known gaussian bell-shaped curve. For further details, the reader is referred to ref. (8).

The assumption of a multivariate normal distribution is very restrictive, as it is practically never fulfilled by clinical data. Much less restrictive assumptions suffice for the application of nonparametric methods as developed and advanced by Wilks (9), Wald (10), Tukey (11), Fraser (12), Kempermann (13), Abt (14), Ackermann (15), and others. These methods only suppose continuous clinical quantities. Sometimes, the additional assumption is made that the domain (the multidimensional region of variation) of the quantities is connected and convex. Essentially, the distribution-free methods consist of constructing from the reference data certain multidimensional domains, also called "tolerance regions", which contain a certain proportion, for example 95 percent, of the points representing the reference persons. Furthermore, the region is modified to ensure that future observations fall into the constructed tolerance region with the given probability, provided that the observation units are drawn from the reference population. In addition, computational rules are derived which enable one to decide whether a given data set belongs to the above-mentioned tolerance region.

A comparison of the parametric and the nonparametric methods reveals advantages and disadvantages of both procedures. On the one hand, the assumption of a multivariate normal distribution leads to particularly simple and easy-to-handle formulas. Parametric data analysis can therefore be done rapidly with an electronic computer. The construction of nonparametric tolerance regions is considerably more cumbersome with respect to programming effort and computing time. On the other hand, clinical data often show striking departures from a normal distribution. If a parametric analysis is done, this not only leads to incorrect decisions and erroneous conclusions but also leaves the practitioner in doubt as to whether he can trust the results of his computations, as the derivation of the computing rules is mathematically sound only in the case of a multivariate normal distribution.

One can raise the question at this point, of whether the data can be transformed before the data analysis in such a way that the transformed data would show a multivariate normal distribution although the original data did not: one could then do a parametric analysis with the transformed data. To obtain multivariate normality in the transformed data requires as a necessary first step the examination and (if need be) normalizing transformation of all the individual clinical chemical quantities, considered here as marginal variables. Theoretically, such normalizing transformations can always be found. Although finding such normalizing transformations is necessary in this context, this step may not be sufficient. Indeed it is possible for all the individual components of a transformed data set to be normally distributed without the transformed data following a *multivariate* normal distribution. This means that it would be impossible to find a suitable transformation with the required qualifications. Nonetheless it is reasonable to assume that the application of transformations which lead to normally distributed individual components also leads to a fairly good approximation to the theoretical multivariate normal distribution.

The search for normalizing transformations is a task which has to be tackled with close regard to the data. Clinical measurements often display a positive skewness, i. e. they are unsymmetrically distributed with the mode of concentration at lower values and a thinning out at higher values. In this case the logarithmic transformation: $y_i = \log(x_i + a)$ often leads to a sufficiently symmetric distribution, if the constant a is appropriately chosen. A more general attempt would be:

$$y_i = \frac{(x_i + a)^b - 1}{b},$$

which converges to the above logarithmic transformation as b tends to zero. One can try to choose the constants a and b in such a way that an approach as close as possible to a normal distribution results for the transformed values y_i. There are a wealth of other possibilities for choosing transformation formulas. If the given sample is of a moderate size there is no way of deciding with statistical means which of them is the best. Furthermore, the determination of the free constants in the formulas can be troublesome. The reader is referred to Box and Cox (16) and Hinkley and Runger (17). One is led to the question, therefore, of whether the time-consuming and inefficient search for appropriate transformations cannot be dispensed with. In the following it will be shown that a transformation, called the "X-transformation" here, which was applied and discussed by van der Waerden in another context (18), can be utilized for multivariate data assessment.

2 The X-Transformation of van der Waerden

Van der Waerden was led to the X-transformation in an attempt to find an alternative to the well known t-test for the comparison of two means. The t-test suffers from a loss of power if the distribution of the observed quantities deviates from a normal distribution. This loss is even more serious if the data set contains undetected outlying observations. It is preferable therefore to use a distribution-free test procedure, of which the "X-test" invented by van der Waerden is an example. The transformation which he applied in performing the X-test had been mentioned by others before, especially in psychological investigations. We shall call the transformation, for the sake of brevity, the "X-transformation". The principal idea which leads to the X-transformation is based on the following line of argument:

The goal is to find a transformation of the form: $y_i = T(x_i)$ which will produce normally distributed quantities y_i from the observed quantities x_i. This problem is usually solved in two steps. The first step consists in finding a transformation of the form: $u_i = S(x_i)$ which produces *uniformly* distributed quantities u_i. If we denote by $P(u \leqslant t)$ the probability that the quantity u is less than or equal to the real number t, then we say that the quantity u is uniformly distributed in, for example, the interval (0,1), if we have:

$$P(u \leqslant t) = t \text{ for all } t \text{ with } 0 < t < 1.$$

Once the transformation S has been found, the second step consists in computing y-values from the uniformly distributed u-values by using the formula:

$$u_i = \Phi(y_i) \text{ or, solved for } y_i: y_i = \Phi^{-1}(u_i)$$

Here Φ stands for the cumulative distribution function of the standardized normal distribution function with expectation zero and dispersion one, i. e. the area function under the bell-shaped gaussian. Φ^{-1} denotes the inverse of Φ. The computation of the inverse function Φ^{-1} can be done by simple approximation formulas (see below). The y_i are then distributed according to a standardized normal distribution. This means that we have:

$$P(y \leqslant t) = P(\Phi^{-1}(u) \leqslant t) = P(u \leqslant \Phi(t)) = \Phi(t)$$

since the u_i are uniformly distributed.

As can be seen, the essential difficulty of the transformation problem consists in finding the transformation S with the property that it maps the x-values into uniformly distributed u-values. This difficulty is avoided by the X-transformation. At first the x-values are transformed into rank numbers in the usual fashion. These ranks are mapped into the open interval (0,1) by the scaling down: $r_i = R_i/(n + 1)$, where the R_i denote the usual ranks. To the resulting r-values the inverse integral function Φ^{-1} can be applied:

$$X_i = \Phi^{-1}(r_i)$$

The resulting values will be called "X-values" here – in remembrance of van der Waerden who constructed the "X-test" with the aid of this transformation. Furthermore, the above-described transformation will be named: "X-transformation". It is summarized in the following steps:

1. Transform the x-values x_i into ranks R_i.
2. Scale the ranks down by: $r_i = R_i/(n + 1)$.
3. Apply the inverse normal integral function Φ^{-1} to the ranks r_i and call the result X_i.

The single steps are illustrated graphically in Figure 1 using, as an example, five cholesterol values to which the X-transformation has been applied. The curve in Figure 1 is the area function under the normal distribution density function. The X-values are obtained by projecting onto the abscissa the points of intersection of the integral curve and the lines parallel to the abscissa at the height of the ranks r_i. Table 1 contains the corresponding numbers.

Note: For the computation of the X-values it is not really necessary to invert the integral function Φ numerically, which would be a little cumbersome. For the present purpose it suffices to use simple approximation formulas such as given by Abramowitz and Stegun (19) or by Patel (20). If, for example, we set:

$$w_i = \sqrt{-\ln(4r_i - 4r_i^2)}$$

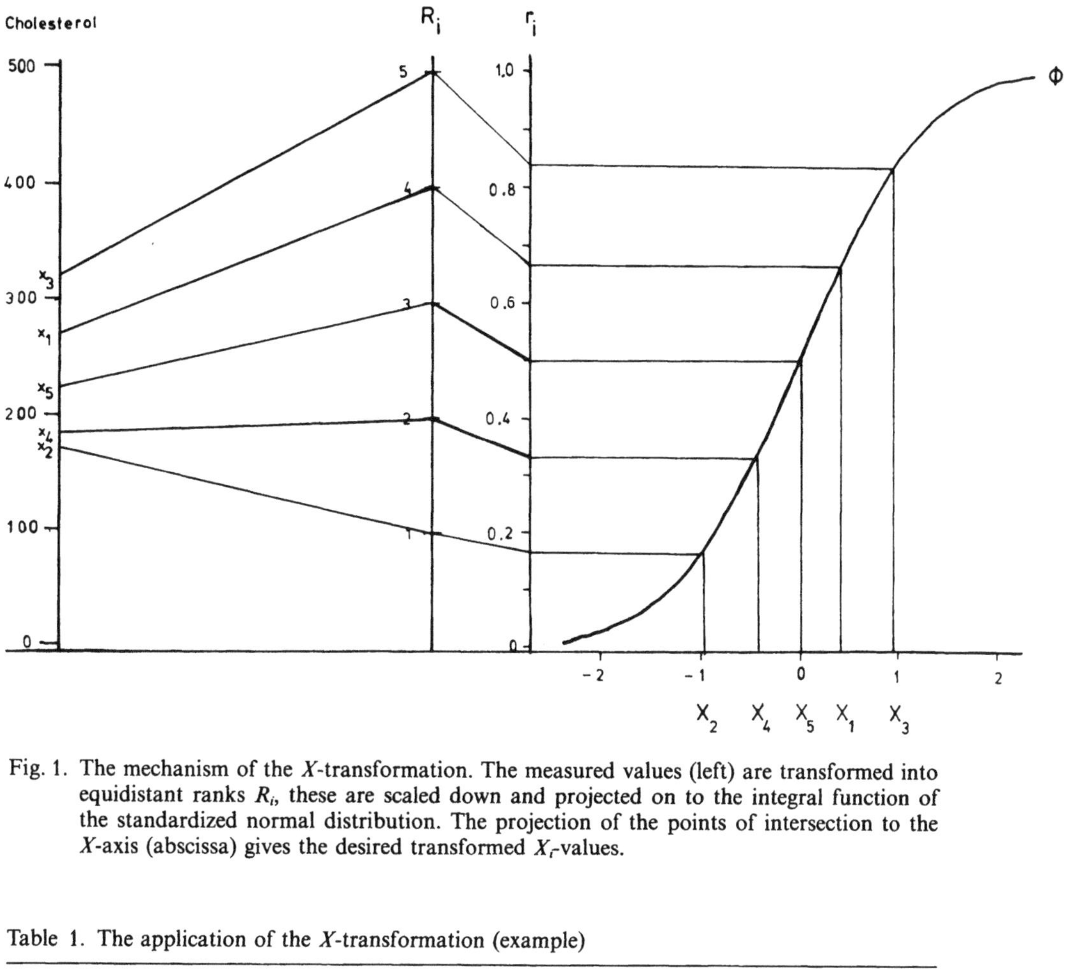

Fig. 1. The mechanism of the X-transformation. The measured values (left) are transformed into equidistant ranks R_i, these are scaled down and projected on to the integral function of the standardized normal distribution. The projection of the points of intersection to the X-axis (abscissa) gives the desired transformed X_i-values.

Table 1. The application of the X-transformation (example)

Measured values Cholesterol	R_i	Ranks r_i	X-ranks
268	4	0.667	0.431
173	1	0.167	−0.967
320	5	0.833	0.967
184	2	0.333	−0.431
229	3	0.5	0.0

then we have, with reasonable accuracy:

$$X_i = \text{signum}(r_i - 0.5)(1.23w_i(1 + 0.0262w_i))$$

This formula was given by Hamaker. See Patel (20).

For the smallest cholesterol value $x_2 = 173$ we have $R_2 = 1$, $r_2 = 1/6$. With the above formula we get $X_2 = -0.968$. The true value is $X_2 = -0.967$.

The X-transformation has the property of converting a continuous quantity, which in principle can take on an infinite number of real values, into a scale of discrete values. The range of the X-transformation, therefore, is a finite set of n real values, the number of which depends only on the size of the observed sample. Only when there are ties in the data, can intermediate values occur. In all cases it follows that there is a loss of information associated with the application of the X-transformation, but this is unavoidable whenever a ranking transformation is applied. The resulting X-values do not depend on the theoretical distribution of the observed quantities. Only the sequence of indices of the fixed X_r-values depends on the particular order of the n observed x_r-values. In this sense the X-transformation leads to a *distribution-free* treatment of the data. In order to distinguish the method of the X-transformation from other multivariate, completely distribution-free, methods it may be called (and has been called in the psychological literature) a *quasi-distribution-free* method. It may be pointed out that the X-transformation is applied to the individual quantities only, i.e. to the *marginal* distributions of the data set, whereas a completely distribution-free method should not depend on the axes of the coordinate system at all.

When the cumulative empirical distribution function is constructed from the computed X-values of a sample, the graphical representation of this function meets, at its points of discontinuity, the curve of the cumulative standardized normal distribution function, provided the values of the empirical distribution function are multiplied by the factor $(n + 1)/n$. This follows immediately from the construction of the X-values. The X-transformation derives its usefulness for the handling of multivariate data vectors from this very property. It is reasonable, namely, to treat the X-values of a data set as if they were observed values with a multivariate normal distribution. The X-values are symmetric to the origin of the coordinate system (provided that there are no ties in the data), so they add up to zero. Hence, the mean values of the corresponding X-values are zero for all marginal distributions. The standard deviations of the X-values are a little smaller than one and tend to one as n becomes infinitely large. Ties in the original data cause slight modifications of the above.

Note: If ranking transformations are applied to the data of a sample of independent observational units then a slight negative correlation results between the ranks of the different individuals. This is also true for the X-transformation. Between the X-values of a clinical chemical quantity which belong to different individuals there is a theoretical negative correlation which amounts to:

$$r(X_i, X_j) = -\frac{1}{n-1} \quad \text{for all } i, j \text{ with } i \neq j$$

The cause of this negative correlation, put somewhat loosely, is that after the ordering of $n - 1$ values there is no free choice for the n-th value. For further details, the reader is referred to van der Waerden (18). As a result, while the observed values of a quantity can be considered to be statistically independent if they arise from different individuals, there appears a slight negative correlation between the corresponding X-values. This correlation, however, is very small and disappears as the sample size n becomes large. Moreover, the correlation would only become apparent if a sample of size n were to be drawn repeatedly, and the correlation determined after the X-transformation had been applied to the data: although theoretically existent in the X-values, the negative correlation has no practical consequence.

2.1 The multivariate application of the X-transformation

The usefulness of the X-transformation in connection with multivariate data sets arises from the fact that the transformed data have many properties similar to those of data following a multivariate normal distribution, as described above. This leads to the assumption that an application of computational procedures tailored to multivariate normal distributions also leads to reasonable and valid results if applied to X-transformed values, even if the underlying distributions of the data are not normal. If this were the case, computations would be dramatically simplified, as data processing and assessment is easy with normally distributed data. This, however, has to be taken with a grain of salt: The prerequisite of the data analysis is a component by component application of the X-transformation, i. e. the individual marginal quantities are submitted to the X-transformation. The resulting data set of X-vectors does *not* show a multivariate normal distribution, not even when the original data are normally distributed. This is so because the transformed values lie on isolated points of a higher-dimensional grid which is defined only by the finite range of the X-transformation. Nevertheless, the transformed data set has the property that all of its marginal distributions can be approximated by the standardized normal distribution. We will say that the data are *quasi-normalized* if submitted to the X-transformation in the above described sense. If from the observational units k-dimensional data vectors are measured, which are denoted by $(x_1^{(i)}, \ldots, x_k^{(i)})$, then the X-transformed data vectors will be denoted by $(X_1^{(i)}, \ldots, X_k^{(i)})$. Here $X_j^{(i)}$ stands for: "the X-value of the j-th quantity in the i-th person".

Careful consideration should precede the application of the X-transformation to the data. The transformation seems to be advantageous and profitable if the practitioner is faced with one of the following two principle problems:

1. k-dimensional data vectors are measured in persons belonging to a homogeneous collective. Correlations between the various measured quantities are to be computed and meaningfully interpreted.
2. k-dimensional data vectors are measured in members of a collective of reference persons and the same data are gathered from several persons who have to be examined for deviations from the reference collective. Furthermore, there may be some continuous regressor variables (such as the age of a person).

There are well worked-out statistical methods for the practical solution of both problems. Most of them are based upon the assumption that the data obey a multivariate normal distribution. One can proceed as follows:

1. The first problem requires as a first step the computation of the matrix C of the (Pearson) correlation coefficients r_{ij} from the data, which are asymptotically unbiased estimates of the true correlation coefficients. The next step consists in

computing the inverse C^{-1} of the correlation matrix C. Then *partial correlation coefficients* can be determined using the formula:

$$r_{ij\cdot k} = \frac{-r^{ij}}{\sqrt{r^{ii}\, r^{jj}}}$$

Here the symbols r^{ij} denote the elements of the inverse C^{-1} of the correlation matrix. The partial correlation coefficient $r_{ij\cdot k}$ provides information as to what extent the i-th and the j-th quantity correlate if all the other variables are held constant (see, for example, ref. (21)). The partial correlation coefficients give a first insight into the complicated pattern of mutual dependences of the different clinical chemical quantities. The correlation structure can be made even more transparent if the correlation matrix is submitted to a model search procedure known as "covariance selection", which was devised by Dempster (22). This procedure consists in testing and accepting certain *hypotheses of independence* which are derived from an inspection of the partial correlation coefficients. In subsequent steps, selected partial correlation coefficients are tentatively set to zero and the admissibility of this operation is tested statistically. In this way a mathematical model of the correlations with an increasing simplicity takes shape. This is continued as long as the mathematical model remains in agreement with the data (the best model being the original matrix itself). The result of this rather abstract analysis is a model of the correlation matrix which, on the one hand, is able to describe the essential correlations between the clinical chemical quantities and, on the other hand, is simple enough to permit an easy interpretation of the correlation matrix. Computer programs to perform the analysis were written by several authors, see ref. (23–25).

An essential precondition for the theoretical applicability of the above sketched procedure is the multivariate normality of the distribution underlying the data structure. If the data are not normally distributed the covariance selection procedure can still be formally executed, but it is unclear how far the results can be trusted. Skew distributions tend to deflate the computed correlation coefficients and thus lead to an underestimation of the internal coherence of the data. Another obstacle is the possible presence of *outlying observations* in the data, i. e. of extreme values which do not belong to the underlying distribution (caused, for example, by errors in the measurements or by dropouts during a data transfer). Such outlying observations bear heavily on the Pearson correlation coefficients, which are not robust. If, prior to the computation of the correlation coefficients, the data are submitted to the X-transformation then the above-mentioned difficulties become practically nonexistent. The reason is that by this means one gets estimated correlation coefficients close to those one would expect if the data could be submitted to a normalizing *continuous* transformation. The correlation coefficients computed with the aid of the X-transformation will, in the rest of this paper, be called "X-rank-correlation coefficients" or simply "X-correlations". They are almost always superior to the usually computed ordinary correlation coefficients. If the data are already normally distributed it is of course unnecessary to compute X-correlations.

2. If the data vector of a person under investigation has to be compared with the data of a collective of reference persons then this is usually done by utilizing

multivariate test quantities. These test quantities, which are usually computed from the original data, can help in the decision whether the person in question deviates significantly from the reference persons. All the measurements are integrated into the computation of the test quantity. When the test quantity exceeds a certain limit the data are considered not to be in accordance with the reference persons. The theoretical background for the compilation of the decision rules is again the assumption that the data of the observational units are random samples from a multivariate normal distribution. In a first step mean values and standard deviations are computed for all clinical chemical quantities for the persons of the reference collective. Moreover, it is necessary to compute the correlation coefficients (or the covariances) between the quantities. Then, from the data set of the investigated person a test value is derived which was given first by Hotelling (26) and which is known as "Hotellings's T^2". If the data are really normally distributed then Hotelling's T^2 is the most powerful test quantity which can be constructed.

As before the assumption of a multivariate normality of the theoretical data structure means a severe restriction. The deviations from normal distributions which are always present in real data can impair the sensitivity of the test quantity and can occasionaly lead to false positive decisions. The skewness of the distributions in the reference data can, for example, result in not rejecting data sets with negative components, which can never be measured values of real concentrations. On the other hand, it is also possible that data vectors will be rejected because they contain large values, although if the skewness of the distributions were taken into account these large values might be consistent with the reference data. All these complications can be illustrated by one-dimensional examples. Whenever difficulties of this kind are to be expected it is advisable to look for transformations which turn at least the marginal distributions into approximate normal distributions. Even here the X-transformation proves useful and convenient if applied to the marginal distributions of the reference data. A problem here is that the data of the investigated person are not also transformed. It will be shown below how this difficulty can be overcome. Another problem which is more serious theoretically is that in applying the X-transformation one leaves the domain of the continuous quantities and has to deal with quantities defined on a discretized scale with a finite range. There is no theory in closed form about how Hotelling's T^2 behaves if it is computed from X-values. Therefore, empirical estimates of the density functions of the test quantity T^2 have to show how the application of the X-transformation stands up in practice.

2.2 The effect of the X-transformation

In the following the results of the application of the X-transformation to clinical chemical data will be described. The analyses utilize concentration measurements of blood serum constituents in healthy and diseased persons. The measurements were performed on the Technicon autoanalyzers SMA 12/60 (sodium, potassium, chloride, total protein, albumin, inorganic phosphorus, cholesterol, urea nitrogen, calcium, creatinine, bilirubin, and uric acid) and SMA 6 plus (iron, copper, magnesium, alkaline phosphatase, acid phosphatase, and glucose). Further, the meas-

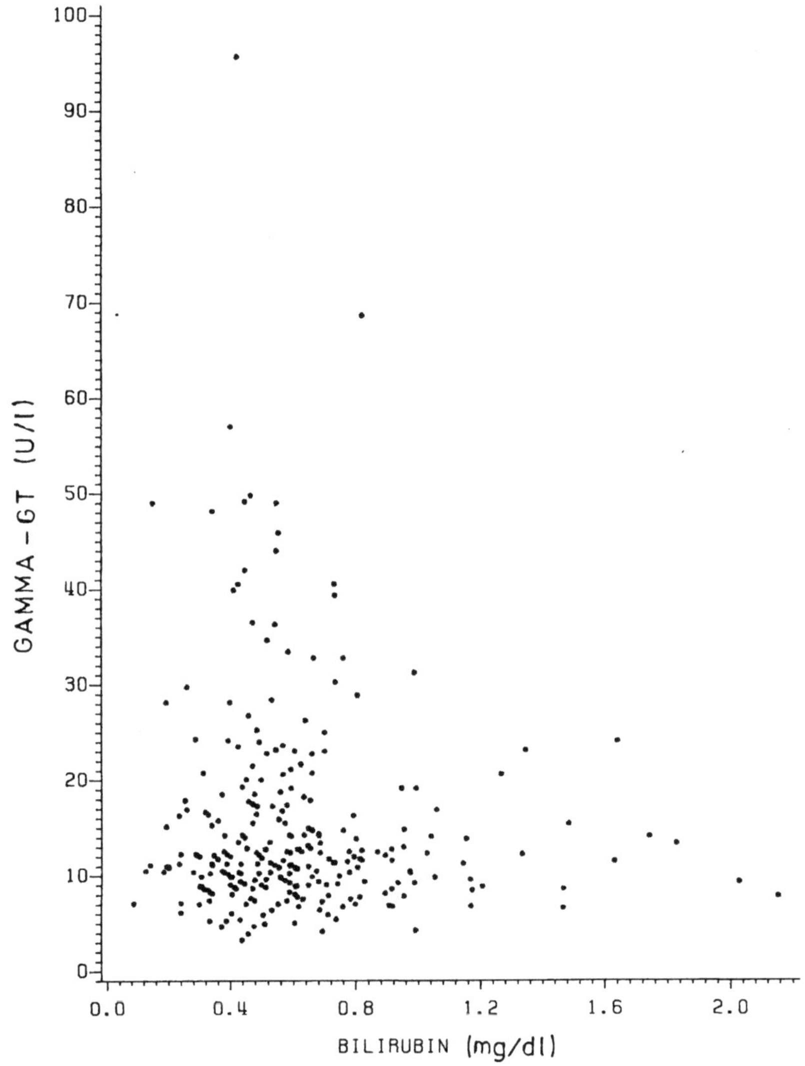

Fig. 2. The statistical correlation between bilirubin and the enzyme γ-glutamyltranspeptidase in 277 male reference persons. The distributions of both quantities are considerably skew. The measurements were corrected for age. The Pearson correlation coefficient is $r = 0.03$.

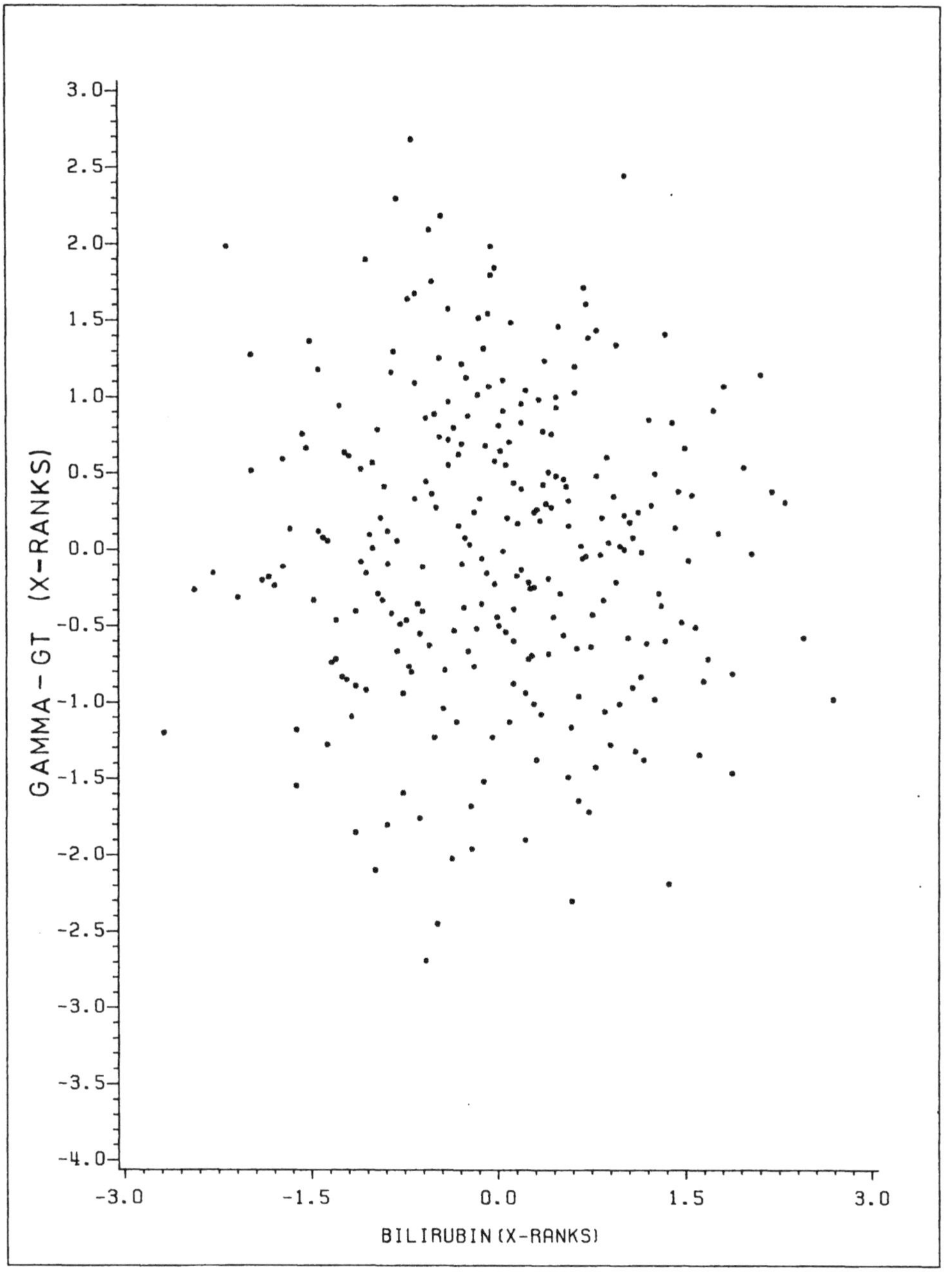

Fig. 3. The statistical correlation between bilirubin and γ-glutamyltranspeptidase after the X-transformation was applied to the measurements. The positions of the transformed data points seem to be in good agreement with the assumption of a bivariate normal distribution. The X-rank-correlation coefficient is $r_X = 0.14$.

urements of the enzymes aspartaminotransferase, alaninaminotransferase and γ-glutamyltranspeptidase were available. The analyses were done in the central laboratory of the Medizinische Klinik of the Universitäts-Krankenhaus Hamburg-Eppendorf. The determination of the enzymes was done according to the recommendations of the Deutsche Gesellschaft für Klinische Chemie with the Netheler & Hinz enzyme analyzer (27). Further details and results are reported elsewhere (28). In all the data of two collectives of reference people (336 males and 519 females), two collectives of patients with primary hyperparathyreoidism (34 males and 67 females,) and a collective of 33 males with cirrhosis of the liver were utilized. Since in the reference collectives some values were missing, the numbers are correspondingly smaller where several values were needed simultaneously. The reference collectives consisted of employed persons who in a routine check-up were classified as healthy and able to work.

The Figures 2 and 3 present the dependence between bilirubin and γ-glutamyltranspeptidase (γ-GT) in 277 males (reference persons) before and after the X-transformation. In Figure 2 it can be seen that the distribution of the two quantities possesses a significant positive skewness. At lower values the points are crowded together, whereas at greater values the concentration of points is much smaller. Since both quantities are age-related the values were corrected using a linear age regression equation. There is no relevant correlation between both quantities. We have $r = 0.03$ and $r_X = 0.14$, with r_X denoting the X-rank-correlation coefficient.

In Figure 3 the skewness of both distributions has completely disappeared after the X-transformation has been applied. The visual impression is that the normalizing procedure has been quite successful. This impression is confirmed by an inspection of graphical representations of other pairs of variables. These pairwise combinations of variables represent two-dimensional marginal distributions of the k-dimensional common distribution. As far as an inference can be made from these marginal distributions, the question of whether the X-transformation is successful in normalizing the data can be answered in the affirmative.

After the application of the X-transformation the data points are placed on a finite grid of isolated positions. This is hardly perceptible in the figures and causes no harm whatever. Another property of the X-transformation is that eventual nonlinear relationships between different quantities are linearized, as far as this is possible. A heteroscedasticity, however, if present in the data, cannot be transformed away, so if a quantity has a dispersion that varies with a second quantity this will probably remain so after the transformation. The X-values are not influenced if a monotonic transformation *precedes* the X-transformation. In this sense the X-transformation is invariant with regard to transformations, or *transformation-free*. This just means that the method is distribution-free in the marginal distributions. For the same reason the X-values have no physical dimension but are pure numbers.

2.3 Structural investigations with the aid of the X-transformation

"Structure" in this context refers to the complicated network of internal relations between the different quantities, i. e. means "correlation structure". This is given initially by the matrix $C = (r_{ij})$ of the correlation coefficients between the different quantities. Little can be learned, however, from an inspection of the correlation matrix C only, since it is not apparent in this matrix which quantities influence which other quantities and what the order of the importance in the correlations might be. A better interpretation is possible if the partial correlations $r_{ij.k}$ are computed and the matrix is submitted to the simplifying covariance selection procedure. As indicated above, this procedure consists in setting appropriate partial correlation coefficients to zero, which is unavoidably associated with an increase in the entropy inherent in the data. The procedure used selects the partial correlation coefficients to be set to zero in such a way that the increase in entropy is as small as possible. For further details, see Dempster (22) and Rehpenning (25). The theory of the covariance selection procedure presupposes a multivariate normality in the data. A normalizing transformation of the data is therefore recommended in most practical situations. The X-transformation offers an easy alternative to the search for normalizing transformations. To compute simple Spearman rank-correlation coefficients would not meet the requirements. If X-correlations are used, however, the result is practically the same as if the quantities had been submitted to normalizing transformations.

Figures 4 and 5 present the result of a covariance selection procedure applied to the X-correlations computed from the SMA 12/60 data for reference males and females. In addition to the clinical chemical quantities, the age of the persons was included in the analysis, since many quantities are age-dependent. The figures

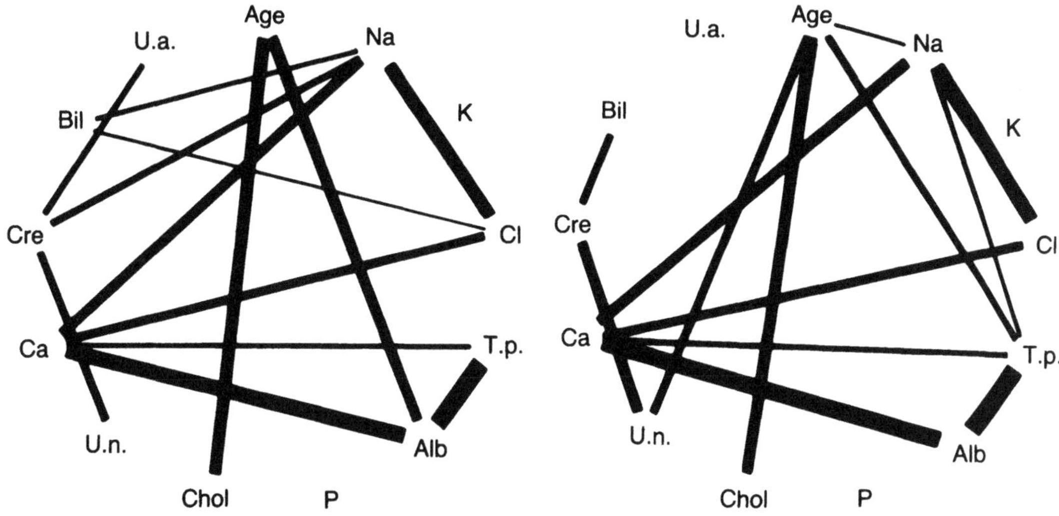

Fig. 4. The internal correlation structure of age and the measurements made with the SMA 12/60 in 252 male reference persons as computed with the covariance selection procedure. Thick connecting solid lines indicate strong correlations. Sodium, albumin, and calcium are correlation centers.

Fig. 5. The internal correlation structure of age and the SMA 12/60 measurements in 436 female reference persons. Age, sodium, total protein, and calcium figure as correlation centers. For details, see text.

contain, in a condensed and easily understandable form an insight into the essential correlations between the variables as derived from the correlation matrix. The lines between the symbols representing the different quantities indicate which of the original partial correlation coefficients survive the analysis, i. e. which partial correlation coefficients are so significantly different from zero that their annihilation would lead to a model inconsistent with the data. The thickness of the lines shows the order of importance of the correlations. In Tables 2 and 3 the computed X-correlations are given for both reference collectives. Here the upper triangles of the tables contain the X-correlations, the lower triangles the additionally computed partial correlation coefficients. On the main diagonal one finds the main diagonal elements of the inverse C^{-1} of the correlation matrix. These numbers are quite useful, since they show how strongly the corresponding quantity correlates with *all the rest* of the other quantities. To see this, let the multiple correlation coefficient of the i-th quantity with respect to the other quantities be $r_{i(k)}$. Then we have:

$$r_{i(k)}^2 = 1 - \frac{1}{r^{ii}}$$

Here r^{ii} denotes the i-th main diagonal element of the inverse correlation matrix C^{-1}. In the correlation matrix of the male reference persons, for example, of all the factors albumin has the greatest value along the main diagonal of the inverse matrix, C^{-1}: $r^{66} = 2.37$. This means that of all quantities albumin is the one which could best be predicted from the rest. If this were done by a linear multiple regression equation then the correlation between the observed and the predicted values would be:

$$r_{6(k)} = \sqrt{1 - \frac{1}{2.37}} = 0.76$$

Tables 4 and 5 contain the corresponding information for the model matrices which were obtained by the covariance selection procedure. They are characterized by a certain pattern of zeros in the partial correlations. It may be noted, however, that the displayed correlation models are already oversimplified, i. e. that they are too simple to be consistent with the data. This was done purposely for the sake of simplicity and for easy comparison. (Note that the statistical significances also depend upon the sample sizes, which are unequal in the different collectives).

A first glance reveals striking common features in the correlation patterns of males and females. The triplet sodium – chloride – calcium constitutes an undecomposable correlating system, as does, essentially, the triplet total protein – albumin – calcium. It is interesting to note that in the latter albumin plays to some extent the role of a *correlation center*. This means that part of the correlation between total protein and calcium is controlled by albumin (a variation in the albumin values induces a parallel running variation in the total protein and calcium values). A particular correlation center is formed by calcium in both collectives of reference persons: the pair total protein – albumin is statistically linked to the pair sodium – chloride via the calcium values. Both collectives show an increase in the cholesterol values with age. The decrease in albumin with age, however, observed in the male reference persons, is not present in the female collective. Since

Table 2. *X*-rank-correlations in reference persons (252 males)

	age	sod.	pot.	chl.	t.p.	alb.	i.ph.	chol.	u.n.	cal.	cre.	bil.	u.ac.
age	*1.66*	−0.10	0.07	0.09	−0.21	−0.39	−0.11	0.44	0.15	−0.20	−0.15	−0.21	−0.02
sodium	−0.01	*1.74*	−0.07	0.51	0.12	0.23	0.11	−0.01	0.16	0.19	0.28	0.13	0.12
potassium	0.02	−0.18	*1.14*	0.03	0.09	0.05	0.02	0.17	0.01	0.16	0.11	−0.09	0.15
chloride	0.06	0.57	0.15	*1.66*	−0.17	−0.06	0.02	−0.03	0.16	−0.17	0.09	−0.13	−0.01
total protein	−0.01	0.07	0.05	−0.14	*1.88*	0.62	−0.14	0.09	−0.03	0.53	0.03	0.21	0.17
albumin	−0.32	0.05	−0.05	0.08	0.43	*2.37*	−0.01	0.01	0.02	0.62	0.15	0.24	0.10
inorg. phos.	−0.13	0.11	0.05	−0.10	−0.15	0.00	*1.14*	−0.06	0.21	−0.01	0.16	−0.06	−0.03
cholesterol	0.45	0.02	0.13	−0.08	0.05	0.11	−0.01	*1.40*	0.17	0.07	−0.00	−0.04	0.17
urea nitrogen	0.16	−0.01	−0.06	0.12	−0.00	0.04	0.20	0.11	*1.24*	0.02	0.31	0.04	0.08
calcium	0.06	0.17	0.17	−0.21	0.18	0.40	0.02	−0.01	0.01	*1.90*	0.14	0.21	0.20
creatinine	−0.11	0.20	0.13	−0.06	−0.10	0.05	0.07	−0.03	0.27	−0.00	*1.35*	0.16	0.26
bilirubin	−0.11	0.13	−0.09	−0.16	0.05	0.03	−0.11	0.02	0.05	0.05	0.12	*1.17*	0.01
uric acid	−0.06	0.06	0.07	−0.03	0.10	−0.09	−0.07	0.15	0.00	0.10	0.23	−0.07	*1.18*

determinant = 0.060685

Table 3. *X*-rank-correlations in reference persons (436 females)

	age	sod.	pot.	chl.	t.p.	alb.	i.ph.	chol.	u.n.	cal.	cre.	bil.	u.ac.
age	*1.39*	0.11	0.05	−0.01	−0.22	−0.14	−0.07	0.40	0.23	−0.15	0.03	0.03	0.10
sodium	0.16	*1.61*	0.04	0.43	0.17	0.27	0.06	0.10	0.17	0.28	0.21	0.18	0.02
potassium	0.05	−0.07	*1.11*	0.04	−0.01	0.08	0.16	0.07	0.13	0.17	0.14	−0.06	0.05
chloride	−0.09	0.51	0.07	*1.51*	−0.14	−0.05	−0.16	−0.09	0.08	−0.11	0.11	−0.08	−0.14
total protein	−0.15	0.06	−0.09	−0.09	*1.83*	0.62	0.17	−0.01	−0.01	0.47	0.04	0.16	0.14
albumin	−0.03	0.08	0.04	0.02	0.48	*1.98*	0.15	0.07	0.07	0.54	0.14	0.20	0.06
inorg. phos.	−0.06	0.12	0.15	−0.21	0.08	0.03	*1.13*	−0.03	0.11	0.14	0.02	−0.03	0.01
cholesterol	0.37	0.05	0.02	−0.13	−0.01	0.07	−0.07	*1.28*	0.19	0.07	0.12	−0.00	0.10
urea nitrogen	0.17	0.04	0.07	0.07	−0.03	0.03	0.13	0.09	*1.19*	0.05	0.26	0.14	0.09
calcium	−0.14	0.24	0.16	−0.18	0.15	0.31	−0.01	0.04	−0.01	*1.68*	0.17	0.11	0.17
creatinine	−0.07	0.06	0.10	0.09	−0.07	0.04	−0.00	0.09	0.18	0.07	*1.22*	0.22	0.17
bilirubin	0.03	0.16	−0.08	−0.18	0.05	0.11	−0.10	−0.09	0.10	−0.06	0.19	*1.17*	0.06
uric acid	0.11	−0.01	0.01	−0.11	0.12	−0.09	−0.04	0.01	0.03	0.12	0.15	−0.01	*1.11*

determinant = 0.104940

Table 4. Reduced correlation model in reference persons (252 males)

	age	sod.	pot.	chl.	t.p.	alb.	i.ph.	chol.	u.n.	cal.	cre.	bil.	u.ac.
age	*1.41*	−0.05	0.00	0.04	−0.24	−0.39	0.0	0.43	−0.00	−0.24	−0.01	−0.02	−0.00
sodium	.	*1.67*	0.00	0.51	0.10	0.12	0.0	−0.02	0.09	0.19	0.28	0.13	0.08
potassium	.	.	*1.00*	0.0	0.0	0.0	0.0	0.0	0.0	0.0	0.0	0.0	0.0
chloride	.	0.56	.	*1.58*	−0.09	−0.11	0.0	0.02	0.04	−0.17	0.14	−0.13	0.04
total protein	.	.	.	.	*1.72*	0.62	0.0	−0.11	0.01	0.62	0.03	0.05	0.01
albumin	−0.26	.	.	.	0.42	*2.19*	0.0	−0.17	0.01	0.62	0.03	0.06	0.01
inorg. phos.	.	.	.	.	.	.	*1.00*	0.0	0.0	0.0	0.0	0.0	0.0
cholesterol	0.41	.	.	.	.	.	.	*1.23*	−0.00	−0.11	−0.01	−0.01	−0.00
urea nitrogen	.	.	.	.	.	.	.	.	*1.11*	0.02	0.31	0.01	0.08
calcium	.	0.24	.	−0.24	0.23	0.39	.	.	.	*1.87*	0.05	0.10	0.01
creatinine	.	0.22	.	.	.	.	.	.	0.29	.	*1.26*	0.04	0.26
bilirubin	.	0.21	.	−0.22	.	.	.	.	.	.	.	*1.07*	0.01
uric acid	.	.	.	.	.	.	.	.	.	.	0.24	.	*1.07*

determinant = 0.113139

Table 5. Reduced correlation model in reference persons (436 females)

	age	sod.	pot.	chl.	t.p.	alb.	i.ph.	chol.	u.n.	cal.	cre.	bil.	u.ac.
age	*1.32*	0.11	0.00	0.07	−0.22	−0.13	0.0	0.40	0.23	−0.07	0.06	0.01	0.0
sodium	0.12	*1.44*	0.00	0.43	0.17	0.17	0.0	0.04	0.02	0.28	0.01	0.00	0.0
potassium	.	.	*1.00*	0.0	0.0	0.0	0.0	0.0	0.0	0.0	0.0	0.0	0.0
chloride	.	0.47	.	*1.32*	−0.03	−0.05	0.0	0.03	0.02	−0.11	0.00	0.00	0.0
total protein	−0.17	0.06	.	.	*1.76*	0.62	0.0	−0.09	−0.05	0.47	−0.01	−0.00	0.0
albumin	.	.	.	.	0.49	*1.86*	0.0	−0.05	−0.03	0.54	−0.01	−0.00	0.0
inorg. phos.	.	.	.	.	.	.	*1.00*	0.0	0.0	0.0	0.0	0.0	0.0
cholesterol	0.38	.	.	.	.	.	.	*1.19*	0.09	−0.03	0.02	0.01	0.0
urea nitrogen	0.20	.	.	.	.	.	.	.	*1.13*	−0.02	0.26	0.06	0.0
calcium	.	0.29	.	−0.22	0.17	0.35	.	.	.	*1.63*	−0.00	−0.00	0.0
creatinine	.	.	.	.	.	.	.	.	0.25	.	*1.12*	0.22	0.0
bilirubin	.	.	.	.	.	.	.	.	.	.	0.22	*1.05*	0.0
uric acid	.	.	.	.	.	.	.	.	.	.	.	.	*1.00*

determinant = 0.190679

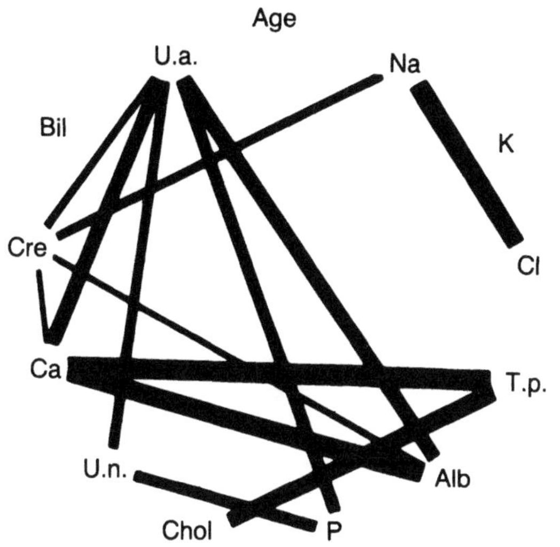

Fig. 6. The internal correlation structure of age and the SMA 12/60 measurements in 33 patients with cirrhosis of the liver. Calcium, creatinine, and uric acid are strong correlation centers. The correlation: total protein – albumin no longer appears.

in the collective of male reference persons in the correlation model there is no partial correlation between age and total protein, the conclusion can be made that the observed negative correlation between age and total protein in the original correlation matrix can be attributed to the decrease of albumin with age. Albumin is part of the total protein. So it is clear that the correlation between albumin and total protein cannot be controlled by other quantities. Hence, it is not surprising that the lines connecting the symbols of albumin and total protein represent the strongest of all partial correlations surviving the analysis in both collectives of reference persons, which is a very satisfying result.

To what extent the correlative structure between clinical chemical quantities can be altered by a disease can be seen in Figure 6, which presents the corresponding graphical structure for the collective of male patients with cirrhosis of the liver. There is no age-dependence of the chemical quantities discernible in the graph. The age dependencies seem to have given way to stronger correlations between the other variables. Uric acid and creatinine have turned into strong correlation centers, whereas the correlation between total protein and albumin – which is particularly strong in the collectives of reference persons – seems to have disappeared completely. A possible explanation for this may be that total protein in these patients is composed to a higher degree of other constituents, such as (presumably) gammaglobulins which, however, also have to be associated with calcium. The *entropy* inherent in the data is much less in the patients with cirrhosis of the liver than in the collectives of reference persons. This can be seen by the much smaller determinant of the correlation matrix (22) – compare Tables 4, 5, and 6. An interpretation of this is that the *overall correlation* is greater in the patients than in the healthy persons – and in patients with cirrhosis of the liver this is particularly so, as can be learned from a comparison with collectives of patients with other diseases (25). The Tables 6 and 7 contain the original and the model correlation coefficients.

It is well known that, in diseased persons, the distributions of clinical quantities tend to skewness moreso than in reference collectives of healthy persons. Therefore

Table 6. *X*-rank-correlations in patients with cirrhosis of the liver (33 males)

	age	sod.	pot.	chl.	t.p.	alb.	i.ph.	chol.	u.n.	cal.	cre.	bil.	u.ac.
age	*1.66*	−0.03	−0.14	0.14	0.08	0.15	−0.18	0.01	0.25	0.11	0.13	−0.40	−0.13
sodium	−0.19	*8.11*	−0.31	0.79	0.12	0.16	0.06	0.23	−0.36	0.11	−0.44	−0.19	−0.15
potassium	−0.23	−0.44	*2.00*	−0.14	−0.06	0.05	0.34	0.08	0.23	0.10	0.13	−0.08	0.22
chloride	0.18	0.89	0.32	*8.38*	0.07	0.19	0.08	−0.06	−0.21	−0.02	−0.28	−0.33	−0.36
total protein	0.05	−0.41	−0.28	0.48	*4.69*	0.58	0.20	0.62	−0.35	0.80	0.01	−0.03	0.07
albumin	−0.15	−0.07	0.17	0.22	−0.18	*6.72*	−0.08	0.49	−0.25	0.79	0.12	−0.33	−0.22
inorg. phos.	−0.15	−0.24	0.31	0.34	−0.03	−0.42	*2.84*	0.25	−0.30	0.20	0.01	−0.24	0.42
cholesterol	0.16	0.49	0.29	−0.47	0.28	0.05	0.09	*3.14*	−0.39	0.71	−0.10	−0.15	0.17
urea nitrogen	0.25	−0.11	0.35	0.16	−0.17	−0.27	−0.52	−0.15	*2.80*	−0.27	0.47	0.06	0.25
calcium	0.09	0.23	0.06	−0.36	0.63	0.72	0.24	0.09	0.19	*10.32*	0.06	−0.28	0.14
creatinine	0.05	−0.38	−0.30	0.21	−0.03	0.47	0.14	0.07	0.32	−0.22	*2.57*	0.03	0.41
bilirubin	−0.37	0.22	0.13	−0.30	0.44	−0.15	−0.24	−0.10	0.00	−0.21	0.17	*2.08*	0.06
uric acid	−0.09	0.46	0.15	−0.44	0.11	−0.35	0.34	−0.08	0.25	0.19	0.51	−0.12	*3.34*

determinant = 0.000208

Table 7. Reduced correlation model in patients with cirrhosis of the liver (33 males)

	age	sod.	pot.	chl.	t.p.	alb.	i.ph.	chol.	u.n.	cal.	cre.	bil.	u.ac.
age	*1.00*	0.0	0.0	0.0	0.0	0.0	0.0	0.0	0.0	0.0	0.0	0.0	0.0
sodium	.	*2.92*	0.00	0.79	−0.02	−0.05	−0.07	−0.01	−0.05	−0.02	−0.44	0.0	−0.18
potassium	.	.	*1.00*	0.0	0.0	0.0	0.0	0.0	0.0	0.0	0.0	0.0	0.0
chloride	.	0.75	.	*2.66*	−0.02	−0.04	−0.06	−0.01	−0.04	−0.02	−0.34	0.0	−0.14
total protein	.	.	.	.	*3.39*	0.63	0.05	0.62	0.03	0.80	0.05	0.0	0.11
albumin	.	.	.	.	.	*4.71*	−0.09	0.39	−0.06	0.79	0.12	0.0	−0.22
inorg. phos.	.	.	.	.	.	.	*1.54*	0.03	−0.30	0.06	0.17	0.0	0.42
cholesterol	.	.	.	.	0.43	.	.	*1.62*	0.02	0.50	0.03	0.0	0.07
urea nitrogen	.	.	.	.	.	.	−0.47	.	*1.35*	0.03	0.10	0.0	0.25
calcium	.	.	.	.	0.48	0.74	.	.	.	*6.16*	0.06	0.0	0.14
creatinine	.	−0.24	.	.	.	0.41	.	.	.	−0.30	*1.73*	0.0	0.41
bilirubin	.	.	.	.	.	.	.	.	.	.	.	*1.00*	0.0
uric acid	.	.	.	.	.	−0.59	0.41	.	0.33	0.47	0.47	.	*2.64*

determinant = 0.007288

a normalizing transformation is especially helpful in interpreting the correlation matrix of collectives of persons with a particular disease. The X-transformation can be considered as a standard tool for handling data whenever correlation matrices have to be compared which come from collectives with differing distributions of the clinical quantities. This is particularly true for a comparison of groups of diseased persons with collectives of healthy persons.

2.4 The role of the X-transformation in multivariate data assessment for diagnostic purposes

In investigating the data vector of a person to be checked – for example in a routine examination – the task is to compare the data set with the data of a reference collective. This is usually done with Hotelling's T^2 test quantity. For the computation of T^2 the mean values of all quantities for the persons in the reference collective have to be computed. (Let the vector of these mean values be denoted by $\bar{\boldsymbol{x}}$.) In a next step the matrix of the variances and covariances must be computed. (This will be denoted by V). If the vector of the measured values of the person in question is denoted by $\boldsymbol{x}' = (x_1, \ldots, x_k)'$ (the apostrophe denotes transposition) then Hotelling's T^2 is given by:

$$T^2 = (\boldsymbol{x} - \bar{\boldsymbol{x}})' V^{-1} (\boldsymbol{x} - \bar{\boldsymbol{x}})$$

If the data in the collective of the reference persons follow a multivariate normal distribution, then the test quantity T^2 can be judged with the aid of the F-distribution. For this, the auxiliary quantity F is computed according to:

$$F = \frac{n - k}{(n - 1)k} T^2,$$

which follows an F-distribution with k and $n - k$ degrees of freedom. Here n denotes the sample size of the reference collective and k the number of measured quantities. If the quantity F exceeds a limit determined by the quantiles of the F-distribution, then the data vector $\boldsymbol{x}$ is considered to be pathological.

If the data in the reference collective do not show a multivariate normal distribution then the above considerations are only approximately valid. The probability of erroneous decisions increases as the departure of the underlying distribution from a multivariate normal distribution increases. It makes sense, then, to submit the data of the reference persons component by component to the X-transformation in order to arrive at an approximation to the multivariate normal distribution in the reference collective. Since the data of the person to be compared with the reference collective are not transformed at the same time it is necessary to attribute appropriate X-values to the components of the data set of that person. This is most easily done in the following way: one compares the measured value of a quantity with those of the reference persons and takes the reference value that differs the least absolutely from the value in question ("nearest neighbor"). One assigns the X-value of the nearest neighbor to the quantity of the person in question. One deals accordingly with the rest of the clinical chemical quantities. We can say then that the person under investigation has been "fitted" by comparison into the grid of the X-values.

Note: Theoretically it would be preferable to integrate the data of the person in question into the collective of the reference persons and to apply the X-transformation to the extended collective. This would be very cumbersome, however, as all mean values, variances, covariances, etc., would have to be recalculated for each new person to be assessed. Moreover, it can be estimated that the X-values calculated in this manner would differ at most by $1/(n + 2)$ from the X-values determined by the method described above. Since this difference is negligible it is more convenient to use the simpler method.

If the values of the person to be investigated are so extreme that they lie outside the range of the reference values then the corresponding X-values would be on the border of the grid defined by the X-values. As a result, the information how far away from the lowest or greatest value of the reference persons the value in question lies, goes lost. This loss of information is characteristic of all distribution-free methods.

The effect of the X-transformation on the values of the patients is shown in Figure 7, for albumin and calcium. On the left side of the figure (part a) the male patients with cirrhosis of the liver were chosen as reference persons and patients with primary hyperparathyreoidism were compared with them. On the right side of the figure (part b), however, the reference collective is formed by the patients with primary hyperparathyreoidism and patients with cirrhosis of the liver were compared with them. That is, the two collectives have changed places from part a

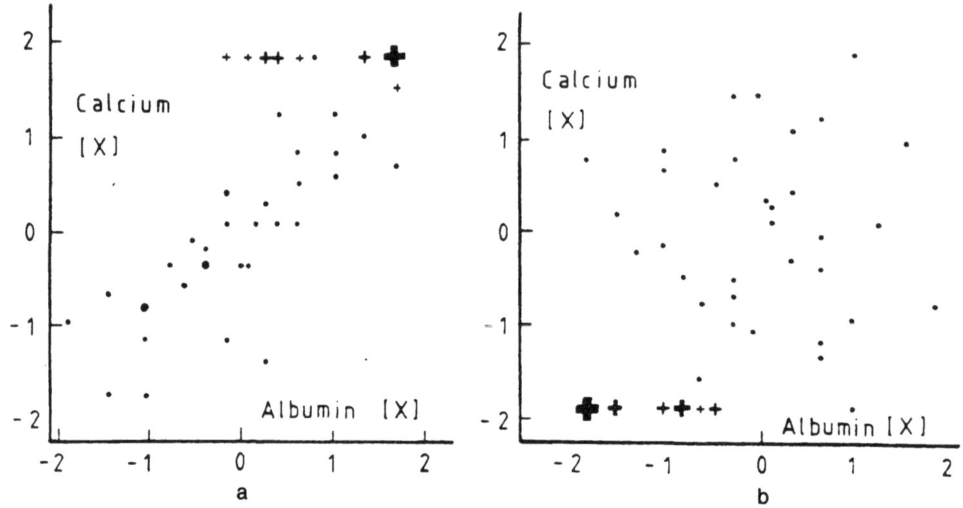

Fig. 7. The effect of the X-transformation on patients' data. Small dots indicate the positions of points representing persons chosen as reference collective, small crosses those of patients to be compared with the reference collective chosen. The abscissa contains the transformed albumin values, ordinate values are the transformed calcium measurements. On the left, patients with cirrhosis of the liver form the reference collective, on the right, reference persons are the patients with primary hyperparathyreoidism. For details, see text.

to part b. Both times the reference persons were indicated by dots. Larger dots imply that several persons are at the same position (caused by ties in the data). The positions of the persons being compared with the reference collective are indicated by crosses, with the size of the cross indicating the number of persons at that position. As a result of the great differences in calcium values between both groups, the points of the persons being compared with the reference collective lie on the border of the X-grid. It becomes apparent, in fact, that the points of the two reference collectives can be described very well by two-dimensional normal distributions. While in the patients with cirrhosis of the liver there is a very strong correlation between the albumin and the calcium values present, this correlation is nonexistent in the patients with primary hyperparathyreoidism. If other combinations of two-dimensional marginal quantities are selected, the impression is confirmed that the application of the X-transformation leads to fairly good normalizations of the two-dimensional marginal distributions. No heteroscedasticity appeared in the transformed data. Moreover, possible relationships between different variables always seemed to be linear after the transformation had been applied.

2.5 The correction for age

Most clinical chemical quantities depend upon the age of the persons. This can be taken into account either by dividing the persons into groups that are homogeneous with respect to age or by correcting the test quantities for age. How the latter is done will be roughly outlined. For further details, see ref. (28). Let the age of the persons be denoted by x_0, and the other clinical chemical quantities by $x_1, \ldots, x_k$. Let the vector of these values be denoted by $\dot{x}$ and the vector of the expected values $\mu_i = E(x_i)$ be denoted by $\dot{\mu}$. Then we have for the matrix $\dot{\Sigma}$ of the variances and covariances between the variables:

$$\dot{\Sigma} = E((\dot{x} - \dot{\mu})(\dot{x} - \dot{\mu})') = (\sigma_{ij})$$

Because of the symmetry of the covariance matrix we have: $\sigma_{ij} = \sigma_{ji}$ for all i, $j = 0, 1, \ldots, k$. The variances and covariances depend on the underlying collective of reference persons. If the influence of age is to be eliminated then the conditional distribution of the quantities must be considered. Assuming that the data can be described by a multivariate normal distribution of dimension $k + 1$, then the conditional distribution with fixed age can also be described by a multivariate normal distribution of dimension k. The covariance matrix of this conditional distribution can be computed directly from the covariance matrix $\dot{\Sigma}$ by dividing this matrix up into submatrices in the following way:

$$\dot{\Sigma} = \begin{pmatrix} \sigma_{00} & \cdots & \sigma_{0k} \\ \vdots & & \vdots \\ \sigma_{0k} & \cdots & \sigma_{kk} \end{pmatrix} = \begin{pmatrix} \sigma_{00} & \beta' \\ \beta & \Sigma \end{pmatrix}$$

Here $\beta' = (\sigma_{01}, \ldots, \sigma_{0k})'$ is the vector of the covariances between the age x_0 and the clinical chemical quantities. Σ is the matrix of the variances and covariances between the quantities disregarding age. The covariance matrix of the conditional distribution with fixed age (which will be denoted by Γ here) can be computed according to (see ref. (8)):

$$\Gamma = \Sigma - \frac{\beta\beta'}{\sigma_{00}}$$

The vector $\boldsymbol{y}$ of the age-dependent expected values follows from the regression equation:

$$\boldsymbol{y} = \mu + \frac{(x_0 - \mu_0)\,\beta}{\sigma_{00}}$$

In the application the unknown expected values must be estimated from the sample. If $\boldsymbol{x}$ denotes the k-dimensional vector of the measured values $x_1, \ldots, x_k$ then the empirical regression equation is:

$$\hat{\boldsymbol{y}} = \bar{\boldsymbol{x}} + \boldsymbol{b}\,(x_0 - \bar{x}_0)$$

The elements b_i of the vector $\boldsymbol{b}$ of the empirical regression coefficients are computed in the usual fashion from the data of the reference collective:

$$b_i = \frac{\sum_{j=1}^{n} (x_0^{(j)} - \bar{x}_0)\, x_i^{(j)}}{\sum_{j=1}^{n} (x_0^{(j)} - \bar{x}_0)^2}$$

It follows that the regression equations can be written as:

$$\hat{\boldsymbol{y}} = \sum_{j=1}^{n} \left(\frac{1}{n} + c_j\right) \cdot \boldsymbol{x}^{(j)}$$

The index j refers to the different persons of the reference collective. The c_j are fixed numbers which fulfil the equation: $\Sigma\, c_j = 0$. From the above equations the covariance structure of the regression vector $\hat{\boldsymbol{y}}$ can be calculated. If this covariance matrix is denoted by VAR ($\hat{\boldsymbol{y}}$) then straightforward intermediate steps yield:

$$\mathrm{VAR}\,(\hat{\boldsymbol{y}}) = \left(\frac{1}{n} + \frac{(x_0 - \bar{x}_0)^2}{(n-1)\, s_{00}}\right) \mathrm{VAR}\,(\boldsymbol{x})$$

Here s_{00} denotes the empirical variance of the values of x_0 and VAR ($\boldsymbol{x}$) the covariance matrix of the vector $\boldsymbol{x}$ with fixed values of x_0, i. e. the covariance about the regression. This covariance matrix is theoretically given by the matrix Γ (see above). The sample estimate may be denoted by G. If x_0 and $\boldsymbol{x}$ are the age and the measured clinical chemical quantities of a person not belonging to the reference collective but with the same underlying distribution of values then it follows that the deviation: $\boldsymbol{x} - \hat{\boldsymbol{y}}$ has the covariance structure:

$$\mathrm{VAR}\,(\boldsymbol{x} - \hat{\boldsymbol{y}}) = B_n^2\, G$$

with:

$$B_n^2 = \frac{n+1}{n} + \frac{(x_o - \bar{x}_0)^2}{(n-1)\, s_{00}}$$

Therefore, the modified text quantity for the examination of the data vector with respect to its deviation from the reference collective in the age-corrected form is:

$$V^2 = \frac{(x - \hat{y})' G^{-1} (x - \hat{y})}{B_n^2}$$

This test quantity, which has been denoted by V^2 to distinguish it from the uncorrected T^2 by Hotelling, has the same distribution as Hotelling's T^2. This means that the F-distribution can be used to assess a person's data vector. If the quantity:

$$F = \frac{n - k}{(n - 1)\,k} V^2$$

exceeds an appropriate quantile of the F-distribution with k and $n - k$ degrees of freedom then the data vector is considered to be significantly different from those of the reference collective.

Note: The theoretical distribution of the test quantity presupposes that the data stem from a multivariate normal distribution. Moreover, it is assumed implicitly that each data testing procedure is associated with a new random choice of the reference collective. Practically, however, once the reference collective is chosen it is often kept for further analyses. Only the persons to be assessed arrive in a random fashion. In this case it is better to compare the test quantity V^2 with a chi-squared distribution with k degrees of freedom. The distribution of V^2 is then a noncentral χ^2-distribution with a noncentrality parameter that depends on how much the sample estimates of the means and the covariances differ from the true values (see also Graybill, ref. (29)). Since the estimates converge to the true values with increasing sample size n, the noncentrality parameter tends to zero. It is reasonable and convenient therefore to pass to the central χ^2-distribution with k degrees of freedom. The χ^2-distribution, by the way, also figures as a limiting distribution if the assessment is made via the F-distribution for large n. The test quantity V^2 will therefore be compared with the χ^2-distribution with k degrees of freedom.

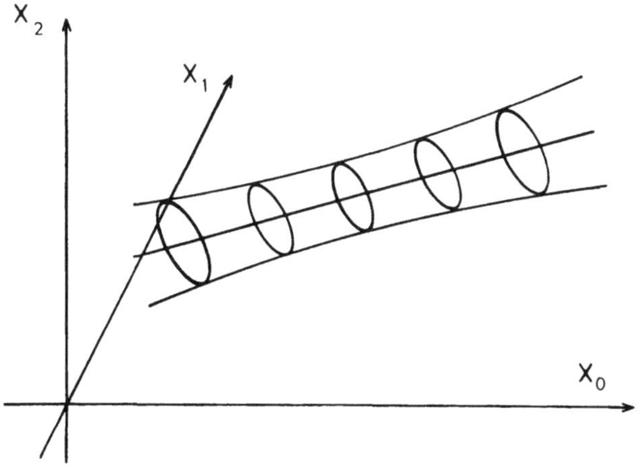

Fig. 8. Hyperboloidal reference region for two clinical-chemical quantities x_1 and x_2 with the regressor variable x_0. The cross-sections parallel to the x_1-x_2-plane are ellipses with parallel axes. The eccentricity of the ellipses is an increasing function of the correlation between x_1 and x_2. The regression line of x_1 and x_2 on x_0 connects the midpoints of the ellipses.

Figure 8 is a graphical presentation of how the constellations of two correlating quantities, x_1 and x_2, which also depend on age, would be positioned about the regression line in the three-dimensional space spanned by the coordinates x_0, x_1, x_2, if the values were submitted to the V^2-test and passed the test as insignificant. These points form a three-dimensional domain bordered by a hyperboloidal surface. This domain can serve as a "normal region" or a "reference region" since it contains (for example) 95% of all possible constellations of values which are consistent with the reference collective. If this hyperboloidal domain is intersected by planes parallel to the $x_1 - x_2$-plane then the sections are ellipses with parallel axes. The eccentricity of the ellipses is determined by the partial correlation between x_1 and x_2. The centers of all the ellipses lie on the regression line expressing the dependence of the x_1 on x_0. The hyperboloidal domain has its smallest diameter at an x_0-value corresponding to $\bar{x}_0$. Conditions are similar in higher dimensions, where a graphical presentation is not possible. The interior of the $k + 1$-dimensional hyperboloid is the set of all points for which the test quantity V^2 does not exceed a certain limit.

In the above considerations a multivariate normality is mandatory for the data. Since this supposition is nearly always violated by real data the application of the X-transformation is likely to improve the analysis. Here one has the choice of applying the transformation only to the measurements or also to the age of the persons. In the following applications both the data and the age of the persons were submitted to the X-transformation. This was not necessary, but it was done with an eye to linearizing possible nonlinear dependencies on age. The test quantity was denoted by V_X^2 to indicate that it was computed from X-values instead of directly from measurements.

Since the X-transformation is a discretizing transformation it follows that the test quantity V_X^2 can also only take on a finite set of values. To calculate the exact distribution of this quantity would seem to require a tremendous amount of computation, which would be hardly feasible (this distribution depends on all the correlation coefficients!). But keeping in mind the construction of the quantity from quasi-normalized values, one can expect the test quantity behave as if it had been computed from normally distributed values.

2.6 The empirical distribution of the test quantities

Figure 9 shows the distribution of the test quantity V^2 as it was computed from the data of two collectives of reference persons. For comparison the density function of the χ^2-distribution with 12 degrees of freedom is also presented. The computations are based upon the SMA 12/60 measurements in the collectives of 252 male and 436 female reference persons. After the computation of the mean values and the covariance matrices the data were read into the computer once again by the program control and were compared with the collective from which they came. All computations were done using a FORTRAN-program on the central computer in Hamburg's university computer center. The estimations of the probability density function of the test quantities were done using normally distributed "kernel functions". This method, if carefully applied, is preferable to the often-used estimation by histograms. In this method of density estimation small gaussian density functions

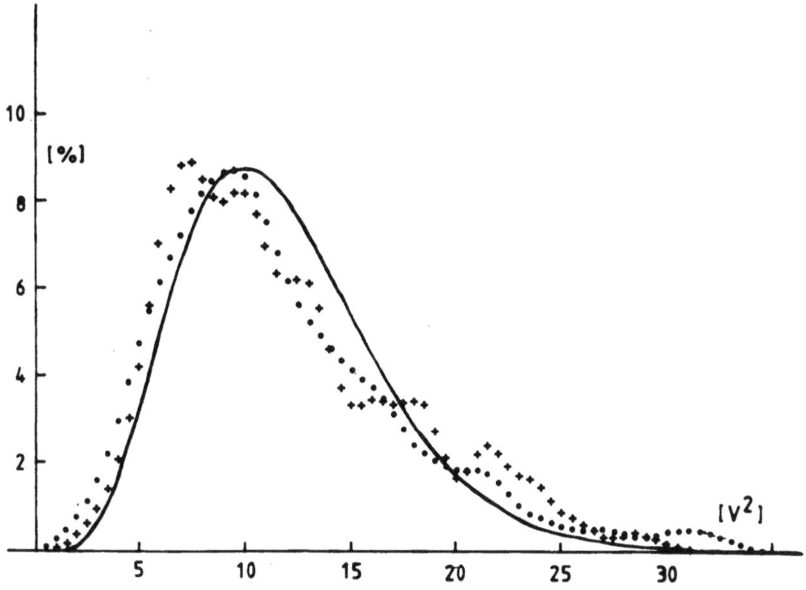

Fig. 9. The empirical density function of the test quantity V^2 when computed from the original data in a comparison with the χ^2-distribution with 12 degrees of freedom. The solid line represents the density function of the χ^2-distribution, the broken lines the density estimates of the test quantity for the reference collectives of 252 healthy males (+ + +) and 436 females (• • •). There are small side maxima.

("kernel functions") are attached to all measured values (here: computed V^2-values). These density functions are centered at the corresponding values and are so scaled that the areas under the curves are $1/n$. Then all these kernel functions are added up to yield the density estimation. Since a critical point is the choice of the widths of the kernel functions (the distances between the points of inflection), an interactive procedure is advisable. After an appropriate choice of the kernel functions the smooth estimations shown in Figure 9 resulted. The theoretical density function of the χ^2-distribution with 12 degrees of freedom is shown as a solid line. It is apparent that the empirical distributions of the test quantity V^2 can be approximated very well by the χ^2-distribution. Only at higher values in both collectives are small side maxima to be found. An explanation for this could be that the assumption of a linear age dependence of the chemical quantities is not fulfilled by the data. There could also be some inhomogeneity in the reference collectives. The latter is the most plausible explanation here because in some of the reference persons a significant increase in the bilirubin values was determined which could not be accounted for.

In Figure 10 it can be seen how prior application of the X-transformation to the data affects the distributions of the test quantities. All other conditions resulting in the curves in Figure 10 were held constant, with only the data of the reference collectives being submitted to the X-transformation. The agreement between the empirical distributions of the test quantity V_X^2 and the χ^2-distribution is now excellent

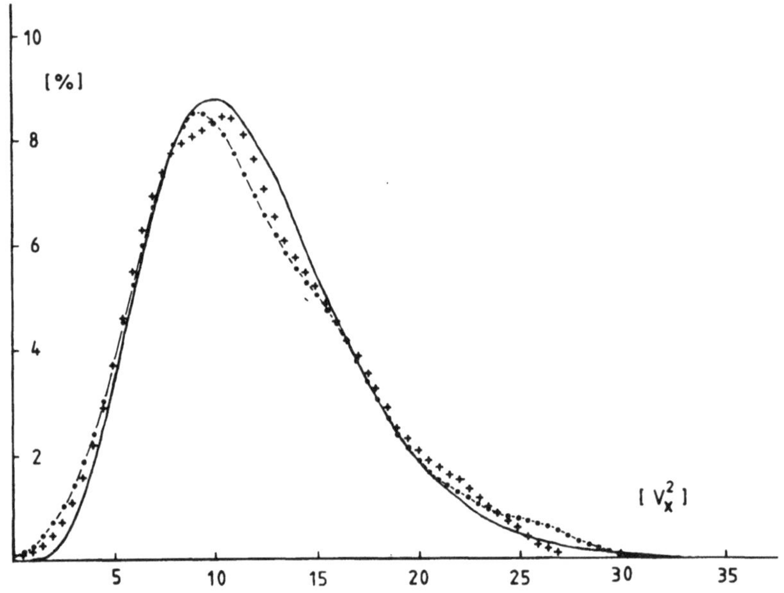

Fig. 10. The empirical distribution of the test quantity V_X^2 computed from the transformed data for the reference collectives (+ + + 252 males, • — • — 436 females) and the theoretical χ^2-distribution with 12 degrees of freedom (solid line). The agreement between the theoretical and the empirical distributions is now excellent.

in both reference collectives. This is strong evidence for concluding that the X-transformation leads to an approximate normalization of the data. Furthermore, it is confirmed that the χ^2-distribution can be used to assess the significance of the test quantity V_X^2.

It is an essential advantage of the X-transformation that collectives with skew distributions of the clinical quantities can be used as reference collectives once the X-transformation has been applied to them. This can be seen in Figure 11, where the empirical estimation of the probability density function of the quantity V_X^2 is compared with the χ^2-distribution for several collectives where patients with the same disease were chosen as reference persons. The computational procedure was the same as above. The resulting empirical distributions can be represented approximately by the χ^2-distribution with 12 degrees of freedom, even if not as well as in the collectives of healthy persons. It must also be taken into account that the sample sizes are much smaller in the groups of patients.

The graph in Figure 12 demonstrates how the groups of patients can be separated from the collectives of the reference persons by the aid of the test quantity V_X^2. The broken lines represent the density estimations for the test quantity V_X^2 on the assumption that the groups of patients were compared with corresponding groups of the same sex of healthy reference persons. We see that almost all of the test values are in the critical region ($V_X^2 > 21.0$), which implies a pathological data vector. The number of falsely negative results is small. At the same time, however, it is apparent that a discrimination between different kinds of diseases is not

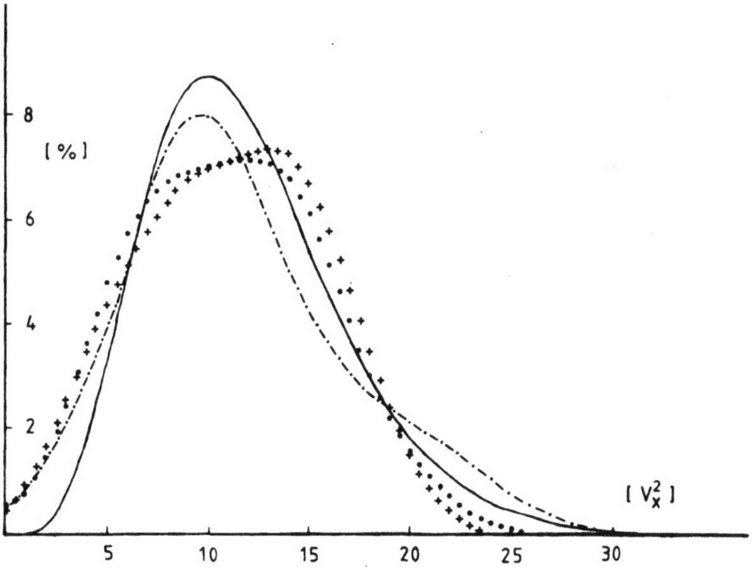

Fig. 11. The comparison between the χ^2-distribution (solid line) with 12 degrees of freedom and the empirical distributions of the test quantity V_X^2 if several collectives of patients are chosen as reference (• • • 34 male and — • — • 67 female patients with hyperparathyreoidism; + + + 33 male patients with cirrhosis of the liver). There are now greater discrepancies between the curves but the χ^2-distribution can still serve as an approximation.

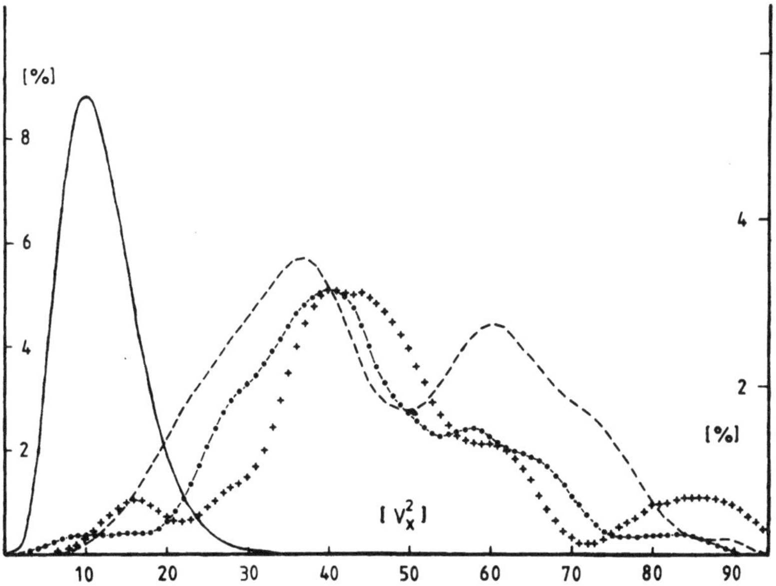

Fig. 12. The empirical distribution of the test quantity V_X^2 computed from X-transformed patient's data (— • — • 34 male and — — — 67 female patients with hyperparathyreoidism; + + + 33 male patients with cirrhosis of the liver). Reference collectives were the healthy males and females. Most values of the test quantity exceed a critical limit. A discrimination between different kinds of disease is not possible in this way.

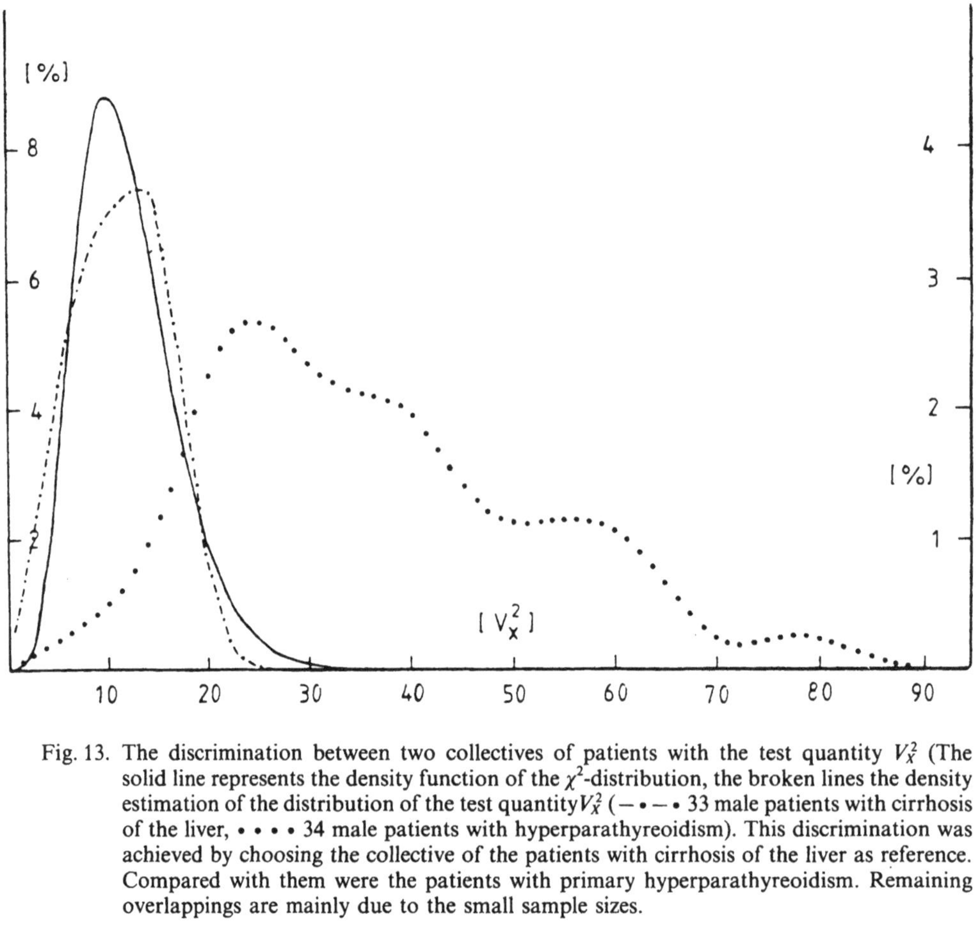

Fig. 13. The discrimination between two collectives of patients with the test quantity V_X^2 (The solid line represents the density function of the χ^2-distribution, the broken lines the density estimation of the distribution of the test quantity V_X^2 (— • — • 33 male patients with cirrhosis of the liver, • • • • 34 male patients with hyperparathyreoidism). This discrimination was achieved by choosing the collective of the patients with cirrhosis of the liver as reference. Compared with them were the patients with primary hyperparathyreoidism. Remaining overlappings are mainly due to the small sample sizes.

possible in this way. The case is different, however, if the reference collective of healthy persons is replaced by a collective of patients with a fixed disease and the patients in question are compared with that collective with the aid of the test quantity V_X^2. An example is given in Figure 13, where the collective of male patients with cirrhosis of the liver is compared with the corresponding collective of patients with hyperparathyreoidism, which serves as the reference collective here. The distribution of the test quantity points to a far-reaching distinction between the two collectives of patients. Quite similar pictures are obtained if other combinations of groups are chosen. Still remaining overlappings between the distributions of the test quantity and the χ^2-distribution can partly be attributed to the small sample sizes of the patient collectives when they are chosen as reference collectives, since small sample sizes restrict the range of the X-transformation.

It may be remarked that an attempt to distinguish groups of patients with a particular disease from the reference collectives of healthy persons is successful with the untransformed data, too. The test quantity V^2 then takes on even much higher values than the quantity V_X^2. There is no computational discrepancy in the assessment of the patients, however. In our computer calculations one weakly significant V^2-value – of a female patient with primary hyperparathyreoidism – was converted into a V_X^2-value which lay close to the critical limit. A separation between the different collectives of patients is also possible with the untransformed data. Yet the results are not in agreement with those obtained after applying the X-transformation to the data. It may be surmised that the results obtained with the X-transformation are the more reliable.

2.7 Deviation patterns and group-conformity behaviour

The comparison of the test quantities V^2 and V_X^2 obtained from patient data is a rather summary procedure for examining the influence of the X-transformation on the data structure. It would seem advantageous, therefore, to consider subsets of the quantities as well, i. e. to pass to marginal distributions. Here a concept may come in useful which will be named "group-conformity behaviour". We will say that a member of a group of patients with a fixed disease behaves *in conformity with this group* with respect to a particular clinical chemical quantity if the deviation of his value from the age-specific expected value estimated from the reference collective lies in the same direction as the mean value of his group. To take an example: Let the albumin-value of a patient with cirrhosis of the liver be below the average value of healthy persons of the same age. We then say that this patient behaves in conformity with the group of patients with cirrhosis of the liver since patients with that disease generally have reduced albumin values. This concept was generalized to combinations of quantities in the following way: From the 12 clinical chemical quantities under consideration all possible subsets containing four quantities were formed. Then the test quantities V^2 and V_X^2 were computed. Then, for both V^2 and V_X^2, the particular subset for which the test quantity was maximal was called *maximal combination*. These maximal combinations were determined for all patients with a fixed disease. The maximal combinations are characteristic for the patients and their disease. After that it was determined how often all possible pairs of clinical quantities appear in these maximal combinations. From the frequencies thus obtained *graphical structures* were derived which demonstrate disease-specific *deviation patterns* based upon collective-conformity behavior in the data. (This was done separately for the quantities V^2 and V_X^2, of course.) The graphical structures are set up in the following way (Figs. 14–19). The different clinical chemical quantities are symbolized by small circles arranged along the circumference of a larger circle. "Plus" and "minus" signs are put into the small circles to indicate how the values of the patients differ on the average from the corresponding values of the reference persons. The magnitude of the mean difference is qualitatively represented by the thickness of the plus and minus signs. Following the conception of group-conformity behaviour, the frequencies with which the different pairs of clinical quantities appear in the maximal combinations are represented by the thickness of the solid lines connecting the corresponding symbols. This was done only as of a certain level of frequency so as to make the pictures not too difficult to read. The influence of the X-transformation on the data can be judged by comparing the pictures constructed from the transformed and the untransformed data.

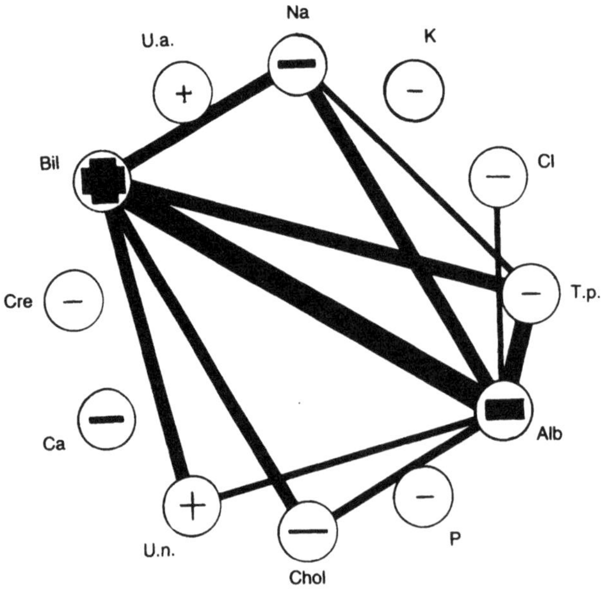

Fig. 14. The disease-specific deviation pattern derived from the measurements of 33 male patients with cirrhosis of the liver. The decrease in the albumin values and the increase in the bilirubin values are predominant. The graphical representation is based upon the concept of "group-conformity behaviour". The untransformed measurements were taken for the analysis. For details, see text.

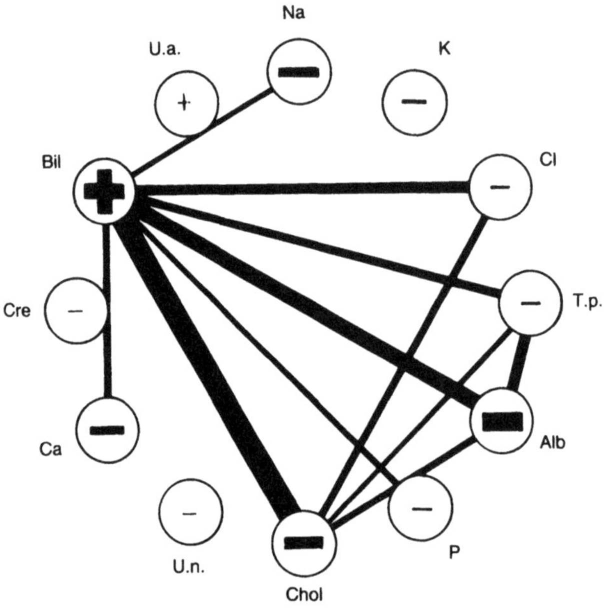

Fig. 15. The disease-specific deviation pattern for the patients with cirrhosis of the liver computed from the *X*-transformed data. The significance of the pairs: bilirubin – albumin and bilirubin – urea nitrogen is reduced while the pairs: bilirubin – cholesterol and bilirubin – calcium turn up as important in the deviation pattern.

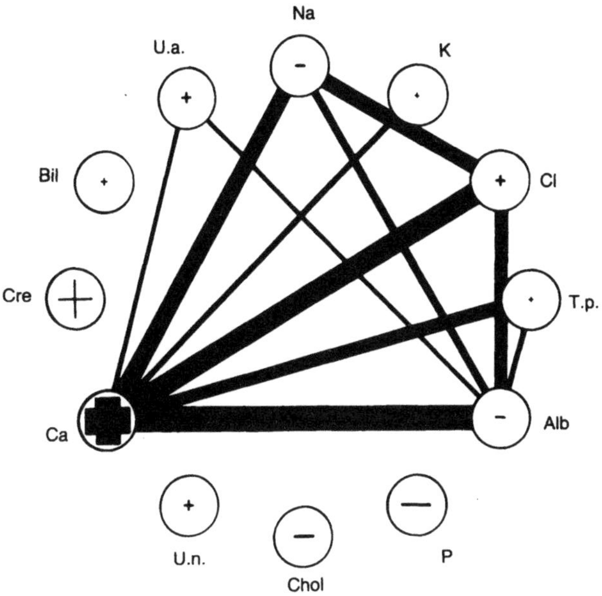

Fig. 16. The disease-specific deviation pattern for the male patients with primary hyperparathyreoidism as computed from the untransformed data. The picture is dominated by the massive increase in calcium values. The pair: calcium – albumin appears most often in the individual maximal combinations. An explanation for this is the positive correlation between calcium and albumin in the reference persons. For details, see text.

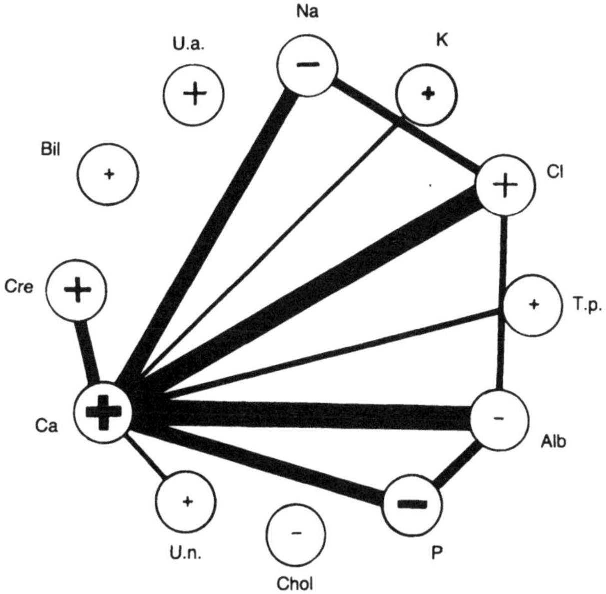

Fig. 17. The disease-specific deviation pattern computed from the *X*-transformed measurements for the male patients with primary hyperparathyreoidism. Calcium, although reduced in its deviation by the transformation, has retained its significance. In contrast with the untransformed values, the significance of anorganic phosphorus is revealed. For details, see text.

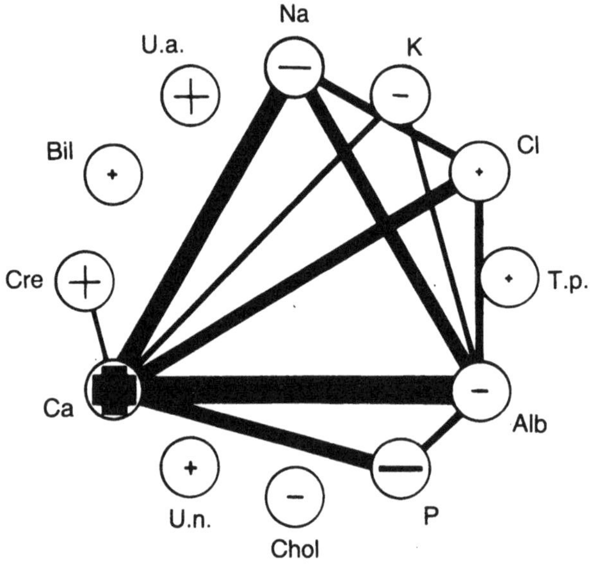

Fig. 18. The disease-specific deviation pattern for the collective of 67 female patients with primary hyperparathyreoidism, computed from the untransformed data. Contrary to the male patients, anorganic phosphorus figures in the deviation pattern. Note the otherwise striking similarity to the male patients.

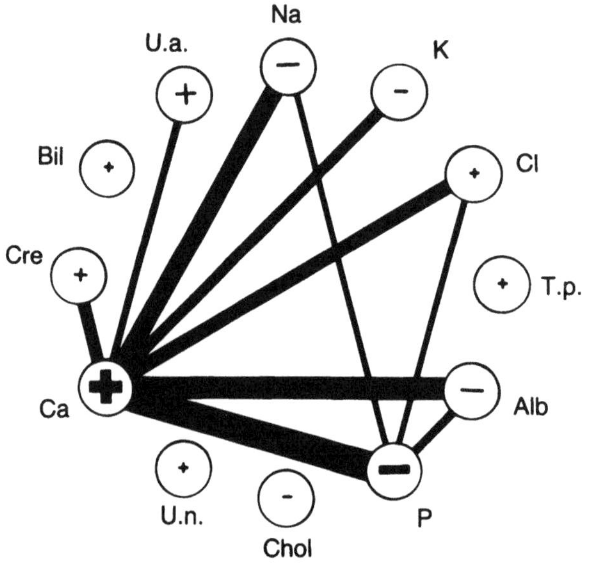

Fig. 19. The disease-specific deviation pattern for the female patients with hyperparathyreoidism, computed from the X-transformed values. The pair: calcium – anorganic phosphorus occurs most frequently in the individual maximal combinations. For details, see text.

Note: The reader might wonder why subsets of just four of the clinical quantities were selected for the analysis, when three or five quantities could have been selected as well. This was actually done, but pictures computed from three measurements did not show enough detail, while the use of five measurements for the subsets consumed too much computing time, hence the selection of four measurements for the subsets.

Figures 14 and 15 present the result of the computations utilizing the data of the patients with cirrhosis of the liver. Figure 14 shows the disease-specific deviation pattern derived from the untransformed data, while Figure 15 contains the corresponding pattern computed from the X-transformed data. There are striking common features in both pictures. Both are dominated by a large decrease in the mean albumin values and a large increase in the bilirubin values. Furthermore, a general reduction of the concentrations of the electrolytes is predominant. There are differences between the patterns, however. In the transformed data cholesterol plays a more central role than in the untransformed data. The opposite is true for the urea nitrogen values, where the deviation even changes sign.

The corresponding deviation patterns for male patients with primary hyperparathyreoidism are presented in Figures 16 and 17. Both pictures are dominated by a massive increase in calcium values. There is a decrease in albumin and in the anorganic phosphorus values in both pictures. Since the decrease in the albumin values is contrary to the expectation with respect to the calcium values (albumin and calcium show a strongly positive correlation in the reference collectives) the pair: albumin – calcium is present in all maximal combinations. As a result, there is a particularly strong connecting line between the symbols representing albumin and calcium. In the X-transformed data the decrease of the anorganic phosphorus values leads to a connecting line between calcium and anorganic phosphorus. This is even more evident in the Figures 18 and 19, where the deviation patterns of female patients with hyperparathyreoidism is presented. At the same time it is apparent that there are strong similarities between the male and the female groups with hyperparathyreoidism. It seems that the disease has essentially the same effect on the clinical chemical quantities in both men and women.

3 Conclusions

The X-transformation has been applied here to clinical data in an attempt to establish conditions in the reference collective which correspond to the situation where the data follow a multivariate normal distribution. If such an attempt is successful, the easy-to-handle formulas available for normally distributed variables can be applied to the data. Here the X-transformation was applied to the marginal quantities of a *homogeneous* group. If the measurements of further persons had to be compared with that group, additional considerations were necessary. The results of the computations show that in the two-dimensional marginal distributions excellent approximations to a corresponding normal distribution are obtained. That the transformed data points are eventually positioned on a finite grid is hardly perceptible and turns out to be no hindrance. Nevertheless, it cannot be expected that a multivariate normalization is always possible. This is simply theoretically impossible, since the transformed data lie on a finite higher-dimensional grid and because there is a slight negative correlation between the values of different persons (which does not matter all that much); there might be practical obstacles as well in the data which cannot be cleared away by the X-transformation. The X-transformation will fail or will not give the desired results in the following cases:

1. If the statistical relations between the different quantities are not monotonic then it is impossible to convert the data into multivariate normally distributed values by a transformation of the marginal quantities. Notwithstanding the difficulty of defining what one really means by a non-monotonic statistical relation this situation cannot be discovered without an in-depth investigation of the data. To discover this it would seem to be best to investigate the graphical representations of the two-dimensional marginal distributions. Another sign of a nonmonotonic relationship can also be that the dispersion of a quantity varies rapidly with the values of another quantity (heteroscedasticity). An example of a nonmonotonic correlation is age dependence of bone density in men. The density of the bone in men increases approximately to the age of 37, and then decreases. Heteroscedasticities can also arise if the correlations between the quantities depend on the age of the persons or if, for example, the ellipsoidal reference regions rotate about the age regression line. In this case is is recommended to form groups of reference persons that are homogeneous with respect to age or to make use of completely distribution-free methods.
2. Difficulties can also be expected if the range of variation of the quantities is not convex. This could occur if some internal regions of the range of variation were avoided by the data, if the distribution had "holes", so to speak. Such behavior, however, can hardly be imagined in connection with real data. Considering the immense topological difficulties of working in higher-dimensional spaces, it is unlikely that such irregularities could be detected in real data. As far as an inspection of the two-dimensional marginal distributions of the above data could reveal there were no indications of such difficulties.
3. If the collective of reference persons in question has more than one center of agglomeration of the values, then the application of the X-transformation does not make much sense. Such multimodal distributions can result, for example, if patients with different diseases are grouped together. In this case we have a mixture of different distributions and the patients should not serve as reference persons.
4. If it is the objective of the practitioner always to compare investigated persons with the same collective of reference persons as would be the case, for example, in a routine check-up of many persons, then it would seem to be better to look for continuous transformation formulas for the normalization of the data. This is admittedly more laborious, but the loss of information connected with the X-transformation can be avoided by using continuous transformations.

The advantage of using the X-transformation becomes apparent if collectives of patients with skewly distributed clinical chemical quantities are to play the role of reference collectives in order to permit a discrimination between different types of disease. We then have the following immediate benefits:

1. The search for appropriate transformations is unnecessary. The X-transformation can be done rapidly with the aid of rational approximations on a computer.
2. A good approximation to a multivariate normal distribution is achieved, if it is at all possible, by marginal transformations. Even if this is impossible, it seems likely that the X-transformed values will behave more like normally distributed values than the original values themselves. Eventual nonlinearities are linearized as far as they are monotonic. Moreover, the influence of outlying observations is reduced.
3. It is likely that the sensitivity of the test quantity V^2 is improved when the X-transformation is applied to a reference collective with skew distributions. At first it would seem as if this were not true. The X-transformation reduces the test quantity if the data contain extremely high or low values, since the X-values of extreme data lie on the border of the X-grid. This would seem to impair the sensitivity of V^2. Practical experience, however, shows that this is not the case. The significance of the test quantity remained basically unaffected, with the exception of one value, as mentioned above. Furthermore, it is known that the superiority of a multivariate data assessment to a combined series of univariate assessments increases, the stronger the correlations between the quantities are. If the correlations between the measured quantities were zero then a multivariate data assessment would be of little use. Unsymmetric marginal distributions tend to deflate the correlations between the variables. This tendency is counteracted by the X-transformation, so that the sensitivity increases. For this it is necessary, however, that the reference collective not be too small so that a sufficiently fine and extended X-grid results.
4. The X-transformation is particularly useful for correlation analyses. The interpretation of partial correlation coefficients or the application of the covariance decomposition technique requires a multivariate normality of the distribution underlying the data structure. Since most clinical quantities are not normally distributed a normalizing transformation cannot be dispensed with and here the X-transformation comes in useful. The results of computer simulations indicate that the X-transformation causes no harm if the data are already normally distributed. Table 8 contains the results of several such computations performed with pseudo-random numbers. Correlation coefficients were computed from normally distributed pseudo-random numbers with a preassigned correlation structure. The correlation coefficients were computed before and after the X-transformation had been applied. The sample sizes were also varied. One can see in the tables that the correlation coefficients agree to approximately two decimal places. The principal advantage of the above method is that the derived correlation matrix is a real (positive definite) correlation matrix. It may be noted in warning that it will not do to compute Spearman's rank correlation coefficients R_{ij} and to convert them with the formula (see ref. (18), page 324): $\hat{r}_{ij} = 2\sin(\pi R_{ij}/6)$ into estimates for the true correlation coefficients. The reason is that the matrix composed of these estimates need not be positive definite – indeed, it often does not even have a positive determinant! This is avoided by the X-transformation.

Table 8. The correlations between normally distributed pseudo-random numbers with a preassigned correlation structure (simulation study)

a) untransformed values				b) each vector of values was X-transformed			
$n = 300$				$n = 300$			
1.0000				1.0000			
0.4039	1.0000			0.4134	1.0000		
0.5052	0.2467	1.0000		0.5116	0.2509	1.0000	
0.4185	0.6066	0.6716	1.0000	0.4201	0.6063	0.6687	1.0000
$n = 400$				$n = 400$			
1.0000				1.0000			
0.4062	1.0000			0.4114	1.0000		
0.5213	0.3614	1.0000		0.5228	0.3612	1.0000	
0.5237	0.6022	0.6778	1.0000	0.5279	0.6015	0.6780	1.0000

A relative disadvantage of the X-transformation is its rather abstract definition. As a result the objection is sometimes heard from the practitioner that he prefers to deal with real data to handling such abstract values. This objection, however, is not to the point: The X-values are auxiliary numbers stored in the computer; the practitioner comes in contact only with the result of the computations, not with the X-values themselves.

The fact that the application of the X-transformation leads to results differing from those obtained from the untransformed data indisputably points to deviations from a multivariate normality in the data. Let us consider, for example, the pair: calcium – anorganic phosphorus, in the male patients with primary hyperparathyreoidism. The analysis reveals, in addition to the strong increase in calcium values, a slight decrease in anorganic phosphorus. Nevertheless the pair is missing in the graphical deviation pattern of the untransformed values (Fig. 16). After the data were submitted to the X-transformation the analysis brings to light not only a stronger decrease in anorganic phosphorus compared with the expected value, but the pair: calcium – anorganic phosphorus is now present in the graphical deviation pattern. This is a very satisfying result, since the quotient: calcium/phosphorus is used for diagnostic purposes in connection with a conjectured primary hyperparathyreoidism. In the female patients with hyperparathyreoidism the pair: calcium – anorganic phosphorus figures as the most important pair in the disease-specific deviation pattern (Figs. 18 and 19). This demonstrates that the truly relevant deviations between the reference collective and the patients are more adequately assessed with the X-transformed data than with the untransformed values.

Since the application of the X-transformation to clinical chemical data presents no difficulties and takes up little computing time the X-transformation can be considered to be a unified and standardized tool for the handling of multivariate data sets in connection with the above-described problems – at the expense, to be sure, of having to ignore certain theoretical qualms.

References

1. Büttner, J. (1977) J. Clin. Chem. Clin. Biochem. *15*, 1–12.
2. Eggstein, M. (1975) Med. Welt *26*, 474–478.
3. Voigt, K. D., Harm, K., Domesle, A. & Rehpenning, W. (1978) Technicon Symposion Frankfurt a. Main.
4. Harm, K., Rehpenning, W., Domesle, A. & Voigt, K. D. (1979) J. Clin. Chem. Clin. Biochem. *17*, 517–522.
5. Rehpenning, W., Harm, K., Domesle, A. & Voigt, K. D. (1979) J. Clin. Chem. Clin. Biochem. *17*, 565–572.
6. Grams, R. R., Johnson, E. A. & Benson, E. S. (1972) Amer. J. Clin. Path. *58*, 188–200.
7. Buret, J., Monfort, F. & Marthos, J. (1975) in: Organisation des laboratoires et interprétation des résultats – Biologie prospective. Comptes rendus du troisième colloque internationale, Pont-à-Mousson 1975, L'Expension Scientifique Française, Paris.
8. Anderson, T. W. (1974) An Introduction to Multivariate Statistical Analysis, Wiley, New York.
9. Wilks, S. S. (1941) Ann. Math. Stat. *13*, 400–409.
10. Wald, A. (1943) Ann. Math. Stat. *14*, 45–55.
11. Tukey, J. W. (1947) Ann. Math. Stat. *18*, 529–539.
12. Fraser, D. A. S. (1951) Ann. Math. Stat. *22*, 372–381.
13. Kempermann, J. H. B. (1956) Ann. Math. Stat. *27*, 180–186.
14. Abt, K. (1977) in: Math. Meth. in der Medizin, Tagungsbericht 8/1977, Oberwolfach.
15. Ackermann, H. (1983) Biom. J. *25*, 351–359.
16. Box, G. E. P. & Cox, D. R. (1964) J. Roy. Stat. Soc., Ser B *26*, 211–252.
17. Hinkley, D. V. & Runger, G. (1984) JASA *79*, 302–312.
18. Van der Waerden, B. L. (1965) Mathematische Statistik, Springer, Berlin, Göttingen, Heidelberg.
19. Abramowitz, M. & Stegun, I. A. (1965) Handbook of Mathematical Functions, Dover Pub., New York.
20. Patel, J. K. & Read, C. B. (1982) Handbook of the Normal Distribution, Marcel Dekker, New York.
21. Cramér, H. (1971) Mathematical Methods of Statistics, Princeton University Press, Princeton.
22. Dempster, A. P. (1972) Biometrics 28, 157–175.
23. Wermuth, N., Wehner, T. & Gönner, H. (1976) Computer Programs in Biomedicine *6*, 23–38.
24. Wermuth, N. & Scheidt, E. (1977) J. Roy. Stat. Soc. Applied Statistics *26*, 88–97.
25. Rehpenning, W. (1983) Multivariate Datenbeurteilung, Springer, Berlin, New York, Heidelberg, Tokyo.
26. Hotelling, H. (1931) Ann. Math. Stat. *2*, 360–378.
27. Technicon (1971) Reagent Data Handbook for the Technicon SMA 12/60 System, Technical Pub. No. THO0160-10.
28. Rehpenning, W., Harm, K., Ringe, J.-D., Bützow, G. H. & Voigt, K. D. (1984) J. Clin. Chem. Clin. Biochem. *22*, 319–329.
29. Graybill, F. A. (1961) An Introduction to Linear Statistical Models I, Mc-Graw Hill, New York, Toronto, London.

Chapter 3.7
Data Driven Interpretation of Laboratory Results in the Context of a Medical Decision Support System

P. D. Clayton, R. S. Evans, T. A. Pryor, R. M. Gardner, P. J. Haug and H. R. Warner

1 Introduction

In this chapter, we will describe a routinely used and effective computer based system for assisting the clinician in the role of decision maker. We will first discuss the philosophical issues of such systems, then describe in detail how the system functions. We will also discuss the integration of laboratory data for decision-making purposes and describe our experiences with, and evaluations of, the system.

1.1 Philosophical issues in automated decision assistance

The volume and variety of clinical laboratory tests has substantially expanded during the last decades. This increased utilization has caused two types of problems for clinical pathologists: 1. they must assume increased management and production responsibilities, and 2. they must play a more active role in assisting their clinical colleagues to appropriately use, and interpret, the results of laboratory tests.

To manage the increased volume of tests performed in the laboratory, to provide more rapid results reporting, and to insure adequate records for the financial aspects of providing health care, most modern laboratories have turned to computer based laboratory information systems. In such systems, many of the high volume analytic machines have been directly interfaced to the information system. Although these systems have measurably improved the management and production of clinical laboratories, the main impact of these sytems on improved patient care to date has been to improve the rapidity with which test results for the correct patient can be reported and to improve methods of quality control.

It is the goal of a new generation of computer programs to go beyond the management functions which are currently provided by laboratory information systems. These new systems can transfer the expertise (knowledge of laboratory tests) of trained clinical pathologists, as well as test results, to the physicians who are directly involved in patient care. In other words, computer programs have been developed and are continually emerging which can substantially assist the practicing physician to interpret (convert numbers to diagnostic or therapeutic information) the meaning of laboratory test results for individual patients. Examples of the algorithms and mathematical models which have been developed to interpret laboratory test results, and thereby increase the efficacy and clinical impact of laboratory data, are described in the preceding chapters of this book as well as in reviews by Beck, Meier, and Rawnsley (1) and Henry (2). By using traditional programming techniques, these algorithms have been, or could be made, available in currently available commercial laboratory information systems.

In contrast to these traditional "hard-coded" programs in which the algorithmic logic or expertise is embedded directly in the computer instruction codes, another approach to providing expert consultation has emerged during the 70's and 80's. In this latter approach, the interpretive and clinical expertise (knowledge) is stored separately from the computer program which uses this knowledge to interpret clinical data (facts) about a patient. When the clinical expertise is stored in this separate representation, it is known as the "knowledge base" for the system.

There are several advantages to this knowledge-based approach. A single control program (sometimes called the "inference engine") can function across broad areas of medical applications if the appropriate expert rules for multiple subspecialties are contained in the knowledge base. The clinical expertise and interpretive rules in the knowledge base can be reviewed, modified and expanded independently by clinicians who are not required to have programming capabilities. This approach allows the breadth and power of the system to be improved by adding to and modifying the knowledge base in a modular fashion.

When the content of the knowledge base is used to provide consultation and make decisions in specific medical domains, these knowledge-based systems are sometimes called "expert systems". These systems provide the expertise of a specialist in a particular field to users who may not even be aware of their need for an expert consultation. The topic of expert systems is covered in detail in Chapter 3.8 by Trendelenburg.

The difference between these knowledge-based systems and the dedicated, hard coded algorithms found in the traditional computer programs is generally not a matter of functional capability, but approach. "Since there is no one formal definition of [knowledge-based systems] and recent implementations have explored variations on virtually every aspect, their use becomes more an issue of a programming *style* than of anything else. ... Since it is possible to imagine coding any given [computer] in either procedural [i.e. traditional 'hard coded' programming] or [knowledge-based systems] terms ..., in the formal sense their computational power is equivalent. This suggests that, given sufficient effort, they are ultimately capable of solving the same problems. The issues ... are not questions of absolute computational power, but of the impact of a particular methodology on program structure, as well as the relative ease or difficulty with which certain capabilities can be achieved" (3).

The present authors interpret this insightful statement to mean that "there is nothing magic" about knowledge-based systems. In a hard coded program, the programmer must have explicitly foreseen every branching point in an algorithm.

Such deterministic protocols or algorithms can have a positive impact on healthcare. However, traditional programs are most appropriate in noncomplicated, straightforward situations where the logic is well defined. If there are complicated situations, the separate knowledge base approach lets appropriate rules be applied to situations which the knowledge base authors did not explicitly foresee or solve in advance.

Another advantage of having the rules for decision-making in a readable stand alone format is derived from the realization that the decision logic may contain an author's moral or ethical judgements or personal biases. In knowledge based systems, these biases remain exposed for scrutiny and discussion by the clinical, rather than the programming community. This independent representation of medical knowledge may also contribute solutions to the as of yet unanswered dilemmas regarding the responsibility and liability associated with the use of decision-making systems.

In spite of these additional advantages of a knowledge based system, it has generally been recognized that the power which leads to superior performance of a computer system comes from the richness and content of the expert knowledge contained in the program rather than the language or style in which it is programmed.

We are of the opinion that there are additional considerations which must be kept in mind as one contemplates a system for the interpretation of clinical data: 1. the way in which the expert logic is activated will have a profound effect upon it's use and acceptance, and 2. some provision must be made to guard against reliance upon a system which may not have the necessary expertise in the domain about which the user is seeking assistance. We feel that the philosophical approach to these issues is extremely important to the success of a useful, productive program for the clinician. Because some expert systems have been developed in a research environment in which the computer science aspects of the project were of more interest than the ultimate clinical application, these two practical issues have not always been fully addressed.

If these issues are neglected, systems which are intrinsically powerful may prove to be quite sterile in a user's environment for the following reasons. First, in order to receive decision-making assistance from the variety of expert systems which are currently (1988) available, the user may be required to have multiple expert systems or hard coded programs; one for diagnosing diseases in internal medicine, another for assessing strokes, a third for assessing rheumatoid arthritis, a fourth for interpreting abnormal bilirubin results... etc. If the user is well enough qualified in a field to know which of the expert systems or programs s/he should turn to, or even more importantly, to know when to consider the need for additional consultation, there is a good chance that the system will not surpass the user's or the consultant's expertise in a specific area. In other words, there are currently only a limited number of systems which contain sufficient knowledge to make decisions in a wide variety of medical specialties or with the capability to support a variety of types of decisions e. g. diagnoses, alerts, management, etc. None of the currently available systems has a comprehensive medical knowledge base.

A second problem occurs when the user approaches the computer and asks for assistance with a specific problem. This is generally referred to as a *goal driven* mode for activating the expert consultant because the user wants to solve a particular problem (What is the diagnosis? Is it all right to administer this drug to my patient? What does this laboratory test mean?). After entering facts about the patient, the user discovers (perhaps by way of a nonsensical answer) that the system does not have expertise in the area concerning the user's goal. An inappropriate response in one area often destroys the credibility of the system although it may be highly proficient in other areas.

The third problem arises because one goal of computer systems is to assist the physician in domains where his/her performance is ordinarily very good, but in which some infrequent but important errors do occur, for example monitoring drug prescriptions. This goal can be accomplished in the critiquing mode in which the computer constantly monitors the physician's decisions and issues a warning or an alert whenever certain decisions are questionable. If the physicians must constantly ask if it is all right to prescribe a drug when, in fact, their unassisted decisions are correct most of the time, they may ultimately tend to avoid asking for assistance from the computer when the yield is so low, unless it is extremely easy to receive this assistance.

In order to address these issues of practical application, we have designed a system in which three tasks were accomplished: 1. the decision-making system is imbedded within a comprehensive clinical information system, 2. the knowledge base, though separate from the decision-making program, is constructed in a very modular, procedural representation, and 3. the appropriate decision logic is activated whenever data (facts) referenced in the decision criteria are stored in the clinical information system. This mode of activating the expert logic is known as a *data driven* activation strategy. This approach to activating decision logic insures that those decisions for which sufficient rules and facts exist in the computer, will be evaluated whenever the data are stored. With this philosophy, the user receives consultation in the form of likely diagnoses, interpretation of laboratory tests, alerts, or contraindication warnings without asking for assistance. The problems associated with an incomplete knowledge base are only partially solved with this approach; the program gives advice when it is able to speak authoritatively, but remains silent when it does not have any knowledge or facts about the problem.

2 Functional Implementation of the HELP System

For the last decade, we have been using a computer-based system for automated medical decision-making which is interfaced to a commercially developed computerized clinical laboratory system as well as other computerized subsystems of the hospital. The hospital based expert system is known as HELP (Health Evaluation through Logical Processing) (4) and a key feature of the system is the integration of administrative data processing functions as well as collection of clinical data from many sources to form a comprehensive clinical database. The resulting system thus has test ordering, results reporting, and charge capture capabilities as well as automated interpretation of clinical data, alert generation, and diagnostic functions.

There are three major advantages for an approach in which the decision-making system is integrated with an on-line clinical database. As we have discussed, one of the most important is the fact that the decision-making capabilities can be automatically evoked whenever new data are added to the patient record.

A second advantage is that the entire computerized medical record (not just information from the clinical laboratory) can be used to provide clinical facts which are referenced in the decision logic. Thus, a patient with a negative sputum culture might still be identified as having a hospital acquired pneumonia because of radiology findings or information provided by respiratory therapy technicians.

The third advantage is that such a comprehensive collection of clinical data becomes the primary repository which the clinical users use to review patient information. Recognition of the importance of this format for data storage and display insures that all involved personnel have vested interests in keeping the data accurate and up to date.

In the sections which follow, we shall describe the overall functional design of the HELP system, then discuss specific issues relative to the inference capabilities and the data driven logic activation mechanism.

As can be seen in Figure 1, the system essentially consists of three main parts: a comprehensive clinical patient database, a separate knowledge base which contains expert logic, and an interpreter which controls the evaluation of the expert knowledge.

Those elements in the upper half of Figure 1 (data collection programs, the long term patient file, the current clinical patient file, the reporting functions and the link to the financial system) are standard components which may be found in some

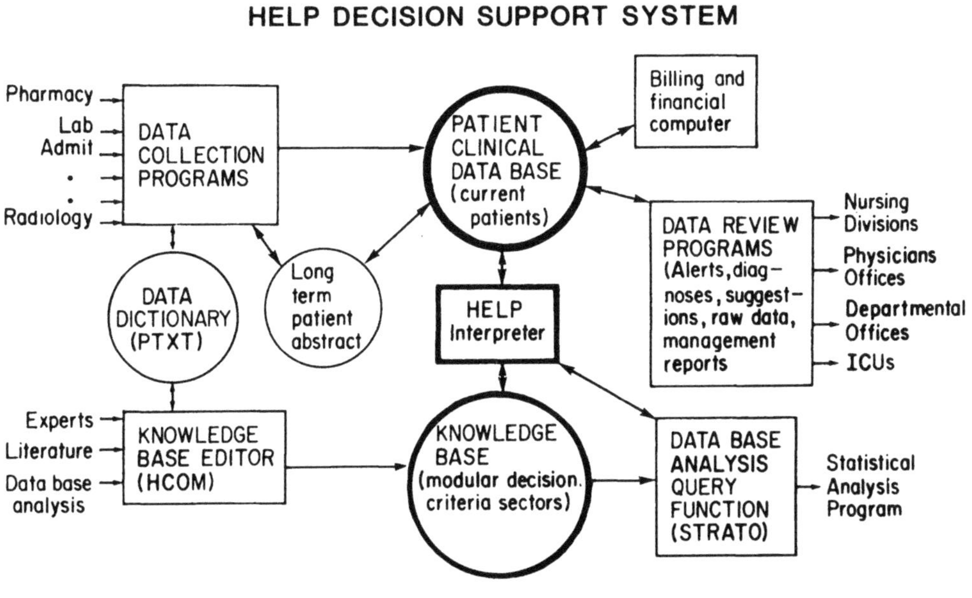

Fig. 1. An overview of the components of the HELP system.

currently available hospital information systems. The central clinical data base integrates data from, and communicates with, computer systems in ancillary departments in order to allow test ordering and results review throughout the hospital.

We have chosen to use the Tandem computer for the central system because it is easily expandable and has built in hardware and software redundancy which help to insure that the system is always operational. In our 520 bed hospital, we presently have 361 terminals and 86 printers attached to a central system which is composed of a cluster of eight computers that function in an integrated mode which allows each computer in the cluster to be backed up by another computer in case of failure by a single unit. Each of these computers in the central cluster has access to all of the system data files. In addition to this central cluster, there are multiple (17 presently) microprocessors attached to the central system which support independent data collection activitiy or act as signal processors.

The long term file contains on-line data for all previously admitted patients and consists of abstracts of clinical and demographic information likely to be useful if a patient is readmitted. The clinical data base contains all data gathered during the current admission. After a patient is released from the hospital, the patient's record is stored in archives which are available for statistical assessment.

In order to facilitate the use of the patient specific information which is contained in the clinical database, it must be stored in some uniquely identifiable form. In our system, clinical data are stored in a coded format which is defined using a data dictionary. The data dictionary is necessary in order to allow those who write programs to acquire patient data, build the logic contained in the expert knowledge base, or construct report formats to accurately reference specific data which may be stored in the patient data base. We have chosen a coded format for the stored data in order to facilitate the rapid retrieval of data which are referenced in the decision-making logic and to allow the use of hierarchical relationships among medical terms. Other developers of knowledge-based systems have used uniquely defined symbols. As long as the user is insulated from the necessity of actually seeing the codes, the distinction between codes and symbols is somewhat artificial. In our case, the codes are retranslated to display the medical terminology to those who review the data.

The elements of the lower half of Figure 1 represent the additional features necessary for a decision support system (knowledge base editor, knowledge base and the HELP interpreter). The expertise contained in the knowledge base can be obtained from opinions of experts, the medical literature, or statistical experience represented in the patient database. The knowledge is stored as compiled "HELP frames" which contain the logic necessary to make a specific decision. The medical knowledge base supports a variety of decision-making models (IF...THEN...rules, patient specific likelihood scores for ranking possible diagnoses, query for "important" missing data, etc.) and allows the medical expert to enter criteria using a high level language contained within the knowledge base editor.

When new results are stored in the patient record or a specific block of the knowledge base is otherwise activated, the compiled logic in the appropriate HELP frames is evaluated by performing the necessary queries to the patient database to see whether the data specified in the logic exist for the patient in question. The frames themselves contain the logic which determines how they are to be evaluated.

Title: Hospital Acquired Pneumonia (16.16.10).

Author: Peter Haug

Message: "Patient may have Hospital Acquired Pneumonia."

Declare Variables:admission^time **as** ADMIT TIME,

late^purulent^sputum as gram^stain^WBCs **or** resp^tx^sputum **where** gram^stain^WBCs is SPUTUM **and** (MODERATE NUMBER OF WBCS or NUMEROUS WBCS) and resp^tx^sputum **Is** PURULENT SPUTUM **from** admission^time + 3 days **until NOW,**

purulent^sputum **as** gram^stain^WBCs **or** resp^tx^sputum where gram^stain^WBCs **Is** SPUTUM and (MODERATE NUMBER OF WBCS **or** NUMEROUS WBCS) **and** resp^tx^sputum **Is** PURULENT SPUTUM

neutropenia **as** white^count **LE** 2.0 where white^count **Is** WHITE BLOOD CELL COUNT,

surgery **as** OPERATIVE RECORD,

late^positive^chest^xray **as** PULMONARY INFILTRATE/CONSOLIDATION from admission^time + 3 days until **NOW,**

positive^chest^xray **as** PULMONARY INFILTRATE/CONSOLIDATION

late^positive^sputum^culture **as** SPUTUM and BACTERIA **from** admission^time + 3 days **until NOW,**

positive^sputum^culture **as** SPUTUM and BACTERIA,

previous^admission **as** LAST ADMISSION **from** 30 days ago,

early^negative^chest^xray **as** early^chest^xray **and not** early^positive^chest^xray **where** early^chest^xray **Is** CHEST XRAY DATA **from** admission^time **until time of** positive^chest^xray **and** early^positive^chest^xray **Is** PULMONARY INFILTRATE/CONSOLIDATION **from** admission^time **until time of** positive^chest^xray.

Logic: **If** (purulent^sputum or neutropenia) **and** (positive^sputum^culture **or** positive^chest^xray) **and** (previous^admission **or** early^negative^chest^xray) **then conclude,**

Else If (neutropenia or late^purulent^sputum) **and** (late^positive^sputum^culture **or** late^positive^chest^xray) **then conclude,**

Else If surgery **and** positive^chest^xray **and** (**time of** positive^chest^xray - **time of** surgery GE 3 days **and time of** positive^chest^xray - **time of** surgery **LE** 10 days) **then conclude.**

Evoke: **If** purulent^sputum **or** positive^chest^xray

Fig. 2. A frame which contains the decision logic for identifying certain classes of patients which may have a hospital acquired infection.

An arithmetic statement can be used to perform tasks ranging from Boolean logic to calculation of a discriminant function. Chronologic statements can be used to retrieve the time of a specified event so that time may be used for data limitations or action flags. Existence statements use the presence or absence of a piece of data, rather than the value as the basis for logical manipulations and/or calculations. Data retrieval statements are used to search the clinical database for specific items within specified time limits. These search items may also trigger the evaluation of additonal HELP frame modules or ask for missing but necessary data.

The logic contained in the frame depicted in Figure 2 contains criteria used to recognize patients who may have a nosocomial pneumonia. The bold face terms in the frame are syntactical and structured commands within the HELP language. The top three headings of the frame give the title by which this frame is known in the knowledge base, the author, and the message which will be displayed if the logical criteria contained within the frame are satisfied. The *logical criteria* in this case are a series of IF...THEN... rules. In order to properly qualify the terminology which is used in the logical rules, the terms (variables/symbols) used in the rules are *declared.* These declarations refer back to terms (denoted by capitalized words) which are defined in the data dictionary. A powerful part of these declarations is the use of chronological constraints as well as qualification regarding the value for a variable. The procedural logic is written in a slightly structured, but understandable, language. The data driving triggers are listed under the *evoke* heading. Whenever the laboratory reports a purulent sputum or a radiologist reports a chest radiograph with evidence of infiltrates or consolidation, the frame is evaluated.

When the necessary criteria for a decision are satisfied, a new data string which reflects this result, is stored in the patient record. The storage of one result may activate other decision logic as well as specified reporting mechanisms. This approach is sometimes called the blackboard method; when a new decision is written to the clinical database (blackboard), it may data drive (evoke) other decisions which depend on this new conclusion.

In contrast to the self contained logic frame shown in Figure 2, Figure 3 shows the algorithmic flow chart which is an alternative representation for this decision.

As explained earlier, this logic could be imbedded directly in a hard coded conventional computer program. The differences are that the frame representation can be easily understood by clinical personnel with authorship or review responsibilities, the frame can be independently added to, or subtracted from, the knowledge base, and the data driven evoking mechanisms for the system will automatically access this logic whenever data referenced by either of the evoking criteria are stored for a patient. This concept illustrates how the system can modularly grow in breadth and detail. Of course, for meaningful results to emerge, this logic must be linked to a database in which microbiology results, radiographic findings and respiratory therapy procedures are routinely entered.

There is a wide variety in the types of rules in the knowledge base; for example, to identify a patient with a reportable disease such as Giardia, the decision logic only needs to look for the presence of Giardia in the microbiology test result. Other types of frames are used to calculate probabilistic likelihood scores which are used to rank the most likely diagnoses for a patient. Still other frames ask for data which are missing in order to conclude certain alerts or diagnoses.

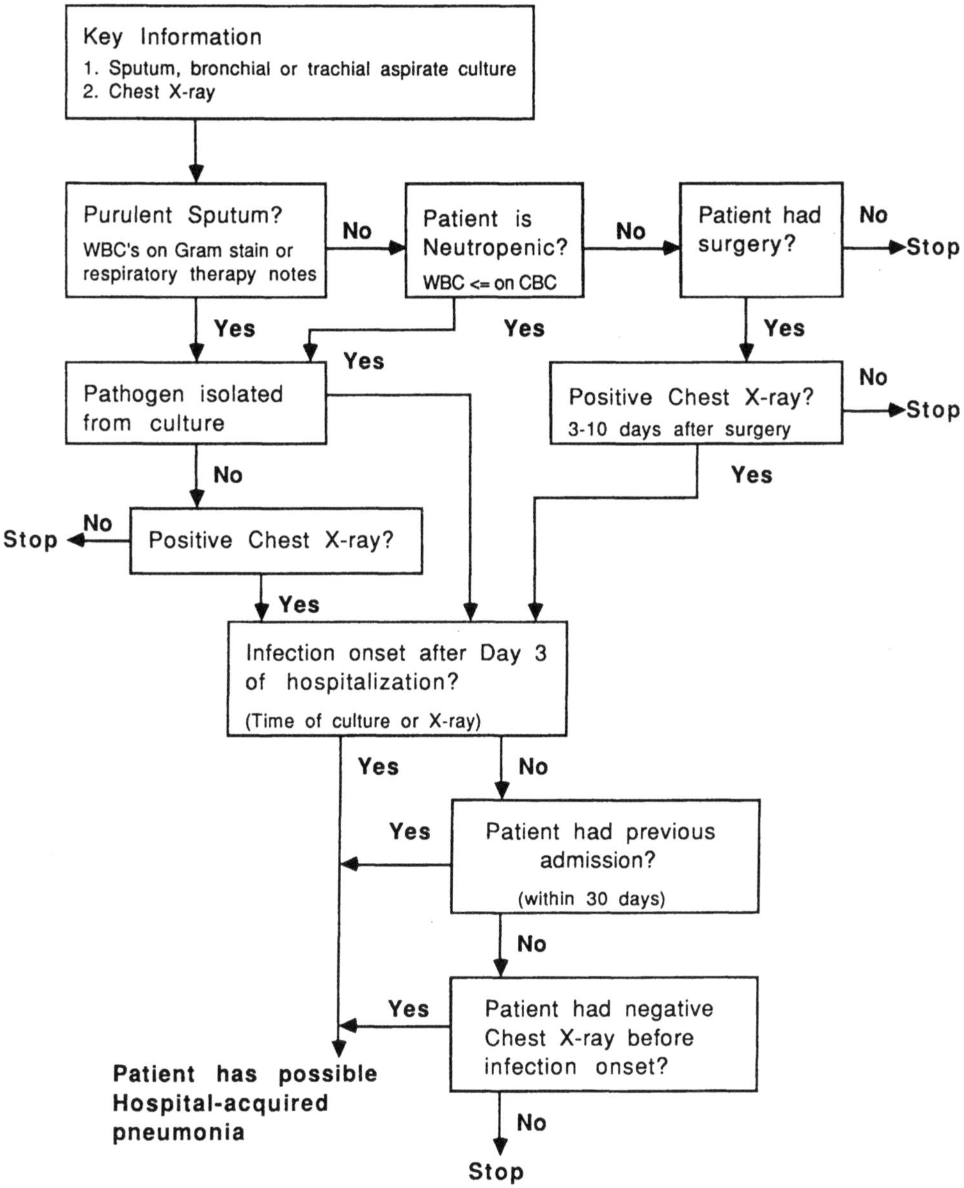

Fig. 3. A conventional algorithmic flow chart representation of the decision logic for recognizing potential nosocomial pneumonias.

2.1 Interface to the clinical laboratory system

Whenever a patient is admitted to the hospital, moved to a new room, or discharged from the hospital, the central computer system notifies the laboratory computer system of these changes. The laboratory system maintains its own database of test results for internal integrity, quality control and operational functions, but results are reported by transferring the data to the central information system and storing the data in the comprehensive clinical data base. When a laboratory test result is sent by the laboratory machine to the central computer, it is temporarily held in a holding file (called a spooler). A translation program then takes the test results, one at a time, and translates them from the laboratory file format into the hierarchical codes used in the central database. Transferring a laboratory result to this central computer activates the charge capture mechanism in the central machine and triggers the appropriate expert logic. The blocks of this expert logic which are activated depend upon the type of test result which has been transferred. Thus, the communication between the systems is two-way; the laboratory system receives information from, and returns its results to, the central hospital system.

2.2 Experience with the decision-making aspects of the system

We currently make decisions in the many areas of medicine. The following list is not meant to be exhaustive, but rather to illustrate the variety of tasks which are addressed by the system:

Diagnosis: Pulmonary disease, anemias, obstetrics, multi-organ failure index, hospital acquired infection, ...

Patient Management: antibiotic usage, diet, obstetrics, ...

Data acquisition: intelligent history, radiology findings, ...

Contraindications/alerts: pharmacy, radiology, clinical laboratory, blood gas, ...

Test result interpretations: Blood gas, hemodynamics, clinical laboratory, ECG, pulmonary function, electron microscopy, ...

Protocol management: Cardiac arrythmias, ...

Departmental management: infectious disease, respiratory therapy ...

Rather than dwell upon all of the specific types of individual decisions that the system is capable of making, we shall describe several of those applications in which laboratory data are used in combination with other sources of clinical data. These types of decisions cannot be easily generated in a stand alone laboratory information system because they rely on the comprehensive nature of the clinical database.

Data driven decision logic is used as the basis for our hospital's infectious disease monitoring program (5). Criteria have been stored in the knowledge base which can be used to identify patients: 1. with hospital acquired infections, 2. not receiving antibiotics to which their pathogens are susceptable, 3. who could be receiving less expensive antibiotics, 4. who receive prophylactic antibiotics longer than appropriate, 5. with infections at normally sterile sites, 6. with reportable diseases, and/or 7. with an antibiotic resistant microorganism. A report containing these alerts is automatically generated every day in the infectious disease department.

Surveillance personnel using computer screening for two months identified more hospital acquired infections than those who used traditional surveillance methods, while requiring only 35% of the personnel time (6). During the same two month period, the computer screening identified 37 patients (in a 520 bed hospital) not receiving appropriate antibiotics and 31 patients who could have received a less expensive, yet equally efficacious, antibiotic. A one month study demonstrated that the computer can automatically identify patients without evidence of infection who are receiving prophylactic antibiotics for an excessive period of time.

The infectious disease knowledge base is also used to generate a respiratory therapy infection monitoring system. This program uses the knowledge base to identify patients who have conditions to which a respiratory therapist should be made aware and also for tracking pneumonias and bacteremias which may have potentially been induced by the respiratory therapy. A daily report identifies patients with positive or pending tuberculosis, mumps, rubella, or hepatitis tests, and/or patients with hospital acquired pneumonias or bacteremias who have also received respiratory therapy. The computer reporting programs also identify which of these patients have a common pathogen and determines whether the same respiratory therapist (technician) worked with two or more of the patients with common organisms during a specific time period.

One of the best received and appreciated applications is the generation of pharmacy-laboratory alerts (7). This application illustrates the strengths of an integrated system with decision-making capability. When drugs are prescribed, the pharmacist enters these prescriptions into the computer. This entry activates decision logic which is based upon a combination of current medications as well as laboratory results. If the prescription is for a diuretic, a group of HELP frames which reference the use of diuretics in their logic are evaluated. One of these frames ascertains whether the prescribed drug is a potassium sparing diuretic and whether the patient's present serum potassium level is within normal limits. If both of these criteria are not met, the computer suggests to the pharmacist that a potassium supplement may be advisable.

If a drug which can potentially impair kidney function (e. g. Gentamicin) is prescribed when the serum creatinine or BUN levels are already high, the pharmacist is alerted that a different drug may be preferrred. If a laboratory test to evaluate kidney function is not requested within 48 hours after the drug is prescribed, the pharmacist is also alerted.

After the pharmacist verifies that the suggested contraindication is valid, the prescribing physician is notified. In approximately 95% of these instances, the physician changes the prescription. In our hospital population we find that 4% of the drugs and 2% of the patients receive pharmacy related alerts. A large fraction of these alerts involve pharmacy-laboratory interactions. A study which estimated the costs associated with stay-extending contraindications showed that the entire pharmacy surveillance expert system was cost effective by a four to one margin. A second study showed that those patients with abnormal laboratory values came back into the normal range significantly faster if the physician or nurse was notified by the HELP system. Based upon these formal evaluations, as well as the broad acceptance of the system which has occurred as physicians have learned how to use the system, we feel confident that expert systems will play an expanding role in the proper utilization of laboratory results.

3 Summary

In domains where the types of data which are to be interpreted are relatively constrained (as in the case of specific laboratory test results), our modular, data driven approach can be very productive and well received by the clinical recipient of the data. The computer rarely surpasses the knowledge of an expert in the field of specialty; most of the alerts to experts come as a result of lack of communication, imperfect memory, oversight or multiple decision-makers caring for the same patient (8). In such cases, most of the alerts are immediately recognized as valid, so the need for elaborate explanations is not a high priority.

On the other hand, a non-specialist is alerted to the need for additional investigation, tests or collaborative support, by the fact that a reminder or diagnosis that s/he had not previously considered, appears. In other words, for the expert, a data driven system provides unceasing oversight in high volume, low yield situations where a small number of mistakes may uncommonly occur for reasons which are not related to the lack of knowledge of the provider. For the non-specialist, the system serves more to alert that the patient may have problems in a domain for which the physician needs additional support. In the present state of the art, we do not think that total reliance on the computer contained knowledge is the ultimate source of this additional support; providing the awareness of the need may be the most important contribution. Once you know that you need help, it's usually obtainable.

In a discussion about how computer systems have failed, Friedman and Gustafson made the following observation. "The great majority of computer applications to medicine, attempted to date, have been excessively modest in scope. Where in other fields the computer has been utilized to perform tasks previously incomprehensive to mankind, in healthcare delivery we have been satisifed to merely duplicate the physician. In mathematics, physics, banking, space exploration, etc., the computer is routinely called upon to perform tasks that all mankind, working 24 hours a day from creation, could not begin to duplicate, but in medicine our measure of success is diagnostic accuracy approaching a skilled clinician, ECG analysis which is substantially correct or historical data acquisition which saves the physician 5 minutes per patient. If our timidity were matched in other fields, it is very unlikely anyone could have justified the expense or the efforts necessary in these successful efforts. The disappointing impact of computer technology on medicine may have been caused by our inability ... to do more than emulate the efforts of an individual physician." (9)

Although some of the unfavorable comparison to the accomplishments which computers made in physics may be explained by the amount of variability in symptoms, progression of disease, individual response to disease and therapy, and to the combinations of problems which sometimes concurrently exist in a patient, the bottom line is that physicians do use logical processes to diagnose disease and manage patients. Data driven knowledge based systems do have the potential of doing something that no human being can logically do: scrutinize every piece of data which is collected for a patient and bring to bear the expertise of specialists regardless of the location of the patient or the time of day at which the data are collected.

As the content of the knowledge base becomes richer and more comprehensive, such systems appear to be capable of tremendously improving the quality, and perhaps the cost, of medical care. Currently, the methods for such accomplishments exist; the need is for more expert physicians to logically examine their thought processes and transmit their expertise to a computer compatible format. Algorithms, rules, and expertise are difficult to elicit and capture for any type of computer based representation. Medicine has traditionally been regarded as an art rather than a science; eliciting the decision-making logic tends to require the transformation of the field into a science. We foresee the day when medical knowledge and expertise will be shared among developers of information and decision-making systems. No single group will be able to adequately capture this knowledge across all subspecialties of medicine for all types (diagnostic and mangement) of decisions.

It is due to the ease of modifying and adding to this vast and currently incomplete medical knowledge base that most current developers feel that the knowledge based systems approach (a separate, easily manageable knowledge base) is superior to the hard coded program approach.

We recognize that the HELP system which we have developed will certainly not be the only system which is capable of providing these services; however, the systems which ultimately successfully emerge must incorporate the two basic principles which have been pivotal in the development and acceptance of our system: 1. integration into a comprehensive clinical database, and 2. data driven strategy for evoking the decision logic. It is still too early to tell whether the successful systems of the future must also have the modular, procedurally oriented frame structure which has characterized the knowledge base of our system. The main advantage of the modular approach is the way in which the system can expand gradually. Because data exist which satisfy the logical requirements for a specific decision, the decision is generally easily validated and believable. Those decisions which are produced are generally specific, valid and timely even though the expert logic may not exist for all possible interpretations or alarms.

In all cases, the ultimate decision-making responsibility remains with the human clinician. Many have wondered about the legal aspects of disagreeing with a computer consultant or placing the blame if the computer gives poor advice which is heeded by a healthcare provider. In our experience, the physician users generally feel that the risks which are avoided by the routine use of such a system substantially outweigh any potential legal disadvantages of such an approach.

References

1. Beck, J. R., Meier, F. A. & Rawnsley, H. M. (1981) Mathematical approaches to the analysis of laboratory data. Prog. Clin. Pathology *8*, 67–100.
2. Henry, J. B. (ed.) (1979) Clinical Diagnosis and Management by Laboratory Methods. Saunders, Philadelphia.
3. Davis, R. & King, J. J. (1985) The origins of rule-based systems in AI. In: Rule-based Expert Systems (Buchanan, B. G. & Shortliffe, E. H., eds.) p. 21, Addison Wesley.
4. Pryor, T. A., Gardner, R. M., Clayton, P. D. & Warner, H. R. (1983) The HELP System. J. Med. Systems *7*, 87–102.
5. Evans, R. S., Gardner, R. M., Bush, A. R., Burke, J. P., Jacobson, J. A., Larsen, R. A., Meier, F. A. & Warner, H. R. (1985) Development of a computerized infectious disease

monitor (CIDM). Comp. Biomed. Res. *18*, 103–113.

6. Evans, R. S., Larsen, R. A., Burke, J. P., Gardner, R. M., Meier, F. A., Jacobson, J. A., Jacobson, J., Conti, M. & Hulse, R. K. (1986) Computer surveillance of hospital acquired infections and antibiotic use. JAMA *256*, 1007–1011.
7. Hulse, R. K., Clark, S. J., Jackson, J. C., Warner, H. R. & Gardner, R. M. (1976) Computerized medication monitoring system. Am. J. Hosp. Pharm. *33*, 1061–1064.
8. McDonald, C. J. (1976) Computer reminders, the quality of care and the nonperfectability of man. N. Engl. J. Med. *295*, 1351–1355.
9. Friedman, R. B. & Gustafson, D. H. (1977) Computers in clinical medicine, a critical review. Computers in Biomedical Research *10*, 199–204.

Chapter 3.8
Expert Systems in Clinical Chemistry

Ch. Trendelenburg

1 Introduction

Medical problem complexes have always played an important part in the development of so-called expert systems, that is knowledge-based systems. The use of expert systems is spreading rapidly in scientific and industrial areas and one result of this will be that these systems will increasingly be adopted in practice in the medical field. Taking as example PRO.M.D., a Prolog System for decision support in the field of clinical chemistry and laboratory medicine, the properties and possible uses of so-called expert systems will be explained, taking particular account of the special requirements for laboratory medicine problem complexes.

The name "system for decision support" is a far more exact and better one as regards type of use and fields of application but compared with the name "expert system" it is only rarely used. Due to the typical separation made in expert systems of knowledge base and inference engine such systems are also referred to as knowledge-based systems.

The two aspects explained below are the main factors for predicting a further spread of knowledge-based systems in the area of medical problem complexes:

Improvement of availability: For medical problems, the use of expert systems depends to a high degree on their actual availability. Restrictions in this respect result in particular from the fact that expert systems, due to their complexity, were developed mainly on powerful minicomputers and mainframes. In addition, the systems developed so far often lack essential properties, such as implementation of the time concept essential in medical diagnostics. Now that powerful development languages are available on microcomputers of the so-called industrial standard (IBM-PC/AT or compatible versions of other manufacturers), it has become possible to develop expert systems which can run on computers of this type, which have meanwhile become widespread generally and specifically in laboratories. Analogous to the development of text-processing programs or programs for processing relationally structured data, a rapid spread of expert systems of ever increasing capacity and efficiency is now to be expected because the possibility of loading these systems on microcomputers made in larger numbers and a more effective feedback of the user's wishes will result in further improvement. Increasingly powerful versions will be developed which will approach the performance of the expert systems which can be loaded on for example minicomputers and in some areas even exceed their performance. Another point to remember is that the field of laboratory medicine in particular is closely linked to the industrial development of laboratory instruments and apparatus and in the near future this may be expected to result in an increased use of corresponding expert systems.

Knowledge explosion: Another reason for increased adoption of such systems in the medical field is the ever more rapidly progressing knowledge explosion which makes it impossible for the individual to always have at his disposal in optimum form, as is expected of the physician, all the relevant data for individual diagnostic and therapeutical decisions, and consequently more efficient systems for decision support than are used today will have to be employed. To do away with a widespread misunderstanding right at the beginning it is expressly emphasized here that the actual decision which is finally made will always be taken by the physician and not by a system of any kind. The system employed in partial areas simply supports the physician in making his decision in fundamentally the same manner as for instance reading a textbook. The fact that this is done in a manner related more exactly to the specific case and particularly efficient may possibly lead to a reintegration of medical action again, something which has been lost to some extent by the splitting up of medicine into increasingly specific subdivisions.

2 Structure and Mode of Operation of Expert Systems

A review of structure, mode of operation and greatly varied fields of use of expert systems will be found in excellent and easily understandable form in Goodall (1985), Waterman (1986) and Buchanan (1986). A few important characteristics will be briefly emphasized below, based on the observations of the above authors. In addition, a possibility of program-technical implementation of expert systems using the program language Prolog will be illustrated. To begin with, reference is made to Figure 1 which shows in simplified form the human communication process. It is clear from this Figure that even with relatively small interactive communication an answer to a question can be obtained if adequate common context (same language, similar degree of possession of expert knowledge, etc.) is present. In the symbolic notation of expert knowledge described below this common context in the interaction with an expert system is represented to a first approximation, no doubt very imperfect, as in a mirror and it is then certainly possible for a subject-specific dialog with concluding diagnostic indications to result when employing a suitable system.

2.1 Characteristic properties of expert systems

The block system (Fig. 2) shows the typical components of an expert system. In every expert system there are components for knowledge acquisition and components for deduction in interactive use with the possibility of explaining the formation of the consultation-specific conclusions. The knowledge base can be considered

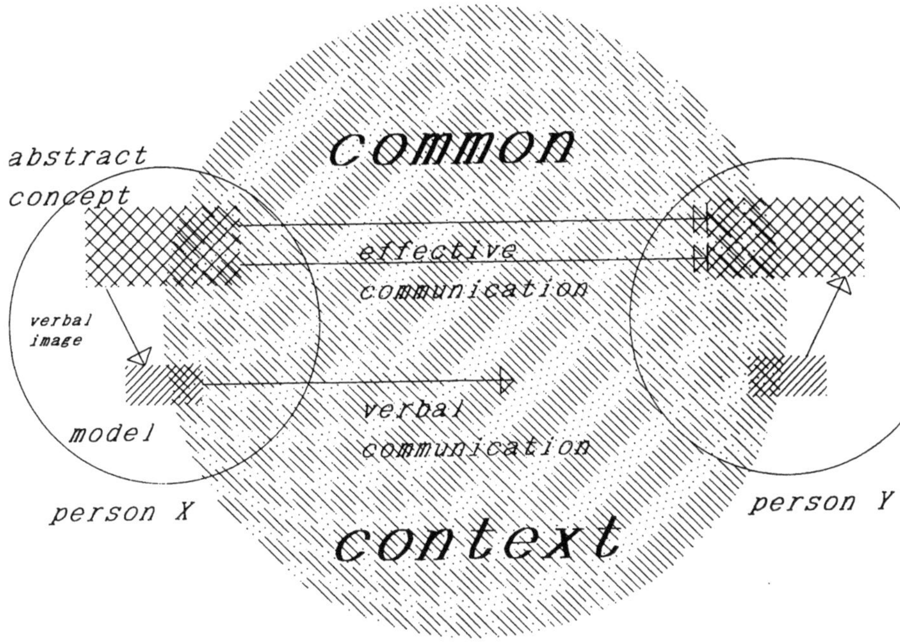

Fig. 1. The human communication process.

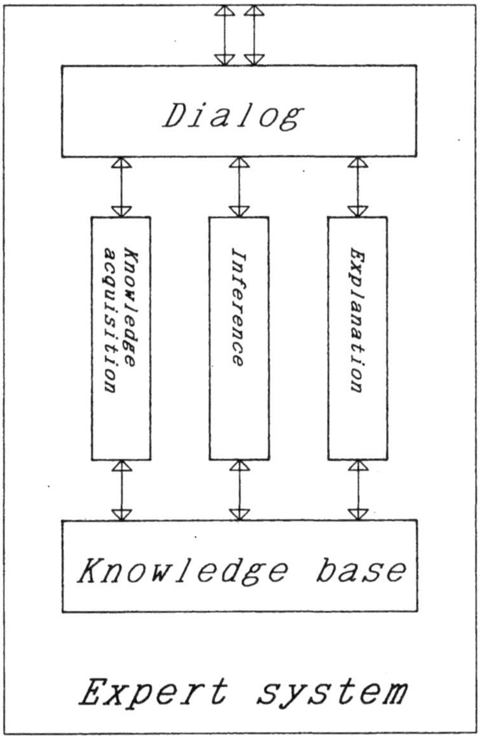

Fig. 2. Typical components of an expert system.

isolated from the system. Of great significance is the dialog component which must be present to build up the knowledge base and for interactive use. The separation of the user interface from the interface influencing the knowledge representation is becoming increasingly pronounced in more recent systems.

2.1.1 Separation of knowledge bases and inference mechanism

Expert systems differ from conventional programs, which of course can contain substantial knowledge components or decision aids, typically in particular in that they do not contain the knowledge representation in the program itself but in external so-called knowledge bases or rule systems. This characteristic is even used for a practice-oriented definition of the concept (Goodall 1985). Admittedly, there are always great interactions between the knowledge base and the inference engine; nevertheless, an expert system is typically suitable for different question complexes if in the individual uses the strategy of the inference process is substantially comparable. The knowledge components in these knowledge bases are for example in the form of numerous rules regarding the structure premise/conclusion, and a gradual easily performable modification and extension of the knowledge components can also be regarded as a typical characteristic of an expert system. Thus, in contrast to conventional programs knowledge components can be removed or added without appreciably disturbing or influencing the execution of the program; this only changes the quality of the decision support. In recent developments of expert systems a further difference has appeared, that is the possibility of representing the knowledge in a form which although syntactically restricted nevertheless corresponds largely to natural language.

2.1.2 Capability of symbolic inference

Symbols are defined as character chains for identification of concepts and material behaviour of reality (Waterman 1986). Such symbols can be combined to set forth the logical relationships between the symbols. To solve a given problem an expert system manipulates these symbol structures in a manner which is more heuristic than mathematical. An expert system works here effectively in a very narrowly defined area containing difficult problem complexes. This is why the rules of an expert system are necessarily frequently of high complexity due to their individual form or simply due to their great number. On the basis of the symbolic notation the rules can be formulated in a form of expression approximated to natural language.

2.1.3 Capability of explaining the inferences

An expert system can itself draw conclusions about its own mode of operation and thus optimize this mode of operation. Thus, in a rule-based system at any time the actual chain of inferences can be checked for correctness, consistency and plausibility of the investigated hypothesis. This makes it possible to derive from the system and represent explanations and justifications adapted to the individual situation of the consultation process. This is done by specific rules in the knowledge base which are higher-ranking than the other rules and contain the knowledge on logical and expedient modifications of the decision process (metaknowledge). A requirement for future systems in this connection is that these higher-ranking rules themselves be formulated as independently as possible from the actual knowledge bases so that the separation thus achieved in turn makes possible application of the higher-ranking rules to different knowledge bases.

Knowledge bases and the expert systems employed with them must be subjected to extensive trials before they can achieve a high degree of error-free applicability. Even then, as with human experts, they cannot be expected to be absolutely free of error. This however also applied to complex conventional programs with the essential difference that such programs did not have any explanation and justification properties and consequently were far less able to detect errors than expert systems and due to the lack of modularity of the knowledge components were also far less able to remedy any errors. In more recent developments, by deriving relevant knowledge bases from large collections of examples this ability of the expert system to learn through user feedback will be even further optimized.

2.2 Prolog (programming in logic) as efficient implementation language

LISP and Prolog are mainly used for implementing knowledge-based systems. Conventional languages, such as BASIC and Pascal, are becoming increasingly less significant for this purpose although an expert system can certainly be implemented in such a conventional language. A comparison of LISP and Prolog cannot be part of this work, although as a matter of interest it is mentioned that, for whatever reasons, it would appear at present that in America LISP is greatly favoured whereas in Europe and Japan Prolog is preferred.

Prolog differs in very many properties greatly from conventional programming languages. One basic requirement for Prolog to be easily learnt and applied in its appropriate form may be that learnig this language does not require any knowledge of another programming language or that a readiness exists to abandon the procedural arrangement of the statements and instructions, as is essential and cogent with conventional languages, in favour of the definition of logical declarations. This property of Prolog of itself seeking the method for fulfilling these declarations, if necessary even by backtracking and new ramifications in the case of several possibilities, is one of the main characteristics of Prolog distinguishing it from conventional program languages. Further essential characteristics depend on a built-in dynamic data base, an efficient implementation of recursive notation and rapid linking and rapid comparison of occupied or unoccupied variables, as well as the ability to process lists and structures in almost any conceivable form. It is also possible to employ for representing knowledge relationships the so-called operator concept which permits the compilation of interpretable simple sentences and statements and which is explained in detail in the representation of the language concept for the formulation and maintenance of the rules. All these properties are exceedingly helpful for the so-called symbolic programming and thus for the development of expert systems.

Another very import factor the further spread of Prolog has proved to be that in the meantime efficient implementations of the language on small computers of the so-called industrial standard have meanwhile become available. This applies in particular to Prolog-2 (Expert Systems International, Oxford) which was used in the development of the system described below for decision support in laboratory medicine. The most important properties of Prolog-2 going beyond the language scope described by Clocksin and Mellish (1985) or by Bratko (1986) are briefly enumerated here: module concept with definable range of variables, library procedures, partial compilation, virtual access technique for large data base components, automatable or automated indexing and hash access techniques in the area of data base functions and support for use of so-called screen windows which are necessary for example for creation of an easily understandable user guide. In the development of a knowledge-based system described below Prolog-2 has proved exceedingly suitable for achieving these objectives. The declaratively symbolic notation does indeed extremely simplify the implementation of ideas and concepts during the development and the improvement of the system compared with conventional programming.

2.3 A brief review of medical expert systems

In the following a brief review of medical expert system is given (Pohl and Trendelenburg 1986). Medical consultation systems such as MYCIN (Shortliffe 1976), INTERNIST (Pople 1975), CASNET (Weiss 1978) and PIP (Pauker 1976) were developed in the 1970s for consultation on specific medical problems. These were then followed by skeleton expert systems such as EMYCIN (van Melle 1984) and EXPERT (Weiss 1979) which were independent of a specific knowledge base. The inference engines in these systems were generalized to enable them to be used for different domain knowledge bases. Thus, the knowledge base of INTERNIST could be translated successfully into an EXPERT model (Goldberg 1982). ATTENDING (Miller 1983 and 1984) was developed in the early 1980s and was the first criticizing system. It was given a list of patients' medical problems and the procedure planned and discussed the risks and benefits of the proposed approach and of other reasonable approaches.

Although probabilistic reasoning is implemented in most expert systems it is not required for many decisions in medicine and pure propositional logic would suffice in many cases (Darvas 1978, Szolovits 1978 and Friedman 1983). Scoring inference engines (Puppe 1985) additionally add up the scores of the systems and signs present and conclude a diagnosis only if the total score confirming it exceeds a predetermined threshold value. Fuzzy logic (Zadeh 1965, Forsyth 1984 and Adlassnig 1985) extended classical Boolean logic to real numbers, where 1 represents truth and 0 represents falsity and all fractions between 0 and 1 are employed to indicate partial truth. Certainty factors (Shortliffe 1984) relate premise and conclusion of a causal rule by one single number and represent a mixture of probabilistic and cost/benefit judgement. The Bayesean theorem (Bayes 1753 and Diamond 1979) is widely accepted as a means of predicting the probability of diseases. It requires too much statistical data to permit its broad application. The Dempster-Shafer theory (Barnett 1981 and Gordon 1984) is an extension of the Bayesean theorem and is more flexible in representing and dealing with uncertainty. It requires a hierarchically ordered dictionary and its computational complexity depends critically on the size and structure of the knowledge base.

Since time-oriented data management is one of the main characteristics in taking medical decisions, there is a great need of temporal reasoning in medical expert systems. The Ventilator Management Program VM (Fagan 1979 and 1984) can handle real time data from an intensive care unit. MECS-AI (Koyama 1981) is a consultation system which provides a diagnostic judgement and gives treatment and status of the patient in the past as well as his current status. The Digitalis Therapy Advisor (Gorry 1978) uses a formal model of pharmacokinetics and a variety of qualitative clinical data to construct a patient-specific model to enable conclusions to be drawn regarding the time for the next dose of digitalis. A dynamic mathematical model, as used here for pharmacokinetics, is a powerful general method of representing temporal knowledge. Obermeier (1985) implemented the knowledge-based text-processing system GROK to translate natural language sentences from a patient's history case report to a data structure where the statements are ordered chronically and the representation of the temporal information is interval-based. The statements are organized around domain-specific key events, e. g. accident, admission or surgery, which have to be recognized by the system. A limited number of key events provides the hooks defining the temporal structure of the domain-specific texts.

Meta-level knowledge (Davis 1984) is knowledge about knowledge. This concept enables an expert system to reason about its own reasoning. This means that a program cannot only use its knowledge directly but can also examine it, abstract it, reason about it and direct its applications. Thus, meta-level knowledge enables an expert system to become as it were conscious of its own knowlege and reasoning.

3 Requirements Made of a Knowledge-Based System for Supporting Laboratory Medicine Diagnostics

Below, taking as example the development of an own Prolog system for supporting medical diagnostics (PRO.M.D.), the typical characteristics and the extensions of significance medically and in laboratory medicine are illustrated. In the field of clinical chemistry and laboratory medicine the use of knowledge-based systems is governed essentially by the actual availability explained at the beginning and by the implementation of specifically suitable inference engines. The inference engine is the actual program core on the specific properties of which the capacity and efficiency mainly depend.

3.1 Implementation of an inference engine suitable for medical question complexes

As regards the quality of inference engines, experience shows that generalized inference engines suitable for many question complexes often tend to give moderate results when applied to specific questions. This makes it clear why so far no standardized generally accepted implementation form has been found for inference engines. For this reason an attempt was made to develop an inference engine particularly suitable for laboratory medicine.

In contrast to program development using conventional languages, in which as comprehensive as possible a primary concept is important, the methods of gradual refinement are expedient. During the development phase by creating and testing small exemplary knowledge bases for the particular current version of the system pronounced interactions between the system properties and the question complex being represented are generated. The declarative nature of Prolog makes it possible in astonishingly simple manner to convert the difficulties arising here in the formulation of the knowledge components to improvement proposals and incorporate them into the system in order to adapt the system characteristics and the notation of the knowledge components in this manner step by step to the peculiarities of the knowledge area to be represented. In this process in particular the aim must be to continuously improve the inference engine with the objective of enabling area-typical relationships, strategies and other peculiarities to be represented in linguistically easily reproducible form in the formulation of the knowledge components.

3.2 Combination of numerical and non-numerical information

The combination of numerical data and non-numerical information, for example clinical information, which is exceedingly important in laboratory medicine diagnostics is either absent completely from many systems or not contained in appropriate form. Thus, as implemented in PRO.M.D., it must be possible for results in the typical two-sided range gradation (in the normal range, somewhat deviating, greatly deviating, extremely deviating) or for results in pure numerical form and also for characteristics of clinical anamnestic data in symbolic notation to be entered equally both in the rule syntax and during the interactive use and to be interpretable by the inference engine. Calculated or derived values are also possible. In addition, if required these particulars can be made more precise as regards the weighting of the final part of the particular rule and as regards the time relationship.

3.3 Weighting

In rule-based systems the dynamic structure of so-called certainty factors has become a standard property. It is possible here to allocate weighting factors to individual hypotheses in individual rules thus assigning in accordance with the sequence of the consultation and number of statements indicative of a hypothesis an increasing weighting factor to the latter by which the further diagnostic strategy can then be influenced. The weighting factors have nothing to do with a statistical description and statistical techniques; they are rather to be understood as subjective measure of significance based on experience. There are various techniques for compiling these factors during the consultation. For PRO.M.D. the Shortliffe (1976) technique was implemented in modified form. Components of the weighting factors supporting a hypothesis (measure of belief) and components contraindicative of a hypothesis (measure of disbelief) are maintained separately throughout the entire consultation, can be called up jointly or separately and are united until the final output of the diagnostic indications. The incorporation of statistically substantiated decision criteria at suitable points in the rule syntax of PRO.M.D. has been prepared and will be added in one of the next configuration levels. Generally, the trend of expert systems is towards the incorporation of decision parameters from model calculations or statistic calculations into the decision within the empirical-heuristic sequence of the consultation at points at which this is possible. This goes as far, as in the case of PRO.M.D., as executing freely definable actions in the form of external Prolog auxiliary programs which can be called up.

3.4 Screen representation and user interface

The screen representation is configured so that a consultation takes place on a (stationary non-rolling) screen form (Fig. 3) with constant logic/spatial divisions. There are fixed screen areas for displaying the information indicated below:

- Program name and program version number
- Current question
- Program state ("input" as request to input or "working" as indication of internal program activities)
- Input area (for numerical input and input of for example file names, or for displaying the concepts selected via the function keys)
- Reference to the selection possible at any time of additional options during the consultation or reference to result output when no further questions are put or the user calls up the final output via the option list.
- Active rule or additional explanations or information on the validity of a rule in the course of application of the inference engine or diagnostic references during or at the end of the consultation.
- Function key allocation to the program operation for input of symptom characteristics and representing and inputting values which are specified as regards their range gradation.

The input characteristics such as "elevated", "positive", "true", "indefinite", etc., is accordingly not by entering the concepts via the alphanumerical keyboard but by selection via the function keys to which the corresponding possible answers have to be reassigned from question to question. This also applies to calling up options by which a great number of auxiliary functions can be used or activated during the consultation.

As in most expert systems, the interface to the user in PRO.M.D. is kept as far as possible in natural language and this is facilitated in many ways by Prolog.

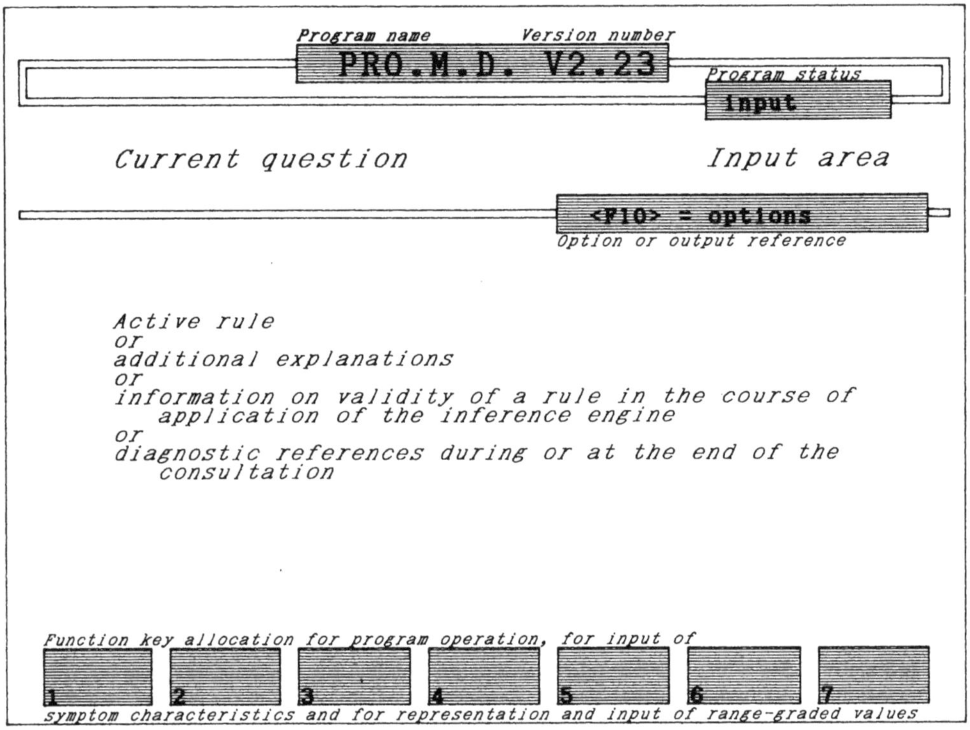

Fig. 3. Screen layout of PRO.M.D.

The entire diagnostic context including all diagnostically relevant information is compiled by the user himself when the knowledge base is created. These characteristics enable the knowledge notation to be implemented in particularly simple manner. This is of particular significance because asking about expert knowledge and the conversion of it to program-adapted form has proved to be the problem complex most difficult to solve in many program systems, requiring in some cases the assistance of an additional specially trained person for this task as mediator between experts and expert system (knowledge engineering). For the objectives and task area of PRO.M.D. the interface to the user has been configured so that the user himself, with appropriate expert knowledge but without specific computer skills, is able to perform on his own the knowledge notation. An essential factor here for this is the natural language structure of the rules because in this case the expert statement of a rule can be derived from the latter without decoding directly and understood.

3.5 Properties of the inference engine

Fundamentally, the sequence of processing the rules of the knowledge base can be either a data-directed one or a goal-directed one. In the goal-directed processing by calling up underlying rules after predefining a hypothesis all information necessary for confirming the hypothesis is interrogated. If a rule is selected whose reply first requires that another rule be called up, then this other rule is called up first, and so on. In data-directed processing the interrogation takes place firstly in the order of the rules but is then greatly influenced in the course of the processing by the individual fact constellation because rules which on the basis of the facts already known are no longer valid are not processed any more. PRO.M.D. has in modified form both properties because firstly the processing is data-directed and in the further course of the consultation may also be goal-directed. This procedure was implemented because in the case of laboratory medicine question complexes usually a symptom-dependent and result-dependent selection from amongst several hypotheses is desired. Specific rule formulations then permit exact examination of a hypothesis in accordance with its weighting.

3.6 Storing and reconsultation

In additon to the actual knowledge base the data of individual patients can be stored and called up at any time to permit interruptions of the consultation. On such an interruption dynamic data on the actual diagnosis process are also stored. When called up again the additional relations are automatically updated taking account of the stored time data and using the date. This possibility of reconsultation has been implemented in this form in PRO.M.D. because it corresponds to the usual diagnostic procedure. Thus, analogously to the course of a diagnosis it is possible for example with an inpatient to clarify indications initially provided by the system, for instance by ordering an examination, and then to continue this process in the course of time so that diagnostic strategies can also be imaged in the knowledge base. Because of these characteristics the system is able to provide indications even when the information items becoming available distributed over a period of time are incomplete. This is an essential difference to many statistically operating methods for diagnosis support in which all the items of information must be available at the same time as is typically not the case until the end of the diagnostic process.

3.7 Imaging of the time concept

In the development of PRO.M.D., from the start the time concept exceedingly important in medical diagnostics has been imaged in the inference engine and knowledge notation. This property is implemented only very rarely in expert systems. If this property is not present the system is restricted to the assessment of static findings and symptom constellations and this is a very serious restriction. Even in its present version PRO.M.D. permits formulation of the diagnostic time concept in a new manner which is very general and can thus be universally employed and with which the diagnostic process can be imaged along both directions of the time axis with pathologically dependent time period variation. This implements the representation of temporal relationships between diagnostically significant pathological events, the so-called key events, relatively to each other and absolutely relatively to a date (Trendelenburg 1985, 1986a).

It is therefore possible to formulate and support in the knowledge base complex time-oriented rules for imaging diagnostically important temporal relationships. This will be made clear by the following hypothetical rule formation which is however syntactically possible (cf = certainty factor).

premise
sign_a was positive with_time_relation 6 to 8
months after start of disease and sign_b is positive today

conclusion
symptom_x becomes pronounced with time relation of 3 weeks after symptom_y with 0.4 cf

This type of representation of time relationships can be taken as a first attempt at a general knowledge notation of the concept of disease for knowledge-based systems.

3.8 Fact input

With uniform question complexes, such as the interpretation of findings patterns, a preliminary input of characteristics for example of laboratory investigations may be desirable. This may also be done by accessing an external file or an external relational data base system. In the simplest form, for this purpose a corresponding rule is simply prefixed at the start of the knowledge base and initiates the interrogation of the laboratory investigations which are known from the start. With this procedure the interactive input only starts when in the course of application of the inference engine questions must be put on additional requirements.

3.9 Display of conclusions and correction

By selecting the option "propositions" it is possible at any desired instant during the consultation to display the entries of the user and the inferences and conclusions of the system. Information on time relationships and the magnitude of the weighting factors are also contained. By selecting the option "update" user entries can be corrected or modified to run through the corresponding effects on the sequence of the inference engine application (inference chain). Thus, questions of the type "What would happen if" are put to PRO.M.D. and the current knowledge base.

3.10 Explanation facility

By selecting the option "why" or the option "explanation", for a given question once or continuously the current rule text is also included in the display as an explanation. On selection of the option "information" PRO.M.D. shows observations made in the course of the application of the inference engine in the context of the rule which has been effective. These options can also be deactivated again by the option "i + e off". In one of the next configuration levels a graduated explanation function will be implemented which will make it possible to display free text for explaining a question or rule on request or in dependence upon the course of the individual consultation.

3.11 Connection to the outer world

Through a specific rule type, the so-called "action rule", when using PRO.M.D. an external Prolog program can be read in and executed in dependence upon the course of the consultation. After execution the external Prolog clauses are again automatically removed from the collection of all the Prolog clauses disposed in the memory. This makes it possible at a suitable point to represent complex information also in graphical manner or make available information for the system by applying suitable computations whenever it is expedient and logical to apply for example an underlying mathematical model or a multivariant statistical theorem. The essential point here is that for such auxiliary functions the PRO.M.D. system itself need not be modified and the auxiliary functions can be created by the user if he has an appropriate version of Prolog.

3.12 Range facility

Typically, laboratory results in a diagnostic process must be used both in numerical and in non-numerical form, i.e. symbolic form. An expert system suitable for laboratory medicine question complexes must allow both these types of representation. A specific range rule permits simple allocation of value classifications such as slightly low, normal, greatly elevated, etc. For a further configuration level of PRO.M.D. extension of the range facility is planned for the dimensions age and sex and for range-typical conditions.

3.13 Indication output

At the end of the consultation the system provides indications and hints for diagnoses or partial diagnoses, stating the weighting factors and the time relationships. It also provides information on other diagnoses or partial diagnoses which are supported by the knowledge base and of which no indications were found in the current consultation in order to draw the user's attention to the diagnostic capabilities which the system has.

3.14 Structure of the knowledge base, rule types

The system permits creation and processing of expert rules from any desired medical areas. The rules have the form shown below:

- "if premise then conclusion"
  or "if premise then conclusion with number cf.".

- A premise is a propositional logical statement with so-called propositions as elements. In other words, a premise consists of one or more propositions which are separated from each other by "or" or "and". An "and" forms a stronger link than an "or".
- Conclusion is a freely selectable designation for the deduction which can be drawn from the premise.
- Number, expressed as certainty factor (cf) in an arbitrary unit, is the certainty with which the conclusion can be deduced if the premise is true.
- A proposition is a condition which must make a statement about the patient, e. g. "urea greater than 12". There are three types of propositions:
  1. Statement of a word (e. g. "infectious"). Such a proposition usually denotes a property to which no numerical value can be assigned.
  2. A construct is the form "word is value". Word is usually an examination or a symptom, e. g. "potassium" or "joint swelling". The admissible expressions for value are "keywords" such as "present", "positive", "negative", numbers and comparative operations, etc.
  3. To express time relationships the two first variants can be modified to "word time ago" and "word was value time ago" respectively. Time is a statement of duration such as "1 day", "2 days", "3 weeks", "6 months". A relative time relationship with respect to a diagnostically significant key event can also be represented using the statement "with time relationship".

  Examples of the propositions:
  "icteric"
  "potassium is greater than 4.2"
  "hbsag was positive 4 weeks ago"

As an example of a rule picked arbitrarily from the context of a large number of rules a hepatitis rule system will be set forth here:

Premise
acute-start-of-disease is
wrong and
hbsag is negative and
hbsag was negative 6 months ago
and
acute-start-of-disease was
present 6 months ago and
(anti-hbe is positive or
anti-hbs is positive)

Conclusion
hbsag-negative-hepatitis-b
with
0.6 cf.

It is of interest here that conclusions can in turn appear as propositions in subsequent rules. This makes it possible to create complex knowledge bases which in interaction with the user specifically request missing information in accordance with the course of the individual consultation.

The system can be asked at any time why a specific question was put; it is also possible to select a mode which explains why a concluding diagnostic indication is given.

The following types of formulation can be assigned to the premises (V) and conclusions (F). Particular attention should be paid to the temporal formulations and the formulations influencing the certainty or weighting factors (cf). Examples have been selected from a knowledge base for supporting serological hepatitis diagnostics:

(V) anti-hav is at least
slightly-elevated
(V) hepatitis-a with-more than 0.4 cf
(V F) anti-hav is positive
(V F) hbsag was positive 1 – 5 months ago
(V F) start-of-disease was acute 3 – 4 months before seroconversion with 0.6 cf
(V) "and" and "or" links of premises ...
(F) only "and" links of conclusions ...
(F) hepatitis-a with 0.7 cf
(F) question-time-of symptom-x
(F) put hepatitis-b to 0.8 cf

3.15 Language concept, rule maintenance

The entire knowledge base can be created in approximately natural language formulation. Prolog greatly supports this through the operator concept by which concepts such as and, or, with, ago, after, days, month, slightly-elevated can be used in the usual linguistic form based on the correct logical declaration.

The Prolog operators are used here to prove the comprehensability of the rules in accordance with the example given in Figure 4.

```
premise: enzyme_x was greater_than 90 units 4 weeks ago    conclusion: disease_y with 0.3 cf

                                   ---units                                           ----cf
                      greater_than--------- --weeks                        ----------with-------
                      ------------------------------ago
         ---------was----------------------------------
premise:-----------------------------------------------
-----------------------------------------------------------conclusion:-----------------------
```

Fig. 4. The use of Prolog operators for rule formulation in PRO.M.D.

By keeping to certain syntactic rules the knowledge notation is thus made directly in Prolog syntax without this being noticed by the user. Even the full-stop at the end of the sentence, which is very important to Prolog because it denotes the end of a logical declaration, can be employed in the usual manner for the user in this type of operator concept application. This thus obviates a serious obstruction in the use of program systems configurable in any manner, that is the conversion of the entry to computer-oriented form.

3.16 Extension of the language scope of PRO.M.D.

Previous uses of PRO.M.D. have shown that the incorporation of statistical data on the validity of individual test methods is of great importance to laboratory medicine question complexes. PRO.M.D. will therefore permit in the next configuration level an extended form of knowledge notation as represented by the way of example below (Pohl 1987). The following example, meant only as an illustration, is intended to show how the financial expenditure, the benefit of an examination and specificity are to be represented in the language.

This gives the following knowledge basis.

Comments	'hypothetical knowledge basis for differential diagnosis of abdominal pain in male patients aged between 5 and 25 years'.
Question	'abdominal pain' costs 20 Pfs.
Question	'age (in years)' costs 20 Pfs.
Question	'right-side lower abdominal pain' + (yes, no) costs 1 DM
Question	'squelching coecum' costs 1 DM
Question	'leucocytes/μl' costs 4 DM
Result	'appendicitis' benefits 2000 DM
Result	'gastroenteritis' benefits 100 DM
Premise	'age (in years)' $>$ 5 and 'age (in years)' $<=$ 15
Conclusion	'boy'
Premise	'age (in years)' $>$ 15 and 'age (in years)' $<=$ 25
Conclusion	'youth'
Comment	'prevalences of the diagnoses'
Premise	'abdominal pain' and 'boy'
Conclusion	'appendicitis' in 55%
Premise	'abdominal pain' and 'youth'
Conclusion	'appendicitis' in 60%
Premise	'abdominal pain' and 'boy'
Conclusion	'gastroenteritis' in 47%
Premise	'abdominal pain' and 'youth'
Conclusion	'gastroenteritis' in 30%
Comment	'symptoms of the diagnoses'
Premise	'abdominal pain' and ('boy' or 'youth')
Conclusion	'appendicitis' with 'right-side lower abdominal pain' with sensitivity 90% specificity 35% and with 'squelching coecum' with sensitivity 15% specificity 8% and with 'leucocytes/μl' $>$ 12000 with sensitivity 80% specificity 40%

Premise	'abdominal pain' and ('boy' or 'youth')
Conclusion	'gastroenteritis' with 'right-side lower abdominal pain' with sensitivity 60% specificity 7% and with 'squelching coecum' with sensitivity 95% specificity 90% and with 'leucocytes/μl' > 12000 with sensitivity 55% specificity 47%.
Comment	'end of the hypothetical knowledge base'

4 Applications

The first application for daily routine we selected was the interpretation of quantitative lipoprotein electrophoretic tests (precipitation method, Immuno AG, Vienna) in combination with values of serum cholesterol and triglycerides. The large knowledge base (Wieland 1985, Trendelenburg and Wieland 1986) incorporating a wide variety of aspects enables the system to perform the tedious task of daily intepretation in the following manner:

- examination of preanalytical influences, such as errors due to storage or erroneous data input of interferences due to drug therapy, e. g. heparin
- detection of errors during execution and of the presence of atypical lipoproteins such as Lp(a) and Lp(x)
- interpretative diagnostic comments for different forms of hypertriglyceridemia, the quality of prebeta lipoproteins, hyperlipemias according to the Fredrickson classification, hypolipemias such as hypoalphalipoproteinemia, Tangier's disease, abetalipoproteinemia, hypobetalipoproteinemia, typical patterns induced by various underlying diseases: diabetes, renal insufficiency, hypo and hyperthyreosis, cholestasis
- assessment of the individual cardiovascular risk, including advice on any further laboratory tests (if required) to confirm the diagnosis
- individual therapeutical proposals for the specific case.

At present a knowledge base is being created for interpreting the thyroid gland findings pattern from the laboratory diagnostic point of view (Keller 1986) incorporating additional data of the sender (e. g. clinical question complex, diagnostically relevant extrathyroidal diseases and conditions, drugs).

Also planned is a knowledge base for supporting diagnostics of consumptive coagulopathy with the resulting challenge to the perfection of the imaging of temporal processes by PRO.M.D.

5 Further Expert Systems for Clinical Chemistry Question Complexes

Three systems will be briefly characterized below as representatives of further expert systems for clinical chemistry uses. No claim to completeness is made.

The system of Ahmad and Wells (1985) is written in Prolog. In the various development stages of the system an attempt is made to unite the declarative statement (a collection of facts with small set of inferencing procedures) and the procedural statement (the bulk of the knowledge represented as procedures for its uses). In a first stage Ahmad and Wells have created a declarative system with notation of the knowledge in rule form and a simple goal-driven, backward-chaining inferencing mechanism. Confidence factors are used to handle inexact reasoning. Use has shown that with this attempt the procedural knowledge elements frequently additionally employed for clinical algorithms and question complexes can only be poorly imaged. For this reason, in a second attempt a more axiomatic approach was chosen in which the procedural information about the next step is stored within the rule itself. This system has been tried for the following knowledge areas: differential diagnosis of hypoglycaemia, classification of hyperlipoproteinemias, clinical comments on digoxin measurements, therapeutic monitoring of theophylline infusion. The present development of the system is pursuing the aim of incorporating the concept of normal values and time series, implementing connections to external program actions and employing so-called blackboard techniques with which the inferences of a relatively large number of expert systems which are related to each other can be combined.

An approach has been adopted by Frick, Sator, Kratzer and Knedel (1986) which was very successful as regards the representation of uncertain knowledge. As implementation language Prolog was chosen together with a very convenient interface to the language C for additional implementation of rapid mathematical capabilities and a graphical user interface. As field of use the diagnosis of blood coagulation disturbances was chosen. The concept of normal and pathological ranges, together with the related problems of the exact description of range transitions, was solved by employing the so-called fuzzy logic. The system supports in a graphical screen dialog guide the selection of a diagnosis best fitting a current value combination. In a converse conclusion it can also represent the most logical test combination for an assumed diagnosis. In contrast to many other expert systems it is intended and suitable for real-time use within the scope of laboratory medical plausibility checks.

In contrast to the systems described hitherto Besset (1986) has used a commercially available system for creating a knowledge base called BABAR for supporting the work sequence and diagnostics in a medical laboratory.

BABAR was created using the expert system Shells M.1 and K.1 of FRAMENTEC. The objective is to create with a rule-based system both medically oriented mechanisms for plausibility checks and indications for diagnosis for a commercially operating laboratory in which the 50 most common analyses are carried out. From these analyses BABAR can give indications of 95 diseases which give specific result constellations, possibly to be further checked, or for example make a further analysis in blood coagulation necessary.

6 Limits and Possibilities of Knowledge Representation by Expert Systems in the Field of Laboratory Medicine

6.1 Limits

The limits on using expert systems are due mainly to the capability and efficiency of the expert system and scope and complexity of the field of use not being matched to each other. Application areas requiring different complexity levels, different types of knowledge representation and consideration of specific underlying models can be processed only with expert systems permitting a corresponding notation of the knowledge. Of course, depending on the objective compromises can be made here; thus, by suitable concept choice within the symbolic notation more complex components of knowledge notation can be satisfactorily imaged or bridged even with inference engines of very simple structure.

A very important factor is the limiting and ability to delineate the problem complex to be processed. Comprehensive differentially and diagnostically complex areas frequently require with the present development state of expert systems and the techniques of knowledge notation, themselves just being devised, years of use and even then there is no certainty of success. A further essential factor to be remembered when using expert systems is to ensure that their application does indeed bring an appreciable advantage in daily use; this is not always the case even with systems which have been developed at great expense with the objective of achieving precisely such an advantage.

6.2 Possibilities

On the whole, the conditions for the use of expert systems in the area of laboratory medicine are favourable at present. Generally speaking, the use of computers in laboratory medicine has already become a tradition to a certain extent and in some laboratory EDP systems in the meantime the first attempts at a configurable pattern detection, findings examination and findings comments are being implemented (Trendelenburg 1984). In the HELP system (Clayton 1985) which spans several specialist fields configuration areas exist which in many aspects correspond to small expert systems and knowledge bases although this applies more to the procedural notation than the declarative. In this system diagnostic comments, therapeutical recommendations and alarm messages in the case of ominous constellations are generated beyond the boundaries of the individual specialized fields.

The use of computers within laboratory medicine for diagnostic question complexes as well also has a certain tradition. By way of example, attention is drawn to the great number of statistical attempts, in some cases very different from each other, for handling multivariant result assessment. A particularly important factor making possible an efficient use of expert systems in the field of laboratory medicine is that some potential areas of use within laboratory diagnostics fulfil the limits and requirements specified above very well. It is precisely in the evaluation of localized specific result combinations, the so-called specialist findings evaluation, that the use of expert systems appears expedient in all cases where additional diagnostically relevant knowledge is contained in the corresponding knowledge base, and here the aim must be to obtain a consensus within the specialized area on the nature and scope of this knowledge notation going beyond the individual scope. Particularly suitable for employing an expert system are those areas of specialist findings evaluation in which the incorporation of clinical data is essential to the diagnostic indications, i. e., the quality of the evaluation greatly depends on this incorporation and on the existence of such data.

6.3 Objectives

At present, the following objectives can be defined for the use of expert systems in laboratory medicine:

- an improvement of the informative content and the technical execution of specialist findings evaluations in correctly selected areas of use
- the start of a new type of knowledge notation and knowledge maintenance going beyond the familiar generalizing representation in books and other publications in order to permit a quicker application of extensive knowledge in a specific manner adapted to the needs of the individual patient.

References

1. Adlassnig, K. P., Kolarz, G. & Scheithauer, W. (1985) Present state of the medical expert system CADIAG-2. Method. Inform. Med. *24*, 13–20.
2. Ahmad, K. & Wells, J. G. (1985) Expert Systems in Clinical Biochemistry: Some Case Studies. In: Computing in Clinical Laboratories: Databases – Data Presentation – Expected Developments, Proceedings of the Fifth International Conference (Trendelenburg; Chr., ed.) pp. 135–136, Kohl, Brackenheim.
3. Barnett, J. A. (1981) Computational Methods of Mathematical Theory of Evidence. Proc. 7th IJCAI, pp. 868–875.
4. Bayes, T. (1753) An essay towards solving a problem in the doctrine of changes. Philos. Trans. R. Soc. London (Biol.) *53*, 370.
5. Bratko, I. (1986) Prolog Programming for Artificial Intelligence. Addison-Wesley, Wokingham.
6. Besset, D. (1986) Medical Diagnostic Using Expert Systems: A Practical Case. In: Artificial Intelligence and Advanced Computer Technology Conference (AI-Europa) Proceedings, TCM Exposition Publisher, Liphook.
7. Buchanan, B. G. (1986) Expert systems: working systems and the research literature. Expert Systems *3/1*, 32–51.
8. Clayton, P. D., Gardner, R. M., Larsen, K. G, Pryor, T. A. & Warner, H. R. (1985) HELP/PATHLAB integration – A decade of experiences using an expert system interface to a clinical laboratory system. In: Computing in Clinical Laboratories: Da-

tabases – Data Presentation – Expected Developments, Proceedings of the Fifth International Conference (Trendelenburg, Chr., ed.) pp. 109–113, Kohl, Brackenheim.
9. Clocksin. W. F. & Mellish, C. S. (1985) Programming in Prolog. Springer, New York.
10. Darvas, F., Futo, I. & Szeredi, P. (1978) Logic-based program system for predicting drug interactions. Int. J. Bio-Medical Computing *9*, 259–271.
11. Davis, R. & Buchanan, B. G. (1984) Meta-Level Knowledge. In: Rule-Based Expert Systems (Buchanan, B. G. & Shortliffe, E. H., eds.) pp. 507–530, Addison-Wesley, Reading.
12. Diamond, G. A. & Forrester, J. S. (1979) Analysis of probability as an aid in the clinical diagnosis of coronary-artery disease. N. Engl. J. Med. *300*, 1350–1358.
13. Fagan, L. M., Kunz, J. & Feigenbaum, E. A. (1979) Representation of Dynamic Clinical Knowledge: Measurement Interpretation in the Intensive Care Unit. Proc. 6th IJCAI, pp. 260–262.
14. Fagan, L. M., Kunz, J., Feigenbaum, E. A. & Osborn, K. K. (1984) Extension to the Rule-Based Formalism for a Monitoring Task. In: Rule-Based Expert Systems (Buchanan, B. G. & Shortliffe, E. H., eds.) pp. 397–423, Addison-Wesley, Reading.
15. Forsyth, R. (1984) Fuzzy reasoning systems. In: Expert Systems – Principles and Case Studies (Forsyth, R., ed.) Chapman and Hall.
16. Frick, H. G., Sator, H., Kratzer, M. A. A. & Knedel, M. (1986) WILFUL – A knowledge based system supporting coagulation diagnostics by means of prolog and fuzzy logic. In: Artificial Intelligence and Advanced Computer Technology Conference (AI-Europa) Proceedings, TCM Exposition Publisher, Liphook.
17. Friedman, R. H. & Frank, A. D. (1983) Use of conditional rule structure to automate clinical decision support: A comparison of artificial intelligence and deterministic programming techniques. Computers and Biomedical Research *16*, 378–394.
18. Gordon, J. & Shortliffe, E. H. (1984) The Dempster-Shafer Theory of evidence. In: Rule-Based Expert Systems (Buchanan, B. G. & Shortliffe, E. H., eds.) pp. 272–292, Addison-Wesley, Reading.
19. Goldberg, R. N. & Weiss, S. M. (1982) An experimental transformation of a large expert knowledge base. Journal of Medical Systems *6/1*, 41–52.
20. Goodall, A. (1985) A Guide to Expert Systems. Learned Information Systems, Oxford.
21. Gorry, G. A. & Silverman, H. (1978) Capturing clinical expertise – a computer program that considers clinical response to digitalis. The American Journal of Medicine *64*, 452–460.
22. Keller, H. (1985) Klinisch-Chemische Laboratoriumsdiagnostik für die Praxis. Thieme, Stuttgart, New York.
23. Koyama, T., Kaihara, S., Minamikawa, T. & Kurokawa, T. (1981) Time-Oriented Features for Medical Consultation Systems. Proc. 7th IJCAI, pp. 910–912.
24. van Melle, W., Shortliffe, E. H. & Buchanan, B. G. (1984) EMYCIN: A knowledge engineer's tool for construction rule-based expert systems. In: Rule-Based Expert Systems (Buchanan, B. G. & Shortliffe, E. H., eds.) pp. 302–313, Addison-Wesley, Reading.
25. Miller, P. L. (1983) Critiquing anesthetic management: The ATTENDING computer system. Anesthesiology *58*, 362–369.
26. Miller, P. L., Blumenfrucht, S. J. & Black, H. R. (1984) An expert system which critiques patient workup: Modeling conflicting expertise. Computers and Biomedical Research *17/6*, 554–569.
27. Obermeier, K. K. (1985) GROK – a knowledge based text processing system. In: Artificial Intelligence and Advanced Computer Technology Conference (AI-Europa) Proceedings, TCM Exposition Publishers, Liphook.
28. Pauker, S. G., Gorry, G. A., Kassirer, J. P. & Schwartz, W. B. (1976) Towards the simulation of clinical cognition: Taking a present illness by computer. Am. J. Med. *60*, 981–996.
29. Pohl, B. & Trendelenburg, Chr. (1986) PRO.M.D. – Towards an optimal inference engine including temporal reasoning in medical diagnostics. In: Artificial Intelligence and Advanced Computer Technology Conference (AI-Europa) Proceedings, TCM Exposition Publisher, Liphook.
30. Pohl, B. & Trendelenburg, Chr. Extension of the language scope of PRO.M.D. (in preparation)
31. Pople, H., Myers, J. & Miller, R. (1975) Dialog. A model of diagnostic logic for internal medicine. Proc. 4th IJCAI, pp. 848–855.
32. Puppe, F., Puppe, B. & Beetz, M. (1985) MED1 – a shell for diagnostic expert systems. Nachr. f. Dokum., *36/1*, 28–32.

33. Shortliffe, E. H. (1976) Computer-Based Medical Consultation: MYCIN. American Elsevier/North Holland, New York.
34. Shortliffe, E. H. & Buchanan, B. G. (1984) A model of inexact reasoning in medicine. In: Rule-Based Expert Systems (Buchanan, B. G. & Shortliffe, E. H., eds.) pp. 233–262, Addison-Wesley, Reading.
35. Szolovits, P. & Pauker, S. G. (1978) Categorical and probabilistic reasoning in medicine. Artificial Intelligence *11*, 115–144.
36. Trendelenburg, Chr. (1984) Von der auftrags- zur patientenorientierten Arbeitsweise bei Einsatz eines EDV-Systems im Krankenhauslabor. Laboratoriumsmedizin *8*, 334–340.
37. Trendelenburg, Chr., Zeller, H. & Krautter, W. (1986 a) The expert system PRO.M.D.: A Prolog-system supporting medical diagnostics. Proceedings 1st International Expert Systems Conference, pp. 153–162, Learned Information Systems, Oxford.
38. Trendelenburg, Chr. & Wieland, H. (1986 b) Routine use of a clinical chemistry expert system with a knowledge base on disorders of the lipoprotein metabolism. Biochimica Clinica *10*, 928–929.
39. Waterman, D. A. (1986) How do expert systems differ from conventional programs? Expert Systems *3/1*, 16–19.
40. Weiss, S., Kulikowski, C., Amarel, S. & Safir, S. (1978) A model-based method for computer-aided medical decision making. Artificial Intelligence *11*, 145–172.
41. Weiss, S. & Kulikowski, C. (1979) EXPERT: A system for developing consultation models. Proc. 6th IJCAI, pp. 942–950.
42. Wieland, H. (1985) Lipoprotein-Quantifizierung nach elektrophoretischer Trennung: Indikation und Interpretation. Mitteilungen der Deutschen Gesellschaft für Klinische Chemie *16/2*, 48–57.
43. Zadeh, L. A. (1965) Fuzzy sets. Information and Control *8*, 338–353.

Chapter 3.9
Single-Point Method and Bayesian Approach for Individualizing Theophylline Dosage

M. Oellerich and B. Schneider

1 Introduction

Frequently used drugs like theophylline, aminoglycosides or digitalis glycosides show a low toxic/therapeutic ratio and considerable interindividual variability of pharmacokinetics. Optimal dosage selection therefore is often difficult for the physician. The therapeutic efficacy of these drugs is well correlated with the respective serum drug concentrations (1 – 5). Hence, monitoring serum drug levels can facilitate individual dosage adaptation. Usually data from therapeutic drug monitoring are only used in a "trial and error" fashion to optimize the patient's dosage regimen. A more rapid and effective way is to apply appropriate pharmacokinetic methods for individual dosage adjustment.

In the present chapter theophylline is used as a model drug to illustrate some features of a single-point method and a Bayesian technique for dose prediction. The selection of the optimum theophylline dosage occasionally causes problems, since this drug has a narrow therapeutic range and its clearance is affected by certain diseases, smoking habits, concomitant drug therapy and diet (1). The

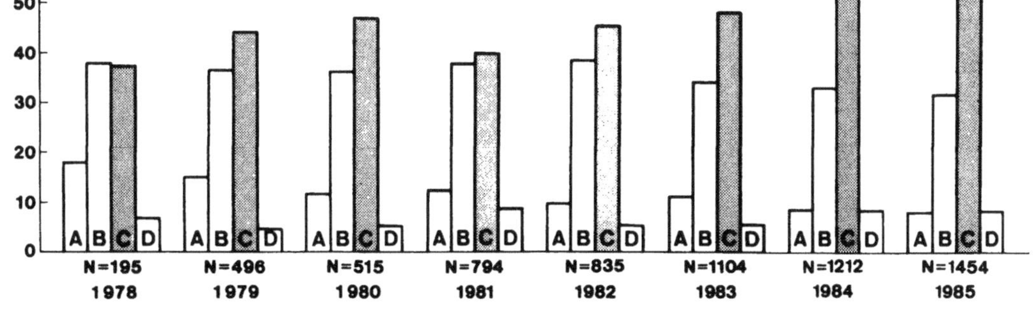

Fig. 1. Frequency distribution (%) of all theophylline serum concentrations determined in samples from patients with obstructive respiratory disease treated at our hospital between 1978 – 1985. Concentration ranges: A: < 2.5 mg/l; B: 2.6 – 7.9 mg/l; C: 8.0 – 20.0 mg/l; D: > 20.0 mg/l.

frequency distribution of theophylline values in patients who were treated for obstructive respiratory diseases at our hospital during the years 1978 to 1985 cleary indicates the difficulty of achieving the optimum dose for this drug (Fig. 1). Only about 37 to 54 per cent of the theophylline levels determined were within the therapeutic range. The consistently high percentage of theophylline concentrations in the subtherapeutic range is noteworthy. In these patients pharmacokinetic dose prediction methods were not routinely used for individual dosage adjustment.

The only use of empirical guidelines based on average population pharmacokinetic estimates does not permit a reliable prediction of the dosage requirement in the individual patient (1, 6). Therefore dosage adjustments must be guided by the observed clinical response of the patient and serum theophylline concentration monitoring. In addition, the direct assessment of individual pharmacokinetic parameters (e. g. clearance) by theophylline serum concentration measurements can be helpful to optimize theophylline therapy. Such pharmacokinetic methods of maintenance dose prediction may be used particularly in the early course of treatment, in order to reduce the time necessary to stabilise the patient.

Various pharmacokinetic methods have been developed to devise individualized dosage regimens. Some techniques require intravenous infusion of theophylline and multiple determinations of serum theophylline concentrations (7, 8). For routine application methods which can work with only one or two theophylline concentrations would be more convenient (9–21). The two-point method of Chiou et al. (18) allows estimation of the individual theophylline clearance from two theophylline serum concentrations obtained during intravenous theophylline ethylenediamine infusion and from the population mean distribution volume. Furthermore single-point methods have been described which are based on the assumption that under certain conditions a linear relationship exists between the required maintenance dose and a given drug concentration after a test dose (9–15). The more flexible Bayesian technique (21, 22) combines *a priori* information on population distributions of pharmacokinetic parameters with data on actually measured one or more serum theophylline concentrations in estimating the individual patient's clearance and distribution volume.

2 Single-Point Method

2.1 Pharmacokinetic basis

It has been observed that steady state concentrations of certain drugs strongly correlate with the respective serum concentrations after a single test dose (23–26). On grounds of theoretical considerations Slattery et al. have put forward the hypothesis that, for many drugs, a linear relationship exists under certain conditions between the mean steady state concentration $\bar{C}_{ss}$ and the concentration C' measured after a single test dose (9).

The mean steady state concentration $\bar{C}_{ss}$ of a drug in serum can be approximated from the maintenance dose D_m, the fraction absorbed F, the distribution volume

V_d, the elimination rate constant K_e and the interval of time (hours) between doses τ:

$$\bar{C}_{ss} = \frac{F \cdot D_m}{V_d \cdot K_e \cdot \tau} \tag{1}$$

Assuming a 1-compartment model and very rapid absorption and distribution, the concentration C' of a drug determined some time t' after administration of a test dose can be established from the test dose D', the fraction absorbed F, the distribution volume V_d, the elimination rate constant K_e and the interval of time t' before measurement:

$$C' = \frac{F \cdot D'}{V_d} e^{-K_e t'} \tag{2}$$

The ratio of the steady state drug concentration $\bar{C}_{ss}$ to its concentration C' determined t' hours after a single test dose is:

$$\frac{\bar{C}_{ss}}{C'} = \frac{D_m \cdot e^{K_e t'}}{D' \cdot K_e \cdot \tau}. \tag{3}$$

A factor ψ can be defined as:

$$\psi = \frac{D_m \cdot \Omega}{D' \cdot \tau} \tag{4}$$

where

$$\Omega = \frac{e^{K_e t'}}{K_e}. \tag{5}$$

If Ω is plotted against K_e at a fixed value of t' a relationship is obtained with a minimum at

$$K_e = \frac{1}{t'}. \tag{6}$$

For the case in which the term Ω remains reasonably constant over the range of K_e values observed in the patient population studied, a linear relationship results for a drug with linear pharmacokinetics between its mean steady state concentration $\bar{C}_{ss}$ and the concentration C' measured after a test dose:

$$\bar{C}_{ss} = \psi C' \tag{7}$$

For a given maintenance dose the mean steady state concentration of a patient can then be predicted from a single concentration measured after a test dose.
Rearrangement of Eqs. 3 and 5 yields:

$$D_m = \frac{\bar{C}_{ss} \cdot D' \cdot \tau}{\bar{\Omega} \cdot C'} \tag{8}$$

where $\bar{\Omega}$ represents the arithmetic mean of the individual Ω values (Eq. 5) from a representative group of test persons. This equation allows the prediction of the maintenance dose D_m for a desired mean steady state concentration $\bar{C}_{ss}$ of a drug.

There is a reciprocal relationship between the maintenance dose and the drug concentration measured after a test dose:

$$\frac{1}{D_m} = \psi' C' \qquad (9)$$

where

$$\psi' = \frac{\bar{\Omega}}{\bar{C}_{ss} \cdot \tau \cdot D'}. \qquad (10)$$

The total body clearance Q_B of a patient is calculated as:

$$Q_B = \frac{D_m \cdot F}{\bar{C}_{ss} \cdot \tau} \qquad (11)$$

Inserting the term for predicting the maintenance dose (Eq. 8) into the equation for calculating the clearance (Eq. 11) yields:

$$Q_B = \frac{D' \cdot F}{\bar{\Omega} \cdot C'} \qquad (12)$$

By use of this equation it is possible to estimate total body clearance of an individual patient provided that the constant $\bar{\Omega}$, the test dose and the drug concentration at a specific time after administration of the test dose are known. The described single-point method for dosage prediction is valid if the following conditions are fulfilled: rapid absorption and distribution of the test substance, stable distribution volume and clearance, linear pharmacokinetics, pharmacokinetics corresponding to a 1-compartment model, elimination rate constants within a specific range, appropriate interval of time between test dose and determination of the serum drug concentration.

2.2 Application for dosage prediction

Based on the theoretical considerations outlined above, a single-point method for prediction of theophylline dosage has been developed (11–12) which showed an acceptable performance in adult ambulatory patients with partly reversible obstructive airways disease (13). This technique will be described in more detail.

2.2.1 Sampling time and population constant $\bar{\Omega}$

For setting up a single-point method for theophylline it was necessary to select the appropriate sampling time t' after an initial test dose for the measurement of serum theophylline concentration and to determine the mean population constant $\bar{\Omega}$. With regard to the performance of the single-point method the sampling time t' is a critical factor. From theoretical considerations Slattery argued that the term ψ'

(Eq. 10) will show minimal fluctuations with changes in clearance and volume of distribution, when

$$t' = \frac{1}{\bar{K}_e} \tag{13}$$

where $\bar{K}_e$ represents the mean elimination rate constant for a given population (27).

To determine the optimal sampling time we evaluated concentration-time curves of 21 adult subjects after oral administration of theophylline ethylenediamine solution (Euphyllin®-Tropflösung, Byk Gulden) (12). A mean elimination rate constant of 0.1158 h^{-1} (range: 0.0540–0.2630 h^{-1}) was observed. These values correspond to elimination half-lives within the usual range for nonsmoking adults of between about 3 and 12 hours. By use of Eq. 13 the optimal sampling time (t') was calculated to be 8.6 hours. Hence, a sample collection time of 8 hours appeared to be adequate and was used in subsequent studies.

In order to determine the mean population constant $\bar{\Omega}$, the individual Ω values for 21 test subjects were plotted against the respective elimination rate constants at a fixed value of t' (Fig. 2). The broken line marks the arithmetic mean $\bar{\Omega}$ (23.7 h) of all the Ω values. The deviation of the individual Ω values from $\bar{\Omega}$ was less than 15 per cent for subjects with elimination half-lives of between three and

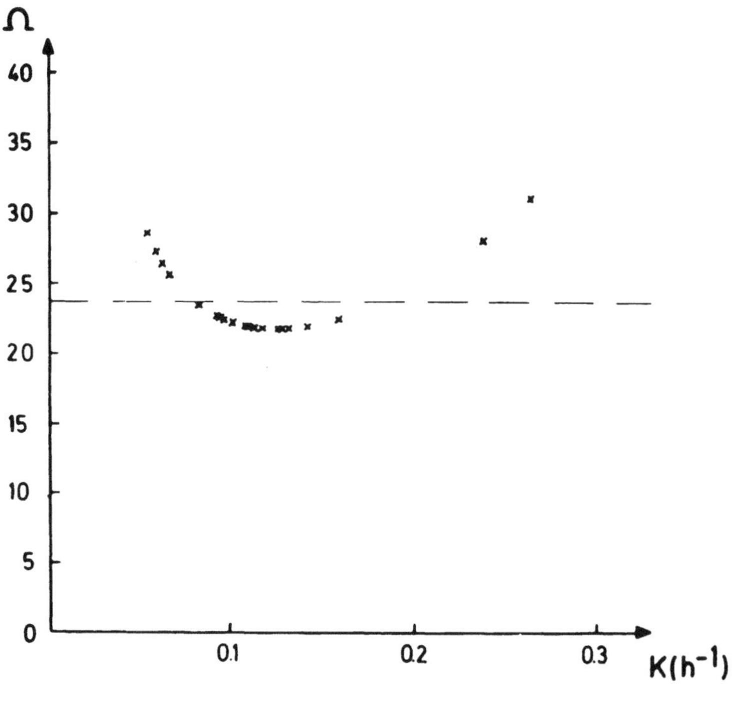

Fig. 2. Dependence of the individual Ω values on the individual elimination rate constant K_e. The broken line marks the arithmetic mean $\bar{\Omega}$ of the Ω values ($\Omega = \frac{e^{K_e t'}}{K_e}$). (From ref. 12).

11 hours. Therefore it was concluded that the obtained $\bar{\Omega}$ value can be used as a constant factor in predicting clearance and maintenance dose for patients with elimination half-lives between 3 and 12 hours (12).

2.2.2 Computer simulations

Using computer simulations, we investigated the influence of sample collection time, absorption and elimination on the absolute clearance prediction error of the single-point method (13). In these simulations 23 (Fig. 3) and 32 (Fig. 4) populations each of 40 subjects, were constructed randomly by use of a single point numerical model and own pharmacokinetic data. The points on the plots represent the mean absolute error of clearance prediction for random subpopulations as a function of the absorption rate constant k_a (Fig. 3) or K_e (Fig. 4).

The computer simulations indicated that in adults a sample collection time of 8 hours would introduce the comparatively lowest overall prediction error (Figs. 3, 4). A mean prediction error of less than 15% was calculated for subjects with elimination rate constants between 0.07 and 0.25 h^{-1}. For subjects with very high elimination rate constants, however, a shorter sampling time would be more ap-

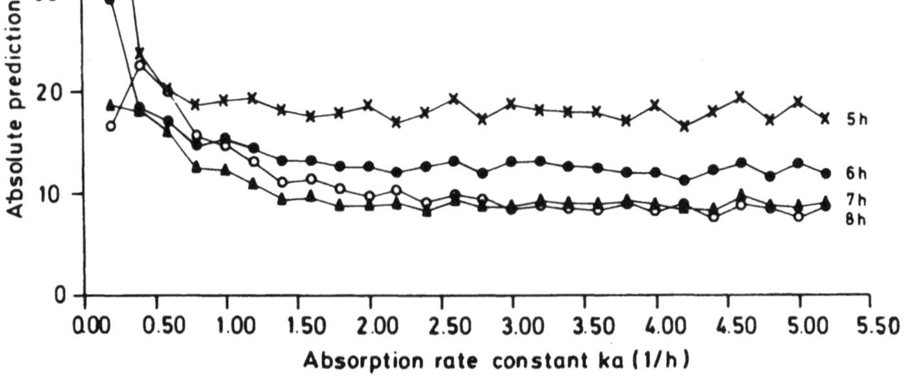

Fig. 3. The relationship between rate of absorption and mean absolute prediction error is plotted for four different sampling times (X = 5h, ● = 6h, ▲ = 7h, ○ = 8h) using the following parameters for the simulation ($K_e = 0.116\ h^{-1}$, $CV(K_e) = 95\%$; $V_d = 0.42$ l/kg, $CV(V_d) = 21\%$; CV (Assay) = 5%; CV (Dose) = 3%; CV (Time) = 3%). No random variation in the absorption rate constant is assumed. Since the observed prediction error remains constant for k_a exceeding 0.8 h^{-1}, it can be assumed that i. v. administration of the test dose will not improve prediction accuracy. (From ref. 13).

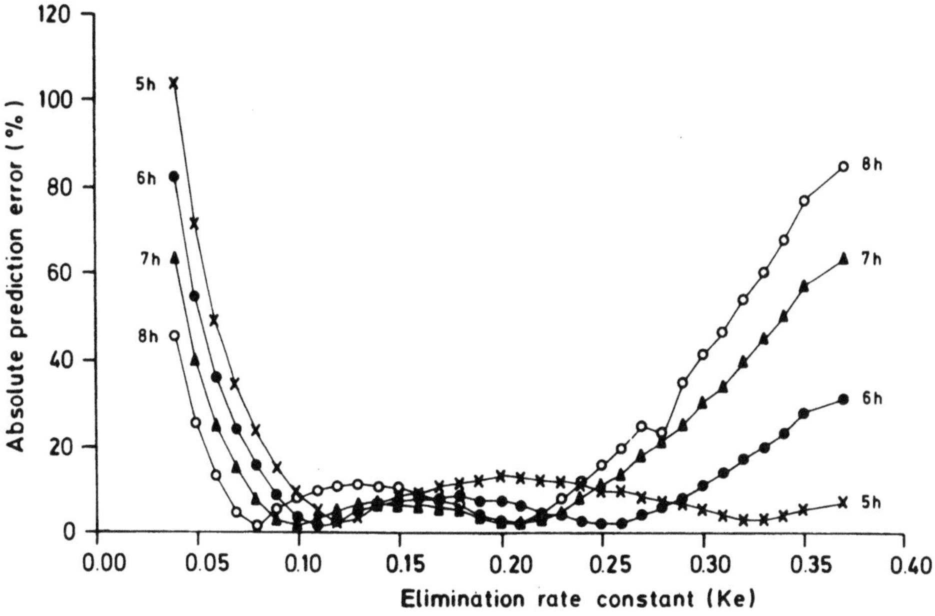

Fig. 4. The relationship between elimination rate constant and mean absolute prediction error is plotted for four different sampling times (X = 5h, ● = 6h, ▲ = 7h, ○ = 8h) with the following parameters used for the simulation ($k_a = 8.56\ h^{-1}$, CV (k_a) = 95%; $V_d = 0.42$ l/kg, CV (V_d) = 21%; CV (Assay) = 5%; CV (Dose) = 3%; CV (Time) = 3%). No random variation is assumed in the elimination rate constants at the given values in order to permit assessment of the mean prediction error in the individual patient. (From ref. 13).

propriate. At absorption rate constants of the oral test dose > $0.8\ h^{-1}$ the prediction error remained relatively constant (Fig. 3). Therefore it appears that i. v. administration of the test dose will not improve prediction accuracy.

From the computer simulations it was concluded that subjects with a serum theophylline concentration in the range of 1.7 to 7.8 mg/l determined 8 hours after the test dose, and corresponding half-lives of theophylline between 2.8 and 12.0 hours, would receive an appropriate maintenance dose with a mean error of less than 15% (13). However, if serum theophylline concentrations outside this range are found 8 hours after the test dose, a larger error in dose prediction can be expected. Such patients would require individual dose titration.

2.2.3 **Nomogram for dosage prediction**

To facilitate the use of the single-point method, a nomogram (Fig. 5) was prepared (12). It is based on a target mean steady-state serum theophylline concentration of 12 mg/l. A standard test dose of 5.6 mg of anhydrous theophylline per kg of body weight is used. The serum theophylline concentration is determined in a sample collected 8 hours after administration of the test dose. The point for the measured theophylline concentration on the center scale is connected with the point for the

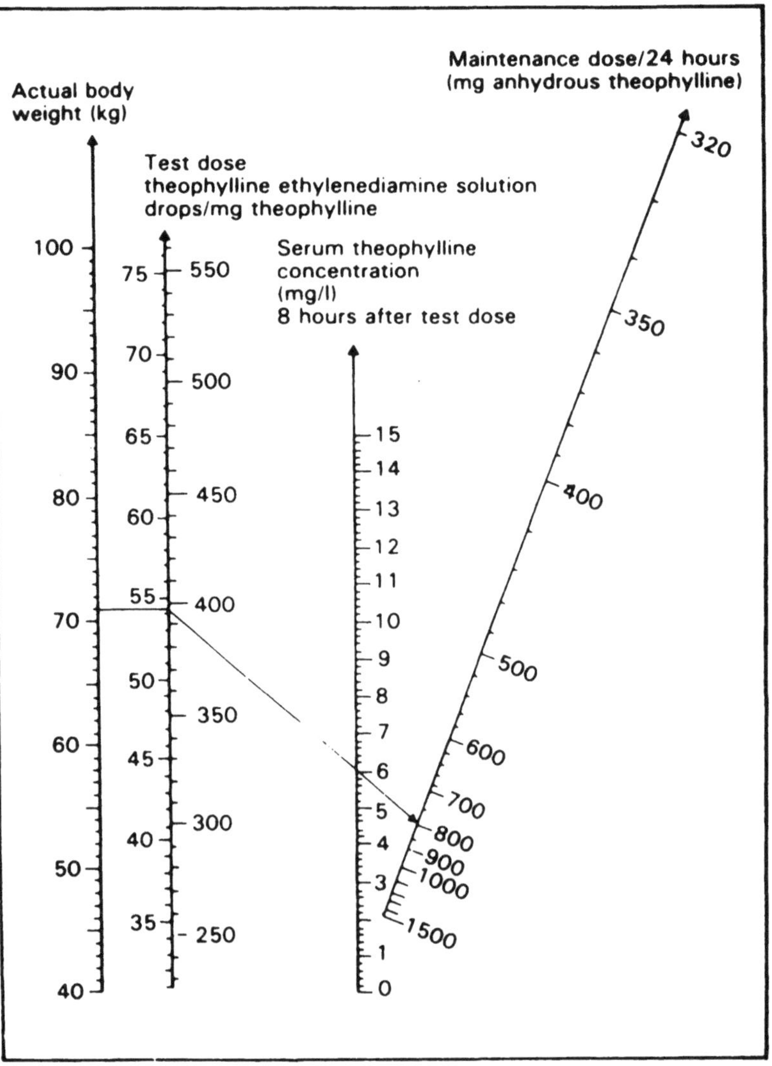

Fig. 5. Nomogram for the prediction of the individual maintenance dose for a mean serum theophylline target concentration of 12 mg/l. This nomogram is valid for adults only. The nomogram has not been evaluated in children. (From ref. 11).

test dose on the left-hand dual scale. The extension of that line intersects the right-hand inclined scale at the point indicating the daily maintenance dose of anhydrous theophylline, which will result in a mean steady-state serum theophylline concentration of ca. 12 mg/l on a twice daily dosage regimen.

For a patient with 71 kg of body weight, the test dose amounts to 397.6 mg of anhydrous theophylline. The theophylline ethylenediamine solution used contains 193.2 mg theophylline monohydrate per ml (175.6 mg anhydrous theophylline). If

1 ml of the solution is equal to 24 drops, then the 397.6 mg of anhydrous theophylline required as test dose correspond to 54 drops of the theophylline ethylenediamine solution. If in our example the serum theophylline concentration (6.0 mg/l) measured 8 hours after the test dose is connected to the given test dose (54 drops) by a straight line, a daily maintenance dose of 800 mg can be read from the dose axis of the nomogram (Fig. 5).

3 Bayesian Method

3.1 Pharmacokinetic parameter estimation

The fundamental difference between "classical" (Neyman – Pearson) statistics and Bayesian statistics lies in the treatment of the population parameters; i. e. the numerical characteristics of the probability distribution for the given sample of measurements or observations within the population of all possible samples. Whereas in classical statistics these parameters are assumed to have a fixed but unknown value for all sample values, in Bayesian statistics the parameters are assumed as random variables varying randomly between the units of the sample or the whole population (interindividual variation). This fundamental difference leads to two different approaches for parameter estimation.

In classical statistics a "point estimator" is wanted which represents optimally the unknown parameter value. It is well known that the value which maximizes the likelihood function is such an optimal point estimator. The likelihood function is the probability or distribution density (for continous random variables) of the observed sample values considered as a function of the parameters.

In Bayesian statistics point estimation is meaningless, as the parameters are considered as random variables and not as fixed values. Therefore one has to estimate the probability distribution or distribution density of the parameters with the knowledge of the observed sample values; or – to be more precise – to revise the *a priori* assumed probability (or density) of the parameters (prior distribution) with the knowledge of the observed sample values. The assumption of such an *a priori* probability is essential for Bayesian methods and has been a point of controversion between statisticians since more than hundred years. This *a priori* probability can be considered as a primary guess about the parameters before the experiment or observations are performed. If nothing is known about the parameters at this stage, a so called "no informative" prior distribution may be used; i. e. an uniform distribution within the range of possible parameter variation. If the results of former experiments or observations are available, the distribution revised by these results (posterior distribution) may be chosen as prior distribution for the new experiment. This approach is called the "empirical Bayes" approach.

After the experiment or observations are performed the primary guess of the parameter distribution is revised by calculating a new distribution of the parameters with the observed values. This distribution is called the *a posteriori* distribution or posterior distribution. The rule for this calculation is "Bayes formula" which was developed by Rev. Thomas Bayes (1702–1761) and presented to the Royal Society 1763 by his friend Mr. Richard Price (1723–1791) two years after his death (28). This formula uses as "mediator" between the *a priori* and *a posteriori* distribution the likelihood.

If we denote the prior distribution (density) of the parameter p with $q(p)$, the likelihood (density) of the measurement c for a given parameter value p with $f(c \mid p)$ and the posterior distribution (density) with $g(p \mid c)$, Bayes formula reads:

$$g(p \mid c) = \frac{f(c \mid p) \cdot q(p)}{h(c)} \tag{14}$$

The denominator $h(c)$ is the "unconditional" distribution (density) of c and is calculated as: $h(c) = \int f(c \mid p) \cdot q(p)\,dp$, where integration is done over the range of p.

The application of this approach to pharmacokinetic parameter estimation is straightforeward:

Suppose that in a pharmacokinetic study with an individual (e. g., patient or volunteer) at times $t_1, \ldots, t_m$ serum concentrations $c_1, \ldots, c_m$ of some compound are measured and these measurements c_i can be additively decomposed in a functional term $c^*(t_i; p_1, \ldots, p_n)$ (depending on n individual kinetic parameters $p_1, \ldots, p_n$) and an error term e_i [residual plasma concentration-error (22)]:

$$c_i = c^*(t_i; p_1, \ldots, p_n) + e_i \tag{15}$$

The functional term may be a feasible expression for the time-course of the serum concentrations; e. g. the open one-compartment function:

$$c^*(t; p_1, p_2) = p_1 e^{-p_2 t} \tag{16}$$

for a bolus injection or the Bateman function:

$$c^*(t; p_1, p_2, p_3) = p_1 (e^{-p_2 t} - e^{-p_3 t}) \tag{17}$$

for oral application, where p_1 is a function of the dose, bioavailability and volume of distribution, p_2 the elimination constant (clearance divided by volume of distribution) and p_3 the apparent first order absorption rate.

Usually it is assumed that the e_i are realisations of independent, normally distributed random variables with mean 0 and variance σ_e^2. With this assumption the c_i are – for fixed parameter values $p_1, \ldots, p_n$ – realisations of independent, normally distributed random variables with mean $c^*(t_i; p_1, \ldots, p_n)$ and variance σ_e^2; so the likelihood is:

$$\frac{1}{(2\pi)^{\frac{m}{2}}} e^{-\frac{1}{2\sigma_e^2} \Sigma (c_i - c^*(t_i; p_1, \ldots, p_n))^2} \tag{18}$$

For prediction purposes one can assume that the parameters as well as the residual variance depend on further parameters or patient characteristics like sex, age or state of illness (22); but for the moment we neglect this dependencies.

It seems reasonable to assume as prior distribution of the parameters p_j a multivariate normal distribution with some assumed mean vector:

$$p^{*\prime} := (p_1^*, \ldots, p_n^*) \tag{19}$$

and covariance matrix S_p. If in a first attempt ["for didactic purposes" see (22)] the parameters are assumed as independent (which is not a feasible assumption) the covariance matrix reduces to a diagonal matrix with the variances σ_{pj}^2 as diagonal elements and the *a priori* distribution is simply:

$$\frac{1}{(2\pi)^{\frac{n}{2}}} e^{-\frac{1}{2}\Sigma \frac{(p_j - p_j^*)^2}{\sigma_{p_j}^2}} \tag{20}$$

Both assumptions lead – according to Bayes formula – to a posterior distribution density whose logarithm is proportional to:

$$\sum_{j=1}^{n} \frac{(p_j - p_j^*)^2}{\sigma_{p_j}^2} + \sum_{i=1}^{m} \frac{(c_i - c^*(t_i; p_1, \ldots, p_n))^2}{\sigma_e^2} \tag{21}$$

It should be remarked that this *a posteriori* distribution is not a normal distribution and the parameters p_j are *a posteriori* dependent in spite of the *a priori* independence assumption. Therefore it seems more reasonable to assume *a priori* dependence between the parameters. In this case the logarithm of the *a posteriori* distribution density is:

$$\ln(g(p|c)) = K - (p - p^*)' S_p (p - p^*) - (c - c^*(p))' I \sigma_e^2 (c - c^*(p)) \tag{22}$$

where p is the vector of parameters, c the vector of serum concentration measurements, $c^*(p)$ the vector of functional terms (expectations) and I the identity matrix; ' means the transpose of a (column) vector or matrix. K is a term independent of p and c.

To get a more comprehensive idea of the *a posteriori* parameters one can use some characteristics of the posterior distribution. Usually the mode of this distribution is used and called "estimate" of the population parameter (22). It should be remarked that this is not a point estimate in the sense of classical statistics but a characteristic of the posterior distribution.

The mode is the value of the vector p which maximizes the posterior distribution $g(p|c)$ resp. minimizes its negative logarithm. As these are nonlinear in p, the mode must be calculated with an iteration procedure (29, 30). If the serum drug concentration measurements c of an individual patient are used in the posterior distribution, the mode can be considered as the best "estimate" of this individual patient's pharmacokinetic parameters (i. e. clearance, distribution volume) from prior population-specific information and actual drug level measurements.

The approximation of the "estimate" to individual patient's conditions can be improved, if individual patient-characteristics (e. g. age, sex, state of illness) are implemented in the estimation procedure. This can be done by assuming the pharmacokinetic parameters p_j as functions of these characteristics x:

$$p_j = f_j(x_1, \ldots, x_r; \theta_1, \ldots, \theta_s) \tag{23}$$

These functions depend on population-parameters θ_k which must be estimated by analogous procedures from the data of a relevant sample of patients.

The Bayes procedure as developed above has two serious disadvantages:

1. The posterior distribution does not belong to the same class as the prior distribution (i. e. normal distribution).

2. The mode of the distribution can be calculated only approximately with iterative procedures and does not coincide with the mean or median of the distribution.

Both disadvantages result from the fact that the pharmacokinetic function $c^*(t, p)$ is nonlinear in the parameters p. This can be avoided if only small variations of the parameters about their means are assumed. In this case the pharmacokinetic function c^* can be developed in a Taylor-series about the mean p^* and the series be cut down after the linear term. This results in:

$$c^*(p) = c(p^*) + C(p^*)(p - p^*) \tag{24}$$

Here $C(p^*)$ is the Jacobian matrix of the first derivatives of c^* to p_j with the mean p^* inserted for the parameters.

Using this approximation the posterior distribution of p is normal (assuming a normal prior distribution) with mean vector:

$$p^*(c) = p^* + \sigma_e^{-2} A^{-1} C'(p^*)(c - c^*(p^*)) \tag{25}$$

and covariance matrix:

$$A^{-1} = (S_p^{-1} + \sigma_e^{-2} C'(p^*) C(p^*))^{-1} \tag{26}$$

Here S_p is the *a priori* covariance matrix and p^* the *a priori* mean vector of the parameters p.

As for a normal *a posteriori* distribution the mode is identical with the mean and median of the distribution, the Bayes "estimates" of the parameters p are given by the *a posteriori* mean $p^*(c)$ which can be calculated by simple linear vector algebra immediately.

3.2 Drug dosing program for theophylline

The theophylline dosing program[1] allows patient-specific pharmacokinetic parameter estimation and subsequent prediction of the drug's concentration-time curve for a given dosage regimen.

Clearance p_1 and distribution volume p_2 are the two pharmacokinetic parameters which are estimated by minimization of the Bayesian objective function (Eq. 21). For calculation of the theophylline clearance and distribution volume the lean body weight (31) is used. Patient-specific data (sex, weight, height, age, smoking history, concomitant diseases, concurrent drug therapy) are used to identify the subpopulation to which the patient belongs.

The subpopulation mean estimates of theophylline clearance p^*_1 can be derived from data collected from various patient populations (1, 32, 33). Initial subpopulation-based estimates of clearance may be generated from an assumed standard clearance of 40 $ml \cdot h^{-1} \cdot kg^{-1}$ for nonsmoking healthy adults by multiplication

[1] A Bayesian drug dosing program (Abbott Pharmacokinetic Systems Theophylline Program) adapted to the IBM Personal Computer XT is available from Abbott Laboratories, Abbott Park, Ill., USA.

with certain factors accounting for specific patient characteristics (33). These factors are for example: 1.6 for smoking, 0.8 for severe airway obstruction, 0.4 for congestive heart failure and 0.4 for pneumonia.

The mean population distribution volume estimate p^*_2may be set at 0.5 l/kg for all patients (21, 34). However, it has to be considered that the mean volume of distribution is slightly larger in premature newborns, adults with hepatic cirrhosis or uncorrected acidemia, and the elderly (1). In all other circumstances the volume of distribution remains relatively unaffected.

It has been suggested to set the subpopulation standard deviations σ of clearance at 50% of the average subpopulation clearance $p^*_1$and that of distribution volume at 20% of the mean population value p^*_2(34). The standard deviation of the ith measured serum theophylline concentration is determined by use of a specific weighting algorithm (34):

$$\sigma_e = Q^{t'} \cdot 0.15 \cdot c^* \tag{27}$$

where σ_e denotes the standard deviation of the measured theophylline concentration c_i, t' is the time in hours between the most recently and an earlier measured serum theophylline concentration ($Q = 1.01$), 0.15 is the relative standard deviation of the measured theophylline concentration reflecting analytical error and error due to the employed pharmacokinetic model, and c^* is the predicted serum theophylline concentration. The time-dependent weighting of measured serum theophylline concentrations is necessary to ensure that the revised pharmacokinetic parameter estimates are primarily based on the most recently measured values (34). Weighting of serum theophylline levels is particularly important when the patient's theophylline clearance changes during the period of observation.

Using the patient-specific estimates for clearance and distribution volume obtained by the Bayesian method, serum theophylline concentrations can be predicted for a given dosage regimen. The required simulations of theophylline serum concentrations are based on the assumption that the kinetic of theophylline can be described by a 1-compartment open linear model with first-order absorption (oral dosing) or zero-order intravenous infusion.

The time course of serum theophylline concentrations after termination of an intravenous infusion is given by

$$C = \frac{k_0}{V_d K_e}(1 - e^{-K_e T}) e^{-K_e t'} \tag{28}$$

where C is the serum drug concentration at the postinfusion time t', k_0 is the rate of drug infusion (amount per unit time), V_d is the distribution volume, K_e is the elimination rate constant, and T is the time at which the infusion was stopped (35).

After an oral dose of theophylline the following relationship between serum concentration C and time t results:

$$C = \frac{k_a F D}{V_d (k_a - K_e)} (e^{-K_e t} - e^{-k_a t}) \tag{29}$$

where k_a is the apparent first order absorption rate constant, F is the fraction of the administered dose D which is absorbed, V_d and K_e are as defined previously.

The equation describing the serum concentration versus time curve following multiple oral dosing of a drug is

$$C_n = \frac{k_a F D_m}{V_d (k_a - K_e)} \left[\left(\frac{1 - e^{-nK_e\tau}}{1 - e^{-K_e\tau}} \right) e^{-K_e t} - \left(\frac{1 - e^{-nk_a\tau}}{1 - e^{-k_a\tau}} \right) e^{-k_a t} \right] \tag{30}$$

where C_n denotes the serum drug concentration at time t after the nth dose during a dosing interval τ, F is the fraction of the administered dose D_m which is absorbed, V_d, K_e and k_a are as defined previously. Equation 30 can be used to predict the serum theophylline concentration at any time during any dosing interval. At steady-state the time course of a drug in serum is described by the equation:

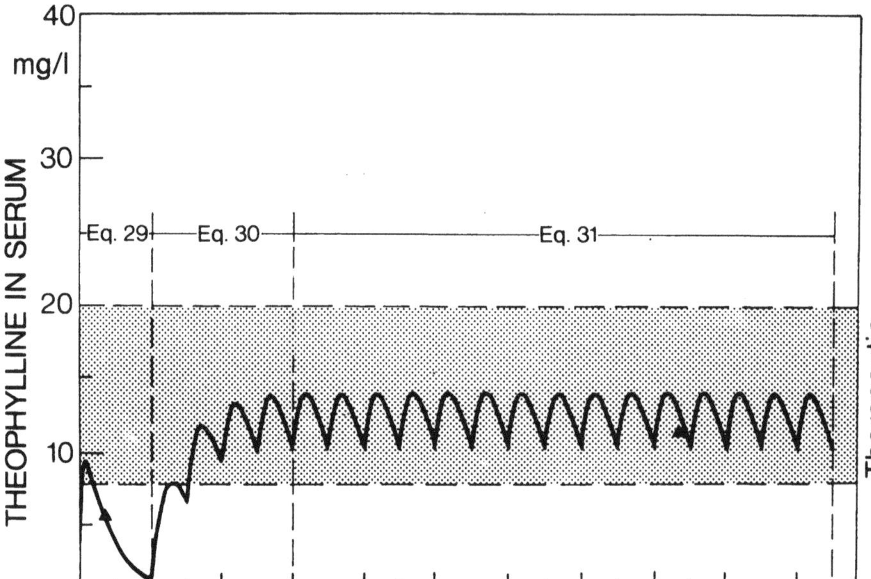

Fig. 6. Predicted time course of theophylline serum concentration in a 25 year old nonsmoking male patient assuming a daily maintenance dose of 800 mg theophylline (PulmiDur®) administered in two equally divided doses. Individual estimates for theophylline clearance (49.6 ml · kg^{-1} · h^{-1}) and distribution volume (0.5 l · kg^{-1}) were obtained by the Bayesian method from patient-specific data and a serum theophylline concentration determined 8 hours after an initial oral dose of 5.6 mg/kg theophylline (Euphyllin®-Tropflösung, Byk Gulden). The represented segments of the concentration-time curve were calculated employing the Eqs. 29, 30 and 31 as indicated (V_d = 26.5 l, based on a lean body weight of 53 kg; K_e = 0.099 h^{-1}, calculated from clearance Q_B and distribution volume V_d: $K_e = \frac{Q_B}{V_d}$; PulmiDur®: k_a = 0.18 h^{-1}, F = 1; theophylline ethylenediamine solution: k_a = 1.5 h^{-1}, F = 1). Triangles represent theophylline serum concentrations measured by EMIT® 8 hours after the initial dose and during steady state 8 hours after the morning dose. (Based on data from ref. 13).

$$C_\infty = \frac{k_a F D_m}{V_d (k_a - K_e)} \left[\left(\frac{1}{1 - e^{-K_e \tau}} \right) e^{-K_e t} - \left(\frac{1}{1 - e^{-k_a \tau}} \right) e^{-k_a t} \right] \quad (31)$$

The equations described above allow the prediction of theophylline serum concentrations as a function of time for a given dose and dosage form. Individual contributions of each dose are taken into account by use of the principle of linear superposition (35).

The application of the described equations is illustrated by an example of a simulated concentration-time curve shown in Figure 6. The marked curve segments between 0 and 24 h, 24 and 72 h as well as 72 and 252 h were calculated by use of Eqs. 29, 30 and 31 respectively.

4 Validation of Methods

4.1 Single-point method

The single-point maintenance dose prediction method (cf. p. 404) has prospectively been evaluated in adult patients with partly reversible obstructive airways disease (13). In these studies the reliability of the single-point method was assessed by comparing predicted and observed mean steady-state serum theophylline concentrations. Computer simulations indicated that on a twice daily dosage regimen with slow release theophylline the best approximation of the mean steady-state concentration would be obtained at a sampling time of 8 hours after the morning dose during steady-state (13). Therefore this sampling time was chosen for the determination of the mean steady-state theophylline concentrations in our studies.

Sheiner et al. suggested to use the prediction error as a measure of the goodness of serum drug concentration prediction (22). The prediction error is defined as the difference between the predicted and the measured serum drug concentration. The accuracy and precision of a dose prediction method are reflected by the average prediction error and the error standard deviation respectively.

Figure 7 shows the prediction error observed in a study with outpatients (13). The mean prediction error is indicated by the horizontal line in the center of each column. Vertical columns represent one standard deviation on either side of the mean prediction error. The broken lines delimit the maximum range within which 68% of error should fall if predictions are to be clinically acceptable. According to Sheiner et al. (22) this range can be derived as follows: The therapeutic serum concentration range of theophylline is usually assumed to be 8 to 20 mg/l. A theophylline concentration of 12 mg/l was chosen as target value. In order to avoid subtherapeutic or toxic concentrations the clinically desirable absolute error would therefore be less than 4 mg/l. In case of a normal distribution of the measured theophylline concentrations about the target value, without bias, about 95% of the observed concentrations will be within 4 mg/l of the target value if the standard deviation of the prediction error is equal to 2 mg/l. The analysis of data from our

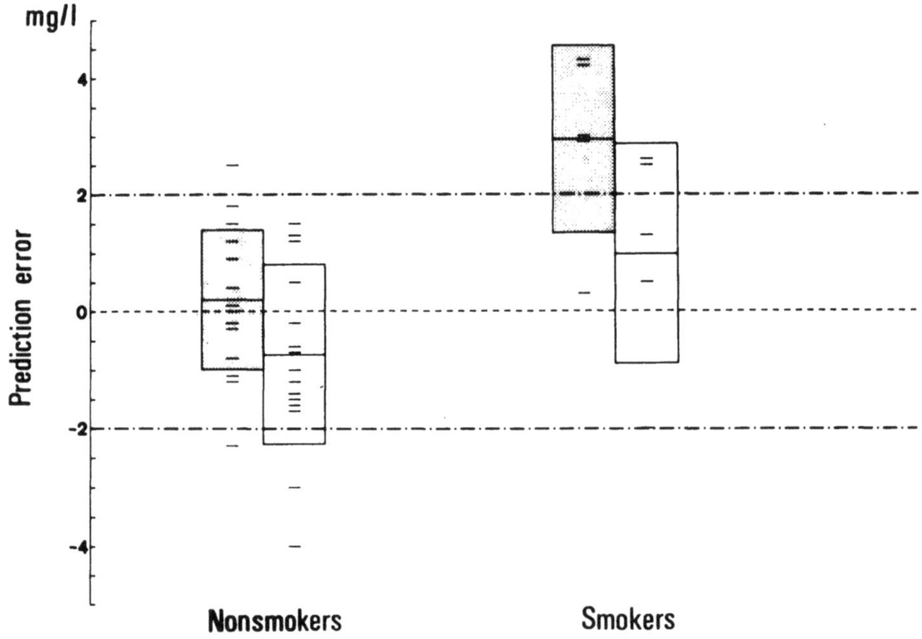

Fig. 7. Prediction errors (predicted minus measured values) for mean steady-state serum theophylline concentrations observed in 17 nonsmokers and 5 smokers with obstructive airways disease. Predictions were made by use of the single-point method (shaded columns) and the Bayesian technique (clear columns). The columns delimit one standard deviation on either side of the mean prediction error, indicated by the horizontal line in each column. The horizontal bars represent the observed individual prediction errors. The vertical broken lines delimit the clinically acceptable range for one standard deviation of the prediction error. (Based on data from ref. 13).

study (13) indicates that in nonsmoking patients the predictive accuracy and precision of the single-point method is clinically acceptable (Fig. 7). Predicted and observed mean steady-state serum theophylline concentrations are compared in Figure 8. In nonsmoking outpatients predicted and measured serum theophylline concentrations were in good agreement. The mean absolute deviation found between predicted and observed theophylline concentrations was 8.2% (range 0.0–21.7%). This is in accordance with the results from a similar study with hospitalized patients, where an absolute deviation of 14.8% (range 5.1–33.1%) was found (12). In 16 out of 17 nonsmokers therapeutic serum theophylline concentrations were attained when maintenance doses were predicted by use of a nomogram based on the single-point method (Fig. 8). The patient with the subtherapeutic theophylline concentration (7.5 mg/l) had received only 70% of the predicted maintenance dose. In this patient the dose was not further increased because of complete relief of symptoms. The required daily maintenance dosage showed large interindividual differences and ranged from 7.6 to 20.7 mg/kg (Fig. 8). These results demonstrate that the single-point method can greatly facilitate the individualization of theophylline dosage.

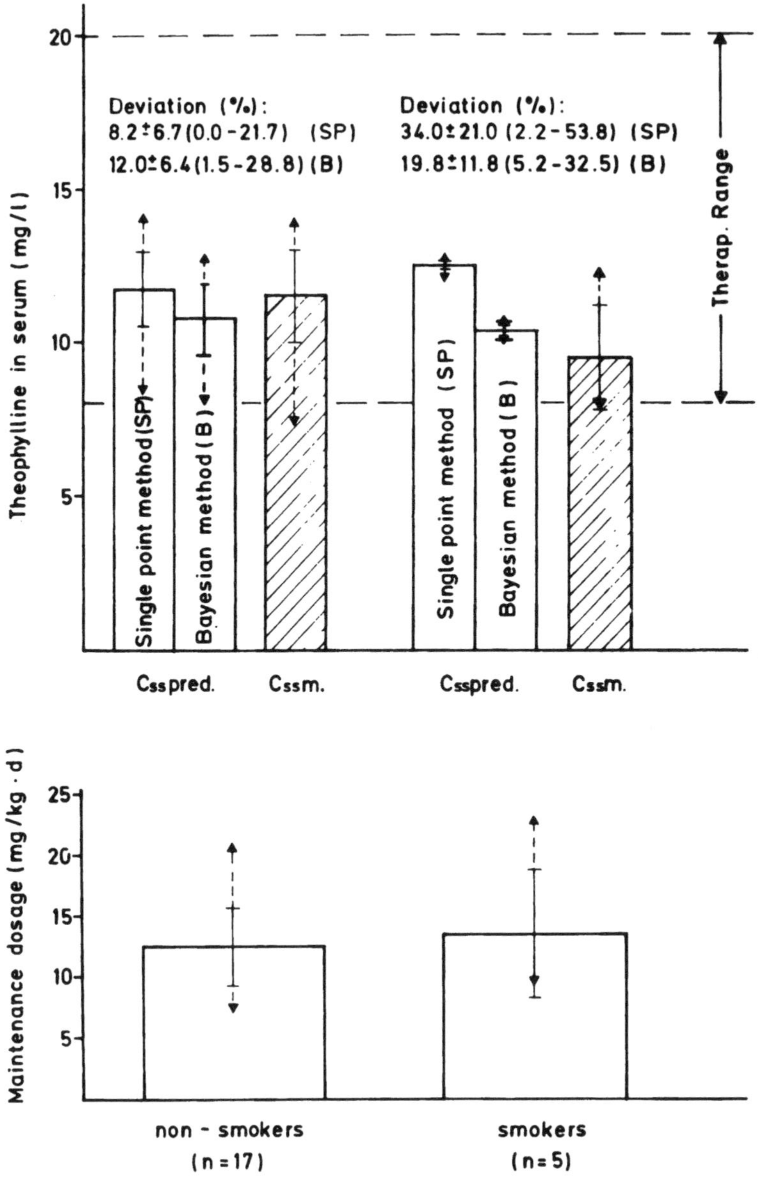

Fig. 8. Comparison between predicted (clear columns) and measured (shaded columns) mean steady-state theophylline serum concentrations in 17 nonsmokers and 5 smokers. Predictions were made by use of the single-point method and the Bayesian technique. The daily administered theophylline maintenance dose is also given. The columns show the mean values with standard deviation (vertical bars) and range (triangles).

$$\text{Deviation (\%)} = \frac{C_{ss}\,\text{pred.} - C_{ss}\,\text{m.}}{C_{ss}\,\text{m.}} \times 100$$

(Based on data from ref. 13).

For smokers the results were less satisfying (Fig. 8). The mean value of the observed steady-state theophylline serum concentrations was significantly lower than that of the predicted values ($p < 0.02$, t-test). The average prediction error showed a substantial bias indicating inadequate predictive accuracy (Fig. 7). The mean absolute deviation between predicted and observed steady-state theophylline serum concentrations was 34% (range 2.2–53.8%) (Fig. 8). Smokers showed a significantly greater difference between predicted and observed steady-state theophylline serum concentrations than nonsmokers (13). Nevertheless, theophylline concentrations within the therapeutic range were achieved in all of the smokers tested (Fig. 8).

Furthermore 17 single dose theophylline serum concentration time curves from smokers were evaluated (13). The average value of predicted clearance was significantly about 9% lower than that of the observed clearance ($p < 0.02$). The mean absolute deviations between predicted and observed clearances were only 11.3% and 9.8% at post-dose sampling times of 8 and 6 hours respectively. Therefore the choice of sampling time was unlikely to have caused the large errors in predicting the steady-state theophylline serum concentrations in smokers. The reasons for the mainly unsatisfactory performance of the single-point method in smokers are not known.

In children the single-point method has not been evaluated so far. The performance of a previously described different single-point method was less than optimal in children (10, 36).

The single-point method attained in nonsmoking adult patients the same predictive accuracy as much more complex procedures such as the Bayesian method (Figs. 7 and 8). It is a major advantage of this method that it can be easily used without a computer in the form of a nomogram. The clinical application of the single-point method is very useful for outpatients in whom rapid assessment of the dosage requirement is desirable and for those in whom repeated return to the clinic for incremental dose adjustment is inconvenient. In addition, the comparison of predicted and measured steady-state theophylline serum concentrations allows assessment of patient compliance. Potential sources of error are the presence of significant nonlinearity in theophylline clearance, inappropriate timing of dose and sampling and rapid intraindividual changes of clearance or absorption. It is a disadvantage of the single-point method that it can only be used if the drug concentration prior to the test dose is zero.

The results of various studies have shown that the single-point method permits reliable predictions of theophylline maintenance dosage in adult nonsmoking patients (12, 13). The use of this method in cigarette smokers, however, may result in an underestimation of the dosage requirements. Further prospective studies are needed to evaluate the clinical utility of the single-point method and to gain more practical experience with this procedure.

4.2 Bayesian method

The Bayesian method was validated retrospectively on data of previous studies (13, 37) and in a prospective study (38).

Figure 6 shows as a typical example the prediction of theophylline serum concentrations in a 25 year old patient. Individual estimates for theophylline clearance and distribution volume were obtained by the Bayesian method from patient-specific data and a serum theophylline concentration determined 8 hours after an initial dose. The represented time course of theophylline serum concentration was simulated assuming a daily maintenance dose of 800 mg theophylline administered in two equally divided doses of a sustained-release preparation (PulmiDur®, Astra Chemicals). The measured mean steady-state theophylline serum concentration (11.5 mg/l) agreed well with the predicted value (12.7 mg/l).

In Figure 7 the prediction errors obtained by the Bayesian method are compared with those observed with the single-point method in the same outpatients. Predictive accuracy and precision of the Bayesian method were clinically acceptable in nonsmoking patients. The Bayesian method was associated in these patients with a somewhat greater predictive error than the single-point method. However, the Bayesian method unlike the single-point procedure yielded a still satisfactory predictive accuracy also in smokers. Predicted and observed mean steady-state theophylline serum concentrations were mainly in good agreement (Fig. 8). The mean values of predicted and measured theophylline serum concentrations did not differ significantly ($p > 0.05$, t-test). The average absolute deviations between predicted and observed theophylline serum concentrations were 12.0% (range 1.5–28.8%) in nonsmokers and 19.8% (range 5.2–32.5%) in smokers.

The sampling time for the measurement of theophylline serum concentrations should be carefully selected to obtain meaningful Bayesian estimates of clearance and distribution volume. The evaluation of 10 single dose theophylline serum concentration-time curves in healthy subjects showed that the deviation between predicted and observed values for clearance and distribution volume strongly depends on the sampling time (38). As expected the prediction error for the distribution volume was lowest if theophylline serum concentrations determined shortly after the end of the intravenous theophylline infusion were used in the Bayesian estimation of this pharmacokinetic parameter. Conversely estimates for theophylline clearance showed a comparatively higher accuracy when predictions were based on theophylline serum concentrations determined during the elimination period about 8 to 12 hours after theophylline administration. With predictions based on two serum theophylline concentrations determined at postinfusion times of 0.25 and 12 hours, the mean absolute deviations between observed and predicted values for both, clearance and distribution volume, were less than 10%.

Within certain limitations the evaluated Bayesian method appears to be suitable for dosage predictions in adult patients. The performance of this procedure has not yet been sufficiently validated in children. In comparison with the single-point method the Bayesian method is much more flexible. Dosage predictions can be made by use of the Bayesian method at any time during theophylline therapy. The Bayesian technique, however, requires a computer (e.g. IBM-PC) to perform the necessary calculations. Potential sources of error are rapid intraindividual changes in clearance or absorption of theophylline and the presence of significant nonlinearity in theophylline clearance (37, 38). Prospective studies are in progress to further evaluate the reliability and clinical utility of this procedure.

References

1. Hendeles, L., Massanari, M. & Weinberger, M. (1986) in: Applied Pharmacokinetics Principles of Therapeutic Drug Monitoring (Evans, W. E., Schentag, J. J., Jusko, W. J., eds.) pp. 1105–1188, Applied Therapeutics, Spokane.
2. Neuenkirchen, H., Wilkens, J. H., Oellerich, M. & Sybrecht, G. W. (1985) Nocturnal asthma: effect of a once per evening dose of sustained release theophylline. Eur. J. Respir. Dis. *66*, 196–204.
3. Zaske, D. E., Bootman, J. L., Solem, L. B. & Strate, R. G. (1982) Increased burn patient survival with individualized dosage of gentamicin. Surgery *91*, 142–149.
4. Moore, R. D., Smith, C. R. & Lietmann, P. S. (1984) Association of aminoglycoside plasma levels with therapeutic outcome in gram-negative pneumonia. Am. J. Med. *77*, 657–662.
5. Reuning, R. H. & Geraets, D. R. (1986) in: Applied Pharmacokinetics Principles of Therapeutic Drug Monitoring (Evans, W. E., Schentag, J. J., Jusko, W. J., eds.) pp. 570–623, Applied Therapeutics, Spokane.
6. Gotz, V. P., Russell, W. L. & Lopez, L. M. (1983) Evaluation of the "condition correcting factor method" of estimating theophylline clearance. Ther. Drug Monit. *5*, 103–107.
7. Horai, Y., Ishizaki, T., Sasaki, T., Watanabe, M. & Kabe, J. (1982) Individualized aminophylline therapy in patients with obstructive airway disease: oral dosage prediction from an intravenous test dose. Eur. J. Clin. Pharmacol. *23*, 111–121.
8. Pancorbo, S., Sawchuk, R. J., Dashe, Ch. & Schallok, M. (1979) Use of a pharmacokinetic model for individualizing intravenous doses of aminophylline. Eur. J. Clin. Pharmacol. *16*, 251–254.
9. Slattery, J. T., Gibaldi, M. & Koup, J. R. (1980) Prediction of maintenance dose required to attain a desired drug concentration at steady-state from a single determination of concentration after an initial dose. Clin. Pharmacokin. *5*, 377–385.
10. Shapiro, G. G., Koup, J. R., Furukawa, C. T., Pierson, W. E., Gibaldi, M., Fuquay, D. & Bierman, C. W. (1982) Individualization of theophylline dosage using a single serum sample following a test dose. Pediatr. *69*, 70–73.
11. Oellerich, M., Ranke, C., Schmidt, G. & Sybrecht, G. W. (1983) Optimised theophylline dosage by estimation of clearance using a simple pharmacokinetic method. Brit. J. Clin. Pract. (Suppl.) *23*, 37–41.
12. Ranke, C., Schmidt, G., Oellerich, M. & Sybrecht, G. W. (1983) Optimierung der Therapie mit Theophyllinpräparaten bei obstruktiven Ventilationsstörungen. 3. Vorhersage der Theophyllin-Clearance mit einer einfachen pharmakokinetischen Methode und Vergleich der Bioverfügbarkeit verschiedener Theophyllin-retard-Präparate. Prax. Klin. Pneumol. *37*, 99–108.
13. Wilkens, J. H., Neuenkirchen, H., Sybrecht, G. W. & Oellerich, M. (1984) Individualizing theophylline dosage: evaluation of a single-point maintenance dose prediction method. Eur. J. Clin. Pharmacol. *26*, 491–498.
14. Goldstein, R. S., Allen, L. C., Thiessen, J. J., Michalko, K., Dayneka, N. & Woolf, C. R. (1986) Daily maintenance dose of a long-acting theophylline from a single theophylline serum level. Chest *89*, 103–108.
15. Koup, J. R., Sack, C. M., Smith, A. L. & Gibaldi, M. (1979) Hypothesis for the individualisation of drug dosage. Clin. Pharmacokin. *4*, 460–469.
16. Grambau, G. R., Johnson, C. E., Conlon, P. F. & Weg, J. G. (1985) Reliability of theophylline clearance in determining maintenance intravenous aminophylline therapy. J. Clin. Pharmacol. *25*, 381–383.
17. Ball, W. D. (1982) Unstructured two-point estimation of one-compartment linear pharmacokinetic parameters. J. Clin. Pharmacol. *22*, 326–334.
18. Chiou, W. L., Gadalla, M. A. F. & Peng, G. W. (1978) Method for the rapid estimation of the total body drug clearance and adjustment of dosage regimens in patients during a constant rate intravenous infusion. J. Pharmacokinet. Biopharm. *6*, 135–151.
19. Vozeh, S., Kewitz, G., Wenk, M. & Follath, F. (1980) Rapid prediction of steady-state serum theophylline concentration in patients treated with intravenous aminophylline. Eur. J. Clin. Pharmacol. *18*, 473–477.
20. Rizzo, A., Mirabella, A., Ferrara, G., Mangiacavallo, A., Bonanno, A. & Bonsignore, G. (1986) Early estimate of theophylline clearance during intravenous infusion. Eur. J. Respir. Dis. *68*, 291–297.
21. Gilman, T. M., Muir, K. T., Jung, R. C. & Walberg, C. B. (1985) Estimation of theo-

phylline clearance during intravenous aminophylline infusions. J. Pharmaceut. Sci. *74*, 508–514.

22. Sheiner, L. B., Beal, S., Rosenberg, B. & Marathe, V. V. (1979) Forecasting individual pharmacokinetics. Clin. Pharmacol. Ther. *26*, 294–305.
23. Cooper, T. B., Bergner, P. E. & Simpson, G. M. (1973) The 24-hour serum lithium level as a prognosticator of dosage requirements. Am. J. Psychiatry *130*, 601–603.
24. Cooper, T. B., Simpson, G. M. (1978) Prediction of individual dosage of nortriptyline. Am. J. Psychiatry *135*, 333–335.
25. Montgomery, S. A., Mc Auley, R., Montgomery, D. B., Braithwaite, R. A. & Dawling, S. (1979) Dosage adjustment from simple nortriptyline spot level predictor test in depressed patients. Clin. Pharmacokin. *4*, 129–136.
26. Brunswick, D. J., Amsterdam, J. D., Mendels, J. & Stern, S. L. (1979) Prediction of steady-state imipramine and desmethyl-imipramine plasma concentrations from single-dose data. Clin. Pharmacol. Ther. *25*, 605–610.
27. Slattery, J. T. (1981) Single-point maintenance dose prediction: Role of interindividual differences in clearance and volume of distribution in choice of sampling time. J. Pharm. Sci. *70*, 1174–1176.
28. Bayes, Th. (1763) An essay towards solving a problem in the doctrin of chances. Philos. Trans. R. Soc. London *53*, 370–418.
29. Horowitz, D. L. & Homer, L. D. (1970) Analysis of biomedical data by time-sharing computers. I. Non-linear regression analysis. Project No. MR 005:20-0287, Report No. 25, Naval Medical Research Institute, National Naval Medical Center, Bethesda, Md. 20014.
30. Peck, C. C. & Barrett, B. B. (1979) Nonlinear least-squares regression for microcomputers. J. Pharmacokinet. Biopharm. *7*, 537–541.
31. Wissenschaftliche Tabellen Geigy, Teilband Somatometrie und Biochemie (1982), 8th. ed., pp. 9–12, Ciba-Geigy, Basel.
32. Jusko, W. J., Gardner, M. J., Mangione, A., Schentag, J. J., Koup, J. R. & Vance, J. W. (1979) Factors affecting theophylline clearances: age, tobacco, marijuana, cirrhosis, congestive heart failure, obesity, oral contraceptives, benzodiazepines, barbiturates, and ethanol. J. Pharm. Sci. *68*, 1358–1366.
33. Powell, J. R., Vozeh, S., Hopewell, P., Costello, J., Sheiner, L. B. & Riegelman, S. (1978) Theophylline disposition in acutely ill hospitalized patients. The effect of smoking, heart failure, severe airway obstruction, and pneumonia. Am. Rev. Respir. Dis. *118*, 229–238.
34. Peck, C. C. (1984) in: Pharmacokinetic Basis for Drug Treatment (Benet, L.Z., Massoud, N., and Gambertoglio, J. G., eds.) pp. 349–356, Raven Press, New York.
35. Gibaldi, M., Perrier, D. (1982) Pharmacokinetics. 2nd ed., Marcel Dekker, New York.
36. Koup, J. R. (1982) Single-point prediction methods: a critical review. Drug. Intell. Clin. Pharm. *16*, 855–862.
37. Hurley, S. F. & Mc Neil, J. J. (1988) A comparison of the accuracy of a least squares regression, a Bayesian, Chiou's and the steady-state clearance method of individualising theophylline dosage. Clin. Pharmacokin. 14, 311–320.
38. Stein, U., Oellerich, M., Sybrecht, G. W. & Schneider, B. (1988) Evaluation of a novel Bayesian method for individualizing theophylline dosage. J. Clin. Chem. Clin. Biochem. 26, 405–414.

Chapter 3.10
Discriminant Analysis: A Special Procedure for the Detection of Alcoholism

E. Hansert and D. Stamm

1 Introduction

In connection with a study on the detection and exclusion of alcohol abuse based on clinical chemical and hematological findings it became clear that the procedures commonly used for multivariate analysis of measurements yield unsatisfactory results (1). A new procedure was therefore developed and its effectiveness demonstrated. The procedure is described here in detail and discussed at length because it appears to be appropriate for use in many different medical situations requiring discriminant analysis.

To understand the development of the procedure it is helpful to be familiar with the medical problems involved in detecting and excluding alcoholism on the basis of clinical chemical and hematological studies. These problems are therefore outlined briefly before the experimental design is described and the development of the statistical procedure discussed.

2 The Problem of Detecting and Excluding Alcohol Abuse on the Basis of Clinical Chemical Parameters

Alcoholism is a common illness; the prevalence among patients admitted to medical wards of hospitals for reasons other than drunkenness is between 10 and 15 percent.

By alcoholism is meant here excessive consumption of alcohol leading to a disturbance of physical, mental and/or social function. This definition is consistent with the 1952 definition of the World Health Organization and with the term "alcohol-related disabilities" used by Edwards et al. (2). The development of alcohol dependence is intentionally not discussed in this contribution.

Because once consumed alcohol can reach virtually all body cells, alcohol abuse leads to metabolic, functional and sometimes even irreversible anatomical changes in numerous organs. The symptoms of these changes are extremely diverse, variable and, frequently, uncharacteristic. Therefore, intensive efforts have been made to develop methods for the detection and exclusion of alcoholism.

There is still an urgent need for a test or a combination of tests for the detection of alcoholism that can be carried out easily and at any time both in hospitals and in physicians' offices.

In past years many studies have been reported in which combinations of laboratory findings were used to detect and exclude alcohol abuse. For reasons discussed elsewhere (3), the results of these studies cannot be compared, and some of them are even contradictory.

Considering the importance of detecting and excluding alcoholism, we felt a prospective study would be justified if the following prerequisites could be met:

1. Valid outside criterion available for alcohol abuse
2. Thorough standardized examination of the subjects possible
3. Reliable laboratory investigations possible for the clinical chemical and hematological parameters to be assessed, with the steps in the investigations standardized and subject to monitoring
4. A suitable statistical procedure available for the multivariate assessment of the laboratory results

Prerequisites 1, 2 and 3 had already been met at the beginning of the study. The necessary statistical procedure was developed especially for the study because due to the structure of the data the available procedures appeared unsuited and preliminary statistical analyses with them had yielded poor results.

In the following section prerequisites 1, 2 and 3 are described to the extent necessary for an understanding of the new statistical method. The alcoholism study itself is described in detail elsewhere (3, 4).

3 Description of the Alcoholism Study

Subjects: The subjects in the study were 82 men with a definite diagnosis of alcoholism (A). Because the clinician is not usually interested in distinguishing between alcoholics and healthy individuals but rather between alcoholics and patients with other disorders, 70 patients with other disorders but clearly without alcohol abuse or alcoholism were selected from 2 hospitals and examined with the same procedures as the alcoholics.

Outside criterion: Alcoholism was diagnosed with the Munich Alcoholism Test (MALT) (5, 6).

Examinations: The following standardized investigations and examinations were performed: past history, alcohol questionnaire, physical examination and neurological examination. Complete data were obtained for each patient.

Laboratory investigations: All of the steps in the clinical chemical and hematological studies (7) were standardized and subject to ongoing quality control procedures (8); a detailed description is given there in Sections 4 and 5.

The following quantities were investigated:

1. Hematological studies (EDTA venous blood)

hemoglobin	reticulocyte count
hematocrit	leukocyte count
erythrocyte count	differential blood count
MCV	thrombocyte count

2. Capillary blood

β-glucose

3. Clinical laboratory investigations (serum)

total protein	cholesterol
protein electrophoresis	triglycerides
bilirubin	creatinine
uric acid	urea
sodium	GGT
potassium	GOT (AST)
calcium	GPT (ALT)
chloride	ChE
phosphorus, inorganic	AP
magnesium	GLDH
iron	Ck
copper	

4 Problems in Applying Discriminant Analysis to Clinical Chemical Measurements

In medicine, discriminant analysis is used to support the diagnostic process with statistical arguments. The statistical arguments relate to the certainty with which particular values of particular variables can be assigned to one of the diagnostic categories in question. In the case of two diagnostic categories, which is the only one of concern to us in this contribution and for which "alcoholism" ("A") and "non-alcoholism" ("NA") are used for illustration, this certainty depends on the two criteria *sensitivity* and *specificity*:

(1a) Sensitivity: Probability of assignment to A if A holds

(1b) Specificity: Probability of assignment to NA if NA holds,

or expressed as *efficiency*:

(1c) Efficiency = sensitivity + specificity
(For more on this type of definition see (1).)

Now let us assume that the whole process by which the values of the variable are obtained can be considered a two-step process in which the probability of A in the first step is characterized by p (see (1)) and the probability of NA by $q = 1 - p$.

Then the criteria (1a), (1b) and (1c) need to be weighted to obtain the expected "hit rate." To get estimates of these criteria, "initial samples" of subjects are needed; for a discussion of the problems involved in such estimates see (9). A particularly difficult problem is that of specifying *p*. Usually this is done on the basis of subjective estimates, and as is well known, the values proposed by different experts often vary widely. Furthermore, it is important to realize that the probability of an individual with "A" being examined at a given institution is not necessarily identical with the prevalence of "A" in the general population.

The different discriminant analysis procedures differ in the kind of discriminant regions for which the values of the criteria (1a), (1b) and (1c) are specified. In the case of two categories (A and NA), the discriminant regions are obtained by dividing the whole range of values into the two regions "positive test results" (D^+) and "negative test results" (D^-). The dividing line between them is defined by the discriminant function(s). A positive test result is assigned to the category "A" and a negative one to the category "NA" with a certain risk. In Section 7 we will discuss the problems related to the abrupt change in the assignment when the dividing line is crossed. The power of, or information provided by, particular values specified for sensitivity and specificity or for efficiency can be seen in analogy to a statistical test, as has been discussed in detail in (1). If a discriminant analysis procedure is used that does not involve prevalences, then depending on the question asked we are interested in:

(2a) Maximum efficiency,

(2b) Maximum specificity when a minimum level of sensitivity is specified, or

(2c) Maximum sensitivity when a minimum level of specificity is specified.

Every statistical procedure for assigning observed values to a category is based on the differences in the distributions of the variables studied in the different categories. Before using a discriminant analysis procedure it is therefore useful to get a clear picture of the type of differences in distribution that will be described.

To illustrate, we will briefly look at linear discriminant analysis (after R. A. Fisher, e.g. (10)), which is still often considered the method of choice. It was developed from the concept of a multidimensional normal distribution, whose mathematical expression depends only on the first and second moments, i.e. on the expected values, variances and covariances (or product-moment correlation coefficients). If there is a normal distribution associated with each of two categories and the two distributions differ only in the expected values, in other words, if the distribution that holds for one category is obtained from the distribution that holds for the other by a simple shift within the multidimensional range of values formed by the variables, one obtains discriminant regions that are separated by a hyperplane, i.e. by a surface that is defined by a linear function. Here again, only the first and second moments of the initial samples are needed for a mathematical expression of this linear function. Used independent of this very specific prerequisite that the data be normally distributed, the result reflects a situation in which certain normal distributions are hypothesized for the initial samples, and the quality of the result then depends on how close the hypothesized distributions are to the actual ones.

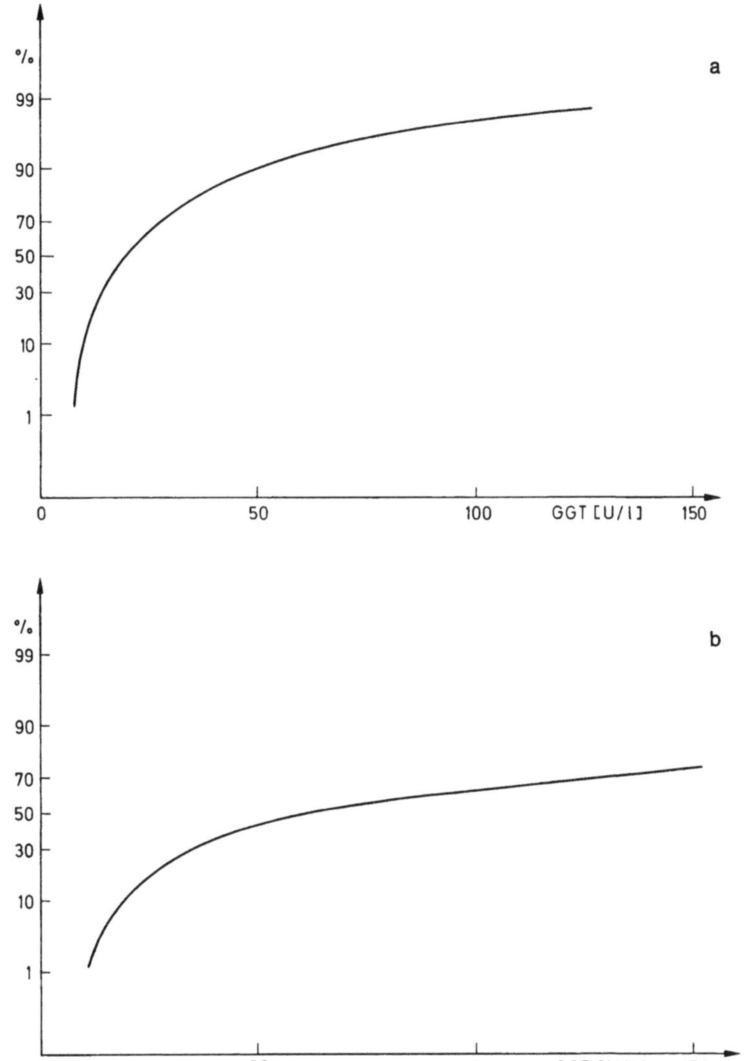

Fig. 1. (a) Smoothed curve showing distribution of GGT values for the non-alcoholics studied; values plotted on probability paper.

(b) Smoothed curve showing distribution of GGT values for the alcoholics studied for the range up to 150 U/l; values plotted on probability paper.

With our initial samples, we found that the hypothesized distributions were quite different from the actual ones, and the results obtained with linear discriminant analysis were therefore unsatisfactory, as the following figures make clear.

Figure 1a shows the empirical distribution function, plotted on probability paper, for the GGT values found for the sample of non-alcoholics studied ($n = 70$). The results up to 150 U/l have been plotted; only one result was higher. The individual

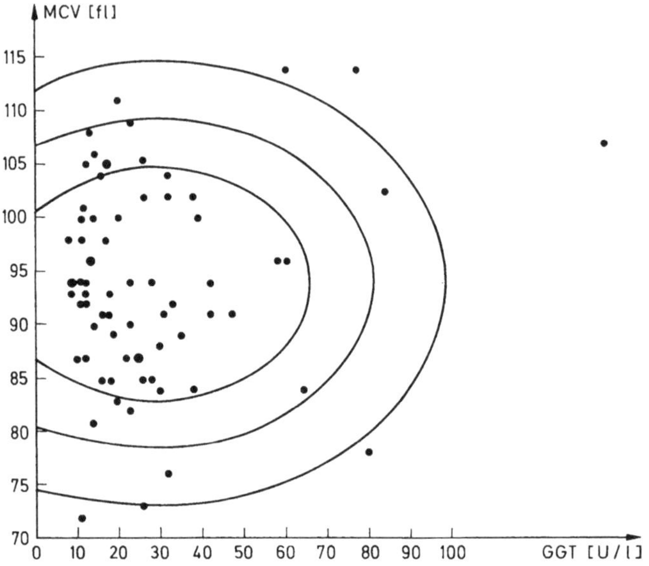

Fig. 2. GGT and MCV values for the non-alcoholics and three ellipses of constant probability density of the two-dimensional normal distribution whose central point and variances are the same as in the sample (theoretical coefficient of correlation = 0); 50%, 25% and 8.2% of the probability, respectively, would lie outside of the ellipses.

dots are extremely close to the smoothed curve. Figure 1b provides the same information for the sample of alcoholics studied (n = 82). Although approximately 30% of the values are above 150 U/l, the distributions for the two samples have surprisingly similar shapes. The marked deviation from a normal distribution is evident. (If results are plotted on logarithmic paper, there is also a deviation from the normal distribution, but it is of a different type and not so marked.)

Figure 2 is a graph of the GGT U/l and MCV fl results for the sample of non-alcoholics. To provide a better visual impression of the type of distribution, several ellipses of constant probability density are included of a two-dimensional normal distribution with the same expected values and variances as in the sample. The value 0 was selected for the coefficient of correlation because that in the sample is −0.011, i.e. roughly 0. We would expect 50% of the values to lie within the innermost ellipse, but in fact the figure is 70%. We would expect 17% between the middle and outermost ellipses, but in fact the figure is only about 5%. A large part of the probability includes negative GGT values, which obviously makes no sense.

Thus linear discriminant analysis yielded a useless result for these two variables: Although for GGT alone a level of efficiency of 150.6 was attainable (sensitivity = 63.4, specificity = 87.1), the combination of GGT and MCV yielded a level of efficiency of only 142.2. Table 1 shows the results for this pair of variables and for two others. In Section 5, after the method of simple discriminant regions has been described, we shall come back to the question of the relationship between the shape of the distribution and characteristics of discriminant regions.

Table 1. Comparison of the results obtained with linear discriminant analysis (LA) and with simple discriminant regions (SR) for three pairs of variables

Combination of variables	Maximum efficiency LA	SR	Order
GGT/MCV	142.2*	158.7	2
GGT/GOT	135.8	158.1	2
GGT/creat	142.3	157.8	1

* The actual value is 142.4. As a result of numerical inaccuracy, one A and one NA near the discriminant line were classified incorrectly.

5 Decision Limits and Simple Discriminant Regions

It was now clear that either linear discriminant analysis is unsuited to identifying differences in the distributions associated with the categories A and NA, which the graphs in the previous section strongly suggest, or the differences in the distributions are not by themselves great enough to enable the use of clinical chemical parameters in alcoholism/non-alcoholism diagnosis.

The need for a different and possibly better suited type of difference between distributions was thus evident. We decided to begin our search with the following observation: Clinicians are used to thinking in single variables and looking at each individual variable to see if it yields a "pathological" or a "normal" value in the person being evaluated. As a rule this process involves a "decision limit" that separates the pathological values (with reference to a particular diagnosis) from the non-pathological ones. In the terminology of discriminant analysis, this means that for a particular quantitative variable there is only one type of discriminant regions, i. e. the values above and the values below a decision limit.

Shouldn't it be possible to transfer this very simple type of argumentation to any number of dimensions via logical connections of the decisions for the individual variables? The first question we had to answer was what type of discriminant regions this would yield, and the second, even more important question was whether such a simple procedure could actually yield useful results. The first question was easy to answer; the second required some experience with real data before it could be answered.

We begin by discussing the problem for the case of two variables, and as in the previous section we use the two variables GGT and MCV. The best decision limits in the sense of maximum efficiency (when the prevalence p is not known) are 45 U/l and 95 fl, respectively (see ref. 4).

The discriminant regions for these two variables taken alone are therefore:

$D_1^+ = D^+(\text{GGT}): X_1 \geq 45$ U/l
$D_2^+ = D^+(\text{MCV}): X_2 \geq 95$ fl

By making use of the two different types of logical connectives we can obtain the following two systems of discriminant regions:

$$(3\,a)\quad \begin{aligned} D^+(\text{GGT,MCV}) &= X_1 \rightarrow D_1^+ \text{ and } X_2 \rightarrow D_2^+ \\ D^-(\text{GGT,MCV}) &= X_1 \rightarrow D_1^- \text{ or } X_2 \rightarrow D_2^- \end{aligned}$$

$$(3\,b)\quad \begin{aligned} D^+(\text{GGT,MCV}) &= X_1 \rightarrow D_1^+ \text{ or } X_2 \rightarrow D_2^+ \\ D^-(\text{GGT,MCV}) &= X_1 \rightarrow D_1^- \text{ and } X_2 \rightarrow D_2^- \end{aligned}$$

As we found in our applications of the program described later, neither type of D^+ (GGT,MCV) yields maximum efficiency; rather, this is achieved with the following definition:

D^+ (GGT,MCV): $X_1 \geq 18$ U/l *and* $X_2 \geq 90$ fl

This is the best efficiency attainable with discriminant regions of the types (3a) and (3b). The maximum is found by first looking for the maximum for each of the two systems (3a) and (3b) given a particular combination of variables and then selecting the larger of the two maxima. In addition, the combination of variables is varied. In contrast to the use of linear discriminant analysis, this procedure yields an encouraging increase in efficiency for the combination of GGT and MCV.

The generalization suggests itself:

Let us assume we have *m* variables $X_1, X_2, \ldots, X_m$, each initially with a preassigned decision limit (for purposes of illustration we assume that these are all upper decision limits; if not, then for the relevant dimensions the "$>$" is replaced by a "$<$"):

$$(4)\quad \begin{cases} D_1^+: X_1 \geq x_1 \\ D_2^+: X_2 \geq x_2 \\ \cdot \\ \cdot \\ \cdot \\ D_m^+: X_m \geq x_m \end{cases}$$

By every possible logical connection with "and" and "or" of the statements $X_i \rightarrow D_i^+$ (X_i provides a value from D_i^+) an *m*-dimensional interval is obtained ($1 \leq i \leq m$). In this way $2^m - 1$ subregions can be formed out of which the discriminant region D^+ can be composed in a suitable fashion. The two extreme possibilities are:

- connect all D_i^+ with "and"; i. e. consider a combination of values for m different variables to be "positive" if and only if each value is the same as or greater than the associated decision limit x_i
- connect all D_i^+ with "or"; i. e. consider a combination of values for m different variables to be "positive" if and only if at least one value is the same as or greater than the associated decision limit x_i.

For the numerous possibilities between these two extremes, a kind of symmetry among the variables was required, which can also be expressed as invariance with respect to the order in which the variables are listed or measured. This led to the definition of $T_{(k)}$, the *simple discriminant region of the* k^{th} *order* that is formed by combining all regions of the types defined in Eq. (4) in which the connection "and" occurs at least k times. In other words, the "simple discriminant region $T_{(k)}$ of the kth order" is defined as follows:

(5) $T_{(k)}$ = all combinations of values of m given variables in which at least k values are the same as or greater than the associated decision limits

The two extreme cases described earlier are thus equivalent to $T_{(m)}$ and $T_{(1)}$.

We thus have m types of discriminant region systems that are all defined by a single m-dimensional point the coordinates of which are exactly the decision limits specified for each value. For a given combination of variables the solution to all three optimization problems (2a), (2b) and (2c) can be found as follows: Specify a value of k ($1 \leq k \leq m$) and then find the optimum for this value by varying the decision limits. Then from the m optimums found, select the best one.

Table 2. Variable combination GGT, MCV, GOT; decision limits for maximum efficiency

	Decision limits			Maximum efficiency		
Order	GGT [U/l]	MCV [fl]	GOT [U/l]	Sensitivity (%)	Specificity (%)	Efficiency
1	≥ 43	≥ 112	≥ 19	66.7	87.1	153.8
2	18	90	24	92.6	68.6	161.2
3	18	90	11	85.2	80.0	165.2

By way of illustration we use the variable combination GGT, MCV, GOT. Table 2 shows the decision limits for maximum efficiency for the orders $k = 1, 2, 3$.

The most efficient regions of positive results are:

For the order $k = 1$:

GGT ≥ 43 U/l *or* MCV ≥ 112 fl *or* GOT ≥ 19 U/l

For the order $k = 2$:

GGT ≥ 18 U/l *and* MCV ≥ 90 fl

or

GGT ≥ 18 U/l *and* GOT ≥ 24 U/l

or

MCV ≥ 90 fl *and* GOT ≥ 24 U/l

For the order $k = 3$:

GGT ≥ 18 U/l *and* MCV ≥ 90 fl *and* GOT ≥ 11 U/l

In the next section we outline a scheme by which such results can be applied in practice.

Using the variable pair GGT, MCV we look once again at the differences between the procedure used in linear discriminant analysis and that involved in simple discriminant regions. To aid us in this comparison, in Figure 3 we drew in

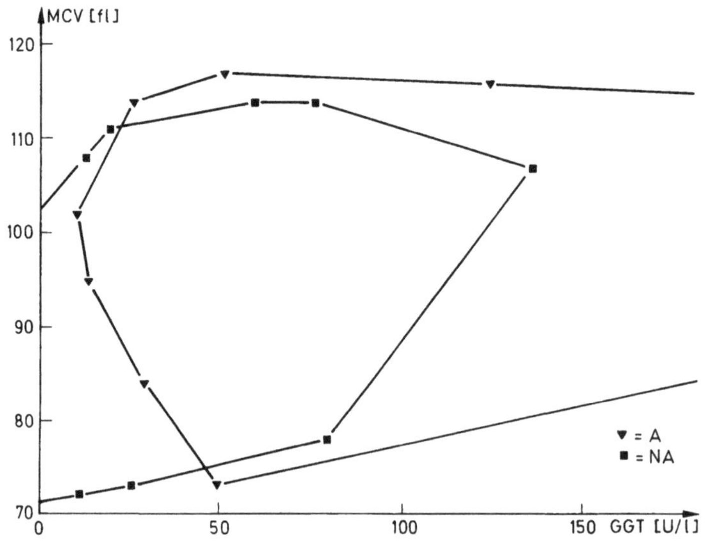

Fig. 3. Convex hulls for the points representing the quantities GGT and MCV (with the exception of GGT values above 150).

the "convex hull" for both the NA and the A samples, i. e. the smallest convex polygon that surrounds all of the points (for the sample of non-alcoholics a single point representing an extremely high GGT value was excluded and for the sample of alcoholics the points representing high GGT values). Next, in Figure 4a we drew in the dividing line resulting from linear discriminant analysis and the points thus falsely classified. Finally, in Figure 4b we marked the most efficient simple region $T_{(2)}$ and the points thus falsely classified. It is evident that the most efficient $T_{(2)}$-region fits the form of the region of the sample of alcoholics particularly well – thus the relatively high sensitivity, whereas the straight line goes right through the middle of both sets of points (which in this case clearly makes little sense – see Table 1).

6 The Computer Program for Finding Simple Discriminant Regions

The computer program developed to solve problems of types (2a), (2b) and (2c) is based on the same principle as the assignment procedure described in Section 5. The givens are two initial samples, one of alcoholics and one of non-alcoholics, a combination of m variables and an m-dimensional decision point x. Included in the definition of a decision point is that for each of the coordinates the direction is specified in which a value must exceed the coordinate to be considered "positive" for that variable. The program compares the coordinates for each subject in the

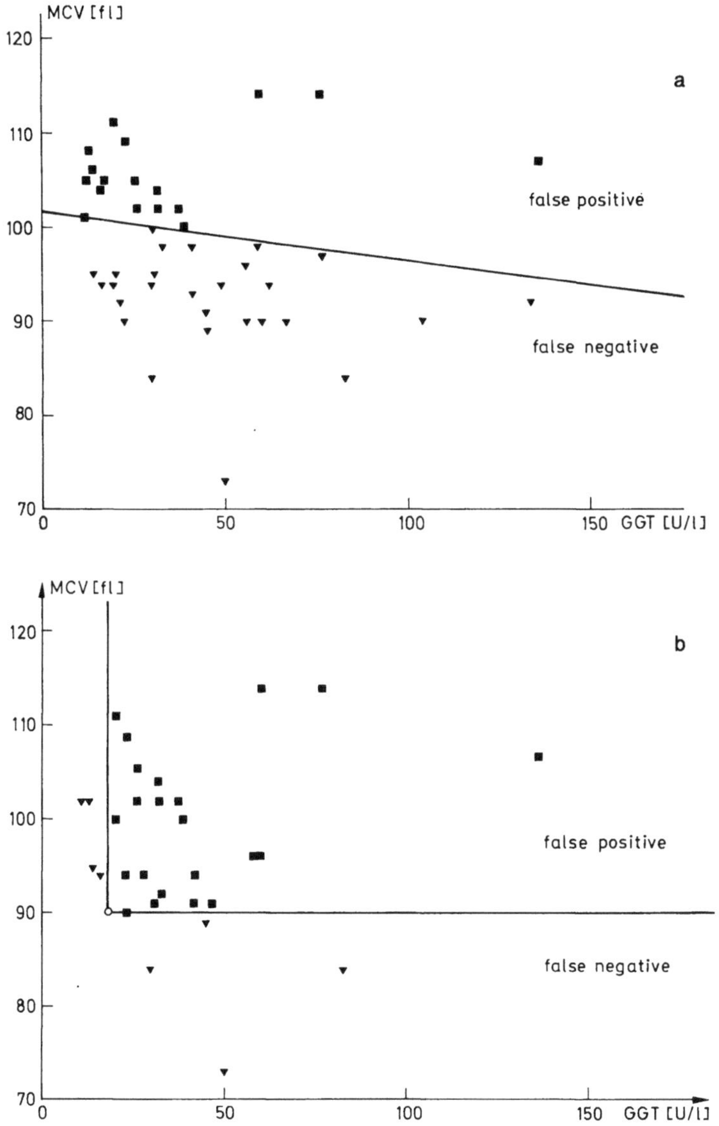

Fig. 4. (a) Straight line (linear discriminant function) for the two quantities GGT and MCV separating the false positive and false negative points.

(b) Optimal discriminant region of the order $k = 2$ for the two quantities GGT and MCV indicating the false positive and false negative points.

initial samples with those of the decision point and determines which are "positive" and which are "negative." This process is programmed as a checking procedure almost free of numerical error. Connecting these comparisons produces the number of positive coordinates for each subject. If the order k of the discriminant region has been specified ahead of time, then after the number of positive coordinates has been determined, the subject can be classified as "positive" or "negative." In this way, for given initial samples, a given decision point and a given order one obtains the figures for sensitivity, specificity and efficiency for the initial samples (at present counting is done without weighting with a value p for prevalence, but such weighting could be done without any major changes in the program). The optimum (or a local optimum) is found by varying the decision point within the range of values of the variables while holding k and the combination of variables constant. The procedure is described in detail in (1) and is based on the specification of lattices of decision points.

The program is run as a dialog. We decided to avoid too much automatization. The variable combination, the lattices of decision points, the order of the discriminant region and the requirements for sensitivity, specificity and efficiency that appear on the screen can be changed while the program is running. The user can get an idea from what he or she sees on the screen about what might be the most useful way to proceed with the analysis.

Two types of lists can be printed out: (a) those decision points that meet criteria for sensitivity and/or specificity or a criterion for efficiency specified in advance, and (b) for a given decision point, the coordinates of all points in the initial samples, with each set of coordinates being marked to indicate the type of "positiveness"; a frequency list can also be included showing all possible types of "positiveness" and the frequency of each type in the two initial samples. Assuming enough computing time is available, a single run of the program is sufficient for a given variable combination, order and optimization task. Printouts of the first type enable listings of discriminant regions with particular characteristics, e.g. the series of nested regions discussed in Section 7.

As to estimating how good optimal discriminant regions obtained from initial samples can be in practical applications, reference is made once again to the literature (9, 10). In the following a scheme is proposed for the way in which a simple discriminant region can be used which is nothing more than a geometric scheme for the logical connections discussed in Section 5 and which enables the whole complex problem of multidimensional regions to be reduced to examining a number of one-dimensional regions.

The variable combination GGT/MCV/GOT/creatinine, which in the form of a third-order region proved to be the most efficient combination of four variables, serves here as an example. The optimal one-dimensional decision limits are:

GGT	$\geq$ 18 U/l
MCV	$\geq$ 90 fl
GOT	$\geq$ 19 U/l
creatinine	$\leq$ 73 µmol/l

The sensitivity of the third-order region with this decision point is 81.5% and the specificity 87.1%; the efficiency is thus 168.6. Figure 5 shows the scales for each of these four variables, with the orientation of the creatinine values being from right to left. The positive one-dimensional regions are shaded and further indicated by arrows. A four-dimensional point is considered positive if at least three of its coordinates are positive. The example given in the figure is the positive point [30, 17, 94, 66]. Hence the determination of whether the point found for a

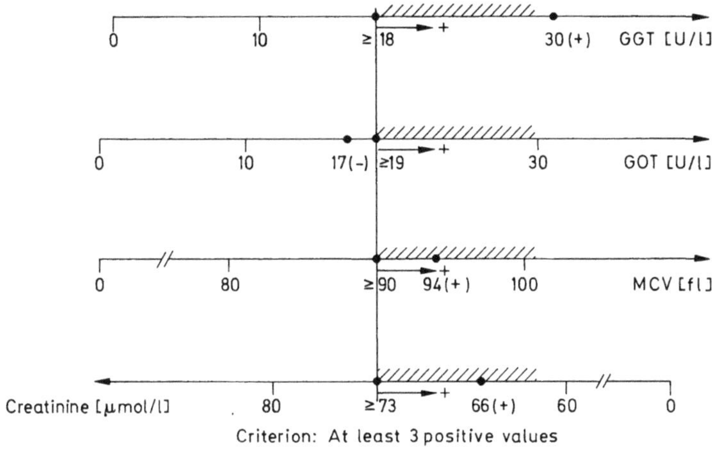

Fig. 5. Example of a decision scheme for the logical combination of four decision limits; the positive region for each quantity is shaded. The example used is the point 30 U/l, 17 U/l, 94 fl, 66 μmol/l, which is positive.

given subject should be considered "positive" or "negative" requires no calculations at all. Furthermore, the diagnostician sees which variables have led to the decision, and this can be of value clinically. In addition, more detailed information of possible clinical value can be made available, as mentioned earlier in connection with the types of computer printouts available; a detailed example is given in ref. 4, Table 3–5.

We have already pointed out that a statistical problem of estimation is involved in such figures based on initial samples. This is doubly true when the initial sample is made up of subsamples of different origins. We have given special attention to this problem and will report on our findings in detail elsewhere. Here we state only that for the two hospitals mentioned in Section 3 and the optimal discriminant region given above, the levels of efficiency were 169.3 and 167.5. At the least, this shows that the discriminant region can be used in the two participating hospitals. However, this does not necessarily mean that it is optimal in both. And in fact there is a much better third-order discriminant region for one of the hospitals, the efficiency being 184.6, if the decision limit used for MCV is 95 fl and that used for GOT is 14 U/l.

7 "Nesting" of Discriminant Regions

Arguments against classifying clinical test results simply as "positive" or "negative" are (a) that in the transition area between the two test regions the assignment to one category or the other is more or less arbitrary, and (b) that when the test result does not fall in the transition area possible additional information is lost. In other words, where in the region a test result lies cannot be without importance. A test result in a region where there has never been an NA result but there have been several A results points much more strongly to a classification as A than simple "positiveness" (but is this justified?). For a formal justification of such a decision see the subsequent discussion. However, it is clear that both criticisms, namely that a judgement of results in the transition area is arbitrary and that a judgement in other areas provides little information, are merely two sides of the same coin: If a space that contains results has a continuous geometric structure, the representation of an evaluation (= assessment) as a continuous function should also be possible.

A way of meeting this need for information and the need for finer and finer distinctions in the assessment would appear to be the introduction of nested regions: Positive regions D_i^+ and negative regions D_i^-

(5a) $\ldots D_i^+ \supset D_{i+1}^+ \supset D_{i+2}^+ \supset \ldots$

and

(5b) $\ldots D_i^- \subset D_{i+1}^- \subset D_{i+2}^- \subset \ldots$

are constructed in such a way that on the one hand a D^+ with a higher subscript provides greater certainty about the correctness of an assigment to A and on the other a D^- with a lower subscript provides greater certainty about the correctness of an assignment to NA.

We now explain this concept with several examples. A welcome characteristic of the method of simple regions is that finding such series of regions (5a) and (5b) turns out to be a natural byproduct of the procedure. In our view, this in turn shows that the procedure is a natural description of differences in distributions (which can also be supported by purely mathematical arguments).

Let us first look at the variable pair GGT/MCV, which we have already used frequently by way of illustration. Figure 6a shows a number of second-order regions. These regions are defined and the associated sensitivity and specificity are given in Table 3. The regions were drawn according to the geometric scheme proposed in Section 6. In Figure 6b the corresponding two-dimensional regions have been marked in, which makes the characteristics (5a) and (5b) immediately visible. The region yielding maximum efficiency, D_3^+, was used as the initial region and has therefore been indicated with a heavy line in Figure 6a (the choice of maximum efficiency as a starting point seems obvious but is nevertheless arbitrary). Let us assume that the point [20, 93] is measured. With reference to the region of maximum efficiency it is positive. But if we consider rejecting it as a case of NA, things do not look very good because the region D_3^+ still contains 31.4% of the NA subjects. The situation is quite different if the point [40, 93] is obtained. This point lies in the region D_6^+, a region that contains only 10% of the NA subjects.

This way of arguing has parallels in the way statistical tests are applied in practice. It is considered "good statistical practice" (11) to distinguish between specifying a test level and considering a critical level, although no statistical theory has yet been developed to support this. Specification of a test level is important in situations where "yes-or-no" decisions must be made that may have far-reaching consequences, whereas consideration of a critical level implies that one is more interested in supporting a hypothesis than in making a final decision (see also (12)).

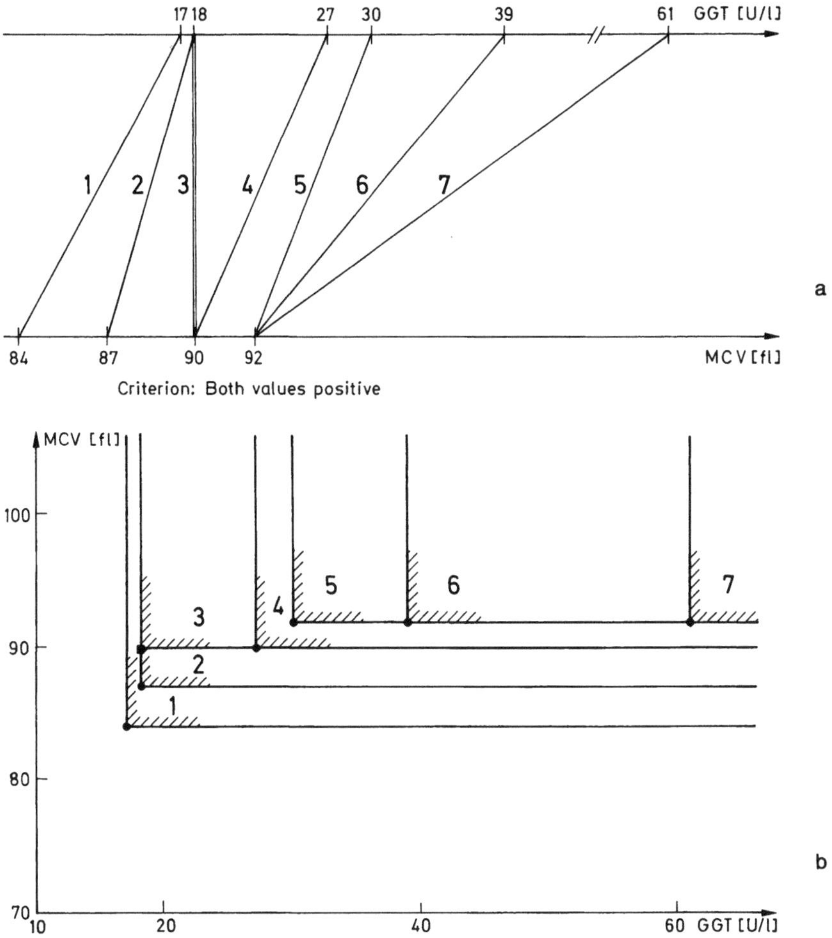

Fig. 6. (a) Decision scheme analogous to that in Figure 5 but for the quantities GGT and MCV only, with several "nested" discriminant regions of the order $k = 2$.

(b) Geometric illustration of the discriminant regions used in Figure 6a.

Table 3. Numerical values for the series of discriminant regions ($k = 2$) defined in Figures 6a and 6b

	GGT [U/l]	MCV [fl]	Sensitivity (%)	Specificity (%)
D_1^+	≥ 17	≥ 84	93.8	45.7
D_2^+	18	87	91.4	60.0
D_3^+	18	90	90.1	68.6
D_4^+	27	90	76.5	80.0
D_5^+	30	92	66.7	85.7
D_6^+	39	92	56.8	90.0
D_7^+	61	92	45.7	97.1

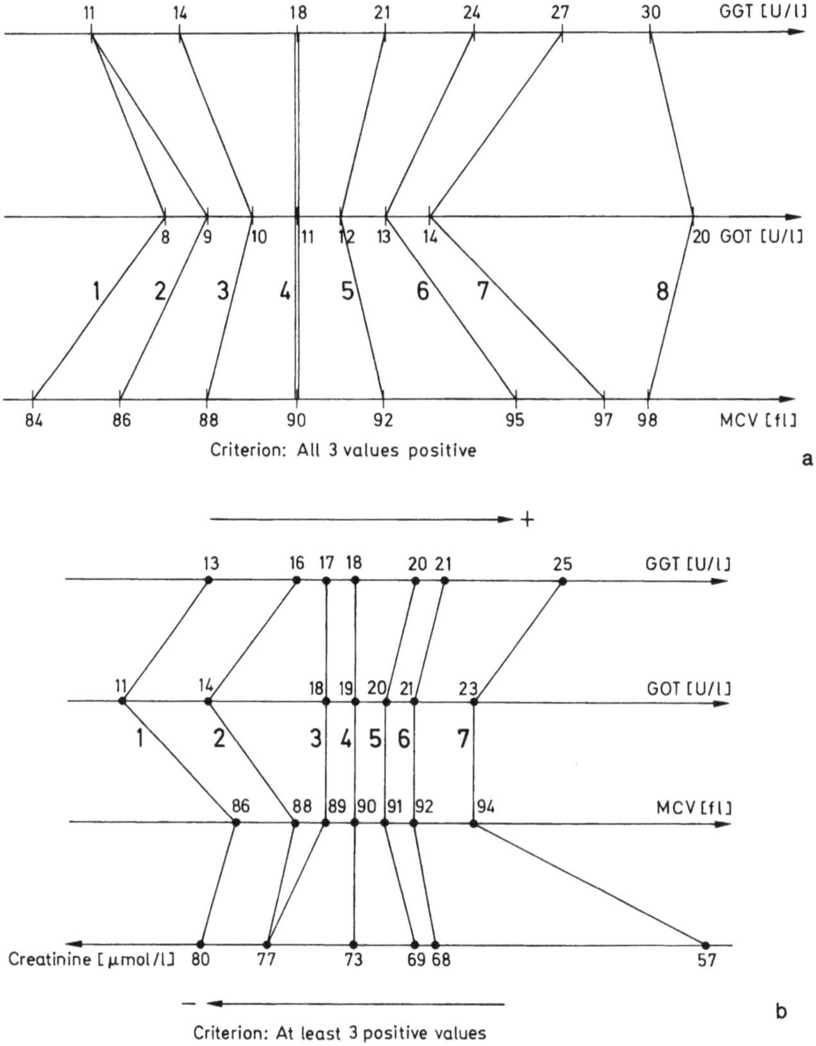

Fig. 7. (a) Decision scheme for the quantities GGT, MCV and GOT, with several "nested" discriminant regions of the order $k = 3$.

(b) Decision scheme for the quantities GGT, MCV, GOT and creatinine, with several "nested" discriminant regions of the order $k = 3$.

The latter corresponds to the typical situation in science and probably also in diagnosis, where a tentative decision is the rule and a final one the exception.

Figure 7a/Table 4a and Figure 7b/Table 4b provide further examples, with three and four variables respectively.

It should be pointed out that in these examples a certain selection has been made among all possible regions: If a practical application is planned, this selection will of course be made with the application in mind.

Table 4a. Numerical values for the series of discriminant regions ($k = 3$) defined in Figure 7a

	GGT [U/l]	GOT [U/l]	MCV [fl]	Sensitivity (%)	Specificity (%)
D_1^+	≥ 11	≥ 8	≥ 84	97.5	30.0
D_2^+	11	9	86	95.1	45.7
D_3^+	14	10	88	90.1	67.1
D_4^+	18	11	90	85.2	80.0
D_5^+	21	12	92	70.4	88.6
D_6^+	24	13	95	58.0	94.3
D_7^+	27	14	97	50.6	97.1
D_8^+	30	20	98	32.1	100

Table 4b. Numerical values for the series of discriminant regions ($k = 3$) defined in Figure 7b

	GGT [U/l]	GOT [U/l]	MCV [fl]	Creat [μmol/l]	Sensitivity (%)	Specificity (%)
D_1^+	≥ 13	≥ 11	≥ 86	≤ 80	100	44.3
D_2^+	16	14	88	77	93.8	68.6
D_3^+	17	18	89	77	84.0	77.1
D_4^+	18	19	90	73	81.5	87.1
D_5^+	20	20	91	69	69.1	90.0
D_6^+	21	21	92	68	61.7	94.3
D_7^+	25	23	94	57	42.0	100

8 Discussion of the Procedure

1. The procedure described here is suitable for all types of variables with at least an ordinal scale, for the only assumption made is the existence of a "greater than" or "less than" relation among the values on the scale.

2. The procedure places no requirements at all on the type of distribution. Each system of discriminant regions is defined by a single point. The coordinates of this point are treated as decision limits, but the individual decisions are connected logically, not considered separately. In mathematical terms, each discriminant region is a compilation of multidimensional intervals and (in the current version) no variable is any more important than the others. The system of discriminant regions that is based on $T_{(1)}$ is comparable to the regions on the basis of which distribution functions in mathematical statistics are defined. Therefore, the other systems of discriminant regions considered here can be regarded as generalizations that describe elementary properties of distributions.

3. The procedure described here is a special case of a general type of discriminant analysis in which an optimal region is sought by joining several subregions of a particular kind. For other examples see refs. 13 and 14.

4. The measures that determine the certainty with which a subject is classified are actually 100 minus sensitivity in percent and 100 minus specificity in percent. This is due to the analogy with statistical test theory discussed in (1). Consequently, the prerequisite for a high level of certainty in rejecting a subject as an alcoholic is a discriminant region with high sensitivity (alcoholism as the null hypothesis), and conversely, for a high level of certainty in assigning a subject to the category alcoholism the prerequisite is a discriminant region with high specificity (non-alcoholism as the null hypothesis). Only if the discriminant region has high efficiency is it an adequate basis for both types of decision.

5. Because the procedure of simple discriminant regions is not dependent on any prerequisites about the type of distribution, as with all procedures having this property, it cannot be optimal with reference to any particular family of distributions. But it does have the mathematical advantage that for a given family of distributions the theoretically optimal simple regions can, at least in principle, be calculated. For the family of two-dimensional normal distributions we did calculations for two examples and found that the difference in efficiency between the optimal discriminant region and the optimal second-order simple discriminant region was less than 5% (relative difference in efficiency). This could indicate that even for normal distributions simple discriminant regions are a useful alternative to linear regions.

6. The reasons why we find this procedure so attractive are:

- Good results can be obtained on a very simple basis.
- A subject can be assigned to one of two diagnostic categories without any calculations being necessary.
- The reasons for the assignment are immediately evident from the values of the variables involved.

In each case the clinician can thus recognize (simply by inspecting the decision diagram) which variables had values that led to, for example, an assignment to the category A. In other words, the clinician can follow the process of assignment, if necessary arrange for additional information to be obtained, and possibly draw conclusions relating to treatment.

9 Discussion of the Results

Of the 152 men studied, 82 were alcoholics (A) and 70 non-alcoholics (NA). Significant differences between the two groups in the distribution of the results were found for the following clinical and hematological parameters:

1. *At the 0.1% level*

GGT	creatinine
GOT (AST)	MCV
urea	

2. *At the 1% level*

GLDH	AP
GPT (ALT)	

From these eight quantities, those combinations of two to six quantities were selected that discriminated best between the alcoholics and the non-alcoholics.

When *conventional decision limits* are used, the following holds:

For the *alcoholics*, in 85.2% of the subjects two or more of the following five variables have values in the regions given in parentheses:

GGT	(≥ 28 U/l)	MCV	(≥ 96 fl)
GOT (AST)	(≥ 18 U/l)	creatinine	(≤ 66.3 μmol/l)
GPT (ALT)	(≥ 22 U/l)		

For the *non-alcoholics*, in 95.7% of the subjects two or more of the following five variables have values in the regions given in parentheses:

GGT	(< 28 U/l)	creatinine	(> 66.3 μmol/l)
GOT (AST)	(< 18 U/l)	urea-N	(> 5.0 mmol/l)
MCV	(< 96 fl)		

It is therefore possible to detect or exclude alcohol abuse in men on the basis of clinical laboratory findings.

By *optimizing decision limits* the diagnosis criteria sensitivity or specificity or efficiency can be improved even more. The optimized decision limits for the six quantities listed above are shown in Table 5.

In looking at papers on markers for alcohol abuse published over the last several years, a realistic picture is obtained only in those studies where alcoholics are compared with other patients who are not alcoholics, not when they are compared with "healthy" subjects.

The differences found in the results in different studies are not due solely to different methods or different outside criteria in patient selection. Rather, these differences are certainly due in large part to differences in the statistical methods used to evaluate the results, because the forms of the distributions were not taken into consideration.

The statistical procedure described in this contribution does not place any requirements on the type of distribution of the results. It yields realistic multivariate decision limits. But practical application of the procedure has also shown that complete separation of alcoholics and non-alcoholics in patient populations, at least based on our data, is not possible with any available procedure. The procedure described here is suitable for multivariate analysis of measurements made on either continuous or ordinal scales. Because there are no requirements regarding the type of distribution, it has a wide range of application. As the example at the end of Section 6 shows, each time before the results are used a check must be made of whether the situation to be assessed has been adequately represented in the initial samples.

Table 5. Decision limits and diagnostic criteria for the detection and exclusion of alcoholism in men

Parameter (Reference interval) [Unit]	Goal		
	Detection of alcoholism by excluding non-alcoholism	Detection of non-alcoholism by excluding alcoholism	Optimization of the number of correct assignments of A and NA
	Required specificity: 95%	Required sensitivity: 95%	Optimization of efficiency
	Decision limits (multivariate optimization)		
GGT (6–28) [U/l]	≥ 27	18	26
GOT (2–20) [U/l]	≥ 15	10	10
GPT (2–20) [U/l]	≥ 11	9	9
MCV (80–96) [fl]	≥ 97	97	93
Urea-N (2.1–7.8) [mmol/l]	≤ 2.9	4.3	6.4
Creatinine (44–97) [μmol/l]	≤ 62	66	67
Number of positive findings	≥ 4	4	5
	Assignments made		
Correctly identified	A 67%	NA 74%	A 83%, NA 89%
Efficiency	162	169	171
Falsely excluded (Type 1 error)	NA 5%	A 5%	
Falsely not detected (Type 2 error)	A 33% (false negative diagnosis)	NA 26% (false positive diagnosis)	A 17%, NA 11%

References

1. Hansert, E., Federkiel, H. & Stamm, D. (1984) A new procedure for discriminating between two patient populations using multivariate decision limits: Application in the detection and exclusion of alcoholism based on clinical laboratory findings. J. Clin. Chem. Clin. Biochem. *22*, 791–810.
2. Edwards, G., Orford, J., Egert, S., Guthrie, S., Hawker, A., Hensman, D., Mitcheson, M., Oppenheimer, E. & Taylor, C. (1977)

Alcoholism: A controlled trial of "treatment" and "advice." J. Stud. Alc. *38*, 1004–1031.

3. Stamm, D., Hansert, E. & Feuerlein, W. (1984) Excessive consumption of alcohol in men as a biological influence factor in clinical laboratory investigations. J. Clin. Chem. Clin. Biochem. *22*, 65–77.
4. Stamm, D., Hansert, E. & Feuerlein, W. (1984) Detection and exclusion of alcoholism in men on the basis of clinical laboratory findings. J. Clin. Chem. Clin. Biochem. *22*, 79–96.
5. Feuerlein, W., Ringer, C., Küfner, H. & Antons, K. (1977) Diagnose des Alkoholismus. Der Münchner Alkoholismus-Test (MALT). Münch. Med. Wochenschr. *119*, 1275–1282.
6. Feuerlein, W., Küfner, H., Ringer, C. & Antons, K. (1979) Münchner Alkoholismus-Test (MALT). Beltz, Weinheim.
7. Stamm, D. (1982) A new concept for quality control of clinical laboratory investigations in the light of clinical requirements and based on reference method values. J. Clin. Chem. Clin. Biochem. *20*, 817–824.
8. Stamm, D. (1986) Experience with quality control in clinical chemistry. Z. Lebensm. Unters. Forsch. *182*, 372–384.
9. Choi, S. C. (ed.) (1986) Statistical Methods of Discrimination and Classification, 136 pp., Pergamon Press, New York; Computers and Mathematics with Applications *12* (2) (special issue).
10. Fahrmeier, L. & Hamerle, A. (eds.) (1984) Multivariate statistische Verfahren. In collaboration with W. Häussler et al., de Gruyter, Berlin.
11. Bross, I. D. J. (1971) Critical levels, statistical language, and scientific inference. In: Foundations of Statistical Inference, Proc. Symp. Foundations Statistical Inference, Ontario, Canada, March 31–April 9, 1970 (Godambe, V. P. & Sprott, D. A., eds.) pp. 500–513, Holt, Reinhart and Winston, Toronto.
12. Stegmüller, W. (1973) Probleme und Resultate der Wissenschaftstheorie und Analytischen Philosophie, Vol. 4: Personelle und Statistische Wahrscheinlichkeit, Studienausgabe Teil D, Springer, Berlin.
13. Büttner, J. (1977) Die Beurteilung des diagnostischen Wertes klinisch-chemischer Untersuchungen. J. Clin. Chem. Clin. Biochem. *15*, 1–12.
14. Keller, H. & Gessner, U. (1982) The risk quotient: A tool for interpretative reporting. Paper given at the 5th Colloque Internationale de Biologie Prospective, October 4–9, 1982, Pont-á-Mousson, unpublished manuscript.

Chapter 3.11
A New Distribution-Free, Multivariate Discriminating Method

First experiences gained in the discrimination of patientgroups suffering from coronary heart disease, alcoholism, and primary hyperparathyreoidism

B. Keller and H. Keller

1 Outline of the Method

If two of more distinct groups are to be discriminated by two or more features, methods of discriminant analysis are used (1 – 9). The "classical" linear discriminant analysis (DA) (10) presupposes a multivariate normal distribution. Frequently this type of distribution is not given or its presence cannot be verified, e. g. because of lack of data. This may be the reason why frequently unsatisfactory results are obtained. Therefore different non-parametric methods of DA have been developed (5, 11, 12) by many authors. Our new, quasi "graphic" procedure is based on the following considerations:

If two collectives are to be separated by two characteristics a dissecting straight line (= discriminant function) is to be found, which separates all points so that none or a minimal number of points is wrongly classified (allocated). It is possible, that several straight lines have the same power of separation (we will not enter into the possibility, that a curve-linear function provides the optimal dissection).

We demonstrate our procedure by a model (Table 1): a scatter diagram (Fig. 1) of parameter X and Y shows the values of group A and B. (In Fig. 1 the points are numbered for easier reference). For each group (cluster) the mean values (or the medians or the modes) are calculated (S_A, S_B respectively).

If one or both variates have a discriminating power with respect to the groups then the means (or medians or modes) cannot be identical. The greater the discriminating power, the greater the distance between S_A and S_B will be. Now the following algorithm is applied: from each point of group A a straight line to each point of collective B is drawn. All lines which do not intersect the connecting line ($S_A S_B$) are ignored. For all lines, that intersect the connecting line, the number of incorrectly allocated points is determined, applying the following rule: a point is considered to be incorrectly allocated if it does not lie on the same side of the straight line as its group mean (or mode or median). Figure 2 shows the six possible straight lines

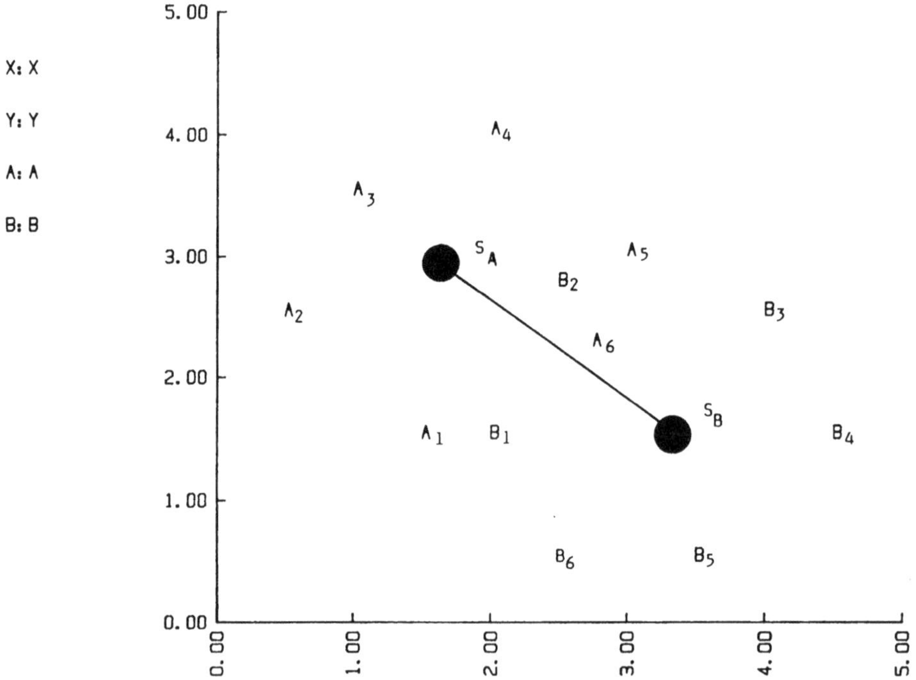

Fig. 1. Model of collectives A and B to be discriminated by parameters X and Y; graphic representation of Table 1. S_B, = centre of collective B.
$(S_A S_B)$ = straight line joining the centres.

Table 1. Model-collective A and B with variables X and Y

	X	Y
A_1	1.5	1.5
A_2	0.5	2.5
A_3	1.0	3.5
A_4	2.0	4.0
A_5	3.0	3.0
A_6	2.75	2.27

	X	Y
B_1	2.0	1.5
B_2	2.5	2.75
B_3	4.0	2.5
B_4	4.5	1.5
B_5	3.5	0.5
B_6	2.5	0.5

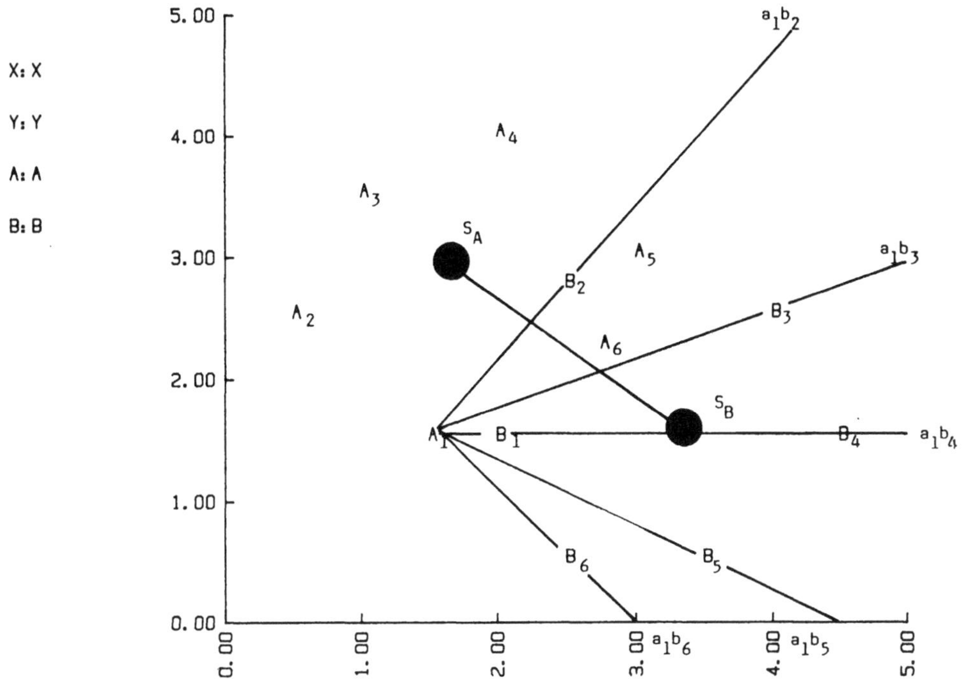

Fig. 2. Straight lines from A_1 to B_1 ... B_6; (S_A S_B) is dissected by (a_1b_2), (a_1b_3) and (a_1b_4).

Table 2. Discriminating effect of different dissection straight lines

Dissection straight lines	False points
a_1b_2:	A_5; A_6
a_1b_3:	B_2
a_1b_4:	B_2; B_3

from A_1 to B_{1-6} (A_1B_1), (A_1B_5) and (A_1, B_6) do not intersect (S_AS_B). The discriminating effect of the three other lines is given in Table 2.

Subsequently for point A_2 all dissecting lines are searched and it is checked whether the discriminating power is better than for the preceding lines. Continuing in this way the discriminating power of all possible lines is investigated. Finally the straight line (or lines) is retained, which provides the smallest number of wrong allocations. In our example this is true for two straight lines (a_1b_3) and (a_6b_1), where just one point is misclassified: B_2 (Table 3).

The optimal straight line is now rotated in such a way, that the defining points are correctly classified, too.

Figure 3 shows the two optimal lines which give only one incorrect classification. The discrimination by the classical linear discriminant analysis is worse, with two incorrect allocations (B_2 and A_6). The distances to the new discriminating line can be used as a new parameter (XY) combining the original parameters X and Y. This

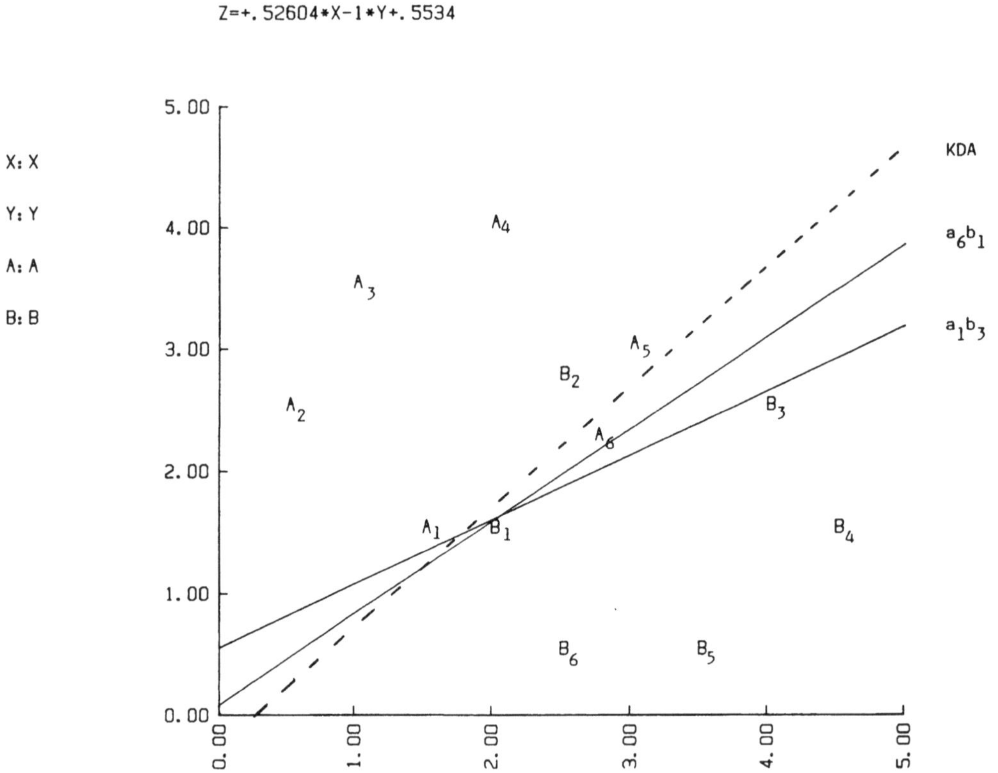

Fig. 3. The optimal dissecting straight lines (a_1b_3) and (a_6b_1) correctly discriminate all pairs except B_2. The dissecting straight line of the "classical" discriminant analysis yields two false allocations: B_2 and A_6.

Table 3. Demonstration of the discriminating steps shown by the display of the calculator. Column A and B correspond to the joining lines, column "false" figures the false allocations of the corresponding dissection straight line

A	B	False
1	2	2
1	3	1
6	1	1

new parameter might be the X-axis of a new scatter diagram where a third parameter forms the Y-axis and one can investigate now, whether some other discriminating line gives a better discrimination.

This new procedure produces lines which are approximately identical with those from the classical linear discriminant analysis, provided that a sufficient number of multivariate normally distributed points is available. In any case, the straight line obtained in this way, is the best possible for the given "learning collective". If more than one optimal discriminating line is found, each one must be investigated to determine whether the addition of a new parameter leads to a better discrimination. It is possible, that important differences occur, although the discriminating lines showed identical power of separation in the first step.

For the application in practice we have developed a program ("OPTIGER") in BASIC for the HP 87XM (see Addendum).

For small collectives – e.g. 12 points in our model – the calculation is performed in a few seconds. If the number of points is considerably larger (> 100), then the procedure takes a few hours.

On the order "determination of the optimal discriminating line" the computer screen displays the individual steps of the procedure. As an example, the discriminating steps of our model are shown in Table 3. Also the classical discriminant analysis can be performed on request by the same program OPTIGER. Less time is necessary for this calculation, e.g. only a few minutes for 100 points.

During the construction of the optimal discriminating line the predominance of false positive over false negative allocations or vice versa is not taken into account. Consequently the diagnostic sensitivity or specificity predominates. Using the OPTIGER program a shifting of the discriminating line is possible in order to improve the diagnostic sensitivity or specificity. Nevertheless in any case a shifting causes a deterioration of the discriminating power.

The efficiency of OPTIGER has been tested using an example of a well known textbook (13) and on some clinical studies from literature. We will demonstrate a bivariate, a trivariate and a tetravariate example.

2 Efficiency Tests

2.1 Textbook model: Discriminant analysis for the assessment of survival in cases of icterus hemolyticus neonatorum (2 parameters available)

In this example the concentrations of hemoglobin and bilirubin in the cord blood of new born babies are determined. It is required to predict the chance for survival by means of these two parameters. Out of 79 infants affected by the hemolytic disease 63 survived and 16 died. The histogram (Fig. 4) shows that with both parameters the groups overlap widely. It is not possible to assess the chance of survival by the hemoglobin or bilirubin level alone.

Another picture shows the scatter diagram (Fig. 5): obviously the survivors are concentrated in the lower right part, and those who died in the upper left.

The calculation of the classical discriminant analysis results in a discriminant function whereby 54 of 63 survivors are correctly classified (and 9 are misclassified) and 14 of the 16 who died are correctly classified, and 2 are misclassified. It is 68 out of 79 (86%) which are correctly classified.

The OPTIGER program finds two straight lines with correct classifications of 60/63 survivors and 14/16 non-survivors. There are only 5 misclassifications, 94%

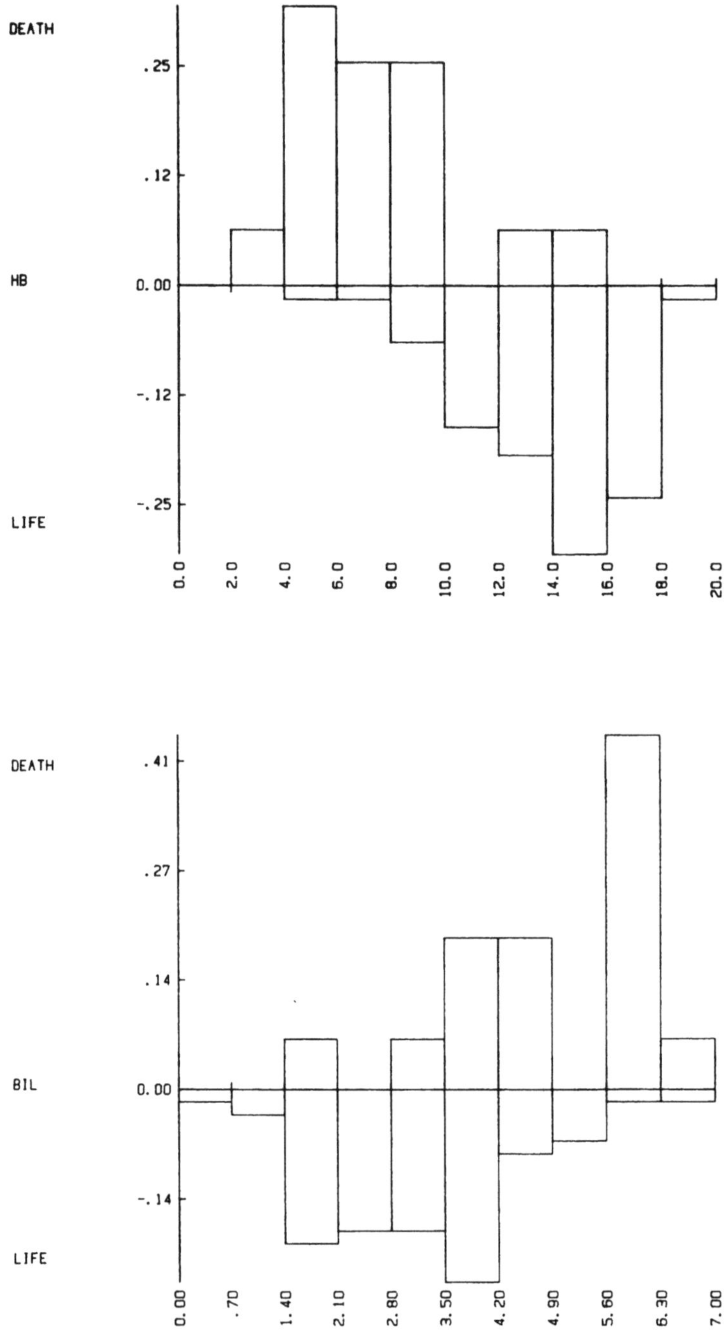

Fig. 4. Histogram of hemoglobin (HB) and bilirubin (Bil) concentrations in cordblood of newborns, suffering of M. hemolyticus neonatorum. Deaths ($N = 16$) top, surviving ($N = 63$) bottom. Both collectives are calculated as 100%.

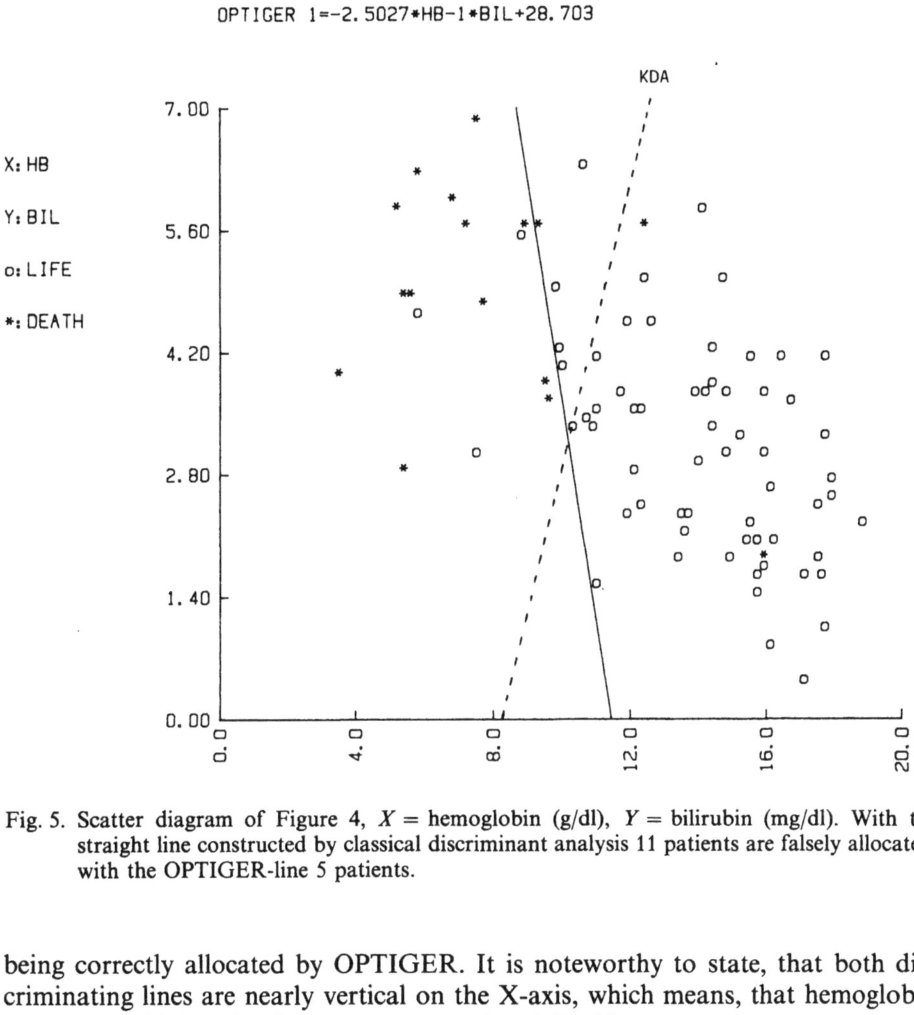

Fig. 5. Scatter diagram of Figure 4, X = hemoglobin (g/dl), Y = bilirubin (mg/dl). With the straight line constructed by classical discriminant analysis 11 patients are falsely allocated, with the OPTIGER-line 5 patients.

being correctly allocated by OPTIGER. It is noteworthy to state, that both discriminating lines are nearly vertical on the X-axis, which means, that hemoglobin has a much higher discriminatory power than bilirubin.

The histogram of the distribution with the new parameter Z_1 of both collectives is shown in Figure 6. By the new discriminant function

$$Z_1 = 2.5 \text{ Hb} - \text{Bilirubin} + 28.7$$

the chances of survival improve with increasing negativity of the resulting value and vice versa.

2.2 Bivariate clinical study: Discriminant analysis for the assessment of the coronary status (8 parameters available)

The lipid status of a group of patients who has had a coronary angiography because of anginale syndromes, has been investigated by Riesen et al. (14). From 110

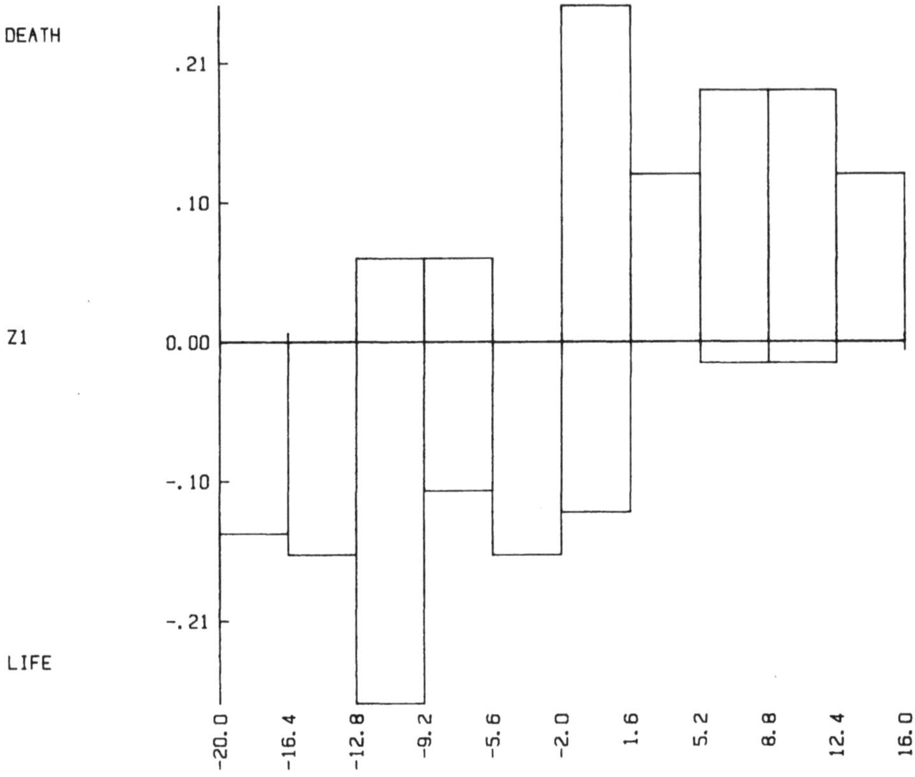

Fig. 6. Histogram in analogy to Figure 4, the function $Z = 2.5\ \text{Hb} - \text{Bil} + 28.7$ is used as parameter.

patients 61 had significant lesions of the coronary arteries, that means they have shown one or more stenoses of more than 50%. These patients are called "patient-group" (PAT). The other 49 patients (with clinical signs of anginale syndromes) did not show a significant coronary arteriosclerosis, they are called "group of reference" (REF). For comparison a collective of 50 blood donors, apparently healthy (GES), is also investigated. In all 3 groups the following parameters were determined: Cholesterol (CH), LDL-Cholesterol (LDL), HDL-Cholesterol (HDL), Triglycerides (TG), phospholipids (PL), apolipoprotein A1 (Apo A_1), apolipoprotein A2 (Apo A_2), and apolipoprotein B (Apo B).

Figure 7 shows the distributions of collective PAT (top) and collective REF (bottom) of four of eight lipid-parameters. Cholesterol is the best discriminating parameter, the trend to higher values of the PAT group is evident. LDL is similar to cholesterol, there is a strong dependency to cholesterol obviously. HDL-cholesterol does not discriminate, there is a complete overlapping of both collectives. The same is true for phospholipids and triglycerides. On the other hand apolipoprotein A_1 and A_2 have a higher discriminating power, each is strongly dependent on the other. Also apolipoprotein B has a significant discriminating effect.

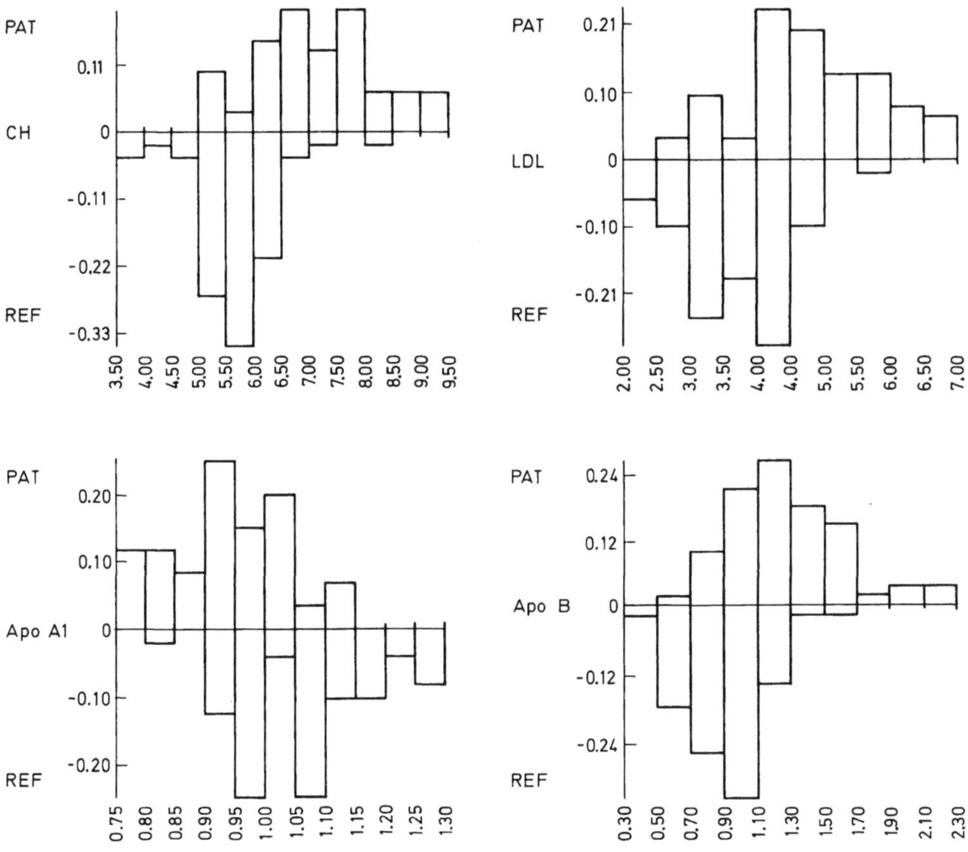

Fig. 7. Histogram of collective "coronary stenoses patients" (PAT) top, and "non-coronary stenoses" (REF) bottom, with the parameters cholesterol (CH), LDL-Cholesterol (LC), Apolipoprotein A_1 ($ApoA_1$) and apolipoprotein B (Apo B). Each collective is calculated as 100%.

Under this conditions it seems logical in a first step, to construct a scatter diagram with cholesterol (X-axis) and Apo A_1 (Y-axis) and to search for the best discriminating line. Figure 8 shows that line corresponding to the discriminating function:

$$Z_1 = 0.14 \text{ Cholesterol (mmol/l)} - \text{Apo } A_1 \text{ (mg/l)} + 0.123.$$

57 of 61 PAT's are located to the right of the discriminating line, corresponding to 4 false negatives. 41 of 49 REF's are on the left side, corresponding to 8 false positives. In other words: 12 of 110 patients are misclassified, corresponding to an efficiency of 89%. The collective GES, additionally recorded into that scatter diagram (Fig. 9) is far from collective PAT, nevertheless there is a certain overlapping with collective REF.

Therefore it is necessary to construct a new discriminating line between the REF and GES group respectively.

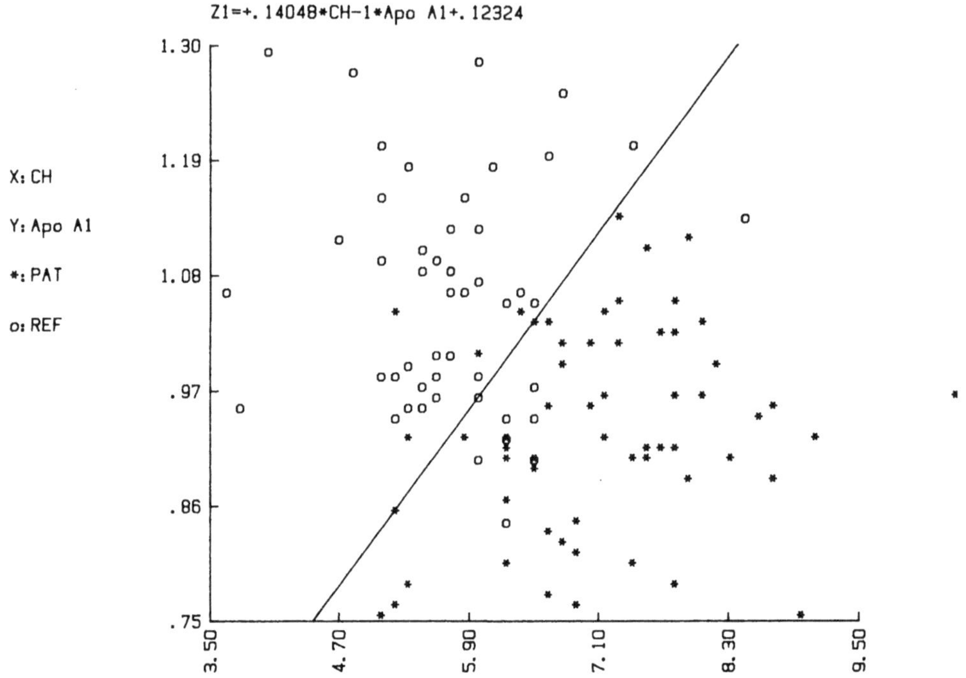

Fig. 8. Scatter diagram of collective PAT (*) and REF (o); X = Cholesterol (mmol/l), Y = Apo A_1 (mg/l); the optimal dissecting straight line is Z_1.

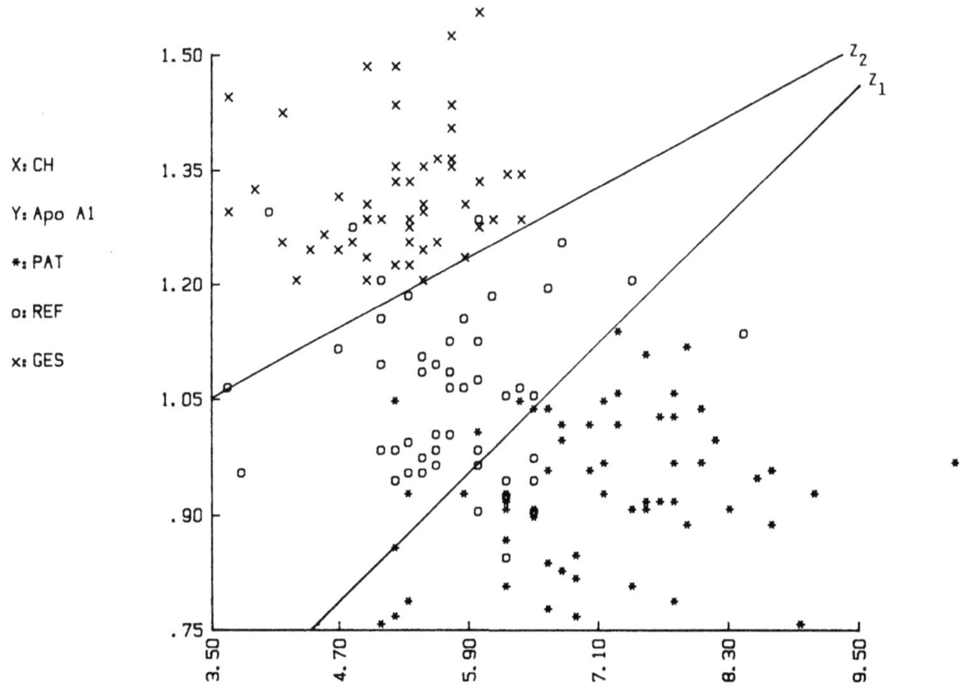

Fig. 9. Like Figure 8, additionally is shown the collective GES (x) and the optimal dissecting straight line Z_2.

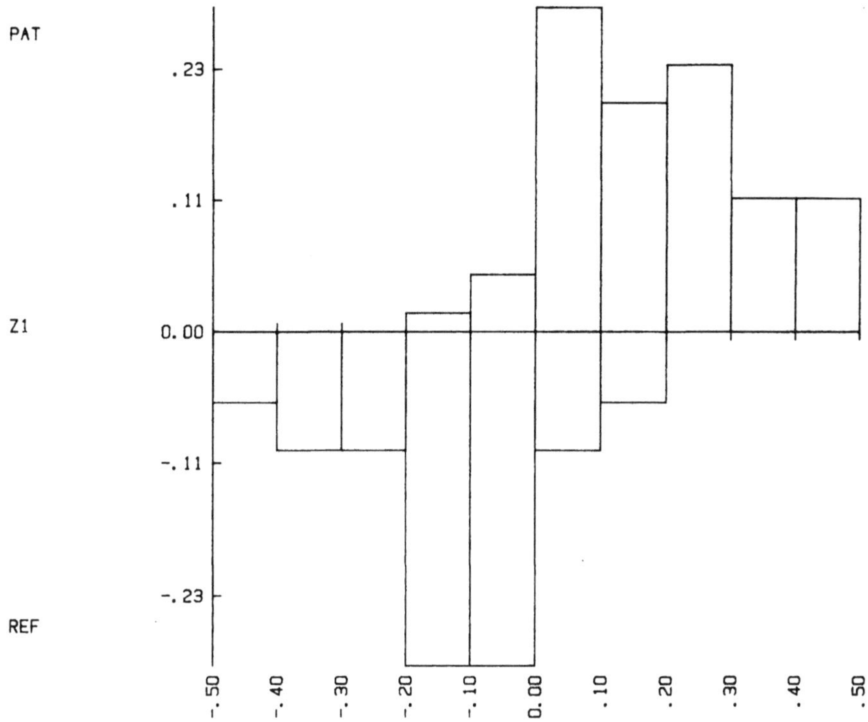

Fig. 10. Histogram of collective PAT (top) and REF (bottom) with Z_1 as parameter.

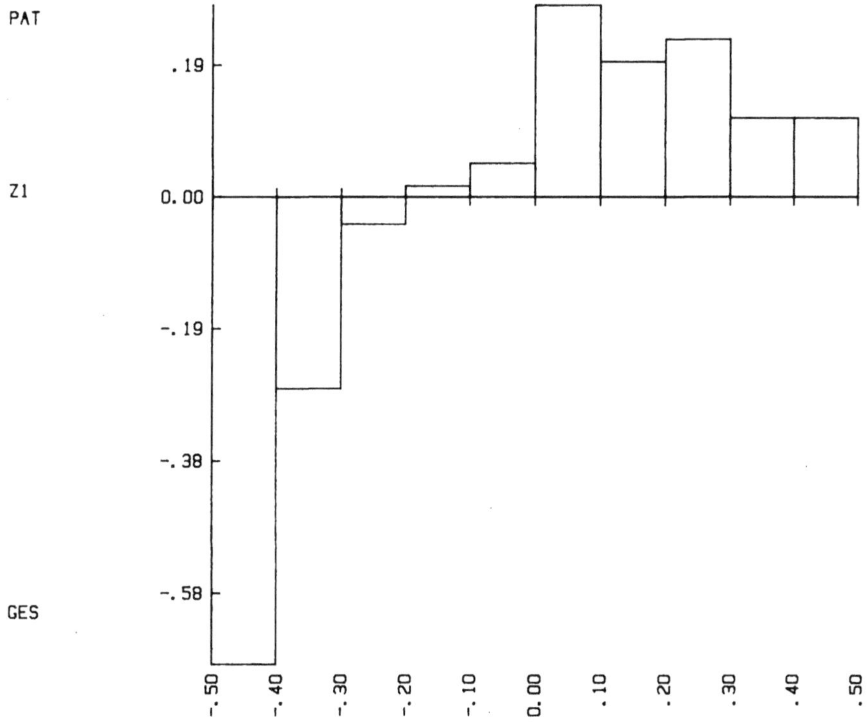

Fig. 11. Like Figure 10 with collective PAT (top) and GES (bottom).

The new line is represented by the function:

$Z_2 = 0.077$ cholesterol (mmol/l) − Apo A_1 (mg/l) + 0.78.

48 of 50 GES are located on the left side, 2 are false positive. 45 of 49 REF patients are on the right side, 4 are false negative. This means, that 84% of collective REF and GES are classified correctly by function Z_2.

Figure 10 shows the distribution of collective PAT and REF by Z_1, Figure 11 for PAT and GES; Z_1 discriminates both collectives completely.

All other 27 bivariate combinations have a lower discriminating power. In Table 4 the results of 10 pairs are demonstrated as example.

It was not successful to increase the discriminating power of the combination cholesterol/Apo A_1 by a third parameter. All multivariate combinations deliver the same number of misclassifications and mostly the same patients are wrongly allocated. The considerable dependency of all lipid parameters seems to be the most important reason for this phenomenon.

Table 4. Discriminating efficiency of 10 bivariate combinations

Parameter *X*	Parameter *Y*	Number of falses	Number of corrects	% Corrects
Chol	Triglyc	32	78	71
	Apo B	23	87	79
	Apo A_1	12	98	89 ⬅
	LDL	22	88	80
Triglyc	Apo B	27	83	75
	Apo A_1	30	80	73
	LDL	24	86	78
Apo B	Apo A_1	22	88	80
	LDL	30	80	73
Apo A_1	LDL	24	86	78

2.3 Trivariate clinical study: Discriminant analysis for the detection of alcoholism (24 parameters available)

The data base originates from a paper by Stamm, Hansert & Feuerlein (15). These authors have determined 24 clinical chemical and 8 hematological parameters in 82 patients with certified alcoholism (ALK). The data are compared with a comparable group of 70 Non-alcoholic patients (ABS), mostly from the same clinical institution. There was no significant difference in age and sex distribution. In our investigation we tried 10 parameters for their discriminating power as monovariates and as multivariates also. Figure 12 shows that GGT has the highest discriminating power followed by the GOT/GPT quotient. The mean of urea and creatinin is

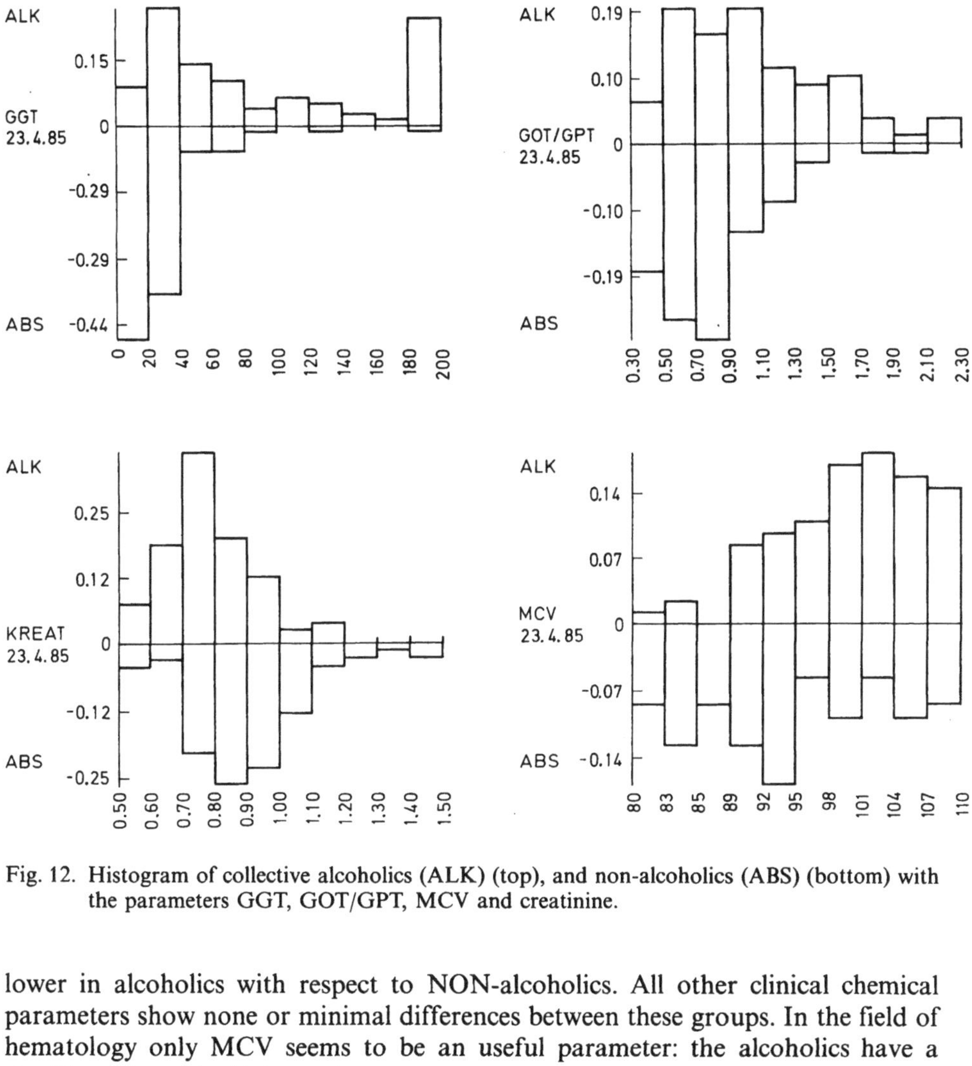

Fig. 12. Histogram of collective alcoholics (ALK) (top), and non-alcoholics (ABS) (bottom) with the parameters GGT, GOT/GPT, MCV and creatinine.

lower in alcoholics with respect to NON-alcoholics. All other clinical chemical parameters show none or minimal differences between these groups. In the field of hematology only MCV seems to be an useful parameter: the alcoholics have a significant trend to higher values.

The pair GOT/GPT versus GGT was the best of all tried bivariate combinations (Fig. 13): 68/82 alcoholics are on the right side of the discriminating line, 14 are false negatives; 53/70 NON-alcoholics are placed on the left side, 17 are false positives. The number of misclassifications is nearly 20%.

If the new parameter Z_3 is combined with MCV (Fig. 14), an additional amelioration results: 70/81 alcoholics are to the left, 54/70 non alcoholics are to the right of the discriminating line. A correct discrimination of 82% is possible by function Z_{3A}. The distribution of the groups by the new parameter is shown in Figure 15. The addition of other parameters e. g. creatinine or urea does not deliver an imporvement with respect to the number of misclassifications. Table 5 gives a review of all combinations checked. With the Z_{3A} function a diagnostic sensitivity of 0.86 and a specificity of 0.77 can be reached. Shifting the line to the left or the

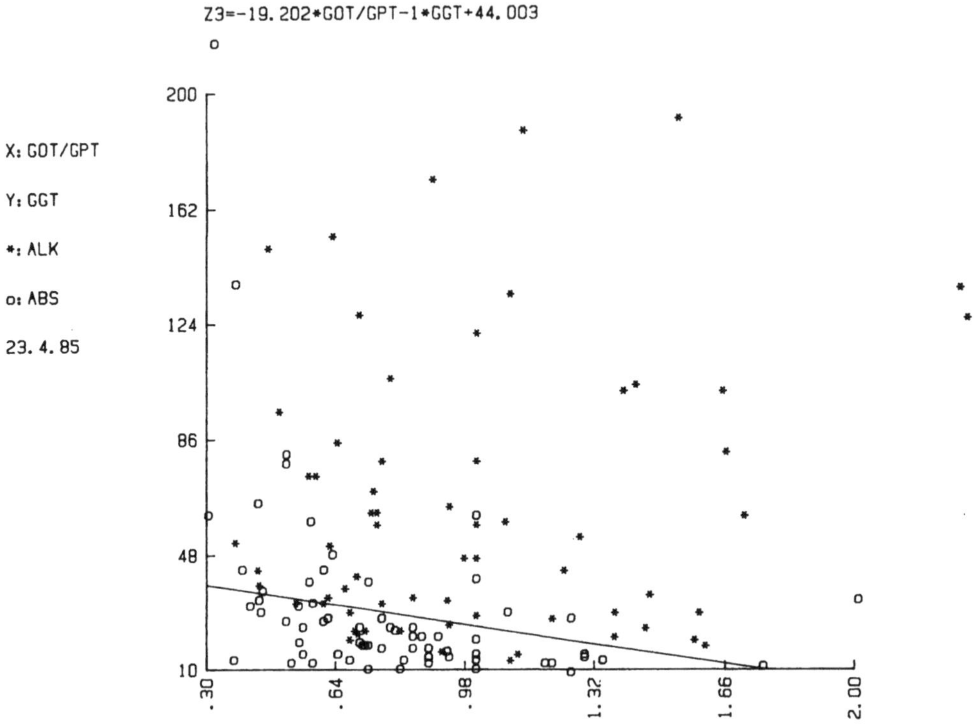

Fig. 13. Scatter diagram of collective alcoholics (*), and nonalcoholics (o); X = GOT/GPT, Y = GGT.

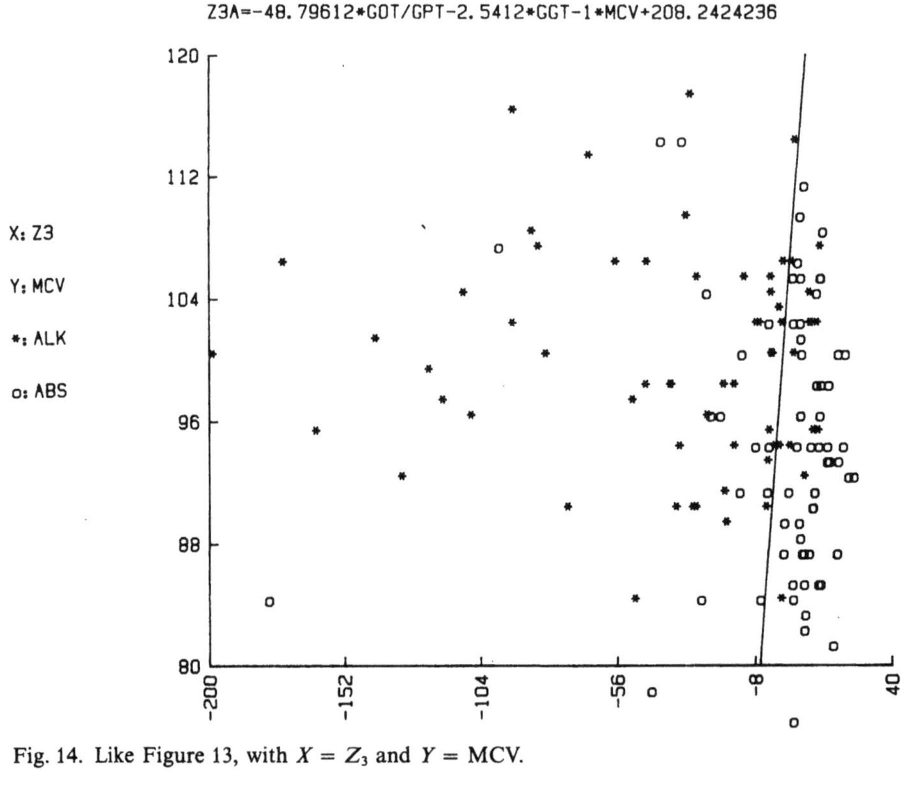

Fig. 14. Like Figure 13, with $X = Z_3$ and Y = MCV.

Table 5. Discriminating efficiency of univariate and multivariate parameters. It is obvious, that parameter Z_{3A} causes the smallest number of false allocations N_f. Additional parameters do not improve

Univariate				N_f
GGT	(discr. 51 U/l)			42
OT/PT	(discr. 0.88)			54
GOT	(discr. 14.5 U/l)			45
Multivariate				
Z_1	GGT *vs* GOT			34
Z_{1A}	GGT *vs* GOT	*vs* Hst		30
Z_{1B}	GGT *vs* GOT	*vs* Hst	*vs* MCV	29
Z_2	GGT *vs* creat.			34
Z_{2A}	GGT *vs* creat.	*vs* MCV		29
Z_{2B}	GGT *vs* creat.	*vs* MCV	*vs* GOT	28
Z_3	OT/PT *vs* GGT			30
Z_{3A}	OT/PT *vs* GGT	*vs* MCV		27
Z_{3B}	OT/PT *vs* GGT	*vs* MCV	*vs* creat.	27
Z_{3C}	OT/PT *vs* GGT	*vs* MCV	*vs* Hst	27
Z_{3D}	OT/PT *vs* GGT	*vs* MCV	*vs* log Hst	27
Z_{3E}	OT/PT *vs* GGT	*vs* Hst		29
Z_4	log GPT *vs* log GOT			32
Z_{4A}	log GPT *vs* log GOT		*vs* log GGT	31
Z_5	OT/PT *vs* MCV			41
Z_6	OT/PT *vs* Hst			39

Table 6. Shifting of the dissecting straight line Z_{3A} by 10 scores in a positive direction causes an increase in sensitivity and a decrease of specifity; and vice versa. In any case the number of false allocations N_f increases when the dissecting straight line is shifted.

Dissecting straight line ± 0

	T	$\bar{T}$			$N_f = 27$
D	70	11	81	sens = 0.86	
$\bar{D}$	16	54	70	spec = 0.77	

Dissecting straight line + 10

	T	$\bar{T}$			$N_f = 33$
D	72	9	81	sens = 0.89	
$\bar{D}$	24	46	70	spec = 0.66	

Dissecting straight line − 10

	T	$\bar{T}$			$N_f = 33$
D	60	21	81	sens = 0.74	
$\bar{D}$	13	57	70	spec = 0.81	

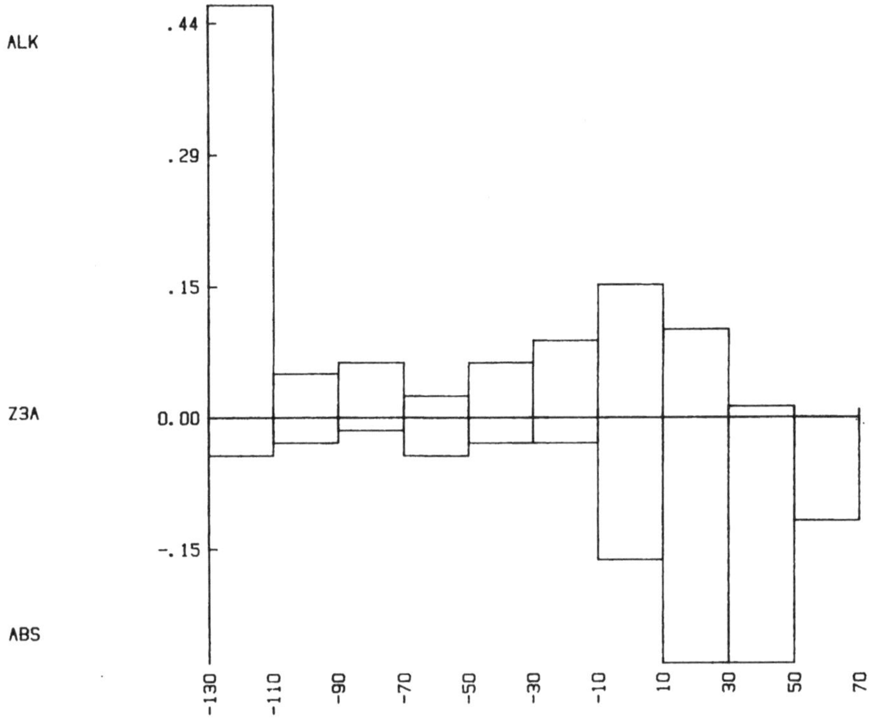

Fig. 15. Like Figure 12, with Z_{3A} as parameter.

right leads to an increase with a corresponding decrease of the dependent probabilities, always in combination with an increase of misclassifications. Table 6 shows the relation between diagnostic probabilities and the shifting of the discriminating line.

2.4 Tetravariate clinical study: Discriminant analysis for the identification of primary hyperparathyreoidism (8 parameters available)

In order to discriminate between the primary hyperparathyreoidism (HPT) and other hypercalcemic disorders, several authors have recommended discriminant analysis (16–20).

Recently a study by Lafferty (Cleveland) (21) was published on this subject. For the discrimination of hypercalcemic patients he used hematocrite, chloride, calcium and phosphate and – for some of the patients – he also used parathormone. We received the complete data base from the author. It contains 100 patients suffering from primary hyperparathyreoidism, verified by biopsy (HPTCL); 31 patients with bone metastases of maligne diseases, 20 patients with pseudohyperparathyreoidism, 4 patients with vitamin D intoxication, 3 patients with thyreotoxicosis and 6 patients

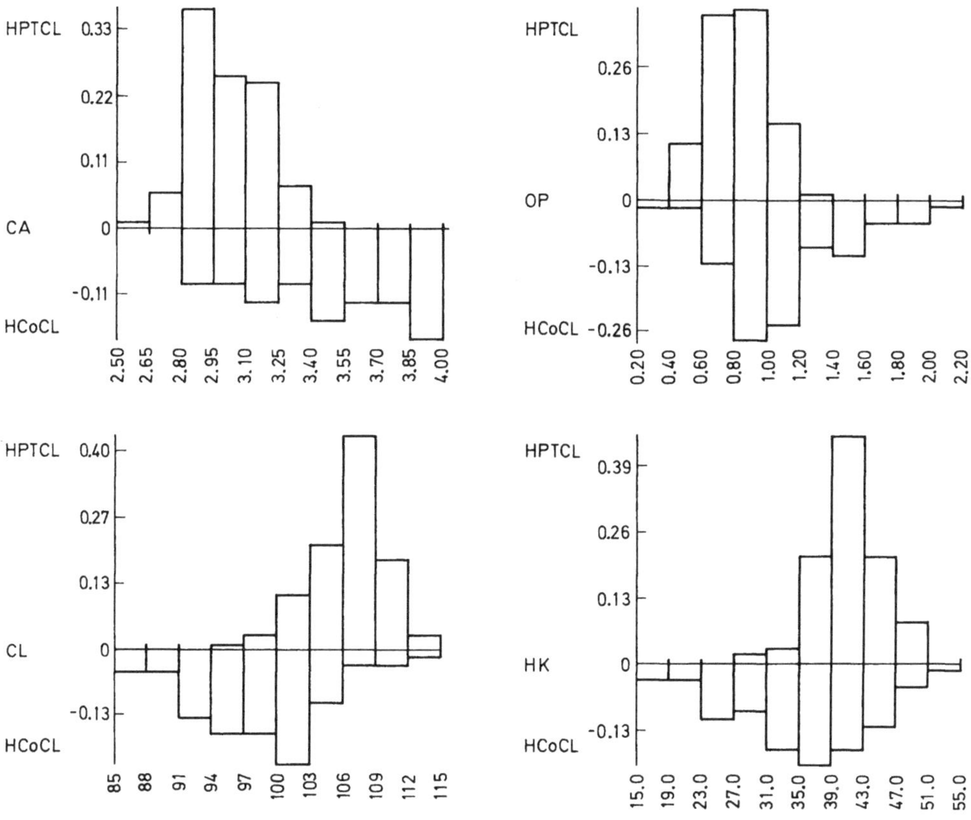

Fig. 16. Histograms of patients with primary hyperparathyreoidism of the Cleveland clinic (HPTCL) (top), and other hypercalemic patients (HCaCL) (bottom): both collectives are calculated as 100%, Ca, Cl and Phos in mmol/l, Hk in %.

with hypercalcemia from other diseases. The histogram of the 4 parameters are shown in Figure 16, where the patients with primary hyperparathyreoidism (HPTCL) are at the top and all the other 64 NON-HPT patients (HCaCl) at the bottom. The calcium concentration in serum is lower in the HPT group compared with the other patients, but there is a broad zone of overlapping.

The same is true for phosphate in serum, where the HPT patients also show lower values.

Within the distribution of the chloride concentrations the HPT group shows a peak, significantly higher than in the NON-HPT group.

Also the mean hematocrit value is higher in HPT patients, again completely overlapping.

Parathormone was not included in our calculations because its concentration was not determined for the majority of patients. Furthermore parathormone is an analytically uncertain parameter with insufficient standardization and insufficient comparability from one laboratory to another. The univariate validity of each

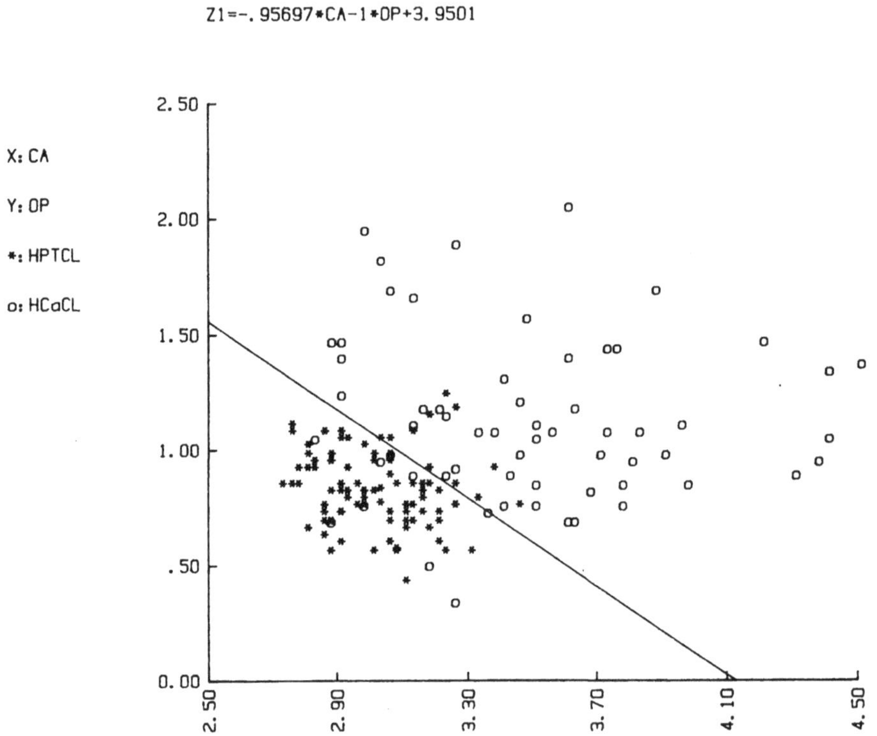

Fig. 17. Scatter diagram of collective HPTCL (*) and HCaCL (o); X = Calcium (mmol/l), Y = phosphate (mmol/l).

parameter does not suffice to solve the diagnostic problem. We therefore tried whether a bivariate combination with 4 parameters and 6 bivariate combinations is possible. Figure 17 shows the scatter diagram of calcium versus phosphate with the optimal discriminating line Z_1, calculated by the computer. The discriminating line allocates 94/100 HPT patients correctly, and 55/64 NON-HPT patients; which means 15 patients were wrongly classified. The discriminating power of the classical discriminant analysis is lower, 18 patients being incorrectly allocated. A new scatter diagram was constructed with Z_1 on the X-axis and chloride on the Y-axis (Fig. 18). With OPTIGER one can find 6 optimal discriminating lines of identical discriminating power: 95/100 HPT patients and 58/64 NON-HPT patients are correctly classified, 11 patients are misclassified. The new discriminant function Z_2 is obviously better than Z_1.

With Z_2 on the X-axis and hematocrit on the Y-axis, the new scatter diagram gives even better discrimination by Z_3 (Fig. 19): now 98/100 HPT patients and 62/64 NON-HPT patients are correctly classified, the number of misclassifications being only 4. The distribution of the "learning group" under Z_3 is shown in Figure 20.

The discriminant power of Z_3 was controlled by two other independent patient groups: 37 samples from patients with verified primary hyperparathyreoidism from

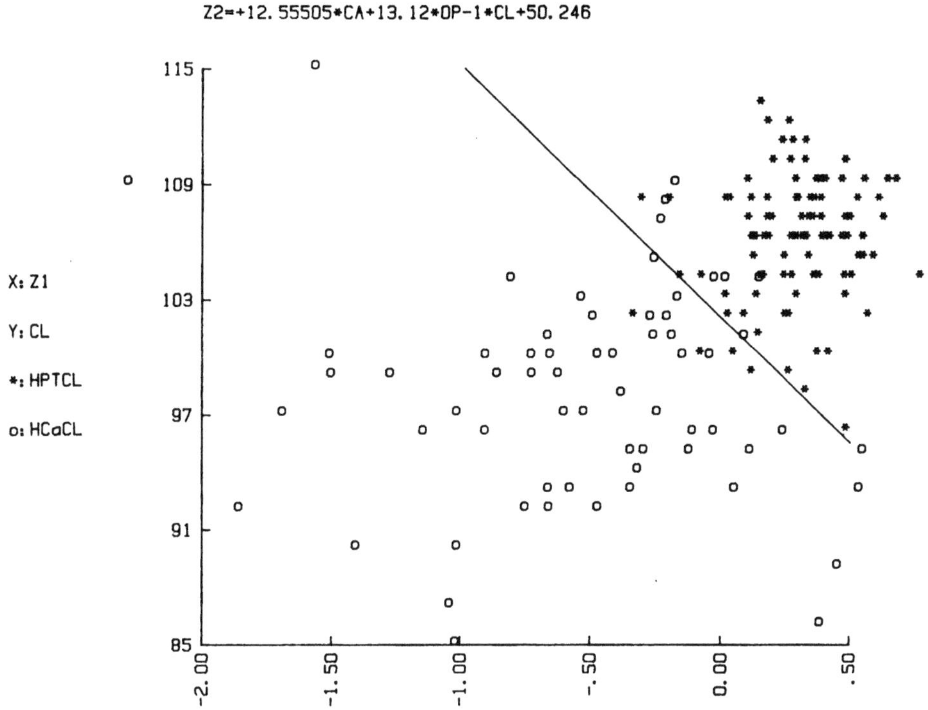

Fig. 18. Like Figure 17, $X = Z_1$, $Y =$ Cl.

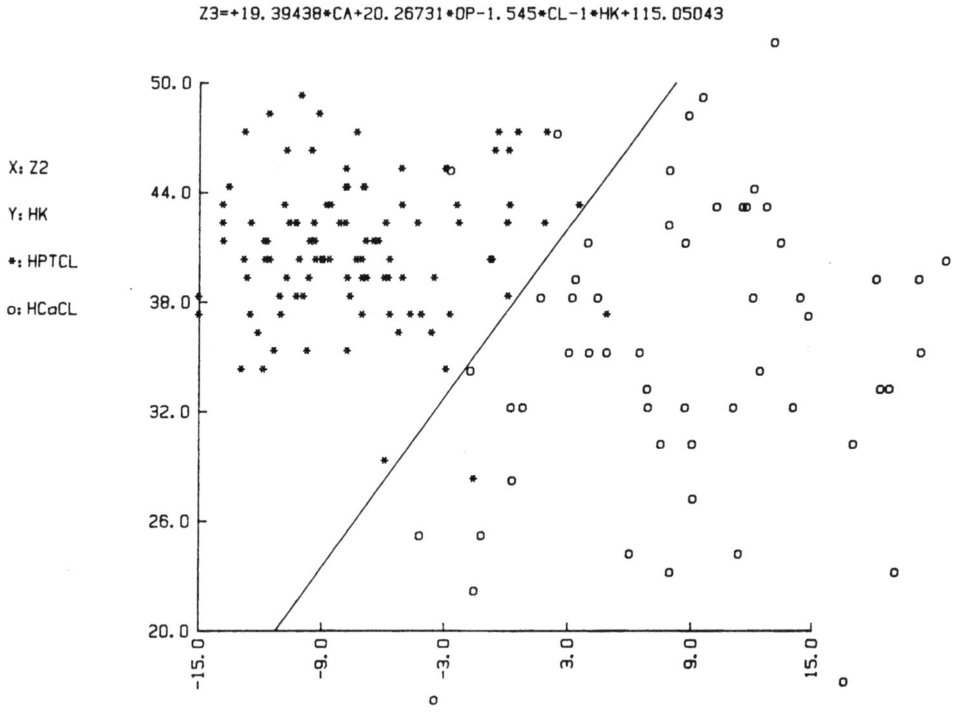

Fig. 19. Like Figure 18, $X = Z_2$, $Y =$ Hk.

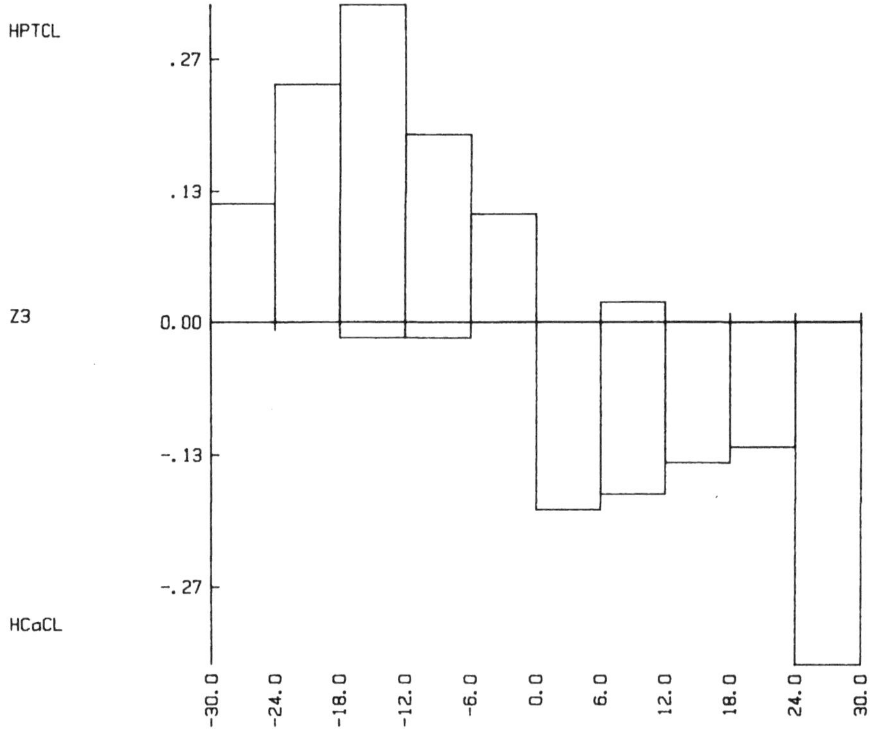

Fig. 20. Histogram of collective HPTCL (top), and HCaCl (bottom), with Z_3 as parameter.

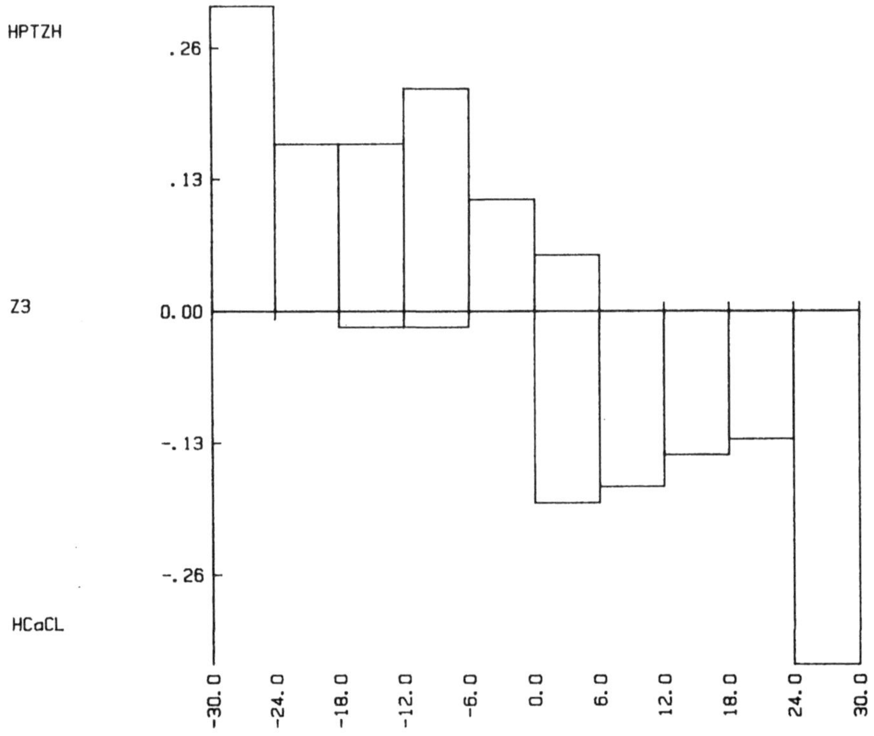

Fig. 21. Like Figure 20, on the top is a patient collective with primary hyperparathyreoidism from the Zürich clinics.

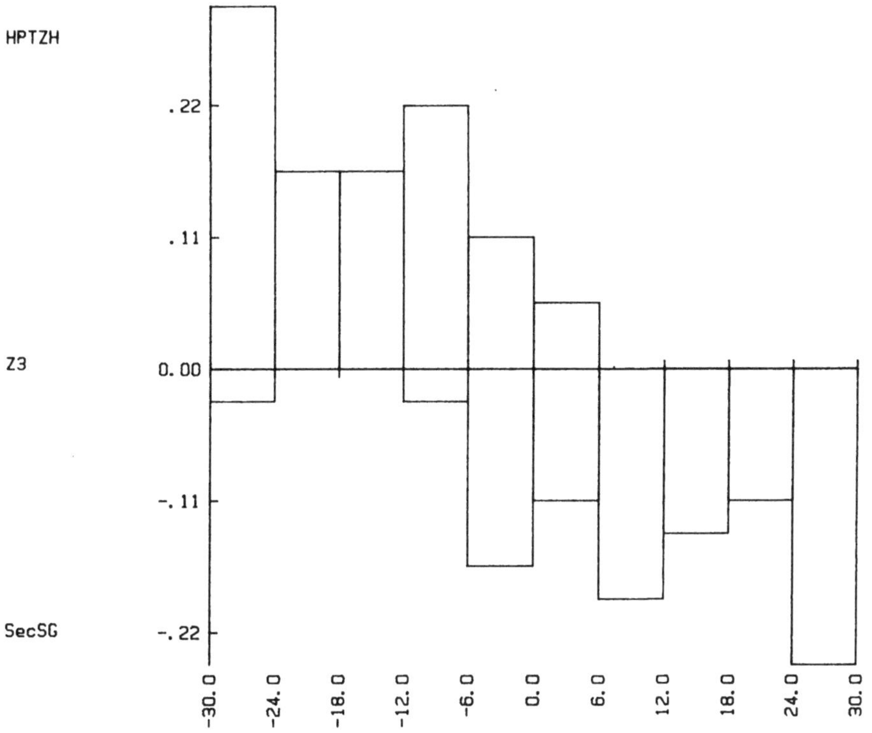

Fig. 22. Like Figure 21, on the bottom is a patient collectiv with secundary hyperparathyreoidism from the St. Gall clinic.

the clinics of the University of Zürich, and 37 samples from the patients of our own hospital, suffering from secondary hyperparathyreoidism. The discrimination of the Zürich HPT patients from the NON-HPT group is nearly as perfect, as that for the learning group (Fig. 21). 35/37 Zürich HPT patients are correctly classified. As expected, the distribution of the patients with secondary hyperparathyreoidism is worse (Fig. 22).

Only 29/37 St. Gall secondary HPT patients are correctly classified.

With the function

$$\begin{aligned} Z_3 = {} & 19.4 \text{ calcium (mmol/l)} + 20.3 \text{ phos (mmol/l)} \\ & 15.5 \text{ chloride (mmol/l)} - \text{Hk} \cdot 100 + 115 \end{aligned}$$

a primary hyperparathyreoidism becomes more probable with increasing negativity of the results, and vice versa.

It should be pointed out that the function Z_3 is valid only for hypercalcemic patients. If normocalcemic patients are to be separated from HPT patients (an illogical problem!) it would be necessary to construct a new discriminating function using an other learning group.

Appendix

1. Denotations

Collective	number of points	points	index	centre
A	n	$A_1, \ldots, A_n$	i	S_A
B	m	$B_1, \ldots, B_n$	j	S_B

2. Algorithms

A point is false with respect to a straight line, if it is not on the side of the centre of its collective

F_g = number of false points with resepct to the straight line g

$F_{\min}$ = minimum of F_g

1 calculate S_A, S_B
2 let $F_{\min} = m + n$
3 $i := 1$
4 $j := 1$
5 determine the straight line $g = (A_i, B_j)$
6 if g does not dissect $(S_A S_B)$, go to 9
7 determine F_g
8 if $F_g \leqslant F_{\min}$, store g and let $F_{\min} := F_g$
9 $j := j + 1$
10 if $j \leqslant m$, go to 5
11 $i := i + 1$
12 if $i \leqslant n$, go to 4
13 rotate the optimal straight line $g = (A_i, B_j)$ by an adequate angle about the center of A_i, B_j
14 list the optimal straight lines

References

1. Lachenbruch, P. A. (1975) Discriminant Analysis, Hafner Press, New York.
2. Solberg, H. E. (1975) Discriminant analysis in clinical chemistry, Scand. J. Lab. Clin. Invest. *35*, 705–712.
3. Anderson, T. W. (1958) An Introduction to Multivariate Statistical Analysis, John Wiley, New York.
4. Ellis, G. & Goldberg, D. M. (1979) Discriminant function analysis applied to laboratory tests in patient with hepatobiliary disease, Comp. Biomed. Res. *12*, 483–501.
5. Flury, B. & Riedwyl, H. (1983) Angewandte multivariate Statistik, Fischer, Stuttgart.
6. Durbridge, T. C. (1984) Applying descriptive discriminant analysis as a visual aid for physicians interpreting biochemical test results, Clin. Biochem. *17*, 321–326.
7. Michaelis, J. (1972) Zur Anwendung der Diskriminanzanalyse für die medizinische Diagnostik, Habil. Schrift, Mainz.
8. Hermans, J., van Zomeren, B., Raatgever, J. W., Sterk, P. J. & Habbema, J. B. F. (1981) Use of posterior probabilities to evaluate methods of discriminant analysis, Meth. Inform. Med. *20*, 207–212.
9. Goldschmidt, H. M. J. & Leyten, J. F. (1985) Medicometries – a new, promising discipline. In: Computing in Clinical Laboratories (Trendelenburg, Ch., ed.) Proc. 5th Internat. Conference, Stuttgart (FRG) June 1985.
10. Fisher, R. A. (1936) The use of multiple measurements in taxonomic problems, Annals of Eugenics *VII*, Pt II, 179–188.
11. Birkenfeld, W. (1978) On step towards a nonparametric discriminant analysis. In: COMPSTADT 1978, Proceedings in Computational Statistics, 3rd Symposium, Leiden, pp. 162–169, Physica Verlag, Wien.
12. Hansert, E., Federkiel, H. & Stamm, D. (1984) A new procedure for discriminating between two patient populations using multivariate decision limits: Application in the detection, and exclusion of alcoholism based on clinical laboratory findings. J. Clin. Chem. Clin. Biochem. *22*, 791–810.
13. Armitage, P. (1971) Statistical Methods in Medical Research, Blackwell, Oxford (repr. 1980).
14. Riesen, W. F., Mordasini, R. & Keller, H. (1983) Chemical relevance of apolipopro-

teins. Proc. Intern. Sympos. New Aspects on Lipoprotein Metabolism, Heidelberg.

15. Stamm, D., Hansert, E. & Feuerlein, W. (1984) Exzessive consumption of alcohol in man as a biological influence factor in clinical laboratory investigations, J. Clin. Chem. Clin. Biochem. *22*, 65–77.
16. Fraser, P., Healthy, M., Rose, N. & Watson, L. (1971) Discriminant functions in differential diagnosis of hypercalcemia. Lancet, 1314–1319.
17. Fraser, P., Healthy, M., Rose, N. & Watson, L. (1976) Further experience with discriminant functions in differential diagnosis of hypercalcemia. Postgrad. Med. J. *52*, 254–257.
18. Transbøl, J., Jorgensen, F. S., Hornum, J. & Keiding, N. (1977) Hypercalcaemia discrimination index: A multivariate analysis of parathyroid function in 107 hypercalaemic patients. Acta Endocrin. *86*, 768–783.
19. Transbøl, J. (1977) On the diagnosis of so-called normocalcaemic hyperparathyreoidism. Acta. Med. Scand. *202*, 481–487.
20. LoCascio, V. H., Vallaperta, P., Adami, S., Cominacini, L., Galvanini, G., Bianchi, I., Ferrari, J. & Scuro, L. A. (1978) Discriminant analysis in differential diagnosis of hypercalcamia. Clin. Endocrin. *8*, 349–356.
21. Lafferty, F. W. (1981) Primary hyperparathyreoidism. Arch. Intern. Med. *141*, 1761–1766.

Subject Index

absorption rate constant 408
accuracy
–, analytical 292
– of clustering 326
– of estimates 107 f
action, medical 278 f
acute myocardial infection 88–96
addressable points 123
age correction 349
alanine aminotransferase *see* GPT
alcohol abuse 425
alcohol dependence 425
alcoholism 3 f, 179, 189, 425–427, 458
–, detection 425 f, 444
–, diagnostics 236 ff
–, exclusion 425 f, 444
algorithm
–, K-MEANS– 287
–, optimization– 287 ff, 297
–, reclassification– 294
alkaline phosphatase 427, 443
alpha error 200 f
ALT *see* GPT
ambiguity 222
analogous 82
analogous and digital data presentation 81–119
analytical variation, effect of 259
animation 124, 129
anterior odds of disease 172 f, 193 f, 197
anterior prevalence of disease 162, 173, 175, 177, 183 f, 190
anterior probability 162, 192
anti-aliased 126, 128
AP 427, 443
aperture 126
apolipoproteins 454
approximation formula 332
artifacts 287
artificial intelligence 205
aspartate aminotransferase *see* GOT
assignment *see also* reclassification
– of patients to clusters 294, 304 ff
AST *see* GOT
attention function scores 265
automated decicion making 367–380
–, use and efficacy 376 f
–, legal implications 379
autoregressive time series models 270

background colour on VDUs 137
backtracking 385
backward probability 222
bar charts
–, detail of 148
–, structure of 148
–, use of colour in 150
barcode 6
BASIC 385
Bateman function 412
Bayes "estimates" 414
Bayes formula 412
– for continuous data 250
Bayes model
–, performance of 250
–, sequential 255
Bayes theorem 191, 247, 255, 387
Bayesian approach 403, 411
Bayesian objective function 414
Bayesian statistics 411
benchmark 130
benefit, patient 255 f
Berger diagram 222 f
beta check 40
beta error 200 f
beta weights 29
bilirubin 451
binary test 221
biological correlation 259
bit-map 127
blood-gas map 69
bolus injection model 96
bootstrap method 255
branching logic 211
business graphics 124, 126

CAD/CAM 129
calcium 462, 464
calculation of the information theoretical quantities 224
calculation printout 43
capital lettering, legibility of 135
carcino-embryonal antigen 179
centours 252
certainty factors 386
chaining 264
channel capacity 223
–, calculation 244
chartjunk 73
charts
–, compared with graphs 147

–, compared with tables 146
–, bar charts 148–151
–, pie charts 147–148
Chernoff faces 14, 70
chloride 462 f
cholesterol 454 f
chromatic aberration, effect on legibility of VDUs 136
class fraction DIS 165, 173
classification of diseases 277 ff
class prevalence of disease 162, 165, 169, 173, 175, 192 f, 199
clearance 406
clinical chemical quantity 349
clinical classes, new 260
clinical laboratory investigations 426 f
cluster analysis 26–30, 260, 277 ff
clustering *see also* partition 3
clustering method 286
–, choice of 286 f
–, hierarchical 287
–, partition method 287
–, Ward's method 287, 289
–, variance criterion 287
clustering technique
–, choice of 264
–, mathematical model of 264
clustering structure *see* stability, validity
clusters
–, assignment of patients to 294, 304 ff
–, number of 9 f
–, similarity between 289 ff, 293 ff
coding of information 14–15, 25, 134 f, 139
Cohen's kappa 323
colour
– as background on VDUs 137
–, effect on legibility 136–137
–, effect on visual acuity 136
– for emphasis 137–138
– as a grouping cue 137
– in tables 143–146
colour coding 139
colour graphics 8
colour luminance values on VDUs 136
colour sensitivity of the human eye 136
colour screens 7
combination of variables 432, 434, 436
communication channel 223, 225
compact disc 130
compartment 92
compartment models 92, 405
compatibility 123 f
– of clustering and reclassification method 294
components of variation
–, analytical 259
–, biological 259
computer conferencing 127
computer interfaces to laboratory systems 375 f
computer networks 5
computer programs 367–376
computer simulations 363
conceptual model 93
conditional probabilities 3, 192
construction sample of patients 255
consumables 122
continuous transformation 362
coronary angiography 453
coronary arteriosclerosis 454
coronary heart disease 3
correction for age 349
correlation
– between analytical errors 259
– of analytical errors, breaking of 259
– between biological and analytical components 259
–, serial 268
correlation center 342
correlation coefficient 233, 258 f, 335, 340, 363
correlation matrix 336
correlation structure 340
cost 121, 123, 129, 255
cost-benefit analysis 241
cost-effectiveness 121, 241
cost function 104
covariance decomposition technique 363
covariance matrix 347, 349
covariance selection 336
CPD curve 170 f
creatine kinase 179, 187, 189
creatine kinase MB 179, 187–189
creatinine 427, 436 f, 440–444
cross validation techniques 255
cumulative findings 49–52, 60
cut-off point 228

DAP program 113 ff
data, time dependent 81–119
data assessment, multivariate 347
data base 7, 37, 49, 56, 60
data dictionary 59
data editing 26
data grouping 15–18, 25
data redundancy 329
data presentation 3–5
–, linear interpolation 83–92
data reduction 279 ff
data storage 6 f
data vectors 335
decision criterion 228
decision limits 431 f, 441
– in combined tests 239
–, conventional 443
–, logical connection 431 f, 441
–, multivariate 239

–, optimized 436, 443
decision making 63, 367–380
decision point 433 f, 436
–, lattice of 436
decision rule 337
decision scheme 228, 240
decision support 65
decoder 228, 231 f
density estimation 353
dendrogram 263
desktop publishing systems 135
deterministic methods 3, 5, 205–217
deviation pattern 357, 361
diabetes therapy control 83 ff
diagnosing, hypodeductive method 279
diagnosis
–, final 303
–, laboratory 302
– theory 278 f
diagnostic
– concepts, current 279 f
– displays 69
– judgement 281, 316, 322
– model 281 ff
– process 241 f
– purposes 364
– statements, reliability 256
– strategy *see* strategy
– time window 188
diagnostic procedure
– along branched logic structures 279
–, exhaustive 279
diagnostics of alcoholism 236 ff
diapositive 124
digital 82
digital and analogous data presentation 81–119
digitising 122, 127 f
direct identification 6
direct search method 104 f
discriminant analysis 15, 280, 425, 427 f, 442, 447
–, linear 428 f, 431–434
–, quadratic 253, 258
discriminant function 254, 428, 435
discriminant function analysis, geometric interpretation of 252
discriminant region 428, 431–433
–, nested 437 ff
–, simple 432–434, 436 f, 442
–, simple, first-order 433
–, simple, k^{th} order 432, 435
–, simple, second-order 433, 435, 438 f, 442
–, simple, third-order 433, 436 f, 440 f
discrimination 354
discrimination information 234
discriminator 161, 163 f, 169, 173, 178 f
discriminator position 162, 164, 167, 173, 178, 195
discriminatory ability 256
disease(s)
–, nosological terms 278
–, taxonomy 277
– terms 277
disease classification 277 ff
diseased persons 161, 164, 169, 177 f
disease-specific deviation pattern 357, 361
displays 63, 66, 69
– effectiveness 74
display techniques, graphic 12
dissimilarity coefficient 260 f
distance
– between objects 283
–, minimal distance partition 287
distribution(s) 167, 169, 178, 194, 198, 411, 415, 429–441
–, convex hull of 434
–, differences in the 428, 431, 438, 442
–, marginal 334, 362 f
– of measurement errors 294 ff
–, multimodal 362
–, multivariate 279
–, normal 428, 430, 442
– of similarity values 289
–, type of 441, 443
–, unimodal 290
–, univariate 279
distribution-free method 329, 362
distribution volume 406
documentation 6
dosage adjustment 403
dose prediction 403
double histograms 166, 168, 170, 172, 174, 176, 178, 180, 197–199
double result list 40
drivers 122
drug dosing program 414
drug monitoring 3, 403
dynamic systems analysis 92

economy, diagnostic 304
editing of data 26
editor 126
effectiveness of graphical displays 74
efficiency 162, 166, 197, 427, 431, 442
efficiency I 197
efficiency II 197
efficiency maximum 428, 432 f, 436
efficiency tests 447
elimination rate constant 405
entropy 220, 225 f
entropy increase 341
equivocation 222

error
–, analytical 292, 296, 315
– in mathematical models 103
–, random 292, 296, 315
–, systematical 292
error rate 258
errors-in-variables methods 29
erythrocyte mean cell volume *see* MCV
estimates
–, accuracy 107 f
–, organo-specific 27
estimation 436 f
evaluation list 37
evaluation sample of patients 255
exhaustive structure 279
external stability 288, 292 ff
expected value 349
expert systems 3, 5, 76, 367–380, 381–402
–, problems 369 f
extreme previous value deviation 40
extreme value check 37
extreme values 296

false negatives 164
false positives 164
falsely negative results 354
filtering of information 12–14, 25
findings
–, microbiological 48, 53, 55
–, preliminary 52, 60
findings report 33–35, 45, 58
first-order absorption 415
flashing images on VDUs 139–140
flat-bed plotter 122
floating point accelerator 126
formulation of a model 92–101
foils 122, 124
frame grabbing 127
frame store 124
fuzzy logic 386, 398

gammaglutamyl transferase *see* GGT
Gaussian distribution, multivariate 263
Gauss-Newton method 104–106
Gestalt principles of grouping 134, 137
GGT 179, 189, 427, 429–431, 433, 435–441, 443 f, 458 f, 461
GLDH 427, 443
glucose tolerance test 100 f
glutamate dehydrogenase (GLDH) 427, 443
glutamate-oxaloacetate transaminase *see* GOT
glutamate-pyruvate transaminase *see* GPT
goodness of fit 106
GOT 427, 433, 436 f, 440 f, 443 f, 458 f, 461
GPT 427, 443 f, 458 f, 461
graphical displays 63
graphical Kernel systems 122
graphical structure 357
graphic display techniques 12
graphics data-base 129
graphics standards 122, 124, 130
graphic computer programme 167, 200
graphs
– compared with charts 147
– compared with tables 146
–, structure and detail of 151–153
–, use of colour in 153–154
group *see* cluster, partition
group-average method 264
group-conformity behaviour 357
grouping of data 15–18, 25
G test 235

hard copy 129
hardware 3, 5
HDL-cholesterol 454
HELP system for medical decicion making 367–380
hematocrit 462–464
hematological studies 427
hemoglobin 451
heteroscedasticity 362
high resolution 129
hierarchical clustering methods *see* clustering methods
hierarchic methods of numerical taxonomy 260
hit rate 428
homeostatic model 270
Hotelling's T^2 337
human computer interaction 127
hypercalcemia 463
hyperparathyreoidism 3, 462
hypotheses of independence 336
hypothetico-deductive principle 279

icon 127
icterus hemolyticus 451
ideal observer 233
identifiability 102, 105–108
identification 6
– of a model 101–108
inaccuracies 122
increase in entropy 341
independence, conditionally 251
independent tests 189
indirect identification 6
inference engine 384
information 219
–, loss of 280
–, missing 304
–, redundand 26
information coding 14 f, 25, 134 f, 139
information decrease by interaction 229
information filtering 12–14, 25

information gain 220, 222
information increase by interaction 229
information pre-filtering 14
information theoretical model 280
– of a diagnostic test 219 f
information theory 3, 219
initial sample 428, 434, 436, 443
input probabilities 221, 223, 238
intelligent interfaces 4
interaction 229
interchanges between clusters 301
interfaces, intelligent 4
internal stability 288 ff
interpolation, linear 82–92
interpolation, nonlinear 82 f, 109–112
interpretation 4
interpretive reporting 205–217

Jacobian matrix 414
joint entropy 221

Kernel functions 353
Kernel method 254, 258
Kernel systems, graphical 122
key board 5
K-MEANS algorithm 287
knowledge base 384
knowledge-based systems 367–380, 381–402
–, problems 369 f

laboratory EDP system 399
laboratory information, psychological and statistical considerations 11–31
laboratory profiles 254
laboratory systems, computer interfaces to 375 f
lactate dehydrogenase isoenzyme 1 (H_4) 181
laser printer 124, 127
LDH isoenzyme abnormalities 26–30
LDH isoenzyme profiles 20–25
LDH isoenzyme report 13, 14
LDL-cholesterol 454
legibility
– of capital lettering 135
– of colour on paper 136
– of colour on VDUs 136–137
– of reverse video 135
light-pen 122
likelihood, likelihood ratio 172 f, 191–193, 198, 280
likelihood function 411
linear discriminant analysis 251, 258
linear interpolation 82–92
line-drawing 122
lipid status 453
lipoprotein test 397
LISP 385
logarithmic transformation 331
logistic regression 253, 259
log likelihood ratio test (G-test) 235

maintenance 129
maintenance dose 404
manpower 129
manufacturer 122
marginal distribution 334, 362 f
Markow chain 268
mark-sense cards 6
matching coefficient 260
material codes 35, 47
mathematic modeling 92 ff
maximal combination 357
MCV 427, 430 f, 433, 435–441, 443 f, 459, 461
mean cell volume *see* MCV
medical action 278
medical report 48 f, 56–58
measurement errors *see* errors
menu 127
metaknowledge 385
microbiological findings 48, 53, 55
microcomputers 14, 122
microfilm 122, 127
minimal distance partition 287
misclassification of objects 305, 310
misclassification rate 255 f, 305
missing information 304
mixture of continuous and discrete variates and Bayes formula 254
model
–, compartmental 92
–, conceptual 93
–, mathematical 92–112
–, minimal 92
–, one-compartment 405
–, two-compartment 98 ff
model errors 103
model formulation 92–101
model identification 101–108
modeling process 93
model validation 109–112
monitoring 68
monothetic classes 260
Monte Carlo methods 286
mouse 122, 126 f
multi-class test 161
multi-media documents 127
multimodal distributions 362
multiple correlation coefficient 342
multiple testing 15
multivariate analysis 205, 247
–, use for interpretive purposes 258
multivariate approach 329
multivariate data assessment 347
multivariate decision limits 239
multivariate normal distribution 329, 347, 351, 361

multivariate profiles 25
multivariate prognostic stratification 267
multivariate reference region 254
multivariate test quantity 329, 337
Munich Alcoholism Test 426
myocardial infarct 70
myoglobin 166

negative test results 428, 435, 437
new clinical classes, construction of 260
nomenclature 161
nomogram 409
noncentrality parameter 351
nonlinear interpolation 82 f, 109–112
nonlinearity in clearance 420 f
nonlinear regression analysis 82
NONLIN program 106 ff
nonmonotonic statistical relation 362
nonparametric methods 329 f
– of discriminant function 254
normal distribution 329, 332, 347, 351, 361
normalization of data 283 ff
normalizing transformation 330
normal pertubation 102
normal range 280
nosology 278
notation, symbolic 384
number fractions 162, 166
numerical taxonomy 258, 260
–, techniques of 260

observational units 335
observer, ideal 233
obstructive airways disease 417
OCR 6
odds 193
one-compartment model 405
one-left-out method 255
on-line connections 6
operating system 5, 60
optimization algorithm 287–289, 297
optimization function 287 ff, 297
–, plot 289, 324 f
optimizing the decision limit of a test 236
optimum input probability 223
optimum subsets 241
organ-specific estimates 27–29
outlier 254
outlying observations 336
output probabilities 221
overall correlation 345

paint box 126
palette 123, 129
parallel combination 229
parallel tests 188
parameter estimation 103–105
parameter reduction 304 ff
parameters, clinical chemical *see also* variables
–, choice of 282
–, histograms of 283
–, normalization of 283 ff
–, transformation of 283 ff
parameters not identifiable 102
parametric method 329 f
parathormone 462 f
partial correlation 336
partial correlation coefficient 259
partition(s)
–, comparison of *see* similarity
– method *see* clustering methods
–, starting 287
Pascal 385
pathognomonic test 222
pathological perturbation 102
patient benefit 255 f
patient-characteristics 413
patient-specific data 414
pattern cognition 277 ff
pattern recognition 277 ff
peaking phenomenon 258
perception 72
perturbation 102
– of input data 288
pharmacokinetics 403
phosphate 462
phospholipids 454
photo-typesetting 127
pie charts, presentation of 147–148
pilot study 175
pixels 123
plausibility check 37, 61
plotter 122
polar diagram 299
polythetic classes 260
population distribution volume 415
population parameters 411
positive definite matrix 330, 363
positive skewness 340
positive test results 428, 434 f, 437
posterior distribution 411
posterior odds of disease 172 f, 192, 194, 198
posterior prevalence of disease 162, 184
posterior probability 249
prediction error 417
predictive accuracy 418
predictive precision 418
predictive value 162, 164–166, 173, 178, 192, 201
–, model of 280
– of correct assignment 306
predictor variate 267
pre-filtering of information 14
preliminary findings 52, 60

pre-optimized starting partition 287
presentation
– of charts and graphs 146–155
– of data 3–5
–, digital/analogous 112–118
– of tabular information 140–146
prevalence 248, 425, 428, 436
– of disease 162, 165, 169, 173, 175, 183 f, 190, 193, 199
prevalence factor 173
preview system 127
principal component 258
principal component analysis 254, 258
prior conditional probability 249
prior distribution 411
prior probability (of disease) 162, 191, 248
probabilistic diagnosis 247
probabilistic reasoning 386
probability
– of correct classification 305 ff
– of disease 162, 191
– of transition 307
probability density function 250
probability notation 162, 191
profiles, multivariate 25
prognostic stratification, multivariate 267
program languages 5
Prolog 381, 387
PRO. M. D. (Prolog system supporting medical diagnostics) 387
prospective studies 420
prospective validation 256
prostatic acid phosphatase 179, 183
pseudo-random numbers 363
pseudohyperparathyreoidism 462
public domain 124
PVneg 164–166, 178, 181, 201
PVpos 164–166, 178, 181, 192

quadratic discriminant analysis 253, 258
QUADROLAB 33, 35, 37, 46, 49, 58
qualitative test, example 227
quality criterion of a test 236
quality control 38, 71, 426
quantitative lipoprotein electrophoretic test 397
quantitive tests 227
quantization 228
quasi-distribution-free method 329, 334
quasi-normalized data 335

Rand's measure 289, 323 f
– corrected for chance 289, 294, 325
– reference curve for 324 f
raster graphics 122
reasoning 386 f
receiver 225
receiver operating characteristic 195
reclassification 294 ff
reclassification algorithm 294
–, compatibility with the classification method 294
recognition of changes 18–20
reconstruction of missing test data 183
recursive notation 385
reduction
– of data 279 ff
– of parameters 304 ff
redundancy 229, 329
redundant information 26
redundant variates 257
reference curve 289, 324
reference data 329
reference group 167
reference interval 4
reference persons 161, 164, 169, 177 f
reference region 352
–, multivariate 254
reference values, normalization 14
referential display 66
regression analysis 28
–, nonlinear 82
regression model 29
regressor variables 335
relational database 7
relative ROC 195
reliability of diagnostic statements 256
remainder list 43–45
replicate measurements, use of 259
reporting, interpretive 205–217
residual errors 106 f
resolution 122, 126
resubstitution method 254
result correlation, striking 41
result unit 34
reverse video, legibility of 135
risk quotients 194, 198
ROC, relative 195
routine check-up 362
sample size 177
sampling time 406
scale 441
–, continuous 443
–, ordinal 441, 443
screen dump 126
screening 190
segregation of effects 20–25
selection procedure 257
semantic information 243
sender 225
sensitivity 164–169, 178, 181, 184, 189–191, 196 f, 201, 256, 307, 329, 363, 427, 436, 442
–, maximum 428
sequence of tests 184, 187
sequential Bayes model 255

sequential combination 229, 232
serial correlation 268
set theory 223 f, 229
short-term hire 126
significance test for grouping structure 288
SIMCA-method 254
similarity *see also* Rand's measure
– between clusters 289 ff, 293 ff
– between objects 283
– between partitions 323
–, Cohen's kappa 323
single linkage method 263
single-point method 403 f
smooth shading 127 f
software 3, 5, 122, 127, 129
software strategy 130
spatial coding 134
Special Interest Group 130
specificity 164–169, 178, 181, 184, 189–191, 196 f, 201, 256, 427, 436, 442
–, maximum 428
spline functions 82
split sample technique 255
stability
–, external 288, 292 ff
–, internal 288 ff
standardization 283
standardized normal distribution 332
starting parameter 305, 310, 312
starting partition 287
stationarity 269
statistical test 428, 438
–, analogy to a 428, 442
–, critical level 438 f
– level 438 f
statistical test quantities 329
stepwise backward selection procedure 257
stepwise forward selection procedure 257
stocastic process 268
storage of data 6 f
strategy, diagnostic 183, 187, 304 ff
–, clinical usefulness 316 ff
–, practical application 316 ff
striking result constellation 41
striking trend 40
structural investigations 340 ff
subpopulation clearance 415
survey plot 167
symbolic notation 384
syndromes 265
synonyms 200
syntactic information 243

tables
– compared with charts and graphs 146–147
–, layout of 140–143
–, structure of 140
–, use of colour in 143–146
tablet 127
target variate 268
taxonomy 277
teaching sample 304, 310
television projection 124
temporal reasoning 387
termination, criterion of 305, 309, 312
test capacity 223
–, calculation 244
test classes 161, 164 f, 167, 169
test combination 228 ff
test data, missing 183
test dose 409
test evaluation quantities 166
testing, multiple 15
test quantity, multivariate 329, 337
test results
–, negative 428, 435, 437
–, positive 428, 434 f, 437
test set 255
test system 221
text segment 49
theophylline 403
theophylline dosage 403
therapeutic drug monitoring 403
therapeutic range 403
three-dimensional graphics 126
three-class test 161, 169, 171 f
thyreotoxicosis 462
thyroid disease 316
thyroid gland findings 397
thyroid stimulating hormone 97, 179, 190
thyrotropin releasing hormone 97
time behaviour 60 f
time depend results 81–119
time series analysis 268
–, medical application of 272
tolerance region 330
training set 255
transformation(s) 283 ff, 289, 296, 298 f, 329 ff
–, influence on classification 283 ff
–, linear 283
–, non-linear 283 ff
transinformation 222, 226
transinformation surface 239
transition area 437 f
transition probability 222 f, 269
TRH test, data presentation 114 ff
trend 268
triglycerides 454
two by two table 163 f, 182
two-class test 161, 163 f, 169, 171, 175, 178, 184
two-compartment model 98 ff
typographic coding 134

ultrametric dissimilarity coefficient 264

uncertainty 220
uniformly distributed quantity 331
univariate approach 329
urea 427, 443 f
utility 256, 307, 309, 311 f

validation
– of a model 109–112
–, prospective 256
validity 219
– of clusterings 287
– of single clusters 295
variables
–, choice of 282
–, normalization of 283
–, number of 282
–, transformation of *see* transformation
–, weighting of 283
variance criterion *see* clustering method
variates, redundant 257
vector diagram 300
video 126
video camera 127
video disc 130
visual acuity, effect of colour on 136
visual coding of information 134 f
vitamin D intoxication 462

Ward's method 287, 289
weights to variates, assignment of 263
working list 37
workstation 127

X-transformation 3, 329, 331 ff

CLINICAL BIOCHEMISTRY

PRINCIPLES · METHODS
APPLICATIONS

Series Editors
H. Ch. Curtius
M. Roth

1 Mass Spectrometry
Editor Lawson

2 Data Presentation/Interpretation
Editors Keller, Trendelenburg

3 Human Plasma Lipoproteins
Editors Fruchart, Shepherd

4 Laboratory Organization/Automation
Editor Vonderschmitt

5 Toxicology/Drug Analysis
Editors Brandenberger, Maes

- Stable Isotopes

- Nuclear Magnetic Resonance in Clinical Biochemistry

Further volumes in this series are in preparation.

W DE G

Walter de Gruyter
Berlin · New York

Bei Fragen zur Produktsicherheit wenden Sie sich bitte an:
If you have any questions regarding product safety,
please contact:

Walter de Gruyter GmbH
Genthiner Straße 13
10785 Berlin
productsafety@degruyterbrill.com